The Geological Society of America

Reviews in Engineering Geology
Volume III

Edited by Donald R. Coates

1977

CONTENTS

PART 4. ENGINEERING GEOLOGY AND HIGHWAY ENGINEERING

12 Utiku landslide, North Island, New Zealand .... *Martin L. Stout*   171

13 Engineering geology of the Woodstock rockslide, New Hampshire .... *Brian K. Fowler*   185

14 Relationship between morphology, hydrology, geotechnics, and vegetation on an old northern Ohio landslide .... *Hazel F. Krist and Murray R. McComas*   197

15 Engineering geology of the slope in stability of two overconsolidated north-central Texas shales .... *Robert G. Font*   205

16 Engineering geology of multiple landsliding along I-45 road cut near Centerville, Texas .... *Christopher C. Mathewson and James H. Clary*   213

17 Three major California freeway landslide areas .... *F. Beach Leighton*   225

PART 5. ENVIRONMENTAL PLANNING

18 Slope-stability studies in the San Francisco Bay region, California ..... *Tor H. Nilsen and Earl E. Brabb*   235

19 Landslides in West Virginia .... *Peter Lessing and Robert B. Erwin*   245

20 Landslides at Sardis in western Turkey .... *Gerald W. Olson*   255

Index 273

Introduction

DONALD R. COATES

In the United States it is finally being recognized that landslides constitute more than a local hazard. This new awareness is manifested by investigations such as those being conducted by the U.S. Geological Survey in the San Francisco Bay region and by the West Virginia Geological Survey in urban areas of that state. The U.S. Flood Disaster Protection Act of 1973 that includes the landslide process within guidelines of insurance clauses is also an indication of current landslide awareness. The purpose of this volume is (1) to update significant information about landslides and present new case histories and (2) to refocus previous works into new syntheses and insights. This volume is the culmination of a four-year series of evolutionary events and represents contributions not only from the authors but also from government agencies, universities, and consulting firms. Great credit is also due the more than 60 reviewers who toiled diligently to make the manuscripts accurate, up-to-date, and highly readable.

The history of this volume started in Minneapolis at the November 1972 annual board meeting of the Geological Society of America's Engineering Geology Division. Plans were made to publish a book in the case histories series on the subject of slope stability. These plans did not materialize. In October 1975 at the annual board meeting held in Salt Lake City, a decision was made to change the topic of the proposed volume to landslides, and I was selected as editor.

The incubation period has now materialized into a volume with 20 chapters written by experts representing a wide variety of interests and backgrounds. Many factors guided me in making the selection of authors and topics. I wanted to obtain as wide a geographic spread as possible to show the universality of landslide phenomena. I believed it was also important to have representation from governmental agencies (6), universities (8), and private consulting firms (6). This blend has produced chapters on Canada, New Zealand, the Philippines, Turkey, and the United States. Landslides that occurred in the states of Alaska, California (3), New Hampshire, New York, Ohio, Oregon-Washington (2), Texas (2), and West Virginia are discussed; thus, many different landslide types are evaluated in greatly contrasting physiographic and climatologic regions. To obtain this spectrum, 10 of the chapters were written by invited authors, and selection of the other 10 authors resulted from voluminous correspondence with many candidates who offered their services in preparing a chapter.

This volume contains more landslide case histories than any publication to date and also has many reviews and integrated analyses of landslide phenomena. Topical features that are presented include (1) climate—arctic, temperate, tropical, arid; (2) terrain—subaerial and subaqueous with extremes of mountainous to nearly flat slopes; (3) lithology—bedrock, regolith, sediments; (4) landslide types—rock avalanches, rock glides, debris flows, debris avalanches, slumps, liquefaction flows, earthflows, and others; (5) landslide size—some of the largest on record as well as smaller ones; and (6) landslide causes—excessive moisture, earthquakes, man-induced. There is also a range of emphases that each chapter addresses to geologists, engineering geologists, engineers, social planners, and management personnel.

Chapters in this volume have been grouped into five parts as a reader convenience and for organizational emphasis. Although this brings into focus the different themes and approaches, it does not imply that subject matter within a specific chapter falls exclusively within a particular part. Most chapters have engineering and environmental aspects, even though not included within those parts of the volume.

Part 1. Overview. The purpose of this part is to provide a fabric and some generalizations of landslides that illustrate the broad sweep of this phenomenon and to enhance the reader's understanding of the remaining chapters. The opening chapter by D. R. Coates is designed to give a preview and some guideposts on landslides regarding their type, habitat, classification, nomenclature, and characteristics. Landslide damages, legal affairs, recognition, prevention and control, and societal impacts are also summarized. Chapter 2 by J. D. Mollard supplies a synthesis of important landslide families in Canada. He divides the country into four dissimilar regions

—the cordillera, the western interior, the St. Lawrence, and the north—and discusses landslides and slope stability problems of each. Engineering aspects of each region are also analyzed.

Part 2. Regional Studies. This part contains those chapters that are primarily devoted to appraisal of landslides on a regional basis. Chapter 3 by K. H. Wyrwoll contains significant data of the Labrador-Ungava region that shed new light on traditional concepts for causes of slope instability in cold areas. He concludes that the external factors that initially created hillslopes contribute more to their instability than frost action, which is the prevailing view for the most common cause of slope failure. Chapters 4 and 5 are regional studies of major rivers. In Chapter 4 L. Palmer discusses landslides of the Columbia River and concludes that they are caused by many different factors including oversteepening of slopes, volcanic loading of the head area, tectonic uplift, and high rainfall. Landslide movement rates range from slow to fast, and liquefaction of clay is one of the dominant flow mechanisms in displacement of earth materials. D. R. Piteau analyzes slope stability of the Fraser Canyon, Canada, in Chapter 5 and shows that 66% of slope failures occur where alluvial fans deflect rivers to the opposite bank and the toe of an unstable terrain is undercut by stream erosion. Regional faulting, climatic conditions, and effects of man are other factors that contribute to slope instability. Chapter 6 by F. J. Swanson and D. N. Swanston documents the relationship that exists in different types of mass movement in the western Cascade Range of Oregon. In this region, creep, slump, and earthflow processes interact to produce most hillslope displacements, and creep initiates failures that lead to more rapid movement. The amount of moisture controls movement rates; type of bedrock and regolith determines the depth of the slip failure.

Part 3. Specific and Local Studies. This part of the volume contains those chapters that concentrate on either a particular landslide or on specific earth materials that have hosted the landslide. P. B. Durgin in Chapter 8 analyzes landslide types that result when granitic rocks are weathered. He illustrates the landslide series of events during the development of corestones, decomposed granitoid, and saprolite that have developed in the granite batholiths of lands adjacent to the Pacific Ocean. J. R. Dunn and G. Banino in Chapter 8 discuss the problems that result from movements in varved clays of glacial Lake Albany, New York. These sediments affect a corridor more than 120 km long in the Hudson Valley where creep and rotational sliding occur from both natural and man-induced causes. One of the largest slides yet recorded is described by B. F. Molnia, R. R. Carlson, and T. R. Bruns in Chapter 9. The submarine slide has a volume of 5.9×10^{11} m³ in a 14×18-km area of the Gulf of Alaska. Analyses of such unstable areas are important considerations if the localities are used for pipeline emplacement or for platform sites in petroleum development. In Chapter 10 J. A. Wolfe supplies information about an immense low-angle landslide on Samar Island, Philippines, where an 18×25-km limestone block moved 5 km on a 1.2% slope. The movement was apparently triggered by an earthquake about 2,000 yr B.P. and affected a rock mass of 1.35×10^{11}-m³ volume—the largest subaerial feature yet identified on planet Earth. C. G. Bock in Chapter 11 describes one of the largest California landslides, but one which has received only scant attention. The Martinez Mountain rock avalanche is an early Holocene landslide in granite that involved a mass of 3.823×10^8 m³. It was earthquake-triggered, and when the ruptured bedrock moved, it became disaggregated producing all material sizes from rock dust to megaliths during its flow over a 7.6-km distance.

Part 4. Engineering Geology and Highway Engineering. The chapters comprising this part specifically address the engineering geology aspects of landslides as they relate to human activities, and especially to highway construction. M. L. Stout in Chapter 12 evaluates a wide range of engineering methods that could be applied to aid in stabilizing the Utiku landslide, New Zealand. The 18-ha slide is affecting State Highway 1 and the Main Trunk Railway. He discusses the feasibility of using such techniques as dewatering, buttressing, shear keys, rock anchors and shear pins, chemical treatment, and vegetation plantings. Seepage in the upper part of the landslide provides the driving force for movement along a failed plane of montmorillonite in the Pliocene siltstones that are predominant in the region. In Chapter 13 B. K. Fowler discusses a 13,000-m³ landslide that affected Interstate 93 in New Hampshire. To stabilize the area, rock benches were cut, a system of high-strength steel tendons were grouted into the toe of the bench, spot rock bolts were installed, and an extensive pattern of rock drains were drilled. Unstable terrain in the Cuyahoga River valley, Ohio, was studied by H. A. Krist and M. R. McComas, and in Chapter 14 they provide data on a typical 200-yr-old slide. Detailed measurement and laboratory analysis of glacial sediments in the 40,000-m³ displaced area show that the landsliding is a function of unstable sediments, their preloaded aspect, and abundant water infiltration. R. G. Font in Chapter 15 discusses three types of slope failure that occur in two Cretaceous shales of the Dallas-Waco area of Texas. Field relationships of rockfalls, mass flow, and slumps are described, and laboratory analyses provide the data from which the engineering characteristics of the materials can be evaluated. He concludes that landslide occurrence is dependent on stratigraphic position, topographic expression, and physical properties of the shales, which are heavily overconsolidated, fractured, and fissured. Multiple landslides that have affected Interstate 45 in another part of Texas are described by C. C. Mathewson and J. H. Clary in Chapter 16. They attributed the landslides to stratigraphic position, shear strength of materials, alteration of terrain during construction, and disruption of surface drainage by the road cut. Laboratory analysis of the Eocene formations provided important data on stability limits of the materials. In Chapter 17 F. B. Leighton provides geologic and engineering information on three case histories of landslides that have affected California freeways. The slides were caused by heavy rains and a surcharge of fill at the heads where rock types and structures were susceptible to failure. Leighton concludes that greater application of geologic information could have reduced highway design modification during grading and minimized width of freeway alignments, thus lowering construction costs.

INTRODUCTION

Part 5. Environmental Planning. The chapters in this part stress the interdisciplinary nature of landslide problems. These problems are not only geology and engineering problems but also societal problems in which all citizens must cooperate for their ultimate solution. In Chapter 18 T. H. Nilsen and E. E. Brabb summarize the extensive five-year study of the U.S. Geological Survey (with the cooperation of the Department of Housing and Urban Development) in the San Francisco Bay region. A variety of maps and reports can be utilized to aid land use planning and management of the fragile hillslopes of the 18,000-km^2 area. Because landslides are a major slope-erosion process in the region and cause millions of dollars of damage annually, the authors warn that additional development of upland areas should not be undertaken without careful evaluation of hillslope stability. Chapter 19 by P. Lessing and R. B. Erwin describe the ongoing landslide investigation by the West Virginia Geological Survey of the seven major urban areas of that state. The study is especially aimed at providing information and advice to homeowners, buyers, and builders. The authors show in a study of 100 landslides from a total of 2,115 in northern West Virginia that, except for rockfalls, most landslides occur by regolith failure such as debris slides, debris avalanches, earthflows, and slumps. The final chapter by G. W. Olson evaluates the extensive landslides, both past and present, at Sardis, Turkey. This area is one of the richest archaeological sites in the world, but it has suffered great landsliding throughout recorded history. Repeated destruction during several millenniums did not act as a deterrent for continued development. As recently as 1970, landslides near Sardis killed 1,100 people and damaged or destroyed 15,000 buildings.

Table of conversions

1 in. = 2.54 cm	1 cm = 0.3937 in.
1 in.^{2} = 6.45 cm^2	1 cm^2 = 0.155 in.^{3}
1 in.^{3} = 16.387 cm^3	1 cm^3 = 0.061 in.^{3}
1 ft = 0.3048 m	1 m = 3.28 ft
1 ft^2 = 0.093 m^2	1 m^2 = 10.764 ft^2
1 ft^3 = 0.0283 m^3	1 mi^3 = 35.314 ft^3
1 yd = 0.914 m	1 mi = 1.094 yd
1 yd^2 = 0.836 m^2	1 m^2 = 1.196 yd^2
1 yd^3 = 0.764 m^3	1 m^3 = 1.308 yd^3
1 mi = 1.609 km	1 km = 0.621 mi
1 mi^2 = 2.589 km^2	1 km^2 = 0.385 mi^2
1 mi^3 = 4.165 km^3	1 km^3 = 0.239 m^3
1 ton = 1.016 tonnes (t)	1 t = 2,204.62 lb
1 acre = 0.405 hectare (ha)	1 ha = 2.47 acres

1 ha = 0.0039 mi^2

PART 1
Overview

1
Landslide perspectives

DONALD R. COATES
Department of Geological Sciences, State University of New York at Binghamton, Binghamton, New York 13901

ABSTRACT

Landslides are abrupt, short-lived geomorphic events that constitute the rapid-motion end of the mass-movement spectrum. The term "landslide" is a misnomer because the process includes gravity displacement of earth material by falling, sliding, and flowing, and most experts also include subaqueous movements. Landslides are also landforms that are nearly ubiquitous in sloping terrain. All that is needed for their formation is a triggering mechanism, such as excessive precipitation, earthquake, or man, to upset natural stability and induce stresses that exceed the shear resistance of the substrate. A new classification is introduced that recognizes the importance various materials play in the production of different landslide types, whether in bedrock, regolith, or sediments. Landslide names also reflect the type of movement and rate, moisture content, shape of the failed surface, and particle size.

There are fewer publications on landslides than on other topics within catastrophic geology, such as earthquakes, volcanoes, and floods. However, the literature is rapidly expanding because of reports required under the southern California grading ordinances and the environmental impact statements mandated in legislation such as the National Environmental Policy Act of 1969. The new awareness and perception are well timed, because landslide damages have been constantly increasing and amount to hundreds of millions of dollars annually in the United States. The escalating costs are the result of increased construction in highways, reservoirs, other service utilities, and urbanization. Double jeopardy occurs in many urban areas where man is forced out of flat areas because of flood hazards and legal prohibitions onto sloping ground that may be landslide prone.

Planning and management strategies for land use in unstable areas are becoming more sophisticated. The Los Angeles grading ordinances of 1952 and 1963 and the San Francisco region five-year investigation by the U.S. Geological Survey are serving as worldwide models. Prevention and control of landslides require cooperation of engineering geologists, hydrologists, geomorphologists, social planners, and the general public. Recognition of a potential problem is paramount in avoiding disaster. My study of sensitive sediments in the Cowanesque Valley, Pennsylvania, is cited as an example of such preparedness as practiced by the Army Corps of Engineers.

A goal of this chapter is to present an overview and synthesis of the art of understanding landslides. Some problems remain unresolved, such as the processes involved in the movement of rock avalanches. However, rapid advances are being made as new frontiers of space and oceans are explored, yielding new insights into landslide mechanics.

INTRODUCTION

Landslides, along with other sudden and short-lived phenomena such as earthquakes, volcanic eruptions, floods, and hurricanes, belong in that aspect of earth science known as catastrophic geology. Such processes produce extraordinary terrain changes for the amount of time the activity lasts. When assessed in geomorphic terms, landslides are included with those agents of earth sculpture known as mass movements. Such agents as mass movements, surface water, ground water, glaciers, wind, and currents and waves are the exogene (surficial) processes that erode the landscape, deposit materials, and create new landforms. When landslides endanger humans and their installations, they become known as hazards; when they cause property damage and loss of life, they are disasters.

Landslides are nearly ubiquitous and occur in all climates, on most hilly terrains, and in lakes and oceans. Some rock types are more landslide resistant than others, but the regolith is especially likely to be landslide prone.

The term "landslide" is a misnomer. There is universal agreement that landslide motion may consist of a fall, a slide, or a flow, or any combination of movements. Furthermore, it is becoming increasingly apparent that landslides constitute an important geomorphic process in the subaqueous environment of lakes, reservoirs, and oceans.

In some terrains landslides may be the most important process operating on hillslopes. For example, in parts of the Hawaiian Islands, landslides cause vertical reduction on hillslopes of approximately 1 m in 1,300 yr (Wentworth, 1943). In the San Francisco Bay region, landsliding is one of the most significant processes of landscape sculpture (Nilsen and Brabb, this volume). Thus landslides can be an effective erosion agent, transporting parts of the hillside to lower levels and into valleys. In valleys the less-viscous flowing water removes the material to a different base-level position.

In the 20th century, man has become increasingly aware of landslide risks. Because of increase in population, urbanization, highways, railroads, dams, and other types of construction, landslide damages in loss to property and lives are reaching staggering proportions throughout the world—hundreds of millions of dollars and thousands of lives yearly. As recently as July 1976, hundreds of people were reported buried by landslides triggered by an earthquake in Indonesia. It is now more imperative than ever that landslide-prone topography be recognized and that appropriate planning and management measures be taken to deal with such problem areas.

Previous Work

When compared with the literature on other geomorphic phenomena, landslide publications are surprisingly few. In his classic work, Sharpe (1939) listed 275 references. However, many of the publication topics, such as mudflows, creep, frost action and solifluction, subsidence, and soil mechanics, do not fall within the discussion of landslides. Similarly, many of the 291 items listed by Tompkin and Britt (1951) in their landslide bibliography are not directly associated with landslide analysis (for example, there are 72 references on soil and rock mechanics methods).

The major works that have proved most influential on landslides include those by Heim (1932), Sharpe (1939), Eckel (1958), and Zaruba and Mencl (1969). In addition to these important books, there are other important publications, many of which are listed in this volume. Since 1969, publications in which discussion of landslides is the dominant theme have averaged about 50 per year (1969, 22; 1970, 32; 1971, 80; 1972, 79; 1973, 58; 1974, 33; 1975, 31; and the first 4 months of 1976, 24). There has always been a "gray" literature on landslides, but two events have caused a great increase in these agency and open-file reports, consulting reports, and other data that are not published in the conventional journals or that do not receive strong peer review. The first major addition to this type of literature occurred after the grading ordinances became established in parts of southern California in 1952, and again after the passage of the 1969 National Environmental Policy Act which mandated environmental impact statements of proposed government construction. For example, some of the chapters in this volume are an outgrowth of work that was originally done as mission-oriented research for a specific project. Much of the new literature on landslides, both published and in the "gray" report area, is concerned with identification of unstable slopes, use of various techniques in the mapping and quantification of landslide-prone topography (both ancient slide areas and potential new threats), and providing benefit-cost analyses as aids in the decision-making process.

Although the volume of landslide literature is not as large as for many other areas of geology, the scope of problems associated with landslides is the equal of most geologic phenomena. A single chapter cannot hope to cover the subject in the detail that would be acceptable to most interested people, because landslides mean different things to the geologist, the geomorphologist, the engineering geologist, and the policy makers and social planners. It is the purpose of this chapter, however, to provide an overview and synthesis for some of the diverse elements that constitute landslides and set the stage for the 20 chapters that fill in many of the details. Such a synopsis shows the necessity for interdiscipline cooperation in landslide studies and indicates that the present generation of contributors is adding important new dimensions to the understanding of this vitally important phenomenon.

TERMINOLOGY

There is no universal agreement concerning the definition of landslide or what range of geomorphic features and processes it encompasses. Terminology has changed during the last several decades, and new words have been added to the landslide nomenclature. For example, the term "landslip" for many years was the preferred term in Great Britain, and in the earliest geologic dictionary (Rice, 1953), "landslip" was the principal indexed term.

The leading definitions for landslide in most common professional literature are the following:

The perceptible downward sliding or falling of a relatively dry mass of earth, rock, or mixture of the two. [Sharpe, 1939, p. 64]

A rapid displacement of a mass of rock, residual soil, or sediments adjoining a slope, in which the center of gravity of the moving mass advances in a downward and outward direction. [Terzaghi, 1950, p. 84]

Denotes downward and outward movement of slope-forming materials composed of natural rock, soils, artificial fills, or combinations of these materials. [Varnes, 1958, p. 20]

Rapid movements of sliding rocks, separated from the underlying stationary part of the slope by a definite plane of separation. [Zaruba and Mencl, 1969, p. 1]

A general term covering a wide variety of mass movement landforms and processes involving the moderately rapid to rapid (on the order of one foot per year or greater) downslope transport, by means of gravitational body stresses, or soil and rock material en masse. Usually, but not always, the displaced material moves over a relatively confined zone or surface of shear. [Gary and others, 1972, p. 396]

Sharpe, Varnes, and Zaruba and Mencl greatly expanded upon their definitions to indicate what they considered to be the full meaning of the term "landslide." They also defined the various types of phenomena included under their concept of landslides, and suggested schemes for classification of those phenomena.

The principal areas of agreement among the authors of this volume on what constitutes a landslide are as follows:

1. Landslides represent one category of phenomena included under the general heading of mass movements.
2. Gravity is the principal force involved.
3. Movement must be moderately rapid, because creep is too slow to be included as landsliding.
4. Movement may include falling, sliding, and flowing.
5. The plane or zone of movement is not identical with a fault.
6. Movement should be down and out with a free face, thus excluding subsidence.
7. The displaced material has well-defined boundaries and usually involves only limited portions of the hillside.
8. The displaced material may include parts of the regolith and (or) bedrock.
9. Frozen ground phenomena are usually excluded as landslides (see Varnes, 1958; Hutchinson, 1968), although Zaruba and Mencl (1969, p. 91) devoted nearly a page to discussion of solifluction.

In the literature the rate of landslide motion is defined as being perceptible, moderately rapid, and rapid. More precisely, the rate of movement has been described as having a lower limit "of ⅕ of a foot per year" (Morton and Streitz, 1972, p. 67) or "of the order of one foot per year" (Gary and others, 1972, p. 396). A more useful approach in the consideration for landslide rate of motion would be to compare the rate with that of adjoining hillslopes. To qualify as a landslide, the mass must move separately and independently at rates much in excess of adjacent slopes. Thus, the key to this approach is to identify those hillslope parts that have segmented from the main part of the terrain and whose motion is independent and is being caused by factors that do not affect other slopes in the same manner.

There is also disagreement in the literature as to whether mudflows should be considered as landslide phenomena. Here they are not included as landslides because most mudflows are associated with sheet wash and other fluvial processes. The only types of mudflow that should be considered within the realm of landslides are those whose debris was largely derived by landsliding on the hillsides—not through the characteristic sheetwash removal of weathered detritus by overland flow into the stream channel. It is not unusual for a landslide to exhibit two or more types of motion. However, for classification as a landslide, there must be some rupture in the head and a pulling away of material from the parental mass.

Landslide terminology has been repeatedly used in the description of subaqueous mass movements, as in the writings of Heim (1932), Jones and others (1961), Moore (1964), Hyne and others (1973), and Molnia and others (this volume). Such extension of landslides is logical, and the increasing number of subaqueous slides, with many man-induced, should be a vital concern to engineering geologists, and planners who are involved in decisions for reservoir construction.

The following types of landslide movement and processes are placed into a composite nomenclature that represents current usage and consists of the most important concepts as described by Sharpe (1939), Varnes (1958), Simonett (1968), Hutchinson (1968), Zaruba and Mencl (1969), and other workers too numerous to mention.

Falls

Among specialists the falling of earth material has generally been included as a landslide phenomenon. Perhaps the rationale is that falling materials may provide sufficient energy to initiate further downslope movement.

Falls are the abrupt free-fall movements of earth materials from cliffs and steep slopes in which the displaced mass retains little integrity after impact. It is implied that considerable volumes of material are involved rather than mere isolated rock fragments loosened by routine weathering. When specific events are described, they are defined in terms of the type of material involved, usually as rockfalls, debris falls, and soilfalls. The movement may range from vertical free fall to rolling, bounding, and ricocheting of materials. When extensive and continuous contact of the hillslope by the earth materials occurs, such movements are not classed as falls.

Slides

This designation is reserved for those mass movements of earth material with displacement along recognized shear surfaces where the ruptured mass moves with some semblance of unitary motion. The nature of the shear surface defines whether the slide is rotational or planar. A slump is a rotational slide in which the earth materials commonly retain coherence and move along a shear plane that is concave upward. Such motion produces a backward tilting of the mass with sinking at the rear and heaving at the toe. Planar or translational slides, those whose movement is parallel to the shear surface, are further defined on such bases as the type of material and its coherence during the movement process. Block glides occur when the displaced bedrock maintains its integrity and geologic attitude. Rockslides occur when the detached bedrock becomes fragmented during rapid movement over the failed surface. They can be of large size, and this landslide type (and the rock avalanche) includes some of the most awesome and spectacular events in catastrophic geology. In unconsolidated materials the term "debris slide" (also called "sheet slide" by Zaruba and Mencl [1969, p. 36]) is used for the relatively dry movement of slides in the regolith. When the materials are unconsolidated and uncemented but contain bedded sediments, the slides developed in such materials are known as slab slides, or failures by lateral spreading (Varnes, 1958). Such slides commonly move on a surface of quick clay, and the involved mass usually consists of some units that become tilted and broken during the motion.

Flows

Landslides in which flow is the dominant transport mechanism involve movements in which the materials act as a viscous mass. They are defined in terms of the nature of the earth materials and moisture content. Greater controversy exists concerning what to include as flows and what constitutes the mechanism of flows than for the other categories of landslides.

Landslides can be complicated and compound events containing many types of motion in a single landslide, ranging from fall to slide and flow. A rock avalanche is an exceptionally large mass of broken bedrock whose terminal movement occurred as a very rapid dry flow. Varnes (1958) and Mudge (1965) used the terms "rockfall avalanche," "rockslide avalanche," and "rock fragment flow" to describe such a process. The version "rock avalanche" is used in this chapter because it is shorter, and it is tacitly assumed that rockfall and rockslide movements are necessary to produce the extraordinary velocity that results in the flow motion. Such usage is consistent with that of Gary and others (1972) and Howard (1973). During transit the rock fragments developed such turbulence that shear surfaces are absent in the matrix. Dry flows can also occur in fine-grained sediments, and their movement must generally be triggered by an earthquake. Such flows assume the name of the most common constituent, such as sand flow or loess flow.

Although many different terms have been used to describe wet flows, only three will be used in this chapter, and their differences are related to type of material, topographic expression, and change in physical state. The original definition as stated by Sharpe (1939, p. 61) is still appropriate for a debris avalanche that has "a long and relatively narrow track, occurs on a steep mountain slope or hillside in a humid climate, and is almost invariably preceded by heavy rains which increase the weight of the unadjusted material and aid in its lubrication." Debris flows differ from debris avalanches in that the scar head is much wider, a larger amount of water becomes incorporated into the earth materials, and the lower viscosity allows for much longer transit into and within stream valleys. Both processes involve earth materials that constitute the regolith on steep slopes. Liquefaction flows (Gary and others, 1972) occur in sensitive stratified sediments where interstitial water has been triggered into separation from the particles so that the entire body moves as a flowing viscous mass. The term "earthflow" is a catchall term and refers to a wide range of unconsolidated materials that have undergone some variation of flowage. A great variety of terms are used in the literature for such features, such as "slipouts," and in this volume (Mollard), "skin flows," and "bimodal flows."

Subaqueous Slides

Underwater sliding mass movements should be referred to as slides and not landslides. Although slides have been recognized for a long time in lake and marine environments, new oceanographic techniques are adding much to their analysis (Molnia and others, this volume). There are at least two common types: flow slides and underconsolidated clay slides. Flow slides occur in sands and silts deposited underwater with low stability and cohesion. Collapse results from stresses that exceed their threshold for maintaining an angle of repose. Underconsolidated clay slides occur in clays with excess porewater pressure. They can move with very little external force on very gentle slopes.

Classification

In referring to schemes for classifying landslides, Terzaghi (1959, p. 88) remarked, "A phenomenon involving such a multitude of combinations between materials and disturbing agents opens unlimited vistas for the classification enthusiast." This certainly has been true. Nearly every conceivable approach has been used in attempts to develop a logical classification of landslides. The reader is referred to Sharpe (1939, p. 10-16) for a review of early classification ideas. Classifications have been based upon one or more of the following parameters: (1) character of the material—bedrock, debris, soil, sediments; (2) water content; (3) type of movement—fall, slide, flow; (4) rapidity of movement; (5) cause of movement—earthquake, excess moisture, man; (6) relation of slide to the sliding surface—planar, rotational; (7) slide location—subaerial or subaqueous; (8) geometry and morphology of resulting topography; (9) relative importance of direct and indirect components of the earth materials; (10) environmental setting; (11) regional and physiographic aspect; (12) size and importance; (13) degree of activity—active, dormant; (14) size of material; (15) mechanics involved; and (16) climate.

The two most widely used systems for classifying landslides (and Sharpe's related mass movements) are those by Sharpe (1939) and Varnes (1958). Their schemes are similar because both are based on the same four parameters: (1) type of movement, (2) rapidity of movement, (3) character of material, and (4) water content.

Some of the more recent classification systems stress the environmental aspects of landslides that must be considered in making land use decisions. For example, Erskine (1973) classified landslides as (1) stabilized landslides—those that show no evidence of recent activity; (2) recently active landslides—those that show recent movement; (3) reactivated landslides—those that started to move again after a period of stability; and (4) active landslides—those that show no signs of stability. Crozier (1973) quantified the geometry of landslide areas to aid in predicting landslide-prone terrain and the types of movement that would most likely occur. In the San Francisco Bay region study (see Nilsen and Brabb, this volume), the U.S. Geological Survey uses a sixfold division of numbers as indices for the degree of relative instability of hillslopes. Czechoslovakia in 1961-1962 carried out a program for landslide recognition and classification for purposes of land management. A total of 9,164 landslides was registered (Zaruba and Mencl, 1969, p. 6), and the landslides were grouped into three categories: (1) landslides with simple rotational surfaces, (2) landslides with planar sliding surfaces and (3) landslides with horizontal translation of pre-existing slide surfaces.

The classification developed for this volume (Fig. 1) is intended to provide a convenient system consistent with the variety of landslide phenomena that are discussed in the 20 chapters. Type of material (bedrock, regolith, sediments) and type of movement (slide, flow, fall) constitute the principal coordinates in the matrix (Fig. 1). Other important considerations include sediment size, integrity of materials, and water content.

CAUSES

Landslides constitute an abrupt and short-lived geomorphic erosional process and can be triggered in many different ways. Most causes involve a loss in support under or in front of the shear surface, or a change in the physiochemical constitution of the earth materials. Three factors influence the stability of hillslopes:

1. The internal properties of the earth materials. These include the type of rock and its structural character. Many features are important, such as the degree of consolidation and cementation; thickness and arrangement of different lithologic units; strike and dip of zones of discontinuity; and size, type, and distribution of internal structures such as faults, joints and other fractures, cleavage, foliation, schistosity, bedding, and so forth.

2. The geomorphic setting and environment. Hillslope characteristics are crucial in determining the size and severity of the landslide. Thus landslides will have different magnitudes depending upon total topographic relief, steepness of hillslope, and shape of the land surface. In addition, the character of vegetation, slope orientation, climate, and antecedent moisture conditions can be determining factors in slope stability.

3. Independent external factors. These causes are usually termed the triggering mechanisms and provide the immediate stress that initiates movement of the mass. Prior to displacement the regolith and bedrock on the hillside are in a state of balance until the critical threshold of their cohesion and stability is exceeded by some force that upsets the equilibrium. Stress that exceeds the boundary stability limit must be unusual, because hillslope stability is the norm, and materials can endure in place for many years prior to failure.

Three of the most important landslide causes will be emphasized in the following sections: excessive precipitation, earthquakes, and human activities.

Excessive Precipitation

Antecedent moisture conditions determine whether large amounts of rainfall will successfully trigger a landslide. If earth materials already contain significant moisture from prior rainfall, the severity of precipitation from a new storm can be less and still trigger landsliding. If other factors are equal, magnitude, intensity, and duration of the storm are all-important factors that can contribute to hillslope instability. Excessive precipitation weakens earth materials by displacing air in pore spaces and fractures, by increasing overburden pressure on potential slide surfaces, and by providing pore-water pressure along the shear surface. Other factors that determine rainfall effectiveness in triggering landslides include the geologic character of the regolith and bedrock and the geomorphic properties of the terrain. Conditions that enhance ultimate instability occur when surficial materials are porous and permeable and are underlain by low-permeability material at depths less than several metres. These conditions are especially prevalent in the loose, friable regolith of crystalline rocks in humid to tropical climates (see Durgin, this volume).

Brazil

Exceptionally heavy rains in 1966 and 1967 ravaged mid-southern Brazil in areas from Rio de Janeiro to a region 50 km west wherein "Landslides numbering in the tens of thousands turned the green vegetation-covered hills into wastelands and the valleys into seas of mud" (Jones, 1973, p. 1). Furthermore, Jones asserted that the 1967 storm "laid waste by landslides and fierce erosion a greater land mass than any ever recorded in geological literature." This storm affected a 25-km-long and 7- to 8-km-wide banana-shaped area.

The initiating forces for these landslides were exceptionally heavy rainstorms, even for Brazil. The storms on January 10, 11, and 12, 1966, created 484 mm of precipitation, and storms of March 26 and 27, 1966, provided an additional 320 mm of precipitation in an 18-hr period. The January 22 and 23, 1967, storm lasted only 3½ hr, but deluged the area with an average of about 240 mm of precipitation. Jones (1973) classified the landslides that resulted from these storms into four groups: (1) slump earthflows, (2) debris slides and avalanches, (3) debris flows and mudflows, and (3) rockfalls and rockslides. The landslides and concomitant floods killed more than 2,700 people (1,000 in 1966 and 1,700 in 1967).

The landslides were numerous, numbering into the tens of thousands, and several of the individual ones, and others in combination, proved exceptionally destructive. The mudflow in Floresta Creek valley alone killed hundreds of people and reached depths of 12 ft (3.6 m). The damage to property and industry throughout the region was stated to be inestimable, and the power plant that served Rio de Janeiro suffered almost total destruction. Nearly 50 debris avalanches and debris flows joined together as a giant mudflow that nearly devastated the power plant. An additional 50 avalanches and flows also ruined nearby facilities. A single rockslide in the Laranjeiras district killed 132 people. In some areas the precipitation set the stage for delayed reaction, as in the Laranjeiras district where the slopes had been weakened. But the most fatal landslide did not occur until February 18, 1967. Other unstable slopes may not fail for years to come.

Asia

Numerous landslides have occurred in Japan, and in historic times several of the rock avalanches have exceeded 100 million m³. Ohsawakuzure slide on the western slope of Mount Fuji is still active and produces frequent rockslides. In March 1961 a massive landslide at Yui Shizuoka Prefecture in central Japan required corrective measures that cost $56 million to keep the railroad route open. Most of the landslides, especially in urban areas, have been triggered by heavy rainfall. In July 1938 Kobe was hit by rainstorms that produced rock mudflows that killed 461 people and destroyed 100,000 houses. Additional disastrous landslides occurred in 1961 and 1967. In September 1945 the Makurazaki typhoon blasted Kure and produced rocky mudflows that resulted in death for 1,154 people. The Kanogawa typhoon of September 26, 1958, produced the heaviest rainfall in the recorded history of Tokyo—392.5 mm in 24 hr. It created 1,029 landslides and land collapses and killed 61 people (Nakona, 1974).

In June 1966 Hong Kong was subjected to a series of rainstorms in which ten to fifteen times the normal intensity was reached. It rained every day for a two-week period, and in 24 hr on June 11–12, 401 mm of precipitation was recorded. This, added to the previous 314 mm of rain from the first of the month, was the trigger that initiated wholesale landsliding. "The most disastrous and spectacular mass movements were generally large-scale and deep-seated landslips involving rotational shearing and slumping" (So, 1971, p. 248). So identified, mapped, and studied 558 of the resultant landslides. Another 144 mass movements that included debris avalanches, boulder falls, rockslides, and washouts were analyzed. "Many of the landslips and the washouts were associated with road sections and slopes artificially modified through construction and cultivation" (So, 1971, p. 248). Bedrock in the area is predominantly volcanic and igneous on which a deep regolith had formed.

Keng (1970) discussed the influence of heavy rainstorms on Taiwan. On this island 364 mm of rain in 24 hr from a typhoon opened a new earth fissure 250 m long, 55 to 75 m wide, and 20 to 30 m deep. In addition a 2-ha rockslide of 240,000 m³ destroyed 42 houses and damaged 8. High water in the adjacent stream had undermined the toe of the hillslope causing the instability in the sedimentary bedrock.

United States

On August 10–11, 1969, Hurricane Camille produced an 8-hr deluge of 710 mm of rain that Williams and Guy (1971, 1973) described as the worst natural disaster in central Virginia history. Property damage in Nelson County alone amounted to more than $116 million, and the combination of debris avalanches and flooding killed 150 people. Most died from broken bones and blunt-force injuries rather than from drowning. Williams and Guy made detailed studies of 186 of the hundreds of debris avalanches and concluded that these landslides (1) followed pre-existing depressions on hillsides generally steeper than 35°; (2) produced head scars at the steepest part of the hill where the convex slope merged with a concave or planar slope; (3) were more numerous on north, northeast, and east-facing slopes; and (4) caused rapid devastating surges of water and sediment in stream channels. The usual dimensions of typical landslide scars were stated by Williams and Guy (1971, 1973) as being 200 to 800 ft in horizontal length, 25 to 75 ft wide, and 1 to 3 ft deep. Some were as short as 20 ft or as long as 1,000 ft and were up to 200 ft wide and 20 ft deep. Complete mapping of 12 representative slides showed the average amount of regolith removed was 88,000 ft³ (2,480 m³).

Causes of landslides in California can be complex, but as Nilsen and Turner (1975, p. 16) pointed out, rainfall can be a significant factor in their development.

The history of landslide activity also indicates that the distribution, amount, and pattern of rainfall exert a strong influence on landsliding. Rainfall in excess of 7 inches (18 cm) ... generally causes large numbers of landslides, particularly if the ground is already wet from previous storms. In general more rainfall is required at the beginning of the rainy season than at the end to produce large numbers of landslides. The pattern of rainfall during the rainy season is more important than the total amount.

Other Areas

A severe storm on February 28, 1966, hit New Zealand with high winds (Beaufort force 9) and rains that averaged 26 mm/h for 8 hr and caused hundreds of landslides. Pain (1969) sampled 50 landslides which he classified and counted as follows: block slumps, 5; block guides, 2; debris slides, 5; debris avalanches, 23; debris flows, 2; and earth flows, 13. The slides all occurred in Jurassic siltstones and sandstones. The Utiku landslide (see Stout, this volume) seems to be an unusual landslide in that its movement is seasonal and largely related to precipitation during the rainy season.

Arber (1971, 1973) discussed various landslides in England and described the largest landslip that has occurred on the eastern coast. In the area known as the Axmouth Landslip, strong gales and heavy spring rains occurred periodically in June 1839 and continued for several months. They apparently so weakened the chalk cliffs that overlie seaward-dipping sands and clay that on December 25, 1839, a major slide with cliffs to heights of 500 ft (150 m) developed in the 4-mi-long, 0.5-mi-wide area (6.4 km, 0.9 km). The landslide, which was subsequently investigated by the famous British geologist Conybeare, formed a new inland cliff 210 ft high, 0.5 mi long and 200 to 400 ft wide (62 m, 0.9 km, 60 to 120 m) and involved at least 150 million ft³ (4.2 million m³) of material. Movement occurred as a slightly rotational shear in clayey materials.

More recently, Lundgren and Rapp (1974) described how high precipitation initiated a landslide in Tanzania and destroyed the water supply of a village. The slide left a scar 300 m long and 10 m deep on the adjacent hillside.

Earthquakes

Probably the most cataclysmic of all landslides are those caused by earthquakes. Severe seismic shocks not only loosen bedrock and regolith but also may produce physical changes in sensitive sediments. For example, fine-grained sediments may undergo liquefaction and flow as a viscous mass. Daly (1926) discussed terrain effects produced by some of the world's most severe earthquakes. The Lisbon earthquake of 1755 affected extensive areas throughout Portugal and caused much landsliding along sea cliffs and in valley walls of rivers.

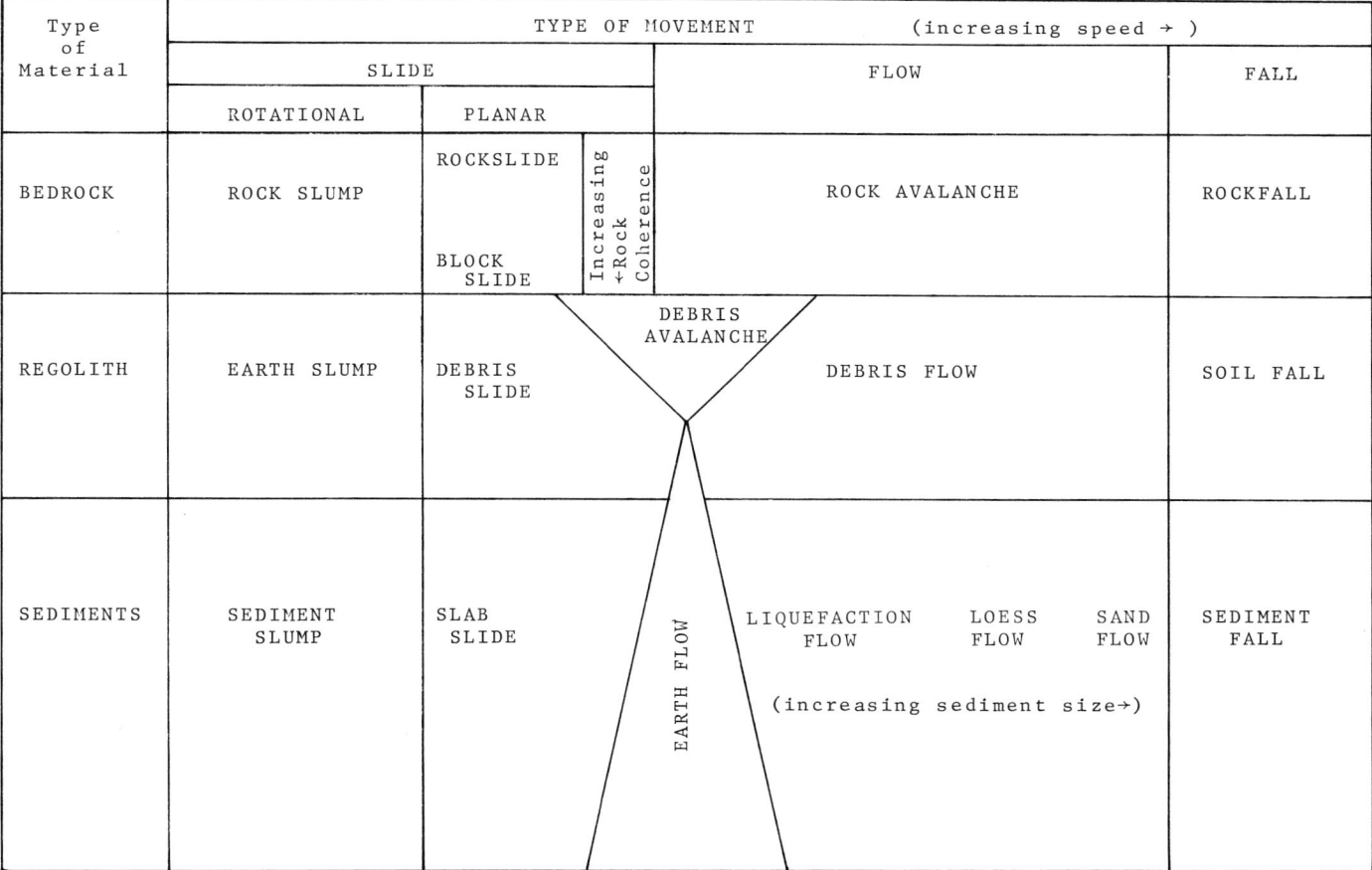

Figure 1. Landslide classification.

According to Daly (1926), the Indian earthquake of 1897 was exceptionally massive and profoundly altered and scarred many valleys, some for stretches of 20 mi. The shock was felt as strong motion in a 160,000-mi² area, and was noticeable in a 1,750,000-mi² area.

United States

1. The New Madrid, Missouri, earthquake of 1811 caused widespread landslides in the loess of river bluffs as far north as the Ohio River. The longest continuous stretch of landslide occurred on the east side of the Mississippi River where for 35 mi (56 km) along the Chicawa Bluffs depressions 100 ft deep (30 m) and 100 ft wide (30 m) were created and were still visible in 1869. Many of the movements were rotational. Fuller (1912, p. 61) stated, "Water soaking downward collected in the more sandy beds above the clays, developing a sort of 'lubricating action' at the clay contact. At the same time it produced, by saturating the base of the upper deposits, a zone of easy flowage and slipping."

2. The San Francisco earthquake occurred at 5:12 a.m. on April 18, 1906. The shock was felt on land covering 175,000 mi² (453,000 km)—north to Coos Bay, Oregon, south to Los Angeles, and east to Winnemucca, Nevada. The earthquake resulted from a series of sharp movements of the San Andreas fault (maximum horizontal motion of 21 ft [6 m]) and lasted little more than a minute, with the sharpest tremor 40 s after the start (Lawson, 1908). The zone of destruction was primarily in a 350-mi-long by 70-mi-wide area (560 × 112 km).

The earthquake opened numerous cracks in surficial materials which during the winter of 1906–07 became sites for extensive slumping. Lawson (1908) divided the landslides into four categories: (1) earth avalanches, (2) earthflows; (3) earth slumps, and (4) earth lurches. The greatest incidence of the earth avalanches, defined as "the slide of dry earth and rock upon precipitous slopes or their fall from cliffs," occurred along the sea cliffs of the coast the morning of the quake. Lawson (1908, p. 386) attributed the earth lurches to the horizontal motion of the faulting, "Cracks and fissures cut the ground up into strips and prisms which lurch toward the stream trench, or, it may be toward an abandoned slough, the lurch being accompanied by a rotation of the prism."

The largest single landslide produced a scar about 1 mi (1.6 km) wide and 0.5 mi (0.9 km) long, but most measured only hundreds of feet. One of the largest earth slumps occurred near San Pablo with a 1,500-ft (450-m) width of scarp that was 50 ft (15 m) high and extended 400 ft (120 m) downslope. The largest earthflow was 2,700 ft (810 m) long, 100 ft (30 m) wide, and 3 ft (1 m) deep. It had moved out of a scarred cavity that was 450 ft (135 m) wide, 900 ft (270 m) long, and 6 ft (2 m) deep, and involved about 90,000 yd^3 (67,000 m^3) of material.

3. In Montana the Madison rock avalanche of August 17, 1959, killed 28 campers. The rocks were dry because it had not rained for 6 weeks, but their structure favored instability, with Paleozoic dolomites dipping 40° toward the valley and a basement Precambrian complex of gneiss and schist.

The essential condition seems to be that a mass of weak rock, held in place largely by friction, was suddenly transformed into a mass which literally flowed into the canyon. Also important is the evidence that this fluidity was produced simultaneously over the slide area. Factors tending to reduce the internal friction or resistance to sliding, or to overcome this resistance by imparting momentum to the mass, are presumably responsible for the sudden change from a solid to fluid behavior. [Hadley, 1964, p. 121]

In addition to the main slide, or rock avalanche, other landslides up to 350,000 yd^3 (267,000 m^3) and rockfalls were common throughout the area. From the time of the earthquake until September 18, 1959, an earthflow 400 to 800 ft (120 to 240 m) wide and 0.5 mi (0.9 km) long had moved a distance of 100 ft (30 m).

Many landslides were rock falls and debris avalanches of weathered bedrock on cliffed slopes, set in motion directly by the main earthquake or by aftershocks. Several were debris slides into Hebgen Lake, apparently augmented by the presence of water-filled material along the lake shore. Many others were massive slumps of colluvial mantle jolted downhill by earthquake shocks. The slumps were commonly marked by fissuring along ridge crests and rarely by compressional folds and thrust faults in surficial deposits at the base of the slope. [Hadley, 1964, p. 107]

4. The Alaska (Good Friday) earthquake of March 27, 1964, killed 114 people and caused significant damage throughout a 50,000-mi^2 (129,000-km^2) area (Hansen, 1965). It registered a magnitude of 8.5 on the Richter scale, and the early shocks lasted 1.5 to 7 min. For a 69-day period after the principal earthquake, more than 12,000 aftershocks of magnitude greater than 3.5 on the Richter scale were recorded, and some were as high as 6.7. The greatest property damage and loss of life occurred in Anchorage where 700 acres with 750 homes, 14% of the city, were destroyed by landslides of the slab-slide type.

Triggering of landslides by the earthquake was related to the physical-engineering properties of the Bootlegger Cove Clay, a glacial estuarine-marine deposit that underlies much of the Anchorage area.... Most of the destructive landslides in the Anchorage area moved primarily by translation rather than by rotation.... [The houses] slid on nearly horizontal slip surfaces after loss of strength in the Bootlegger Cove Clay. [Hansen, 1965, p. 1]

Another type of landslide that the earthquake caused was the rock avalanche of the Sherman landslide which Shreve (1966, p. 1640) has described. The rock mass was shaken loose along bedding planes and down a 40° slope. The sandstone and argillite that compose the bedrock "was pervasively jointed and fractured long before it fell." Shreve discussed the Sherman slide as having similar characteristics to the Blackhawk landslide, and he was especially impressed with the unbruised fragments in the Sherman slide and the jig-saw fit of the shattered rocks.

The Alaska earthquake was also instrumental in initiating extensive subaqueous slides. The Seward, Alaska, waterfront was greatly altered where deltaic-type sediments slid oceanward deepening the original pier area from 20 to 30 ft (6 to 9 m) to 130 to 180 ft (39 to 54 m). The subaqueous slides in the Passage Canal near Whittier, Alaska, created 104-ft (31-m) waves. One of the 52-ft (15.6-m) waves hit Whittier and destroyed the boat harbor, pier, docks, and numerous installations and homes. In Kenai Lake subaqueous slides occurred in the nine deltas, with some slides moving more than 7,000 ft (2,100 m). Backfill waves from the slides occurred to heights of 72 ft (21 m) (McCulloch, 1966).

5. The San Fernando Valley earthquake of 1971 produced more than 1,000 landslides in a 250-km^2 area in the hills and mountains of the adjacent region. The types of landslides include rockfalls, soilfalls, debris slides and avalanches, and slumps. The intensity of ground shaking controlled landslide distribution. Locally abrupt lithologic changes and structural properties were important in determining the size and number of landslides, such as degree of fracturing and faulting (Morton, 1971). Other factors that contributed to landslide development were the presence of pre-existing landslide deposits and the degree of bedrock exposure.

China

Probably the greatest single landslide disaster in recorded history occurred in Kansu Province, China, on December 16, 1920, where a 100 × 300-mi (160 × 480-km) area was devastated, and 100,000 to 200,000 people were killed. The earthquake shock dislodged loess deposits on the steep hillslopes that changed into a pulverized and powdery state and literally flowed into adjacent valleys.

In each case the earth which came down bore the appearance of having shaken loose clod from clod and grain from grain, and then cascaded like water, forming vortices, swirls, and all the convolutions into which a torrent might shape itself. [Close and McCormick, 1922]

Giant scars were etched on the terraced hillsides, and the morphology of valleys was drastically changed by the flood of deposits from the loess flow.

Peru

Another staggering catastrophe occurred in Peru on May 31, 1970. A 7.7 (Richter magnitude) earthquake centered 85 km away caused landslides that destroyed the towns of Ranrahirca and Yungay and killed 21,000 people. The setting of this event was on Mount Huascaran in the Peruvian Andes with elevation of 22,100 ft (6,600 m). Granodiorite formed the bedrock, and the upper slopes contained glacial ice with till and moraines below. The earthquake shock detached a snow cornice near the peak which fell along with rocks in free fall for 2,000 ft (600 m) and crashed into morainic deposits. The snow and ice partly melted on impact, and the entire mass incorporated the unconsolidated morainic sediments and rapidly slid and dropped 9,000 ft (2,700 m) in elevation along a 23° slope, which then changed to 5° in the valley of the Llanganuco River, dropping another 2,200 ft (660 m) in elevation. In all, the slide moved a distance of 14.5 km in 3 min at a speed of 250 mph (400 km/h). Farther downslope the material moved another 50 km at a speed of 15 mph (25 km/h) as a mudflow and highly sediment-charged stream. Part of the movement seems to have been airborne, because at places the rock avalanche demolished vegetation at heights of 60 to 70 ft (18 to 21 m) above the valley, and yet at Yungay a few trees in the valley were undamaged except for some later debris deposited around them (Browning, 1973).

This same earthquake caused thousands of rockslides within a 100-km area of Chimbote, Peru, and the total loss of life from the earthquake, landslides, and floods was estimated to be 70,000 people.

Many other illustrations could be used to document the earthquake-landslide relationship, such as the Himalayan earthquake in December 1840 that

> ... shook loose part of the western spur of Nanga Parbat (26,630 feet), where the Indus has cut a gorge 15,000 to 17,000 feet deep through a great mountain range. The gigantic landslide blocked the river and dammed back the water for 40 miles. The resulting lake reached a depth of over 1,000 feet before it overtopped the obstruction. The water then burst through with such violence that in less than two days the lake emptied. A devastating flood tore down the valley, sweeping away a Sikh army encamped near Attock and carrying destruction for hundreds of miles. [Holmes, 1965, p. 485]

Human Activities

There has been a sharp increase in destruction by landslides from man-induced causes during the 20th century. Not only has the population been expanding but also man's ability to change hillslopes rapidly has produced anthropogene landscapes. The advent of monstrous earthmoving equipment powered by petroleum has permitted massive artificial slope changes throughout the world. In addition the construction of dams (almost universally absent in pre-1900 terrain), the building of highways and more railroads, and the urbanization process have all greatly contributed to man's desecration of natural environments. These changes that can lead to landsliding can be grouped into two principal categories: alteration of hillslope configuration and increase of moisture in earth materials.

Alteration of Hillslope Configuration

Man disturbs the natural equilibrium of stable slopes in a number of ways in various types of construction. Thus any type of activity that increases the hillside gradient, cuts away earth materials, or fills upslope areas can lead to instability and set the stage for a potential landslide.

1. Highway construction is a very common cause of landslides (see Font, Fowler, Leighton, and Mathewson and Clary, this volume). A series of landslides occurred during the construction of a 1.2-mi (1.9-km) divided 4-lane highway near Sears Point, California (Woods, 1958). Excavation of the road cut, 1,500 ft (450 m) long with maximum depths of 75 ft (22 m) was started November 16, 1949. Repeated landsliding occurred during construction which hampered the project. A total of five landslides developed, and the largest had a 400-ft (120-m) toe width. The volume of all landslides totaled 105,000 yd^3 (80,200 m^3). It is believed that the immediate cause of the landslide was the removal of support by the excavation, because aerial photographs show there had been no sliding prior to construction (Wood, 1958). Other conditions that permitted the landsliding were a combination of geologic structure, high water table, and lithologic variations in the sediments. The beds of the semiconsolidated Pliocene and Pleistocene sediments dipped toward the valley, and stability of all earth materials was made precarious by intercalated tuffaceous clays. The landslides were along the dip slopes of these strata.

Nossin (1972) described a series of 104 landslides that occurred between 1966 and 1968 (39 in 1966, 38 in 1967, and 27 in 1968) in the Province of Cosenza, Italy. He analyzed the size of the area that was affected, the cost of treatment to stabilize the slides, and their causes. He attributed their initiation to the building of roads. "In this area the trigger has often been human interference in the slope for purposes of road construction" (p. 606). The rock character of the earth materials was suitable to produce landslides when undercut. The bedrock consists of Paleozoic metamorphics that contain tectonic shearing and crushing structures with clay zones that act as water traps and aid in lubrication of the slip planes.

2. Another way that man affects slope stability is by artificially filling the upslope area so that the excess load creates overburden stress that exceeds the critical cohesion strength of earth materials. In the construction of Lookout Point Dam, Oregon, a trackway was dug high on the slope. The trench was 1,500 ft long, 30 ft deep, and 100 ft wide (450 m, 9 m, 30 m) (Staples, 1957). The 5,000 yd^3 (3,800 m^3) of excavated material was dumped on the hillslope below the trench during August and September 1950. In October the heaviest rains in the history of the area occurred, and from October 26 to 30, 7.67 in. (19.5 cm) of rain fell. The stability of slopes had previously been changed by removal of all trees during logging operations.

> The removal of forest cover produced bad results both due to the removal of roots which lace the soil together and stabilize it, and also because it permitted more free movement of underground water with less storage effect. [Staples, 1957, p. 47]

All of these factors aided in producing two separate landslides, an upper one and a lower one, which apparently were not physically connected. The bedrock in the area consisted of 10 ft (3 m) of medium-grained tuff with porphyritic andesite breccia below. The regolith was 2 ft (0.6 m) thick and consisted of clay with angular rock fragments. The upper slide started heaving in October at a point 80 ft (24 m) below the toe of the fill, and the lower slide started movement on November 22. By December 6, 1970, there had been movement of 5.6 ft (1.7 m). On December 11 the contractors, at the request of the Southern Pacific Railroad whose tracks were being endangered, excavated and removed the bulge in the lower slide which then caused the entire area to slump.

3. Mining operations provide another method whereby man drastically modifies terrain and severely tests the stability of hillslopes. Slate mining was an important cause of the disastrous Elm landslide in 1881 (Heim, 1882). Early reports of the costly Frank landslide of 1903 stated that coal mining was a contributing factor (McConnell and Brock, 1904; Daly and others, 1912), but recent workers conclude that other factors are the dominant consideration (see Mollard, this volume, and his references).

A man-induced landslide caused by the mining of clay at Haverstraw, New York, has been described in the following manner:

The brickyards at the north end of the town were the scene of an earth slide on January 8, 1906, that was accountable for extensive damage and the loss of 20 lives. It is one of the few examples of an artificially produced disturbance of catastrophic proportions on record in the Hudson valley. The cause is ascribable to the overdeepening of a cut in terraced clays, the latter belonging to the series of clay beds which front the river. [Newland, 1916, p. 105]

The clays had been worked back from the Hudson River along a cut that was at least 1 km long and near a village street. The cut did not leave sufficient mass to resist the weight that the street and buildings exerted.

Hamel (1970) described a landslide and rock failure in the Pima Mine, Arizona. The failure zone was 100 ft (30 m) thick, and between June 16 and November 9, 1966, the total movement was 7.2 ft (2.2 m). Another failure occurred between November 29 and December 9, 1966, in an area 70 to 100 ft wide and 600 ft long (30 × 180 m) with movement of 4 ft (1.2 m).

4. Many different activities of man produce an increase in content of water on surface and subsurface materials, such as improper design of drainage ditches, rerouting of channels, watering of lawns, and seepage from septic tanks. The largest landslides, however, result from water-level changes in materials that are hydraulically connected to reservoir water levels. The calamity at Vaiont, Italy, was largely initiated by man-controlled changes in the water level of the reservoir ponded by the spectacular Vaiont dam. On October 9, 1963, a landslide mass of 260 million m³ catastrophically plunged into the reservoir and created giant waves more than 100 m high that overtopped the dam, and 1 km below in the valley the wall of water was still 70 m deep (Kiersch, 1965). More than 2,000 people lost their lives in this event. Both Kiersch (1965) and Jaeger (1969) attributed man's influence as an important factor in this disaster. The changes in reservoir water level altered bank storage conditions, which decreased the slope stability of the rocks. Other factors that contributed to the landslide were the favorable geologic structure which consisted of sedimentary strata that dipped toward the valley. The rocks contain abundant arcuate jointing fractures, and differences in bedding planes are prominent. In addition, heavy rains in August and September 1963 produced increased percolation and helped raise the ground-water level. This recharge increased the bulk density of the rocks above the initial water levels, and the extra weight contributed to a reduction in the shear strength of the materials.

Jones and others (1961) investigated landslides in the Columbia River valley where Franklin D. Roosevelt Lake is located—the reservoir of Grand Coulee Dam. They described 10 different types of landslides in the sample of 321 that were analyzed: (1) recent slump earthflows, (2) recent slump earthflows limited by bedrock, (3) ancient slump earthflows, (4) slip-off slopes, (5) multiple alcoves, (6) landslides off bedrock, (7) talus slumps, (8) landslides in artificial slopes, (9) mudflows, and (10) dry earthflows. Seventy percent of the landslides were slumps in categories 1 and 2.

The principal cause of the landslides in the area was the weakening of sediments by ground water. The sediments generally have a lower shear strength when saturated or partly saturated than when dry. During the filling of Franklin D. Roosevelt Lake the sediments bordering the lake became saturated and many landslides occurred. Apparently, buildup of ground water in certain areas downstream from Grand Coulee Dam has been the principal cause of landsliding there. [Jones and others, 1961, p. 31]

These authors also reported that construction of the dam was hindered by more than 100 separate slides between 1934 and 1952, that the cost from landslides was $20 million, and that "a large part of it could have been saved had presently known geologic facts been available and had engineers applied them in designing and construction work." For example, during the early construction work in 1934 to 37, the landslides were caused by excavation along the toe of riverbank slopes, and some of the later ones were due to the earlier excavation and the overloading near spoil banks.

Additional Landslide Causes

Many landslides are not caused by a single factor but involve a merging of several conditions and events. The Portuguese Bend landslide in the Palos Verde Hills of California is a well-known example of sliding due to several causes. Many slopes throughout this region are unstable and can be classified as landslide-prone topography. Starting in 1956 after a series of home developments on the slopes, the slopes began an obvious accelerated motion. By 1960 the slides had caused damages of millions of dollars, and in 1970 the slides were still continuing to move 0.02 ft/day (0.6 cm). The area is underlain by weakly consolidated sedimentary strata with clay units that can move as a viscous mass when thoroughly wetted. A combination of factors was responsible for the movement. The water level in subsurface materials had been greatly increased by seepage from septic tanks and watering of lawns.

Figure 2 shows a landslide area and the scarred surface that is caused by a combination of factors. This alcove-shaped feature (200 m long, 150 m wide) occurs on the east valley wall of the south-flowing Tioughnioga River, New York. The slide occurs in a topographic reentrant of bedrock walls where glaciolacustrine clay, silt, and fine-grained sands are 21 m thick (Figs. 3, 4). These sediments were deposited behind a barrier to the south which formed during late Wisconsin time at the confluence of the Tioughnioga and Chenango Rivers. The 500,000-m^3 slide is bounded by New York Route 79 on the west and a Broome County road on the east. The following have all played a role in creation of this landslide, which is a sediment slump:

1. During construction of New York Route 79, the road undercut the toe of the landslide-prone terrain.
2. The Broome County road on the upper slope placed an extra load on the head of the slide and allowed water from upslope to drain into the slide area.
3. Most of the movement and terrain collapse occur in the late winter and early spring when freeze-thaw cycles are accelerated, snowmelt is commonly at a maximum, and heavy rains occur.
4. There is instability of the weak sediments on slopes that exceed their normal angle of repose (20%).
5. Because of the interbedded character of the sediments, they have a tendency to undergo piping. Figures 3 and 4 show the emergence of pipes. These conduits transport water that flowed onto the slide area upslope, moved along and wetted bedding-plane surfaces, and re-formed into separate subsurface flow paths in response to hydraulic head changes within the subsurface flow system. Pipes are up to 0.6 m in diameter and extend into the sediments more than 6 m. They develop preferentially in sandy units perched on top of clays.

A variety of other activities by man can also produce landslides. Scott (1931) discussed how blasting for road-grade changes can cause rockslides. At Menton, France, where landslides killed 11 people, a change in land use from growing olive trees to growing carnations was cited as the cause of the landslides (Morton and Streitz, 1972). The earth materials lost their equilibrium when the stabilizing force of tree roots had been destroyed.

There are many other landslide causes, but they can only be briefly mentioned. Snow (1964) discussed the importance of a large landslide in Peru being due to undercutting by the Mantara River. At the outside bend in the river, a 7 million-yd^3 (5.3 million-m^3) landslide created a 331-ft (100-m) dam in the river. It occurred during the dry season, and no seismic activity had occurred in the region. The temporary lakes and terrace deposits produce evidence that along a 175-mi (280-km) stretch of the river, similar landslides occur every 250 yr. Piteau (this volume) also shows the importance of river undercutting in creating landslides.

Pain (1972) studied 40 landslides in New Guinea, and although an earthquake had been the initial cause for disturbing the area, he cited treefall as the triggering mechanism for many landslides. There were three types of landslides: (1) rockfalls, (2) debris avalanches on 40° to 50° slopes, and (3) earthflows on 30° to 40° slopes.

If the environmental setting is susceptible to instability, a host of factors may interact, singly or together, to produce landslides—anything that will cause disturbances or changes in the physiochemical properties promoting stability, including hydration, drying, removal of cement, absorption, drying, shrinkage, and displacement of air or increase in pore-water pressures. Showing the complexity of landslide causes, Brabb and others (1972) concluded:

Degree of slope and nature of the bedrock seem to be the principal factors controlling the distribution of landslides in San Mateo County. However, other factors may be important locally, such as: the relation between the orientation of bedding, foliation, or cleavage and the slope direction; the amount, spacing, and type of jointing and faulting in the rocks; the extent of undermining of bedrock and surficial deposits by streams; the kind and amount of vegetation present; the amount and distribution of ground water and rain; changes in climate; frequency, location, and magnitude of earthquakes; and the activities of man that differentially change the load or increase pore fluid pressure in slopes.

REGIONAL LANDSLIDE STUDIES

Regional studies of landslides take various forms (see Mollard and Part 2, this volume). Baker and Chieruzzi (1959) analyzed landslides throughout the United States and concluded:

Efforts to relate degree of severity of landslides to standard physiographic sections produced encouraging results, although several deviations were noted. The degree of severity was defined as a function of the effect on engineering as opposed to general landslide susceptibility. Both magnitude of moving mass and frequency of occurrence were considered in assigning the measures of severity. [p. 14]

They made further judgments on which physiographic sections contained the most severe landsliding and identified the geologic formations that are especially susceptible to landsliding. In 1961 and 1962 Czechoslovakia carried out a program of landslide identification. Types of landslides and types of areas affected were determined, for example, 59% of the landslides were on agricultural lands and 23% were in forests (Zaruba and Mencl, 1969). Colton (1975) studied landslides in western Colorado by aerial photograph methods and determined that landslide deposits cover 15,600 km^2. Other recent and ongoing investigations include the massive mapping program of the U.S. Geological Survey in the San Francisco region (Nilsen and Brabb, this volume), and the West Virginia landslide study of major urban areas in that state (Lessing and Erwin, this volume). Carrara and Merenda (1976) have developed a landslide inventory method that has been used in a 1,000-km^2 area of southern Italy.

San Juan Mountains

One of the earliest regional studies in the United States was by Howe (1909) in the San Juan Mountains of Colorado. The type of volcanic and sedimentary rocks made this region exceptionally landslide prone, when combined with the ruggedness of the mountains and oversteepening of slopes created during various glacial periods. He concluded:

Figure 4. Closeup view of Figure 3. Diameter of opening is 55 cm. Note mud slurry being discharged through the conduit which can be visually traced more than 6 m into the hillside.

Shreve (1968b, p. 656) took exception to the theory and terminology employed by Kent and contrasted the fluid mechanism of flow with the air-cushion type of lubrication.

Contrary to the suggestion of Kent (1966) fluidization cannot be the primary mechanism of lubrication in landslides of the Blackhawk type, because the geological evidence shows that these landslides slid rather than flowed, whereas fluidized material by its nature flows rather than slides.

Studies of the moon have brought forth remarkable new revelations. Howard (1973) described a rock avalanche on the moon that began on a 2-km-high mountain and moved 5 km on a plain extending from the mountain base. Thickness of the debris blanket ranged from 20 m at the mountain front to only a few metres at the distal end. The lunar material covered a 21-km² area, and its volume was 200×10^6 m³. "Evidently lunar avalanches are able to flow despite the lack of lubricating of cushioning fluid" (p. 1052). And in describing the motion, Howard stated, "Avalanches can behave efficiently like fluids composed of rapidly moving particles" (p. 1055).

Part of the dispute concerns whether the rock fragments were always in contact with the ground surface and whether air blasts were important. Mathews (1960) reported eyewitness accounts of the Madison landslide to the effect that the slide traveled so fast that the nearby winds hit the survivors so hard that "Several had their clothes literally torn from their bodies ... by the hurricane velocity gusts shot out as the slide displaced air in its path" (p. 334). Crandell and Fahnestock (1965), in supporting movement on air, reported that a fragile thermography shelter 1.5 ft high was not damaged by the slide during its movement over the installation. However, Hsü (1975), in a translation of work by A. Heim in 1881, reported that an eye witness to the Elm landslide said, "The debris mass did not jump, did not skip, and did not fly into the air, but was pushed rapidly along the bottom like a torrential flood" (Hsü, 1975, p. 131). In addition, eye witnesses stated they did not feel any gusts of strong air, and a water pipe bored to a 1-m depth on the valley floor was uplifted and transported more than 1 km and deposited in the debris of a lateral ridge.

Hsü (1975) provided an analysis of the different transport mechanisms and adopted many of the ideas of Heim (1932), who used the term "sturzstrom," and provided a new nomenclatural base to explain the phenomenon. For example, Heim (1932) believed the materials flowed like a liquid and referred to their geometrical characteristics as being similar to lava flows and glaciers, but realized the contrasting flow mechanics of these different systems. Drawing heavily on work by Bagnold (1956), Hsü concluded that the motion of rock avalanches is similar to the flow of a mass of cohesionless grains in a fluid medium. The dispersion of the fine debris and pulverized rock dust among the larger blocks provides an uplifting stress during motion. It is this buoyancy of the interstitial fluid material that reduces the effective pressure of the entrained grains. Thus, the exceptional travel distance is related to a reduction of frictional resistance of colliding blocks dispersed in a dust suspension. "One can easily fancy a stream of colliding blocks swimming with terrifying speeds in a sea of small stones and dry rock powder" (Hsü, 1975, p. 136). Such a process could explain the movement of lunar rock avalanches where neither water nor air can be in sufficient amount to enter into the transport mechanism.

Liquefaction Flows and Landsliding in Fine-grained Sediments

The environmental setting for landslides that move as liquefaction flows is vastly different from most other landslide-prone terrain. An important difference is that such movements can occur on nearly level and featureless surfaces. For an understanding of these landslides, or knowledge that would lead to careful management of the terrain, the stratigraphy and sedimentology of the strata are vital. Such movements only occur in fine-grained stratified sediments that contain abundant interstitial water, and that, in a sense, have been "underloaded." Favorable locations where many of these have occurred are in glaciolacustrine and glaciomarine environments (see Mollard, this volume; Dunn, this volume) and in deltaic regions. Various processes can trigger the slides, but most have been caused by earthquakes or by rapid changes in water content or loading. For example, in the Province of Zeeland, the Netherlands, between 1881 and 1946 there were 229 slides varying in size up to more than 2×10^6 m^3, and they were mostly formed immediately following a very high tide period.

There is an entire gradation of landslides that occur in the general family of the slab-slide–liquefaction-flow category. The specific terminology depends on the ratio of foundering blocks of sediments that retained some coherence after movement to the amount of material that flowed in the sediments that underwent liquefaction. The St. Lawrence region provides a wide range of slides (in the Leda clay) whose movement was by liquefaction flow but in which varying percentages of overlying sediments maintained a blocky to broken integrity. Mollard (this volume) provides two case histories in Canada where, for example, the South Nation River landslide material retained some semblance of blocky cohesion after movement. A variety of terms has been used to describe these phenomena, such as quick clays, sensitive clays (sediments), and thixotropic materials.

Characteristics of some of the liquefaction flows are listed in Tables 1 and 2. Other case histories are described by Newland (1916), Varnes (1958), and Dunn and Banino (this volume). One final note should be mentioned concerning the possible importance of liquified sediments and their relation to seismic events. In a historic study Sims (1973) correlated three zones of soft sediment deformation in strata of the drained Van Norman Reservoir, California, with recent nearby earthquakes. If these clues can be used with confidence, such deformed strata can aid in revealing the earthquake record of a region. Toward this objective a feasibility reconnaissance study was made of the 64,700-km^2 area on the American side of the St. Lawrence Lowland (Coates, 1975). The primary objective of the investigation was to determine if landslides and deformed strata of late Quaternary deposits could yield a record of recurrence intervals for seismic activity in the region. It is difficult to obtain a unique solution to the sediment deformation of strata 10,000 yr old. Contorted units and landslides can be caused by many factors other than earthquakes, so these methods need further testing and study and may still have promise for investigations in nonglaciated terrain.

1. St. Thuribe, Canada. This liquefaction flow near Trois Rivieres, Quebec, was described by Dawson (1899). It occurred on May 7, 1898, and the type has been referred to as a bottleneck flow because the large mass of liquefied sediments flowed from a nearly level surface, in this case a terrace, through a very narrow orifice; 3.5 million yd^3 flowed through a 200-ft opening in 3 to 4 hr. It left a scar 1,700 × 3,000 ft and a depression that ranged from 15 to 30 ft deep. "[The] mass of clay must have simulated a liquid body when in motion." (p. 487).

2. St. Jean Vianney, Canada. This new housing site in the Province of Quebec was situated on rather flat terrain and underlain by a 30-m thickness of the infamous Leda clay (a glaciomarine sedimentary deposit that has produced most of the landslides throughout the St. Lawrence region). On April 23, 1971, a few cracks appeared in some of the asphalt streets, two driveways settled 5 in. (Blank, 1971), and half a 40-ft-high hill disappeared leaving a V-shaped hole 200 ft wide, 500 ft long, and 80 ft deep. During the evening of May 4, 1971, the terrain started to move.

The earth simply dissolved to a depth of nearly 100 feet; in the canyon thus formed, a river of liquefied clay—sometimes as deep as 60 ft—flowed at a rate of 16 miles per hour toward the Saguenay River, two miles away. At its widest, the canyon was a half mile across, and it extended for approximately one mile. [Blank, 1971, p. 88]

The liquefaction flow had stopped by midnight of the same day, but the toll was 31 lives and 38 homes which had disappeared into the liquefied material. Cause of the landslide was attributed to the exceptionally heavy rains that had occurred during April, overloading the sensitive clays. Animals at times appear to feel precursors of landslides. Before the Vaiont landslide, it was the cattle that became restive, and at St. Jean Vianney, dogs were the first to become unusually agitated.

3. Ventura, California. Landslides that cover 345 acres or 45% of the producing area in the Ventura oil fields have been described by Kerr and others (1971). In 45 years, from the 1920s to 1969, 61 oil wells were destroyed and 132 were endangered. Although the sliding was generally slow, in January 1968 more than 1 million yd^3 moved 100 ft in a few minutes. Landslide movements can be correlated with heavy precipitation. Cause of the movement is that the montmorillonite clays at depth become thixotropic when wet, and the overlying sandstones move over these disturbed sediments.

4. Youd (1971) described landsliding near Van Norman Reservoir that resulted from the San Fernando earthquake

of 1971. The most damaging slide was a liquefaction flow 1.2 km long that caused $30 million damages. It occurred on terrain that averaged only 1.5° slope and had a maximum slope of 3° at the head. Sand boils were also formed in the area, and the total downslope movement of landslides was only 0.5 to 1 m.

Subaqueous Slides

Sliding of earth materials occurs both in lakes and oceans. Heim (1932) described slides in Switzerland; for example, in Lake Zurich. Hyne and others (1973) discussed the Lake Tahoe landslide (California) which they said was comparable to the largest slides yet recorded. The slide under the lake is 50 m thick and covers 210 km². It consists of poorly sorted coarse sediments, similar to pebbly mudstones. They described its consistency as being more fluid than a slump but less fluid than a turbidity current. It contains a series of disordered mounds that compare with those formed at the Frank, Sherman, and Blackhawk landslide terrains.

Slides in the marine environment range from coherent slump blocks to partially fluidized incoherent slides, and fluidized turbidity currents would be the end member of this sequence. Hutchinson (1968) classified the slides into flow slides and underconsolidated clay slides. Menard (1964) described some marine slides in the Pacific, and Terzaghi (1957) discussed various types and provided examples from the Atlantic. For example, on May 2, 1930, massive slumps at Orkdals Fiord, Norway, broke several submarine cables; the estimated volume of the slumps is 10^7 m³. Menard (1964) reported of other slumps of large size. A slump of August 30, 1935, beyond the mouth of the Magdalena River in

TABLE 1. LANDSLIDE TERRAIN PROPERTIES

Locality	Date	Volume (10^6 m³)	Area covered (km²)	Vertical displacement or fall (m)	Horizontal movement (km)	Runup distance (m)	Velocity (est) (km/h)	Landslide type
Huascaran, Peru	May 31, 1971	10		4,000	14.5		400	Rock avalanche–debris flow
Little Tahoma Peak, Washington	Dec. 14, 1963	10.7		1,890	6.9	90	152	Rock avalanche
Elm, Switzerland	Sept. 11, 1881	12.7	0.6	610	1.4	103	160	Rock avalanche
St. Albans, Canada	Apr. 27, 1894	19.1						Liquefaction flow
Madison, Montana	Aug. 17, 1959	28.3	0.5	400	1.6	130	180	Rock avalanche
Sherman, Alaska	Mar. 27, 1964	28.3		600	5.0	137	185	Rock avalanche
Frank, Canada	Apr. 29, 1903	36.5	2.5	870	4.0	120	175	Rock avalanche
Gros Ventre, Wyoming	June 23, 1925	38.2	0.8	640	1.9	106	164	Rock avalanche
Goldau, Switzerland	Sept. 2, 1806	40		550	1.7			Rock avalanche
Apollo 17, Moon	..	200	21	2,000	5.0			Rock avalanche
Silver Reef, California	Prehistoric	226		790	5.3	46	104	Rock avalanche
Vaiont, Italy	Oct. 9, 1963	260			1.8	240		Rockslide
Blackhawk, California	Prehistoric	283		1,220	8.0	64	120	Rock avalanche
Gohna, India	Sept. 1893	290		1,470	1.6			Rock avalanche
Martinez, California	Holocene	382		2,000	7.6			Rock avalanche
Ticino River, Switzerland	During glacial retreat	500						Rock avalanche
Upper Garhwal, India	Sept. 22, 1893	566		1,520	3.2			Rock avalanche
Sawtooth Ridge, Montana	Prehistoric	650		360				Rock avalanche
Klonsee, Alps	Interglacial	770						Rock avalanche
Tin Mountain, California	Prehistoric	1,795						Rock avalanche
D'Onsoi, Pamir Mtns.	Feb. 18, 1911	2,080	10.3	1,160	3.6			Rock avalanche
Flims, Switzerland	Interglacial	12,000	41.4	1,980	16.1			Rock avalanche
Lake Tahoe, California–Nevada	..	10,000						Subaqueous slide
Saidmarreh, Iran	Holocene	20,000	165.7	1,650	14.5	457	338	Rock avalanche
Samar Island, Philippines	2,000 yr B.P.	135,000						Block glide
Gulf of Alaska, Alaska	Mar. 27, 1964(?)	590,000						Subaqueous slide

Note: Sources include: Browning, 1973; Crandell and Fahnestock, 1965; Harrison and Falcon, 1938; Heim, 1882, 1932; Howard, 1973; Hsü, 1975; Hyne and others, 1973; Kent, 1966; Kiersch, 1965; Mudge, 1965; Newland, 1916; Shreve, 1966, 1968a; Zaruba and Mencl, 1969; and data from this volume in Bock; Molnia and others; Wolfe; and personal observations and calculations.

TABLE 2. LANDSLIDE DISASTERS

	Date	People killed	Remarks
Brenno Valley, Switzerland	1512	600	Rockslide dammed valley; dam broke in 2 yr causing destruction
Tour d'Ai, Switzerland	1584	300	Landslide devastated village of Yvorne in Rhone valley
Mount Conto, Switzerland	1618	2,430	Rockslide
Goldau, Switzerland	1806	457	Landslide destroyed village
Mt. Ida, Troy, New York	1843	15	Sediment slump and flow
Elm, Switzerland	1881	115	Rock avalanche also demolished 83 houses
Trondheim, Norway	1893	111	Liquefaction flow in marine clays
Frank, Canada	1903	70	Rock avalanche destroyed most of town
Kansu Province, China	1920	100,000–200,000	Earthquake caused loess flows
Nordfjord, Norway	1936	73	Rockfall created 74-m wave
Kobe, Japan	1938	461	Rocky mudflows
Kure, Japan	1945	1,154	Rocky mudflows
Yokahama, Japan	1958	61	Rocky mudflows
Madison, Montana	1959	28	Rock avalanche buried campers
Vaiont, Italy	1963	2,000	Rockslide into reservoir created wave that flooded below dam
Anchorage, Alaska	1964	114	Combined toll from landslides and earthquake
Aberfan, Wales	1966	144	Man-made mining spoiled hill; landsliding buried mostly children
Brazil	1966–67	2,700	Combined toll from landslides and floods
Nelson County, Virginia	1969	150	Combined total from debris avalanches and floods
Huascaran area, Peru	1970	21,000	Combined rock avalanche and debris flow buried two cities
St. Jean Vianney, Canada	1971	31	Slab flows buried people and houses

Note: Sources: Blank, 1971; Browning, 1973; Brunsden, 1974; Close and McCormick, 1922; Hansen, 1965; Heim, 1882; Jones, 1973; Kiersch, 1965; Mathews, 1960; Nakona, 1974; Newland, 1916; Williams and Guy, 1971; Zaruba and Mencl, 1969.

Columbia was estimated to contain 3×10^8 m³, and one near Suva, Fiji, in September 1973 was 1.5×10^8 m³. The 1923 Kwanto earthquake caused a subaqueous slide near Sagami Wan, Japan, that Menard (1964, p. 203) described as "the greatest known." Its volume was calculated to be 7×10^{10} m³.

Moore (1964) described what he termed as two giant submarine landslides on the slope of the Hawaiian Ridge northeast of Oahu, Hawaii. The larger one is 160 km long and 50 km wide. Blocky remnants in the area are 8 to 25 km long, 5 to 15 km wide, and have long axes tilted toward the ridge. Individual blocks are up to 2,000 m high. The sole of the slide is 6,000 m below sea level. The second large slide is north of the island of Molokai and is 80 km long and 50 km wide. The most recently studied massive submarine slide is described by Molnia and others (this volume) and occurred in the Gulf of Alaska, with a volume of 5.9×10^{11} m³ in an area 18 km long and 14 km wide.

PLANNING AND MANAGEMENT IN UNSTABLE TERRAIN

Public interest in landslides is generally aroused only when there is clear demonstration of imminent disaster. Such interest is directly proportional to whether one's pocketbook, property, or life will be taxed, lost, or jeopardized. The scientist's role is not completed upon identification and recognition of a landslide, but he must also become involved in the decision-making process and must maintain open communications with engineers when construction activities become necessary, as in either the prevention or the control of landslides. Studies in societal awareness and perception of landslides have lagged behind those of other natural hazards, such as floods (White, 1974). The age-old attitude prevails for most people when considering landslides as a potential hazard, namely, "it can't happen here or to me."

In this section, planning and management policies are combined as the total effort in developing strategies, making policies, and providing for continued maintenance of terrain that has been determined to be unstable and landslide prone. Planning refers to the initiation of studies and programs that lead to the development of policies, whereas management is the implementation, consummation, and administration of the plans and policies. Factors that are involved in the total process include (1) preliminary investigation for recognition and identification of unstable terrain, (2) feasibility study of the benefit/cost ratio for avoidance or construction, (3) legal considerations in the public interest, and (4) comparison and development of appropriate designs for prevention and (or) control of landslides.

Recognition and Identification

Careful investigation of the stability of hillslopes should precede any development on or below them. There is no substitute for prior geologic, hydrologic, geomorphic, and soil-rock mechanics studies. The scale and detail of such work depend on funding and whether regional or local data are required.

Each landslide type has its own distinguishing features that needs interpretation by experts. The reader is referred to Eckel (1958) for a catalog of physical features that various authors suggest can provide clues for landslide identification (also see Mollard, this volume). The expert looks for diag-

Figure 5. Earthflows in Cowanesque River basin, Pennsylvania. These flows were triggered by exceptionally heavy and prolonged rains in 1966. The materials largely consist of glaciolacustrine, fine-grained sediments that were originally piped, but then failed as a landslide.

nostic properties that indicate weaknesses in the regolith and bedrock. These surface symptoms include (1) unusual contour changes and breaks in slope, (2) anomalous vegetation changes, (3) frontal bulges, (4) lateral tears, (5) other remnant scars at head or within the terrain, and (6) aberrant moisture conditions. Although it is difficult to generalize such a wide diversity of landslide characteristics, many are composed of three zones: (1) a breakaway zone marked by scar or depression, (2) a transportation zone where the principal movement occurred, and (3) a deposition zone where the displaced material has piled up and produced a concavo-convex profile.

For years the U.S. Army Corps of Engineers had contemplated construction of a dam on the Cowanesque River in north-central Pennsylvania. Intensive field investigations commenced in 1965, and in 1966 I was assigned the task of mapping the geomorphology of the area that would be affected by the dam and lands peripheral to the reservoir. The primary purpose for this field work was to locate and evaluate those sediments that might prove unstable and cause landsliding or collapse of the dam or of new access roads that would be built along the reservoir.

The east-flowing Cowanesque River has a drainage area of 764 km^2, and during glaciation ice impounded its outlet causing a series of glaciolacustrine and deltaic sediments to be deposited in the proglacial lake. Three types of information were needed: (1) analysis of the sedimentology of materials at the proposed dam site, (2) character of materials and their stability when wetted at the lateral sides of the potential reservoir, and (3) nature of materials where rerouting of peripheral roads was needed. Although I did reconnaissance mapping of the contiguous areas, much of the work was focused on properties of the fine-grained glaciolacustrine sediments because of similarities to the Tioughnioga landslide which I had been observing for years. Thus, knowledge of materials and recognition and identification of landslide-prone terrain were vital to the success of the project. Figure 5 shows recent earthflows in one of the side tributaries of the Cowanesque River. These slides occurred in glaciolacustrine clays that moved in response to exceptionally heavy rains in 1966 that ended a 5-yr drought in the region. Figure 6 shows hummocky terrain of an ancient earthflow type of landslide that at the present is stable. Such pitted topography can be formed by other geomorphic processes as well, obeying the law of equifinality (the development of similar landforms by different processes), so it is very important that the investigator study such features with great care and determine which of the several possible processes actually was responsible for the features. The fine-grained sediments in the Cowanesque basin, therefore, can be a hazard, and in some places they reach thicknesses of 30 m (Coates, 1966).

Another variation on the theme of the law of equifinality must be mentioned. Southern New York has a variety of isolated knobs, hills, and gentle rises in many of the north-south river valleys (Coates, 1974, p. 224–226). A casual observer might interpret these features as identical landforms; however, they have been formed by different processes. Because the terrain is attractive to potential home owners, many houses have been built or planned in the area. In performing slope stability tests on this terrain, it has been necessary to distinguish the nature of the materials, which aids in providing a clue to their genesis (King and Coates, 1973). Although many of these small hills originated as solifluction lobes, at least some probably belong to the landslide category. Figure 7 shows one of these near Blatchley on the east side of Trowbridge Creek, New York. Fabric analyses of sediments within the hill and a buried soil profile with charcoal dated at 280 ± 130 yr B.P. (Geochron Lab No. GX2896) establish that this is a rotational landslide. The major movement probably occurred under periglacial conditions, but undercutting by the stream has periodically reactivated movement in the clay-rich till units.

Costs

Costs involved in areas of potential landsliding include the expense of mapping and surveys that determine the hazard probability. Various systems are in use to aid the planner in policy decisions related to land use in areas of unstable terrain. The U.S. Geological Survey has adopted an inventory classification system (see Nilsen and Brabb, this volume) of mapping hillslopes and assigning value of I–VI for different areas, dependent upon the likelihood of earth movements for that area. Leighton (1976) has developed a method whereby areas are assigned to a point system and ranked according to a series of indices that include (1) adequacy of a landslide data base, (2) landslide stability rating, (3) records of landslides in the area, (4) geologic-engineering codes and standards, (5) implementation and enforcement of codes and standards, and (6) performance records.

After a landslide problem area has been recognized and evaluated, planners must then decide the course of action to take. Such decisions are often related to some type of benefit/cost ratio. Thus the amount of potential damage is weighed against costs that would be involved in attempting an engineering solution. Of course the type of man-made developments in the area and the type of landslide and its probable magnitude and frequency of occurrence are determining factors. If the costs are too great, the area will be avoided or bypassed in new construction activities.

In the United States, damages caused by landsliding are increasing due to the "double jeopardy syndrome" in urbanizing areas. Because of the greater awareness of flooding hazards in low-lying areas (coupled with prohibitive codes for

Figure 6. Hummocky terrain in a terrace-type landform immediately adjacent to the Cowanesque River, Pennsylvania. This is an inactive or "fossil" landslide in glaciolacustrine sediments. Piping played an important role in removal of materials and led to partial collapse and rotation of the slide mass.

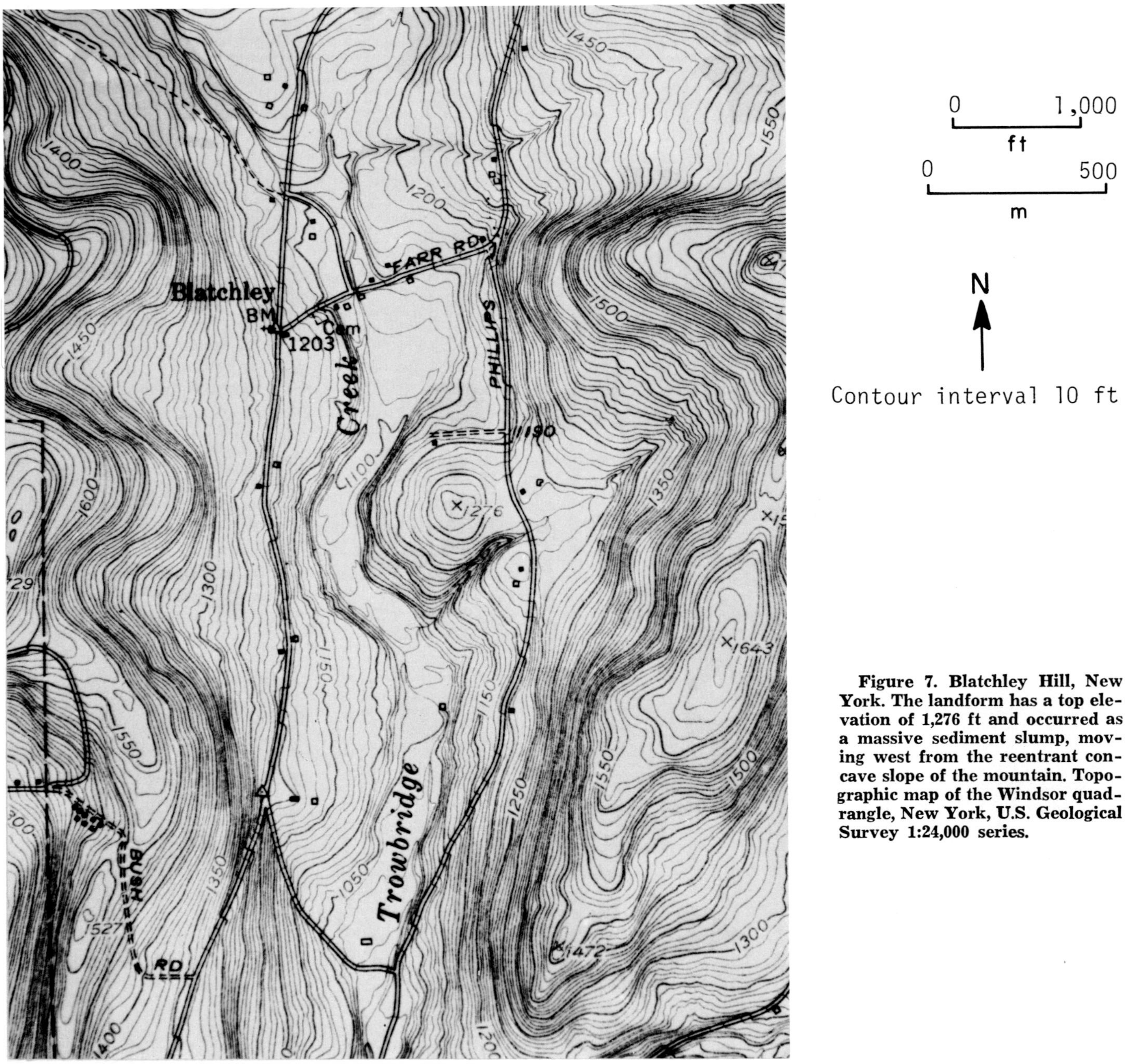

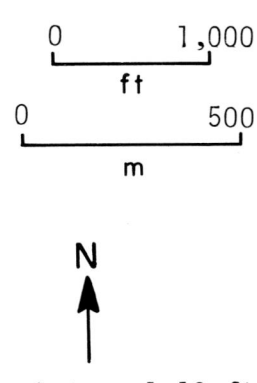

Figure 7. Blatchley Hill, New York. The landform has a top elevation of 1,276 ft and occurred as a massive sediment slump, moving west from the reentrant concave slope of the mountain. Topographic map of the Windsor quadrangle, New York, U.S. Geological Survey 1:24,000 series.

their occupancy), more hillside development is occurring, where landsliding can be the chief hazard. The expansion in facilities, such as highways and other services, has increased the potential for greater costs of landslides.

The direct and indirect costs of landslides to the United States run to the hundreds of millions of dollars each year in damage to highways, railroads, industrial installations, public works, and personal property. [Office of Emergency Preparedness, 1972, p. 87]

One of the first attempts to assess nationally the economic losses from landslides was the study by Smith (1958). Results cited from questionnaires to state highway departments and to railway companies showed the range in losses. States reported the following annual landslide losses for state highways: one state, more than $1 million; three states, between $0.5 and $1 million; one state, between $250,000 and $500,000; five states, between $100,000 and $250,000; six states, between $25,000 and $100,000; and eleven states, less than $25,000. Twelve railways, comprising 22% of the mileage in the United States and 30% in Canada, reported annual landslide costs as follows: one, between $0.5 and $1 million; two, between $250,000 and $500,000; two between $100,000 and $250,000; three, between $25,000 and $100,000; and four, less than $25,000. Such costs did not include expenses related to damage of equipment or lading.

Damages from landslides caused by the Alaska earthquake of 1964 amounted to more than $100 million. Failure of the Bootlegger Cove Clay in Anchorage devastated much of the city, and in one area 750 houses were destroyed. In November 1969 a small landslide occurred between Upper and Lower Silvies Lakes in Alaska. A hillside gave way and swept a new $2.2 million hydroelectric plant into the lower lake and created a power shortage in Ketchikan.

California has more continuing landslide damages than other parts of the United States. Nilsen and others (1976) studied damages that resulted from 335 landslides in Alameda County, California, during the period from 1940 to 1971. Rainstorms during the winter of 1968–69 in Alameda County caused losses of $5,396,700, compared to losses of $25,400,000 for the nine counties of the San Francisco Bay region during the same period. Sonoma and Contra Costa Counties suffered damages of $6,450,000 and $5,200,000, respectively. In Alameda County, damages were largely concentrated on those properties built on slopes steeper than 15%. In addition, "The areas of abundant recent landslides correlate partly with areas of abundant ancient landslide deposits" (Nilsen and others, 1976, p. 1). Costs in San Mateo and Santa Clara Counties were lower—$3,599,018 and $1,899,278—because there are fewer areas of unstable geologic formations and also fewer ancient landslides areas. During this same rainy season, 1968–69, damages in the Los Angeles area amounted to $6.5 million. Alfors and others (1973) estimated costs of landslides in California for the period 1970 to 2000. If land use continues under present growth systems, they calculated total landslide losses will be $9.85 billion. They pointed out, however, that such potential losses could be reduced 99% if stringent grading ordinances, as enforced in some Los Angeles areas, were adopted. On the other hand the use of engineering methods to remedy and control landslides would cost $1 billion, but even this is only 18% of what the total losses would aggregate.

Jones and others (1961) discussed landslide damages in the Columbia River areas, where losses amounting to more than $20 million occurred during the 20-yr construction period of Grand Coulee Dam and its reservoir. It cost $6 million to stabilize a bluff near the left tailrace of the dam. A landslide near Marcus, Washington, necessitated relocation and excavation for the Great Northern Railway at a cost of $239,000. Jones (1973) in another study discussed the great amount of damage caused by landsliding in Brazil, but assigned no monetary units. Nossin's (1972) study of 104 landslides in the Province of Cosenza, Italy, from 1966 to 1968 showed that corrective measures to stabilize them cost about $600,000.

Legal Affairs

One of the most important land use management strategies to minimize damages in landslide-prone terrain is the enactment and enforcement of legislation that prohibits or controls development. The events that led to the first laws regarding control of hillslopes have been described by Jahns (1969, p. 284). A devastating winter storm during January 13 to 18, 1952, caused $7.5 million property damage in Los Angeles, California. "Fresh cuts and fills were scoured and deeply gullied, enormous volumes of rock, mud, and coarse debris were mobilized, and various mixtures of water and solid matter invaded lower areas." The city responded quickly to this event. "Within a short time Los Angeles adopted the Grading Ordinance of 1952, the first regulatory measure of its kind, and thereby assumed a pioneering role in controlling man's modification of natural urban terrain." This monumental legislation required permits for cut and fill projects on hillslopes development tracts and also required soil engineering data. However, very little geologic information was required and no responsibility was assigned to the writer of the reports. These oversights were corrected in subsequent ordinance changes, and by the time of the revised Grading Ordinance of 1963, regulations had become much more stringent. Among other requirements, the 1963 ordinance mandated soils engineering and engineering geology reports during design and construction and the assumption of all legal responsibility. Periodic inspections must be made during construction, and upon completion of the development, the safety must be certified as meeting all present requirements for stability of the earth materials and the structure.

The remarkable success story attributed to these laws has been reported by Slosson (1969). Torrential rains in southern California in 1969 caused $6.5 million in property loss of 1,400 development sites on hillslopes. In making a comparison of these losses, Slosson showed:

(1) From a total of 10,000 sites developed before 1952, with no ordinances in effect, 1,040 sites were damaged and property loss was $3.3 million.

(2) Of the 27,000 developments in the 1952 to 1962 period when only partly adequate codes were in effect, 350 sites were damaged and property loss was $2,767,000.

(3) Eleven thousand developments were built in the 1963 to 1969 period when modern codes prevailed, and only 17 sites were damaged totaling a loss of $182,400. Thus mass-movement property losses per site declined from a high of $330 to only $7 after inactment of the 1963 grading ordinance.

In 1973 Congress passed the Flood Disaster Protection Act which expanded the federal concept of what areas could qualify for insurance coverage. Previous legislation had only included flood disasters, but this new act extended coverage to include mudslides, which in reality includes landslides (see Lessing and Erwin, this volume, for additional information).

Southeastern Brazil has been plagued by many landslide disasters during historic times, and finally in 1955 the government passed a series of laws and decrees to aid in reducing damages in developing areas. The decree of May 15, 1955, required investigations of the stability of sloping land before construction, including effects of any development on land and buildings farther down the slope. The General Office of Transport and Public Works was empowered to scale down the size of developments on slopes that were deemed unstable, and if necessary to condemn any plan that would destroy the equilibrium of the hillside (Jones, 1973). Barata (1969) described additional legislation designed to minimize landslide losses in Brazil. The Forest Law of 1959 made construction work illegal above specified slope levels to preserve the stabilizing influence of vegetation and prevent development across springs. The Law of License of Construction in Uneven Terrain of 1967 regulated construction on steep slopes and elsewhere if the stability of the land was questionable. It demanded that the contractor obtain proof of slope stability before construction was allowed to begin. In Japan, Nakona (1974) described legislation entitled the "Law of Preventing Landslides and Related Events."

Smith (1958) provided information on laws and court cases that involved landslides. Morton and Streitz (1972), in discussing the Portuguese Bend landslide, described how the court ruled that the County of Los Angeles was accountable for landslide motion that had damaged homes. It was held responsible because the excessive fill used to construct a highway had so overloaded the top of the slide that increased movement resulted. The county had to pay damages of $5,360,000 to property owners.

Prevention and Control

Engineering designs and solutions for the remedy of landslide problems occur in such publications as those by Ladd (1935), Terzaghi and Peck (1948), Eckel (1958), Cleaves (1961), Leighton (1966), and Zaruba and Mencl (1969). An annotated bibliography of papers dealing with the subject can be found in Tompkin and Britt (1951). Many data are necessary to reach decisions concerning which approach and method should be used in prevention and control of landslides, including landslide type, magnitude and frequency of movement, characteristics of the regolith and bedrock, topographic setting, and nature of man-made activity or installation. After the costs and legal aspects have been considered, the decision must be reached whether it is most beneficial to avoid the unstable area or to attempt an engineering remedy of the problem.

Avoidance Methods. In some situations complete avoidance or bypassing of the unstable terrain is the best solution for dealing with hazardous slopes. This method has the advantage of being the safest alternative, especially when adjacent stable areas can be used instead. The techniques that can be employed include changing of grade in adjacent terrain, bridging over the slope, and relocation of the development. Bridging may be possible where the slope is steep, the area small, and the rock foundation stable. If bridge supports must be placed on or contiguous to the landslide area, extra caution must be taken so that construction activities of blasting or vibration do not trigger landsliding. Rarely are bridging techniques used if spans must exceed 100 m.

Water Control Methods. Water plays such a significant role in many different types of landslides that techniques for its control and removal in the earth materials are the most universally applied method to prevent landslides and stabilize hillslopes. As in all controls systems the basic principles of landslide control are to (1) decrease the reduction of stress within the system, and (2) increase the shear resistance of the earth materials.

1. Surface water. Surficial water is prevented from entering the landslide area, or is removed from the ground, by techniques such as the following: (a) water from streams, seeps and springs, and sheet wash is diverted by pipes or lined open trenches. (b) Water that collects in depressions is drained away by lined surface trenches. (c) Cracks and other openings are filled with grouting or sealant to prevent water penetration into the regolith or bedrock. Sealants include such impermeable materials as clay, concrete, or bitumens. (d) Slopes are regraded to allow more uniform drainage into ditches and water disposal systems. At times this may include paving the area to allow more rapid runoff and prevent percolation, as was done in the Ventura Avenue oil field of California.

2. Subsurface water. These techniques are used to dewater earth materials and to lower subsurface water levels when appropriate. (a) Galleries and tunnels can be effective in some situations where size and cost are not important factors. They are large laterals that can be subsequently filled with permeable materials to aid in diversion of waters. Some contain infiltration offshoots into more troublesome areas. (b) Horizontal drains and tiles are installed near the surface and are especially effective in soils that are granular, uncemented, and very permeable. (c) Interceptor trenches are usually less than 2 m deep, but depths of 6 m are used under certain conditions. Their use is becoming more rare because of the effectiveness of combinations of other methods, but in special circumstances it may be necessary to use this technique to reach deeper waters that have bypassed surficial or other catchment devices. (d) Vertical drains, holes, and wells take the form of various types of borings that become sumps or collector systems for water removal, commonly below the water table. Water can be transported away by pumping, continuous syphoning, or by subsurface pipes or movement through permeable strata when conditions permit.

Excavation Methods. These methods are some of the first to be used if imminent danger occurs or for slopes that have already failed. Here man-made cuts and deliberate modification of the slope geometry and gradient can take different paths.

1. Removal of slide. This is obviously the most drastic method and can be used only for small slides or those commensurate with the benefit cost ratio. Its advantages are the elimination of the hazard and the guaranteed safety for developments. The disadvantages are that only small slides are capable of removal, and the method is the most costly.

2. Unloading head of the slide. Use of this method depends upon accessibility, but when possible it may be one of the less expensive of the remedial measures. Zaruba and Mencl (1969) report that if 4% of the sliding volume is removed from the head of the slope, the stability index is increased 10%. When feasible the material removed from the head may be used in the regrading process and placed at the foot where it may act as a buttress.

3. Regrading and slope reduction. Here it is vital to know the type of slide and have analyses of soil mechanics of the regolith, because determination of the stable angle of repose is important for the success of the operation. In highway construction this may be critical for calculating right-of-way costs and property acquisition. Thus it is important to determine whether a cut of 4:1 or 10:1 (vertical: horizontal)

should be used to maintain slope stability. The purchase and condemnation for highway right-of-way becomes more costly when shallow gradient cuts are necessary. Another aspect of slope regrading is to eliminate irregularities and depressions that might serve as collectors for surface water, which would add pore-water pressure, increase weight of the potential slide mass, and decrease shear resistance of the mass.

4. Hillside benching. Construction of man-made terraces and berms can relieve stress in certain landslide types. It may be a necessary engineering measure when developments are on steep and long slopes and when the toe of the slide is severed. Whether benches are designed to be horizontal or to slope toward or away from the slope depends upon the nature of earth materials and the length of steepness of slopes. For horizontal and reverse-slope berms, it is necessary to install drains that will divert water away from the cut. On long slopes several berms may be needed to break the slope into separate segments and to prevent movement of the mass as a unit. To be effective benches should be wide and designed to permit remedial and maintenance operations after completion of the cut.

Restraining Structures. This group of engineering structures is designed to be emplaced at or near the toe of the unstable area. One reason for their use is the relative low cost. They also are used as a last resort when the developed site is so near the landslide that something must be done. Such structures to be effective should be used with other methods and generally in materials that are easily drained, because such impediments invariably lead to watertable changes that increase seepage pressures in the distal part. Thus installation of drains is necessary. Such methods are also used in small rotational slides and as catchment for rockfall.

1. Buttresses. These can be composed of rock or earth fill material that is added to the toe of the slide to provide extra weight to increase the shear strength of the original materials. Their most common use is in embankments. Emplacement of the added burden should be done only after soil mechanics studies have shown that the weight will not increase the driving force of the slide. When properly designed, however, as in rotational movements, the added support can feed back into the system and give extra resistance to upslope materials. They are also effective for slip-out types of movements. Most failures occur because the buttress did not extend sufficiently deep or was not coupled to subsurface drainage structures. Their successful use has been described by Cleaves (1961, p. 50) at Belle Vernon, Pennsylvania, and by Robinson and others (1972) in Colorado. After construction on Interstate 70 had begun, the Loveland Basin landslide in Colorado was discovered. By placing 61,775 yd^3 of fill at the toe, the safety factor of the slide was increased to 1.1 at the base and to 1.2 at the top of the slide.

2. Shear keys. These are essentially prisms of compacted fill placed to support only certain sections of the slide. This is a variation of the buttress, but the excavation and insertion of new materials occur in different parts of the slide, such as near the head where shear keys seem most effective. Shear keys were used successfully to stop movement of the Vista Verde landslide, California (Leighton, 1976, p. 285).

3. Retaining walls. There is an almost infinite variety of styles and materials that have been used as a dam or dike to prohibit downslope movement. They have been called walls, cribs, bulkheads, and dikes. Their purpose is preventive rather than corrective, and their construction is generally tied into the original design of the slope cut. Their character is dependent upon the local situation as determined by cost analysis, and their composition may include timber, concrete and grout, stone, metal ribbing, solid fencing, and gabions (wired networks filled with stone rubble). Permeable walls can be more effective with free-draining materials, but all walls need auxiliary drainage components to prevent water buildup. Retaining walls are generally restricted to small areas, at the toe of a predicted instability, and they need to be anchored with tie rods which are locked to stable terrain.

4. Piles. Insertion of some type of pile or vertical plug has occasionally been used in unconsolidated materials, but with no or limited success. The vibration of driving piles may be sufficient to initiate slide motion, and after completion earth materials can still move under or around the piles.

5. Rock bolts. The use of some type of metal dowel or rock bolt in bedrock has become especially popular in the last 15 yr. They work best in rocks that are jointed or bedded, with the plane of discontinuity inclined in the same direction as the slope. They are usually composed of steel and drilled at right angles to the weakness plane. The rock bolts have a wedge or expansion device at the lower end with a plate and washer with nut to serve as anchor. Grouting of the bolt with protective coatings such as polyethylene, bitumen, or cement aids in preventing corrosion, thus increasing its effective life.

Miscellaneous Methods. Many other techniques have been used in attempts to stabilize hillslopes in which the regolith is the principal slide material. Such methods are aimed at increasing the shear resistance by changing the physiochemical properties of the material.

1. Electro-osmosis. This method is most effective for causing accelerated drainage in silty soils. The technique involves driving out pore water by an electric current. The movement of soil water migrates from an anode induced by an electric current to a cathode consisting of a perforated pipe which acts as a well where water can be pumped out. Electro-osmosis cannot be used successfuly in clays that harden, in sands, or in "hard water" where the electrolytes negate the force field. Zaruba and Mencl (1969) reviewed applications of the method and discussed its success in Norway and Canada. For example, during bridge construction in Ontario four anode-cathode rows with electric current of 100 to 150 volts were operated three months. At the end of the period, it had stabilized the slope by decreasing the soil moisture content so that excavation of 1:1 was possible whereas the original slope had collapsed at the shallower gradient of 1:2.5.

2. Grouting. This involves the introduction of cement or chemicals into the soils to harden, stiffen, or release water. Portland cement can be used in granular materials, whereas various chemicals prove more effective in certain soils. For example, sodium silicate can react to form a silicate gel in the interstices of weakly bonded siliceous minerals. Jones (1973) discussed the use of lime as soil treatment in landslide control. The plasticity of clay (especially if it is mont-

morillonitic) can be reduced by the application of lime, $Ca(OH)_2$. By stiffening the clay and eliminating water which can be collected by subsurface drains, this method has been successfully used in California, Oklahoma, Iowa, and Panama. Slides were treated with lime at Ipe, Brazil.

Only a few days after the lime application, the surface of the slide material was firm enough to walk on after a rain, whereas before it was so soft that a man would sink into it 10 or 13 cm. Water issuing from the lower part of the slide prior to the lime application was muddy. After the lime application, clear water emerged from the slide area farther down the slopes.

Two years after the application of the lime to the slide at dike 4 no appreciable movement of the slide had been observed and the slide material was drier and firmer than the material in nearby untreated areas. [Jones, 1973, p. 42]

3. Miscellaneous methods. Several methods have been used in order to temporarily stabilize landslide areas to permit completion of construction projects. One of the most famous cases was insertion of pipes into unstable materials and the freezing of the soil to allow construction of Grand Coulee Dam. Thermic treatment of loess soils has been used in Rumania where the materials were baked and hardened in the vicinity of the bore holes.

SUMMARY

Landslides are almost ubiquitous, occurring in most types of terrain and climate. Although landslides constitute both a geomorphic process and a landform, emphasis is placed on their dynamics. Landslides are classified on their occurrence in bedrock, regolith, and sediments. These habitats are grouped according to type of movement, integrity of material, water content, and sediment size.

The principal causes of landslides are excessive precipitation, earthquakes, and human activities. Regional studies and selected case histories are used to document the style and variations of landslide phenomena. Rock avalanches and liquefaction flows are singled out for special attention.

Planning and management of landslide-prone terrains require the coordinated expertise of geologists, geomorphologists, engineers, and governing officials. Land use policy must be based on evaluation of costs, legal aspects, and engineering ability to prevent, remedy, or control landslide hazards. Wise decisions for appropriate development and stewardship of the land require both the careful evaluation of the site and knowledge of the regional environmental setting.

ACKNOWLEDGMENTS

Great credit is due to John F. Harsh, James T. Kirkland, and Victor E. Schmidt who provided the principal reviews of the manuscript. Their suggestions have been invaluable, but I am fully responsible for any residual misunderstandings.

REFERENCES CITED

Alfors, J. T., Burnett, J. L., and Gay, T. E., Jr., 1973, Urban geology master plan for California: California Div. Mines and Geology Bull. 198, 112 p.

Arber, M. A., 1971, The coastal landslips of south-east Devon, *in* Steers, J. A., ed., Applied coastal geomorphology: Cambridge, Mass., M.I.T. Press, p. 138–154.

——1973, Landslips near Lyme Regis: Geologists' Assoc. (London) Proc., v. 84, pt. 2, p. 121–133.

Bagnold, R. A., 1956, The flow of cohesionless grains in fluids: Royal Soc. [London] Proc., ser. A, v. 249, p. 235–297.

Bailey, R. G., 1971, Landslide hazards related to land use planning in Teton National Forest, northwest Wyoming: U.S. Dept. Agri. Forest Service, 131 p.

Baker, R. F., and Chieruzzi, R., 1959, Regional concept of landslide occurrence: Natl. Research Council, Highway Research Board Bull., v. 21, p. 1–16.

Barata, F. E., 1969, Landslides in the tropical regions of Rio de Janeiro: 7th Internat. Conf. Soil Mech. and Foundation Eng. Proc., Mexico, v. 2, p. 507–516.

Blank, J. P., 1971, The town that disappeared: Reader's Digest, v. 99, Dec., p. 86–90.

Brabb, E. E., Pampeyan, E. H., and Bonilla, M. G., 1972, Landslide susceptibility in San Mateo County, California: U.S. Geol. Survey Misc. Field Studies Map MF-360, scale 1:62,500.

Browning, J. M., 1973, Catastrophic rock slide, Mount Huascaran, north-central Peru, May 23, 1970: Am. Assoc. Petroleum Geologists Bull., v. 57, p. 1335–1341.

Brunsden, D., 1974, Landslides, *in* Brunsden, D., and Doornkamp, J. C., eds., The unquiet landscape: Bloomington, Indiana Univ. Press, p. 41–46.

Carrara, A., and Merenda, L., 1976, Landslide inventory in northern Calabria, southern Italy: Geol. Soc. America Bull., v. 87, p. 1153–1162.

Cleaves, A. B., 1961, Landslide investigations: U.S. Dept. Commerce, Bureau of Public Roads, Washington, U.S. Govt. Printing Office, 67 p.

Close, U., and McCormick, E., 1922, Where the mountains walked: Natl. Geog. Mag., v. 41, p. 445–464.

Coates, D. R., 1966, Report on the geomorphology of the Cowanesque basin, Pennsylvania: Baltimore, U.S. Army Corps Engineers Cowanesque Reservoir Study, 27+ p.

——1974, Reappraisal of the glaciated Appalachian Plateau, *in* Coates, D. R., ed., Glacial geomorphology: Binghamton, State Univ. New York, Pubs. in Geomorphology, p. 205–243.

——ed., 1975, Identification of late Quaternary sediment deformation and its relation to seismicity in the St. Lawrence Lowland, New York: N.Y. State Atomic and Space Devel. Authority, 268 p.

Colton, R. B., 1975, Landslides and related deposits in the western half of Colorado: Geol. Soc. America Abs. with Programs, v. 7, no. 7, p. 1033–1034.

Crandell, D. R., and Fahnestock, R. K., 1965, Rockfalls and avalanches from Little Tahoma Peak on Mount Ranier, Washington: U.S. Geol. Survey Bull. 1221-A, 30 p.

Crozier, M. J., 1973, Techniques for the morphometric analysis of landslips: Zeitschr. Geomorphologie, v. 17, p. 78–101.

Daly, R. A., Miller, W. G., and Rice, G. S., 1912, Report of the Commission Appointed to Investigate Turtle Mountain, Frank, Alberta: Canada Dept. Mines Mem. 27, 34 p.

Daly, R. M., 1926, Our mobile earth: New York, C. Scribner's Sons, 342 p.

Dawson, G. M., 1899, Remarkable landslip in Portneuf County, Quebec: Geol. Soc. America Bull., v. 10, p. 484–490.

Eckel, E. B., ed., 1958, Landslides and engineering practice: Natl. Research Council, Highway Research Board Spec. Rept. 29, 232 p.

Erskine, C. F., 1973, Landslides in the vicinity of the Fort Randall Reservoir, South Dakota: U.S. Geol. Survey Prof. Paper 675, 64 p.

Fuller, M. L., 1912, The New Madrid earthquake: U.S. Geol. Survey Bull. 494, 119 p.

Gary, M., McAfee, R., Jr., and Wolf, C. L., eds., 1972, Glossary of geology: Am. Geol. Inst., 805 p.

Hadley, J. B., 1964, Landslides and related phenomena accompanying the Hebgen Lake earthquake of August 17, 1959: U.S. Geol. Survey Prof. Paper 435-K, p. 107–138.

Hamel, J. V., 1970, The Pima Mine slide, Pima County, Arizona: Geol. Soc. America Abs. with Programs, v. 2, no. 2, p. 335.

Hansen, W. R., 1965, Effects of the earthquake of March 27, 1964, at Anchorage, Alaska: U.S. Geol. Survey Prof. Paper 542-A, 68 p.

Harrison, J. V., and Falcon, N. L., 1938, An ancient landslip at Saidmarreh in southwestern Iran: Jour. Geol., v. 46, p. 296–309.

Heim, A., 1882, Der Bergsturz von Elm: Deutsch. Geol. Gesell. Zeitschr., v. 34, p. 74–115.

——1932, Bergsturz and Menschenleben: Zurich, Fretz & Wasmuth Verlag, 218 p.

Holmes, A., 1965, Principles of physical geology: New York, The Ronald Press Co., 1288 p.

Howard, K., 1973, Avalanche mode of motion: Implications from lunar examples: Science, v. 180, p. 1052–1055.

Howe, E., 1909, Landslides in the San Juan Mountains, Colorado: U.S. Geol. Survey Prof. Paper 67, 58 p.

Hsü, K. J., 1975, Catastrophic debris streams (Sturzstroms) generated by rockfalls: Geol. Soc. America Bull., v. 86, p. 129–140.

Hutchinson, J. N., 1968, Mass movement, in Fairbridge, R. W., ed., Encyclopedia of geomorphology: New York, Reinhold Book Corp., p. 688–696.

Hyne, N. J., Goldman, C. R., and Court, J. E., 1973, Mounds in Lake Tahoe, California-Nevada: a model for landslide topography in the subaqueous environment: Jour. Geol., v. 81, p. 176–188.

Jaeger, C., 1969, The stability of partly immersed fissured rock masses and the Vaiont rockslide: Civil Eng. and Pub. Works Rev., p. 1204–1207.

Jahns, R. H., 1969, Seventeen years of response by the City of Los Angeles to geologic hazards: Geol. Hazards and Public Problems Conf. Proc., Washington, U.S. Govt. Printing Office, p. 283–295.

Jones, F. O., 1973, Landslides of Rio de Janeiro and the Serra das Araras Escarpment, Brazil: U.S. Geol. Survey Prof. Paper 697, 42 p.

Jones, F. O., Embody, D. R., and Peterson, W. L., 1961, Landslides along the Columbia River Valley northeastern Washington: U.S. Geol. Survey Prof. Paper 367, 98 p.

Keng, W. P., 1970, The Sanchuhu rockslide of September 1970, Miaoli, Taiwan: Taiwan Geol. Survey Bull. 21, p. 53–54.

Kent, P. E., 1966, The transport mechanism in catastrophic rock falls: Jour. Geol., v. 74, p. 79–83.

Kerr, P. F., Stroud, R. A., and Drew, I. M., 1971, Clay mobility in landslides, Ventura, California: Am. Assoc. Petroleum Geologists Bull., v. 55, p. 267–291.

Kiersch, G. A., 1965, The Vaiont Reservoir disaster: California Div. Mines and Geology Mineral Inf. Service, v. 18, no. 7, p. 129–138.

King, C.A.M., and Coates, D. R., 1973, Glacio-periglacial landforms within the Susquehanna Great Bend area of New York and Pennsylvania: Quaternary Research, v. 3, p. 600–620.

Ladd, G. E., 1935, Landslides, subsidences and rock-falls: Am. Ry. Eng. Assoc. Bull., v. 37, no. 377, 72 p.

Lawson, A. C., 1908, The California earthquake of April 18, 1906: Carnegie Inst. Washington, v. I, pt. II, p. 225–449.

Leighton, F. B., 1966, Landslides and hillside development, in Lung, R., and Proctor, R., eds., Engineering geology in southern California: Assoc. Eng. Geologists, p. 149–192.

——1976, Geomorphology and engineering control of landslides, in Coates, D. R., ed., Geomorphology and engineering: Stroudsburg, Pa., Dowden, Hutchinson & Ross, Inc., p. 273–287.

Lundgren, L., and Rapp, A., 1974, A complex landslide with destructive effects on the water supply of Morogoro Town, Tanzania: Geog. Annaler, v. 56A, no. 3–4, p. 251–260.

Mathews, S. W., 1960, The night the mountains moved: Natl. Geog. Mag., v. 117, p. 328–359.

McConnell, R. G., and Brock, R. W., 1904, Report on the great landslide at Frank, Alberta: Canada Dept. Interior Ann. Rept. 1902–1903, pt. 8, 17 p.

McCulloch, D. C., 1966, Slide induced waves, seiching and general features caused by the earthquake of March 27, 1964, at Kenoi Lake, Alaska: the regional effects: U.S. Geol. Survey Prof. Paper 543-A, 41 p.

Menard, H. W., 1964, Marine geology of the Pacific: New York, McGraw Hill Book Co., 271 p.

Moore, J. G., 1964, Giant submarine landslides on the Hawaiian ridge: U.S. Geol. Survey Prof. Paper 501-D, p. 95–98.

Morton, D. M., 1971, Seismically triggered landslides in the area above the San Fernando valley, in The San Fernando, California, earthquake of February 9, 1971: U.S. Geol. Survey Prof. Paper 733, p. 99–104.

Morton, D. M., and Streitz, R., 1972, Landslides, in McKenzie, G. O., and Utgard, R. O., eds., Man and his physical environment: Minneapolis, Burgess Publ. Co., p. 64–73.

Mudge, M. R., 1965, Rockfall-avalanche and rockslide-avalanche deposits at Sawtooth Ridge, Montana: Geol. Soc. America Bull., v. 76, p. 1003–1014.

Nakona, T., 1974, Natural hazards: report from Japan, in White, G. F., ed., Natural hazards—local, national, global: New York, Oxford Univ. Press, p. 231–243.

Newland, D. H., 1916, Landslides in unconsolidated sediments with a description of some occurrences in the Hudson Valley: New York State Mus. Bull. 187, p. 79–105.

Nilsen, T. H., and Turner, B. L., 1975, Influence of rainfall and ancient landslide deposits on recent landslides (1950–71) in urban areas of Contra Costa County, California: U.S. Geol. Survey Bull. 1388, 18 p.

Nilsen, T. H., Taylor, F. A., and Brabb, E. E., 1976, Recent landslides in Alameda County, California (1940–71): an estimate of economic losses and correlations with slope, rainfall, and ancient landslide deposits: U.S. Geol. Survey Bull. 1398, 21 p.

Nossin, J. J., 1972, Landsliding in the Crati Basin, Calabria, Italy: Geologie en Mijnbouw, v. 51, p. 591–607.

Office of Emergency Preparedness, 1972, Disaster preparedness: Washington, Exec. Office of President, Rept. to Congress, U.S. Govt. Printing Office (in three volumes), var. pages.

Pain, C. F., 1969, The effect of some environmental factors in rapid mass movement in the Hunua Ranges, New Zealand: New Zealand Earth Sci. Jour., v. 3, no. 2, p. 101–107.

——1972, Characteristics and geomorphic effects of earthquake-initiated landslides in the Adelbert Range, Papua, New Guinea: Eng. Geol., v. 6, p. 261–274.

Rice, C. M., 1953, Dictionary of geological terms: Ann Arbor,

Edwards Bros., Inc., 465 p.
Robinson, C. S., and others, 1972, Geological, geophysical, and engineering investigations of the Loveland Basin landslide, Clear Creek County, Colorado 1963–65: U.S. Geol. Survey Prof. Paper 673, 43 p.
Scott, W. A., 1931, Concrete sidehill viaduct completes Cap Horn cutoff: Contractors and Engineers Monthly, v. 23, no. 1, p. 58–60.
Sharpe, C.F.S., 1939, Landslides and related phenomena: New York, Cooper Square Pub., Inc., 137 p.
Shreve, R. L., 1966, Sherman landslide, Alaska: Science, v. 154, p. 1639–1643.
——1968a, The Blackhawk landslide: Geol. Soc. America Spec. Paper 108, 47 p.
——1968b, Leakage and fluidization in air-layer lubricated avalanches: Geol. Soc. America Bull., v. 79, p. 653–658.
Simonett, D. S., 1968, Landslides, in Fairbridge, R. W., ed., Encyclopedia of geomorphology: New York, Reinhold Book Corp., p. 639–641.
Sims, J. D., 1973, Earthquake-induced structures in sediments of Van Norman Lake, San Fernando, California: Science, v. 182, p. 161–163.
Slosson, J. E., 1969, The role of engineering geology in urban planning: Colorado Geol. Survey Spec. Pub. no. 1, p. 8–15.
Smith, R., 1958, Economic and legal aspects, in Eckel, E. B., ed., Landslides and engineering practice: Natl. Research Council, Highway Research Board Spec. Rept. 29, p. 6–19.
Snow, D. T., 1964, Landslide of Cerro Condor–Seneca, Department of Ayacucho, Peru, in Kiersch, G. A., ed., Eng. Geol. Case Histories No. 5: Geol. Soc. America, p. 1–6.
So, C. L., 1971, Mass movements associated with the rainstorm of June 1966 in Hong Kong: Trans. Inst. Brit. Geographers, v. 53, p. 55–65.
Staples, L. W., 1957, Landslides at north abutment of Lookout Point Dam, Oregon, in Trask, P. D., ed., Eng. Geol. Case Histories No. 1: Geol. Soc. America, p. 43–48.
Terzaghi, K., 1950, Mechanisms of landslides, in Paige, S., Chairman, Applications of geology in engineering practice: Geol. Soc. America Berkey Volume, p. 83–123.
——1957, Varieties of submarine slope failures: Teknisk Ukeblad, no. 43–44, p. 1–16.
Terzaghi, K., and Peck, R. B., 1948, Soil mechanics in engineering practice: New York, John Wiley & Sons, 566 p.
Tompkin, J. M., and Britt, S. H., 1951, Landslides—a selected annotated bibliography: Highway Research Board Bibliography No. 10, 53 p.
Varnes, D. J., 1958, Landslide types and processes, in Eckel, E. B., ed., Landslides and engineering practice: Natl. Research Council, Highway Research Board Spec. Rept. 29, p. 20–47.
Wentworth, C. K., 1943, Soil avalanches in Oahu, Hawaii: Geol. Soc. America Bull., v. 54, p. 53–64.
White, G. F., ed., 1974, Natural hazards—local, national, global: New York, Oxford Univ. Press, 288 p.
Williams, G. P., and Guy, H. J., 1971, Debris avalanches—a geomorphic hazard, in Coates, D. R., ed., Environmental geomorphology: Binghamton, State Univ. New York, Pubs. in Geomorphology, p. 25–46.
Williams, G. P., and Guy, H. P., 1973, Erosional and depositional aspects of Hurricane Camille in Virginia, 1969: U.S. Geol. Survey Prof. Paper 804, 80 p.
Woods, H. D., 1958, Causes of the Sears Point landslide, Sonoma County, California, in Trask, P. D., ed., Eng. Geol. Case Histories No. 2: Geol. Soc. America, p. 41–43.
Youd, T. C., 1971, Landsliding in the vicinity of the Van Norman Lakes, in the San Fernando, California, earthquake of February 9, 1971: U.S. Geol. Survey Prof. Paper 733, p. 105–109.
Zaruba, Q., and Mencl, V., 1969, Landslides and their control: Amsterdam, Elsevier, 205 p.

MANUSCRIPT RECEIVED BY THE SOCIETY SEPTEMBER 7, 1976
MANUSCRIPT ACCEPTED SEPTEMBER 17, 1976

2
Regional landslide types in Canada

J. D. MOLLARD
J.D. Mollard and Associates, Ltd., 815 McCallum Hill Building, Regina, Saskatchewan S4P 2G6, Canada

ABSTRACT

A number of landslide types in Canada are concentrated in physiographic regions and are associated with certain kinds of soil and rock materials, geologic structures, and topographic settings. They include (1) mountain slopes in the Cordilleran region of western Canada, chiefly steeply dipping bedded and foliated rocks; (2) valley sides in Upper Cretaceous argillaceous bedrock, mostly bentonitic marine clay shale, silty shale, and mudstone; (3) river banks and terrace bluffs in the St. Lawrence Lowland region and lower coastal regions of eastern Canada, where postglacial marine submergence and postglacial uplift formed sensitive fine-grained marine deposits; and (4) valley walls, escarpments, and deep thaw basins in the Lower Mackenzie Valley region and adjoining plains of northwestern Canada, where ice-rich permafrost occurs in fine-grained soil and weathered shale materials. Rock avalanches, massive retrogressive slope failures in argillaceous bedrock, and major retrogressive flow slides and earthflows in quick clays are discussed under six headings: (1) morphologic features, (2) stratigraphy and lithology, (3) geologic conditions affecting failure, (4) character and rate of movement, (5) engineering and environmental implications, and (6) case histories. A few other landslide types are discussed but are not presented under these headings.

INTRODUCTION

A high proportion of the destructive landslides in Canada can be divided into a few distinctive types that are characteristic of certain regions. Recurring geologic, geomorphic, geohydrologic, and geotechnical engineering factors are commonly associated with these landslides.

Much of the information discussed in this paper has been obtained from the interpretation of air photos (Brawner, 1964, 1975; Dishaw, 1967; Mitchell, 1976; Mollard, 1952, 1955, 1962, 1972, 1976, 1977). To a great extent, however, the characteristics of landslide terrain evaluated from air-photo studies have been augmented by field reconnaissance and by a search of the geotechnical engineering literature. My own air-photo and field studies were conducted over a 30-yr period from 1947 to 1976 and were made for a variety of route investigations, for existing and proposed recreational developments along the shores of natural lakes, and for new urban developments on marginally stable valley slopes. More than 20 studies dealt with site investigations for dams, bridges, tunnels, and heavy buildings, where slope stability was a concern. Five regional studies involved air-photo mapping and classification of slope-instability features around the shores of proposed and newly created reservoirs behind large power dams.

The landslide types discussed have a significant bearing on economic, engineering, environmental, and land use projects and planning in Canada. The slope failures occur in broad regions that possess characteristic soil and rock materials, geologic structures, and topographic settings. The four regions are (1) the Canadian Cordillera, where steeply dipping bedded and foliated rocks underlie mountain slopes (Fig. 1); (2) parts of the Interior Plains region of western Canada, where a variable thickness of glacial drift mantles Upper Cretaceous argillaceous bedrock, especially bentonitic marine clay shale, silty shale, and mudstone (Fig. 2); (3) the St. Lawrence Lowland region and the lower coastal regions of eastern Canada, where sensitive fine-grained marine deposits were laid down during postglacial submergence (Fig. 3); and (4) the Lower Mackenzie Valley and adjoining plains of northwestern Canada, where permafrost occurs in fine-grained soil and rock materials in a variety of topographic settings (Fig. 4).

Large and generally spectacular types of landslides in these regions include rock avalanches and deep-seated rotational slumps in the Canadian Cordillera, massive retrogressive gravity-slope failures in the Interior Plains, major retrogressive flow slides and earthflows in eastern Canada, and skin flows and bimodal flows in the Lower Mackenzie Valley. Rock avalanches in the mountains, slow retrogressive slope failures in argillaceous bedrock in the Interior Plains, and

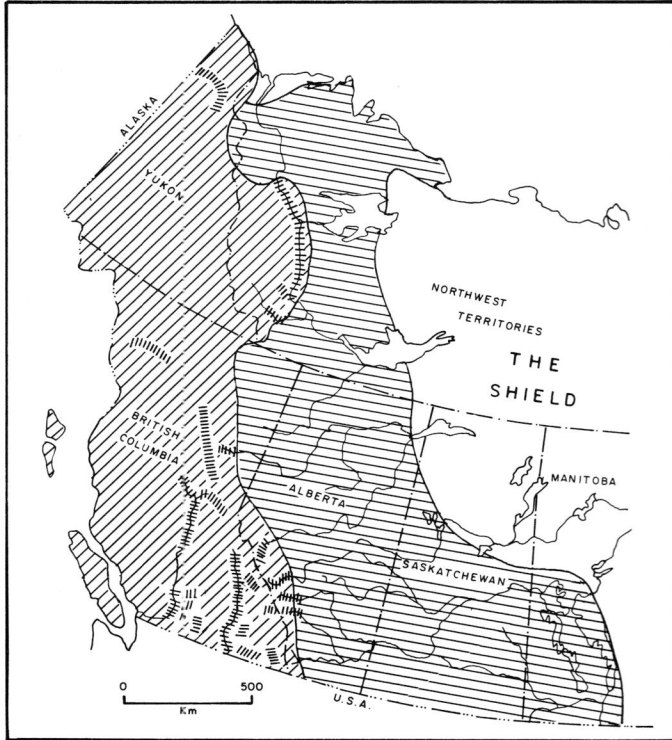

Figure 1. Rock avalanches and other types of landslides in the Canadian Cordillera. Map also shows the adjoining Interior Plains region of western Canada (see Fig. 2). Patterns are as follows: diagonal lines, Cordilleran region; horizontal lines, Interior Plains region; short cross lines, locations where landslides have been studied.

Breakaway scar outlines commonly taper toward the crown (Table 1). The failure debris generally develops a long, thin, glacierlike form called a "debris tongue" or "debris lobe." These tongues are usually long and narrow but may be fan-shaped (Figs. 5 to 9). Lake outlet channels incised into the mass of broken rock detritus tend to be narrow, straight, and steep-walled; rapids and low falls are common features in these channels. Flow morphology features are discernible on the surfaces of debris tongues.

The stream of debris commonly blocks pre-existing drainage courses and forms lakes (Table 1). Distal margins around debris tongues usually show a spatulate outline and a splaying aspect as a result of lateral dilation (Figs. 5 to 9). Shallow, planar (rather than deep-seated, rotational) rupture surfaces are characteristic (Fig. 10). The debris from a rock avalanche may ascend 100 m or more up the opposite valley wall if it is situated within a distance of 1 km or so of the base of the rupture (Fig. 9). In many places a transverse depression occurs between the debris lobe and the toe of the failed slope (Figs. 5 to 8).

Debris cones (Shreve, 1968)—also called "debris mounds" and, locally, "molards" in the French Alps (Gogeul and Pachoud, 1972) — tend to be concentrated in a relatively narrow band along the distal part of the failure debris (Fig. 5). Longitudinal debris ridges, called "debris trains," may be separated by steep-walled, flat-bottomed, longitudinal grooves (Figs. 5 to 7). Scattered angular blocks of broken rock 3 to

earthflows in eastern Canada are discussed here. A few other landslide types, generally treated less extensively in the literature, are also discussed.

ROCK AVALANCHES IN THE CANADIAN CORDILLERA

Morphologic Features

A rock avalanche is defined as very rapid downslope flowage of broken and pulverized rock fragments resulting from very large rockfalls and rockslides (Gary and others, 1972). They are also called "debris streams" (Armstrong, 1976; Hsü, 1975) and are usually referred to as major rockslides by investigators concerned with the mechanics of failure and stability analysis (Cruden, 1976; Krahn and Morgenstern, 1976).

Where the rupture surfaces of major rock avalanches have been recently exposed, they may appear either light (the Frank slide) or dark (the Hope slide) in aerial photographs, depending on the color of the exposed bedrock or vegetation cover (Mollard, 1976). In many places, parallel lineations, the result of rockslide movement, can be seen on the slip surface (Fig. 5). Head scarps, typically low and indistinct, are absent where failure reached the top of a mountain.

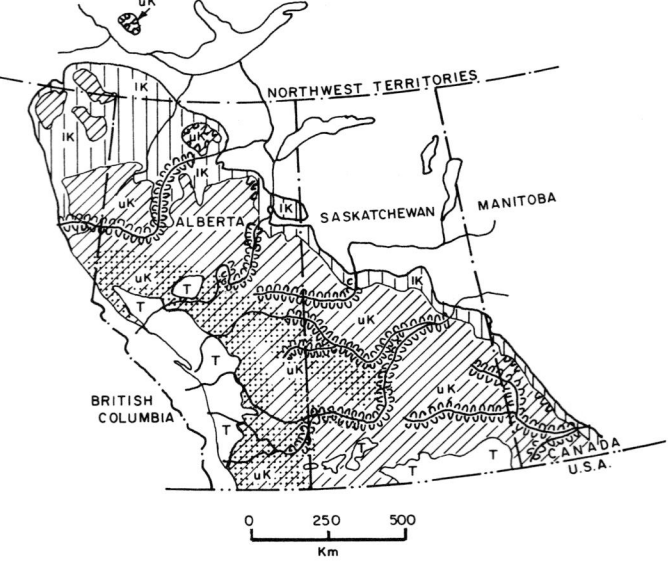

Figure 2. Retrogressive slope failures in argillaceous bedrock in the Canadian Western Interior. Patterns are as follows: lines bordered by U-shaped pattern, segments of major river valleys and upland plateaus along which discontinuous, massive, retrogressive landslides in bentonitic argillaceous bedrock of Late Cretaceous age have been studied from air photos and field reconnaissance; T, Tertiary, nonmarine sedimentary rocks, thinly covered by drift; uK with dot pattern, Upper Cretaceous, nonmarine sedimentary rocks, mantled by drift; uK with diagonal lines, Upper Cretaceous, marine argillaceous sedimentary rocks, in part bentonitic, mantled by drift; lK with vertical lines, Lower Cretaceous, marine sedimentary rocks, mantled by drift.

TABLE 1. COMPARISON OF GEOMORPHIC AND GEOMETRIC CHARACTERISTICS
OF FOUR ROCK AVALANCHES IN THE CANADIAN CORDILLERA

Geomorphic or geometric characteristic	Hope	Maligne Lake	Stalk Lakes	Frank
Valley or cirque glacier erosion and oversteepening of mountain slope behind rupture	W	W	W	P
Long fault scarp near mountain peak (from postglacial rebound)	W	N	M	W
Breakaway scar outline	W	W	W	W
Jumping (rockfall-launching) platform	P	W	W	M
Mound of failure debris below base of rupture scar	M	W	W	M
Transverse trough, or lake, between mountain rupture and debris lobe	P	W	W	W
Subparallel longitudinal debris ridges ("trains")	P	M	W	M
Longitudinal "grooves" (linear to wavy) on debris lobe	P	W	W	M
Splaying of linear features on terminal area of lobe	M	M	W	W
Transverse ridges on debris lobe	M	W	N	M
Transverse depressions on debris lobe	P	W	N	M
Debris cones ("molards"), mostly near distal rim of debris lobe	M	W	W	M
Arcuate outline on distal rim of debris lobe	N	W	W	W
Hummocky (morainelike) surface	W	M	P	M
Large angular blocks of broken bedrock	W	W	W	W
Postfailure, gorgelike channel draining lake	N	W	N	P
Sparse to nonexistent tree cover on failure debris	W	N	W	W
Approximate volume of material lost from upper slope (m^3)	47×10^6	500×10^6	40×10^6	33×10^6
Approximate slope of mountainside in rupture area	30°	25°	20°	40°
Rupture surface inclination	30°	25°–40°	20°–30°	30°–50°
Vertical travel of debris (km)	1.06	0.98	0.70	0.84
Horizontal travel of debris (km)	2.5	5.47	3.0	3.3
Ratio of horizontal to vertical travel	1.6	5.6	4.3	3.9

Note: Legend for geomorphic characteristics: W, well expressed; M, moderately well expressed; P, poorly expressed; N, not expressed.

20 m across are commonly visible in 1:24,000-scale air photos. Faint transverse ridges and fissures may also be evident (Table 1; Figs. 5, 7).

Failure debris commonly extends more than 1 km from the base of the mountain rupture and 2 to 6 km from the top of the breakaway scar (Table 1), giving rise to the expression "excessive travel distance" (Hsü, 1975). The slope at or just below the exposed slip surface generally shows an anomalous bump, which is thought to correspond to the local accumulation of debris and a so-called jumping platform (Hsü, 1975). Such an abrupt change in slope near the bottom of the mountainside probably launches the debris into the air and changes the mode of movement. In any event, a large proportion of the failed rock mass appears to leave the breakaway scar.

Where they are discernible, inclined beds along and below the rupture surface commonly dip at angles of 25° to 40°, with an inclination of about 30° being average and common (Cruden, 1976; Cruden and Krahn, 1973; G. H. Eisbacher, 1976, oral commun.). Foliation and joint planes at the Hope slide strike parallel to and dip in the direction of mountain slopes of about 30° (Mathews and McTaggart, 1969).

Locally, the appearance of transverse ridges on the debris tongue suggests an imbricate internal structure, that is, successively overlapping waves of broken rock debris. These ridges may show compositional banding and debris schlieren (Shreve, 1968) in addition to steep, planar surfaces. Conical debris mounds, closed depressions, a hummocky surface, and a zigzag pattern of interlacing ridges and hollows produces a relief not unlike that of kame-and-kettle topography in glaciated terrain. Debris cones, in particular, resemble small moulin kames and may easily be mistaken for them (Figs. 6, 7).

Stratigraphy and Lithology

Bedrock types underlying shallow, planar mountain ruptures include thick-bedded carbonate rocks (Cruden, 1976), interbedded fine and coarse clastic rocks (Eisbacher, 1971), and metamorphosed volcanic rocks (Mathews and McTaggart, 1969). Usually the rupture surface roughly parallels the original mountain slope. The layered rocks are commonly well jointed normal to the rupture surface (Krahn and Morgenstern, 1976). Moreover, the direction of slip movement is almost always at right angles, or nearly so, to the direction of regional geologic structures, such as anticlines, synclines, and thrust faults (Fig. 10).

Geologic Conditions Affecting Failure

In general, relatively shallow seated rockslides of large volume that develop into rockfalls and eventually into rock avalanches tend to occur in steeply inclined, bedded and foliated rocks, although noteworthy exceptions are known to have occurred [Gogeul and Pachoud, 1972; also, the Barrier

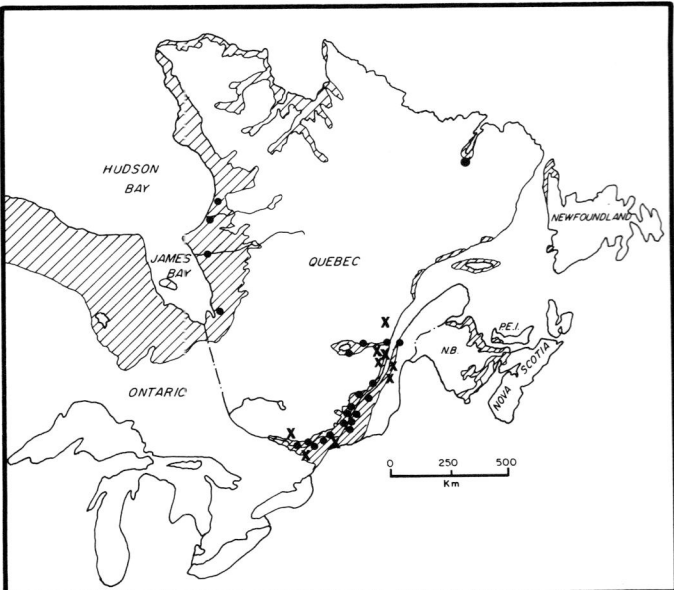

Figure 3. Earthflows in sensitive marine clay in eastern Canada. Solid circles are locations of major retrogressive flow slides and earthflows in sensitive, fine-grained, marine sediments (Chagnon, 1970; Mitchell and Markell, 1974; Mollard, 1976; Parkes and Day, 1975); diagonal lines represent distribution of fine-grained glaciomarine sediments, locally covered by sand; X = epicenters of earthquakes exceeding magnitude 5 for period 1889 to 1960 (Hodgson, 1965; Howell, 1973).

part of the rupture surface; a striking motion and impact of the debris and rapid accumulation of a mound of broken rock on the valley floor, usually some distance beyond a transverse depression that appears to receive very little debris; and a longitudinal flowage accompanied by some spreading of the debris stream. Several theories have been advanced which suggest that in similar terrain the fluid medium transporting the large (10 to 20 m) and small (0.3 to 1 m) rock fragments was compressed air similar to that below air-cushion vehicles (Shreve 1966, 1968), a dense dust cloud (Hsü, 1975; McConnell and Brock, 1904), vaporized interstitial water (Gogeul and Pachoud, 1972), and wet mud (Hsü, 1975). McConnell and Brock (1904) described the accounts of eyewitnesses who reported hearing the sound of rocks breaking away from the mountainside above, after which everything was shrouded in a cloud of dust—the noise of the slide resembling the sound of steam escaping under high pressure. Evidence of air-blast effects was noted at the Hope slide (Mathews and McTaggart, 1969) and also at the Frank slide (McConnell and Brock, 1904; Shreve, 1966). Rapid to very rapid rates of movement are reported in the geological literature—for example, 80 to 330 km/h (Shreve, 1966), 175 km/h (Shreve, 1968), 100 to 133 km/h (Hsü, 1975), 60 km/h (Gogeul and Pachoud, 1972), and 110 km/h (McConnell and Brock, 1904).

slide, Garibaldi, British Columbia (N. R. Morgenstern, oral commun.)]. They also appear prone to develop in places where preglacial stream erosion was followed by Pleistocene glacial erosion, resulting in "overdeepened" and "oversteepened" valleys and in the "daylighting" of steeply dipping, layered rock units (Fig. 11). Old tectonic movements coupled with postglacial isostatic rebound (which induced slip along bedding, foliation, joint, and fault surfaces) have decreased the cohesion and frictional components of shear strength on potential rupture surfaces. In different parts of the Canadian Cordillera, initiating mechanisms of catastrophic rockslides include excessive precipitation (which contributes to high pore pressures), long-continued freeze-thaw cycles (including ice wedging), river and ice erosion at the toe of slopes, and earthquakes of large magnitude (Cruden, 1976; Krahn and Morgenstern, 1976).

Character and Rate of Movement

Failure movement is very rapid and consists initially of translational rocksliding followed by rockfall and rock flow. The following phases, or modes, of colliding rock-fragment movement are suggested (Fig. 11): slide movement with minor fall; a jumping, waterfall-like motion, beginning at a launching ramp or jumping platform upon which some broken rock debris is lodged, commonly on and below the distal

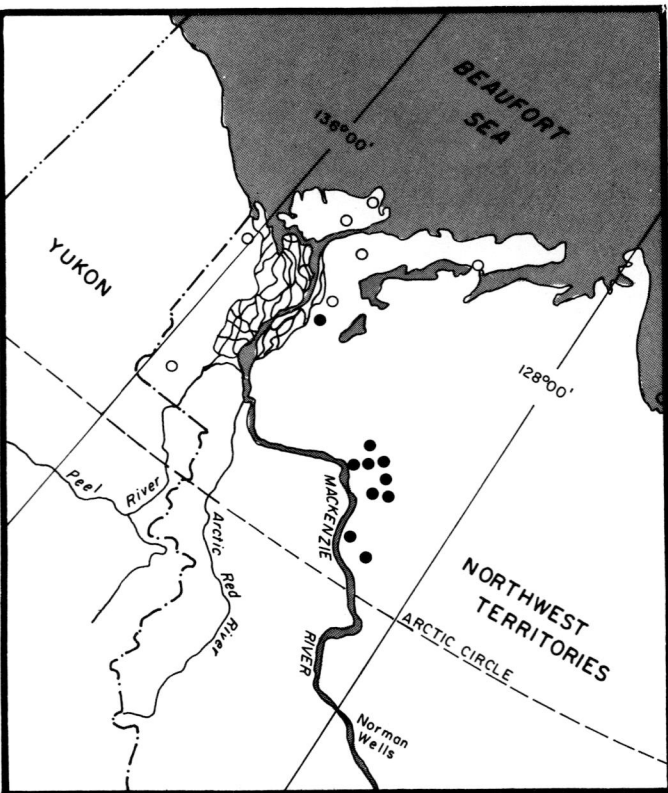

Figure 4. Slope failures in permafrost terrain in northern Canada. Open circles represent bimodal flows; solid circles represent skin flows.

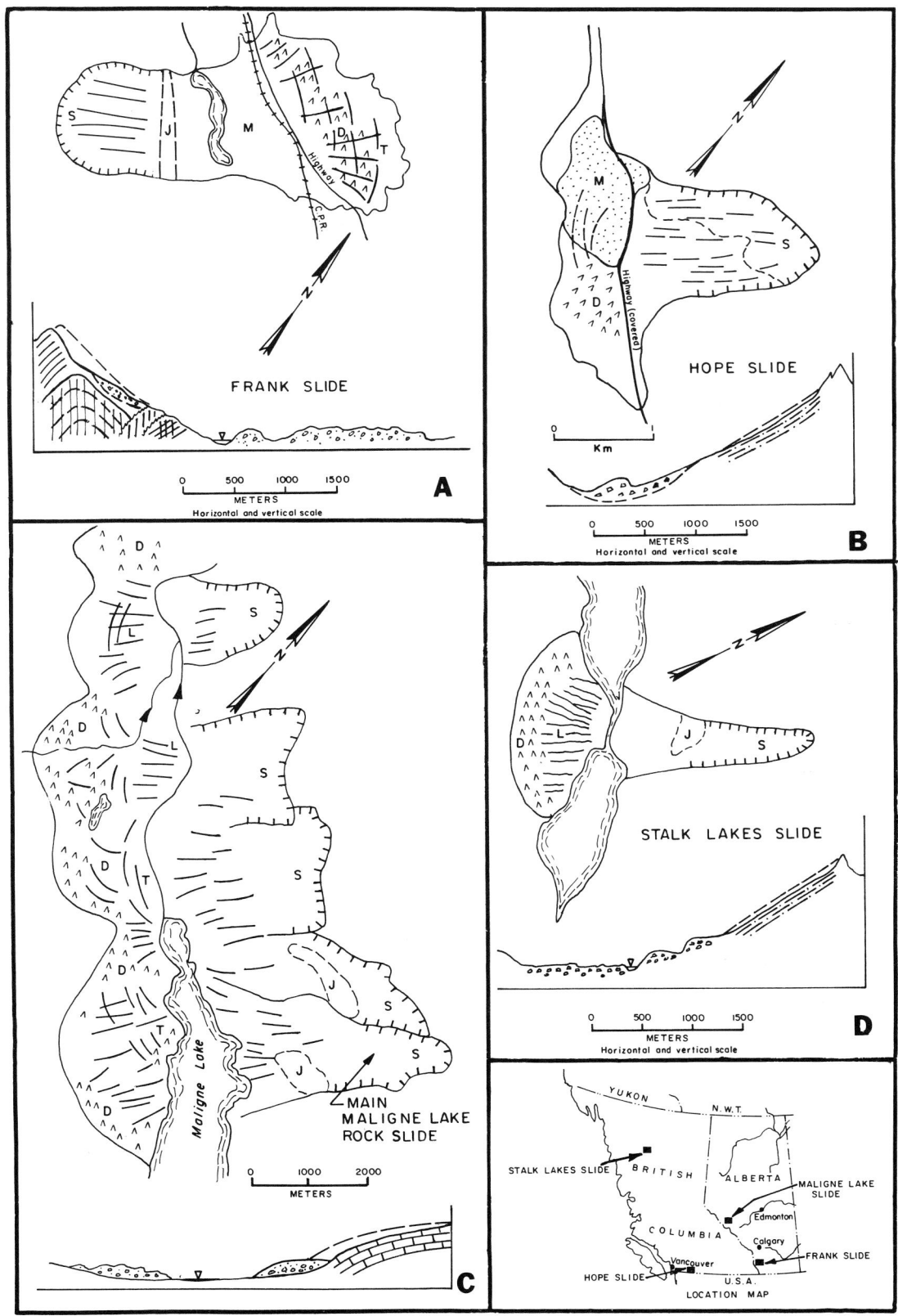

Figure 5. Geomorphic features and cross sections of two postglacial and two historic catastrophic rock avalanches in the Canadian Cordillera. M, area of large mound; J, jumping (airborne launching) platform; L, longitudinal debris ridges ("trains") or longitudinal grooves; T, transverse ridges or fissures; D, debris cones or "molards" (see also small inverted V's); S, rupture surface below rim of breakaway scar. Dashed line above the bedrock rupture surface in the profiles indicates the inferred mountainside before slope failure.

Figure 6. Aerial view of shallow rockslide and rock avalanche (debris stream) crossing Stalk Lakes. Note longitudinal ridges on the debris lobe (A.5 to B.1, 1.5 to 2.0), the debris cones (A.4 to A.5, 1.5 to 2.1), jumping platform (B.3, 1.5), and crest of breakaway scar (B.7, 3.4). Air photo A 12313–403, courtesy of the National Air Photo Library, Ottawa.

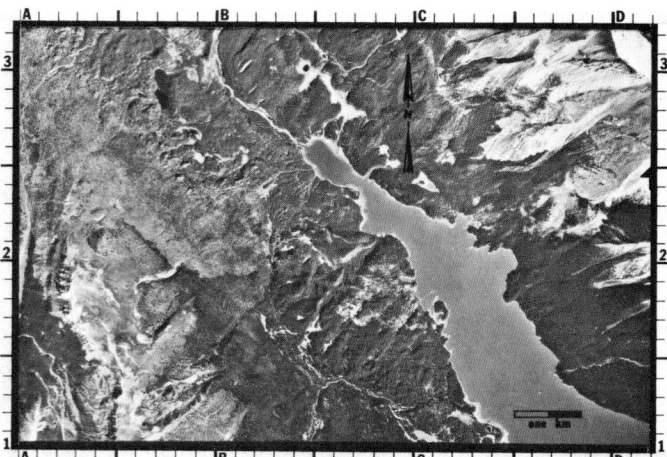

Figure 7. Exposed planar rupture surface (C.8 to D.2, 2.3 to 2.9) and scattered debris cones (A.6 to B.5, 2.0 to 2.7) around margin of failure detritus west of Maligne Lake, Jasper Park, Alberta. Note also the longitudinal ridges and grooves and fan-shaped debris lobe. Although the debris cones resemble moulin kames, farther northwest (Fig. 5C), where the anatomy of these symmetrical, steep-sided, turret-shaped mounds is exposed to view in recent road cuts, the materials composing them lack the characteristic stratification patterns, water-worn pebbles, and varied lithologies usually found in kames (A.M. McCann, 1976, oral commun.). Small cairnlike mounds of broken rock fragments are scattered over the ground surface. Air photo A 13319–192, courtesy of the National Air Photo Library, Ottawa.

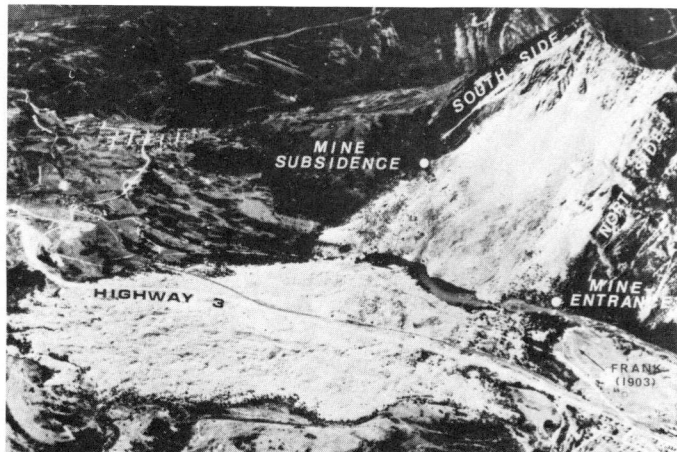

Figure 8. Annotated oblique aerial view of Turtle Mountain and the Frank slide (from the northeast; after Krahn and Morgenstern, 1976).

Figure 9. Oblique aerial view of Hope slide (from the southeast). Oblique air photo courtesy of the British Columbia Department of Highways, Victoria.

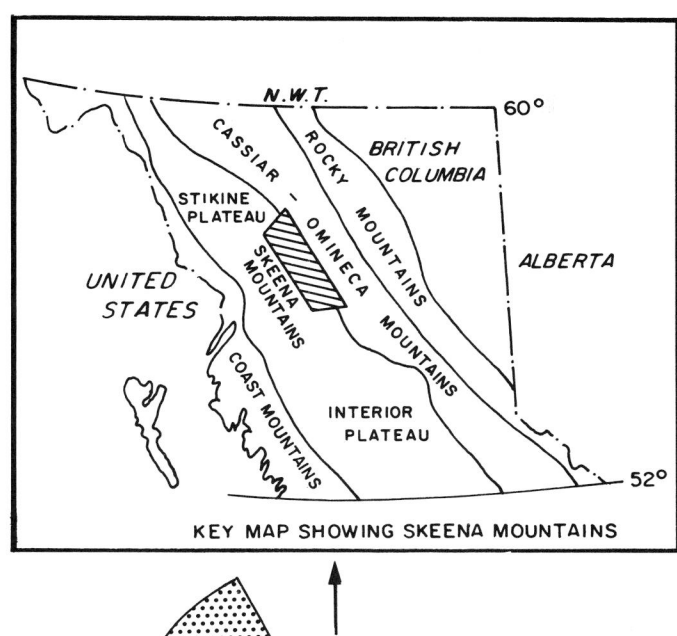

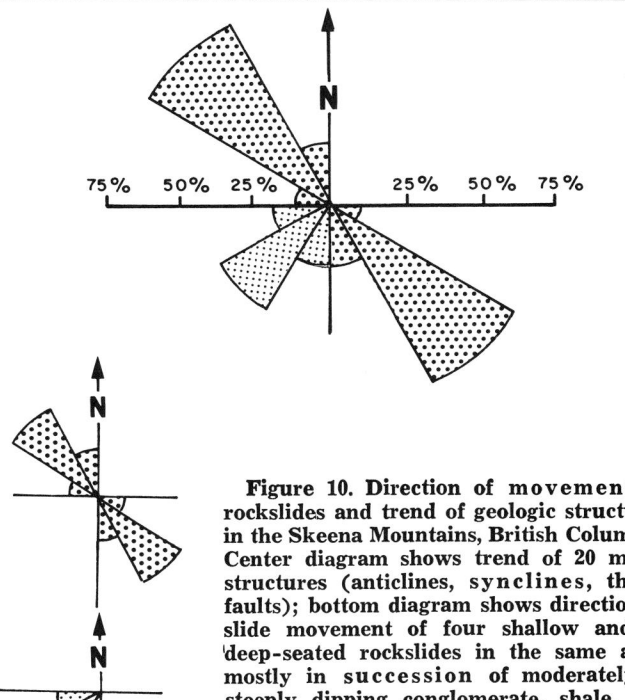

Figure 10. Direction of movement of rockslides and trend of geologic structures in the Skeena Mountains, British Columbia. Center diagram shows trend of 20 major structures (anticlines, synclines, thrust faults); bottom diagram shows direction of slide movement of four shallow and 21 deep-seated rockslides in the same area, mostly in succession of moderately to steeply dipping conglomerate, shale, and sandstone beds (after Eisbacher, 1971).

Engineering and Environmental Implications

It is difficult to predict accurately the factor of safety of existing mountain slopes and even more difficult to predict the time when mountainside failure may take place. Investigators armed with geological maps and high-altitude stereoscopic air photos should try to reconstruct the past sequence of geologic events and geomorphic characteristics that may influence slope stability (Figs. 10, 11; Table 1). A check should be made for dips of nearly 30° in layered rocks oriented parallel to the mountain slope, especially on high spurs; air photos should be studied for long, fresh-looking fault scarps located near the crests of mountain peaks. Thicker snow accumulates and lingers longer into the summer in furrows located upslope of shutterridges formed by displacement on faults traversing the higher slopes of uneven mountain topography, with the displaced part of the ridge "shutting in" the adjacent ravine. Such anomalous-appearing breaks in upper mountainside topography may be associated with postglacial isostatic rebound following the melting of valley or cirque glaciers. The induced differential response to rebound along mountain peaks in a glacier-eroded terrain may follow pre-existing structural weaknesses, such as fault zones. This may have occurred at the Hope slide (Mollard, 1976).

Case Histories

Stalk Lakes Slide (Figs. 1, 6; Table 1). The date of failure and volume of the Stalk Lakes debris stream are unknown. The breakaway scar and debris lobe, however, are sparsely wooded (Table 1), suggesting that rupture may have taken place recently. This mountain rupture was described by Eisbacher (1971). Its identifying physical features are characteristic and generally well expressed: the breakaway scar outline; the jumping platform near the base of the rupture surface; the splaying debris tongue, with conspicuous longitudinal trains and grooves; many debris cones; and a transverse depression beyond the base of the failed mountainside. Figure 6 shows these features particularly well.

Maligne Lake Slide (Figs. 1, 7; Table 1). I first noticed this rock avalanche in the early 1960s when making an air photo study of potential campsites. The main mountain rupture, one of several in this valley (Fig. 5), is located at Maligne Lake in Jasper National Park, Alberta. The date on which this rock avalanche occurred is unknown. This major rock failure has been described by Cruden (1976), who reported that the rupture occurred in thick-bedded limestone, siltstone, and chert inclined toward the valley center at 25° to 40°. He recorded that the rupture surface is about 1 km wide and 1.5 km long and that the huge volume (500×10^6 m³) of debris travelled about 1 km vertically and as much as 5.5 km horizontally. In stereoscopic air photos the rupture surface looks clean and planar (Mollard, 1976).

Striking characteristics of this failure are the numerous conical debris mounds containing angular, slabby fragments (A. Stende, oral commun.) and the manner in which they are concentrated around the arcuate rim of the debris lobe (Fig. 7). Some of the larger cones may possibly be moulin kames, but this interpretation seems unlikely in view of their consistent location with respect to the debris lobe (Fig. 5). Transverse ridges and fissures and faint longitudinal ridges and grooves can be detected. Large angular blocks of broken rock scattered widely over the surface of the main failure can be detected in 1:24,000-scale air photos.

Frank Slide (Figs. 6, 8; Table 1). At 0410 h on April 29, 1903, some 33×10^6 m³ of rock crashed down the east face of Turtle Mountain at speeds exceeding 100 km/h, killing 66 people in the town of Frank. Blocks of limestone and shale as much as 12 m across, mingled with mud, came to rest in the distal rim of the debris lobe more than 1 km from the base of the failed mountainside and 120 m above the valley floor at Frank, Alberta. The rock mass started sliding on a steeply east-dipping flank of the Turtle Mountain anticline, whose crest was close to the crown of the slide scar. The slide probably took place on bedding surfaces, with the orientation of the head scarp and lateral margins controlled by joint sets normal to the bedding (Krahn and Morgenstern, 1976). The rupture surface near the toe of the slide followed a minor thrust fault lying above the Turtle Mountain thrust fault (Cruden and Krahn, 1973), which appears to have been the place where a thick accumulation of rock debris had built up on the lower part of the rupture surface, forming a shelf or platform that launched colliding rock fragments and dust into the air.

Hope Slide (Figs. 1, 9; Table 1). Shortly before dawn on January 9, 1965, a large rockslide took place, and 47×10^6 m³ of broken rock fragments buried British Columbia Provincial Highway 3 to a maximum depth of 75 m. I studied "before and after failure" air photos of the Hope slide shortly after the slide occurred. Failure took place in massive to slightly schistose, green metavolcanic rock (greenstone) containing intrusive sheets of felsite as much as 6 m thick that dipped nearly parallel to the pre-existing 30° slope of the mountainside. Detritus from the 1965 slide buried material from a prehistoric landslide that had occurred at the same site. Geologic conditions contributing to failure are thought to include weakness in or adjacent to the sheets of felsite, including jointing and schistosity parallel to those surfaces (Mathews and McTaggart, 1969).

Creep movements associated with the prehistoric slide, deep glacial erosion and oversteepening of the valley wall in adjoining Eleven Mile Creek Valley, and postglacial rebound may have contributed to the development of a long, conspicuous, fresh-looking scarp and associated snow-filled trough that can be discerned near the top of Johnson Peak. The resulting mountainside adjustment might have caused sufficient slip on potential rupture planes to reduce the angle of shearing resistance from a peak value to some lesser value approaching the residual angle. In any event, till deposits have been observed above the head scarp of the 1965 break-

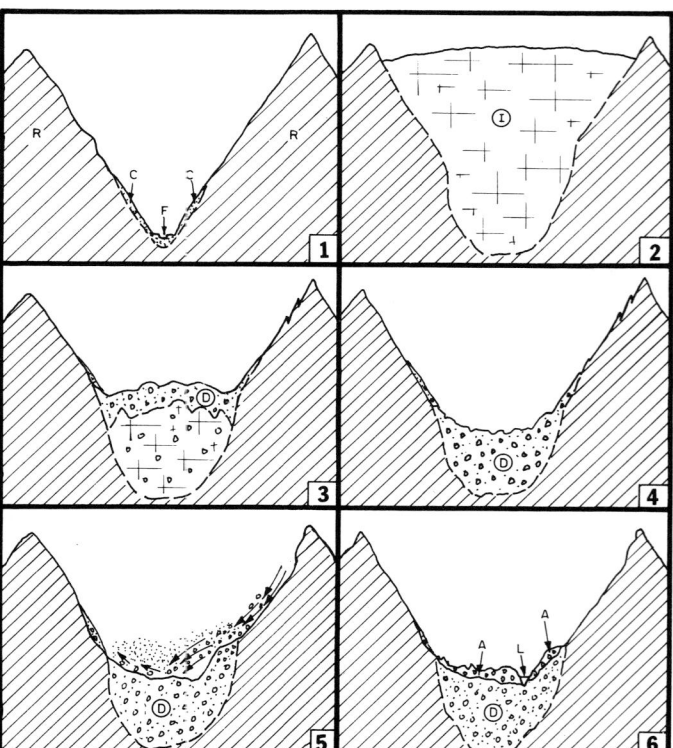

Figure 11. Postulated sequence of geologic events in the formation of rock avalanches. 1, Preglacial valley, more or less V-shaped, with flood plain (F); accumulation of colluvium (C) over steeply dipping rock layers (R). 2, Valley filled with glacial ice (I), which characteristically modifies the lower part of the valley to make it U-shaped; deep glacial erosion of rocks, especially below valley floor. 3, Ablation of valley glacier; accumulation of superglacial drift (surface moraine) (D); differential rebound near mountain top; scarplets formed; talus accumulation. 4, Complete melting of glacier; irregular topography formed on surface of moraine debris (D) in valley bottom. 5, Catastrophic rock avalanche in motion; movement includes components of rockslide, rockfall, and rock flow (see arrows). 6, Present-day features on the debris tongue of the rock avalanche (A); debris-dammed lake (L).

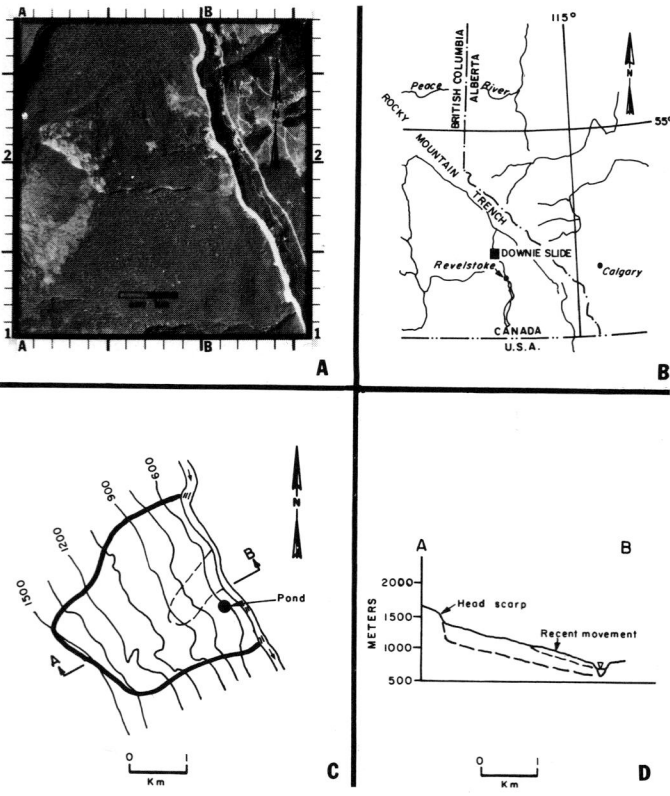

Figure 12. Surface and subsurface characteristics of the Downie Creek slide (A.1 to B.1, 1.8 to 2.5 in A) in the Columbia River valley north of Revelstoke, British Columbia. Creep movements in dashed area of C appear active (after Patton and Hendron, 1974; Piteau and others, 1976). Air photo A 14932–78, courtesy of the National Air Photo Library, Ottawa.

away scar near the top of Johnson Peak (Mathews and McTaggart, 1969), indicating that the valley was formerly filled with ice.

The grooved rupture surface, with two substantially deeper cavities on the northeastern part of the scar, and a pile of coarse rocky detritus in the valley floor exhibit many of the characteristics of a rockslide that successively developed into a rockfall and rock avalanche. Air blast effects, transverse concentric ridges, a hummocked surface, compositional banding, and debris cones can all be discerned in air photos (Mollard, 1976). Broken rock fragments, dirt, and mud—the mud thought to have been derived largely from saturated, fine-grained alluvial sediments on the valley floor upon impact—were driven onto the opposite valley wall, from which they rebounded to flow back up the base of the failed mountain slope (Mathews and McTaggart, 1969). Hodgson (1965) suggested that earthquake tremors on the day of the slide may have triggered failure, but Mathews and McTaggart (1969) expressed doubt that the earthquake was strong enough to be a significant factor.

SOME OTHER TYPES OF LANDSLIDES IN THE CANADIAN CORDILLERA

Eisbacher (1971) stated that in the Skeena Mountains the number of deep-seated, rotational slumps greatly exceeds the number of shallow, planar rockslides that generate rock streams. In the study area he mapped in the Canadian Cordillera, Eisbacher noted that 21 of 25 rockslides belong to the deep-seated, rotational variety of mountainside failure (Fig. 10). They generally show exposed head scarps that are 50 m or more high, with bulging common at the toe.

Downie Creek Slide

The Downie Creek slide (Fig. 12) is a large (1×10^9 m³), deep-seated rockslide whose downslope creep movements conspicuously narrowed the width of the Columbia River and its flood plain. I detected this rockslide during an air photo study in 1961, when searching for granular borrow for a major earthfill dam in the Columbia River valley. Among the features that reveal the huge bedrock failure are conspicuous narrowing of the river channel and flood plain; three sets of rapids in the river opposite the slide; an anomalous-appearing, small, nearly circular pond situated on the side of a steep valley slope; a steep unvegetated head scarp of varying height (30 to 120 m); a distinct breakaway rim; and a series of linear trends in the tree vegetation that appear to be associated with slope movements. Large-scale surface and subsurface exploratory investigations were begun in 1965 (Piteau and others, 1976). Bedrock below the failed mountainside comprises interbedded gneiss and mica schist. The failure surface, which follows a foliation shear dipping 18° toward the river (Patton and Hendron, 1974), lies between 150 and 250 m beneath the mountainside. The nature of surface features in a northeast-trending, U-shaped area near the center of the failed slope (Fig. 12) suggests recent movements. The Downie Creek rockslide has been investigated intensively using different geologic and geophysical exploration methods in the past 10 yr. A novel feature has been the installation of acoustical devices in selected boreholes. The listening devices have been used to monitor the sounds of movements in creeping rock (Piteau and others, 1976).

Nagle Mountain "Linears"

Air photo studies have disclosed conspicuous faultlike scarps, called "linears," near the tops of mountain peaks in different parts of British Columbia. They are worth noting because movements along them may weaken the bedrock in the mountainside downslope and thereby increase the hazard of future slope failures. In 1962 I identified one such area of linears while mapping and classifying former slope instability features in the Columbia River valley near the Mica damsite. The linear ridges are located below the peak of Nagle Mountain (Fig. 13).

Because these distinctive faultlike features occur in mountain valleys that were filled by thick (1 to 1.5 km) ice during Pleistocene time, it is believed that they result from postglacial isostatic rebound movements, possibly along pre-existing fractures (Mollard, 1976). Gneisses and schists ap-

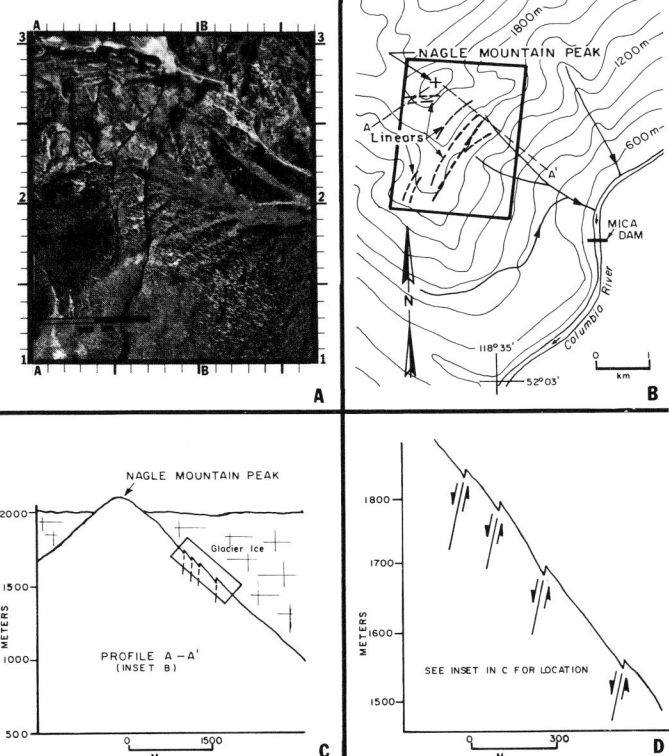

Figure 13. Characteristics of the Nagle Mountain area, British Columbia. A, Air photo showing "linears" (A.4 to A.9, 1.3 to 2.9); see inset in B for location. B, Topography of glaciated Nagle Mountain and location of linear features. C, Section through mountain showing valley filled with glacier ice. D, Detail of linear features in gneiss and schist with steeply inclined foliation. Air photo A 11091-30, courtesy of the National Air Photo Library, Ottawa.

pear to be shattered in the bottom of narrow, V-shaped furrows upslope of narrow, A-shaped ridge features. Rocks beneath the furrows contain zeolite and appear to be highly weathered. The strike of some of the "linears" parallels the gneissic banding. Topographic features resulting from Pleistocene glaciation can be seen near the top of Nagle Mountain, indicating that ice in the Columbia River valley exceeded 1,000 m in thickness. Some of the linear furrows exhibit a distinct splaying aspect (Fig. 13), reminiscent of splay faults, and the furrows support alpine meadow vegetation. Small creek valleys either terminate at the linears or become noticeably smaller downslope from them. Dips on the upslope side of the ridge scarps typically range from 60° to 80°.

These upper mountainside linear features have been studied closely in the field. In 1966 I noticed similar features in air photos when making a study of landslides in the Columbia River valley between Revelstoke and Castlegar, British Columbia.

RETROGRESSIVE SLOPE FAILURES IN ARGILLACEOUS BEDROCK IN THE CANADIAN WESTERN INTERIOR

Morphologic Features

In discussing the origin of the topography caused by landslides in the valley of the South Saskatchewan River, Terzaghi (1955) referred to the riverward "energetic lateral expansion" of shale in valley sides. He stated that the shale probably assumed the character of a "massive gravity creep," causing subparallel ridges and depressions to form in the overlying mantle of glacial drift. Cause of the slope-flattening process would now be interpreted in terms of a reduction from peak to residual shear stress. Because these very large slope failures occur in a retrogressive manner, they are also referred to as "large-scale retrogressive slope failures."

In the Interior Plains region, retrogressive slope failures occur most commonly in Upper Cretaceous bentonitic marine clay shale, silty shale, and claystone, commonly with siltstone and sandstone interbeds. Failed slopes flatten out, as a result of slow creep movements, to gradients as low as 9.5° to 4° (1 on 6 to 1 on 14). Such slope failures are easily identified in air photos owing to their very large dimensions, characteristic locations along valley walls (Fig. 14), and arcuate, elongate ridge-and-depression topography (Figs. 15, 16). They may extend for several kilometres along the sides of a valley and can be many tens of square kilometres in area.

Faintly expressed transverse cracks develop on the upland near and approximately parallel to a high head scarp; in time these cracks become accentuated in the relief by surface runoff (Fig. 17). They have been called "fracture traces" by Scott and Brooker (1968) and are thought to result from stress relaxation phenomena and to be a manifestation of the energetic expansion referred to by Terzaghi (1955) and Peterson (1958), the swelling of overconsolidated clay shales described by Hardy (1957), and the rebound following valley erosion and unloading determined by Matheson and Thomson (1973).

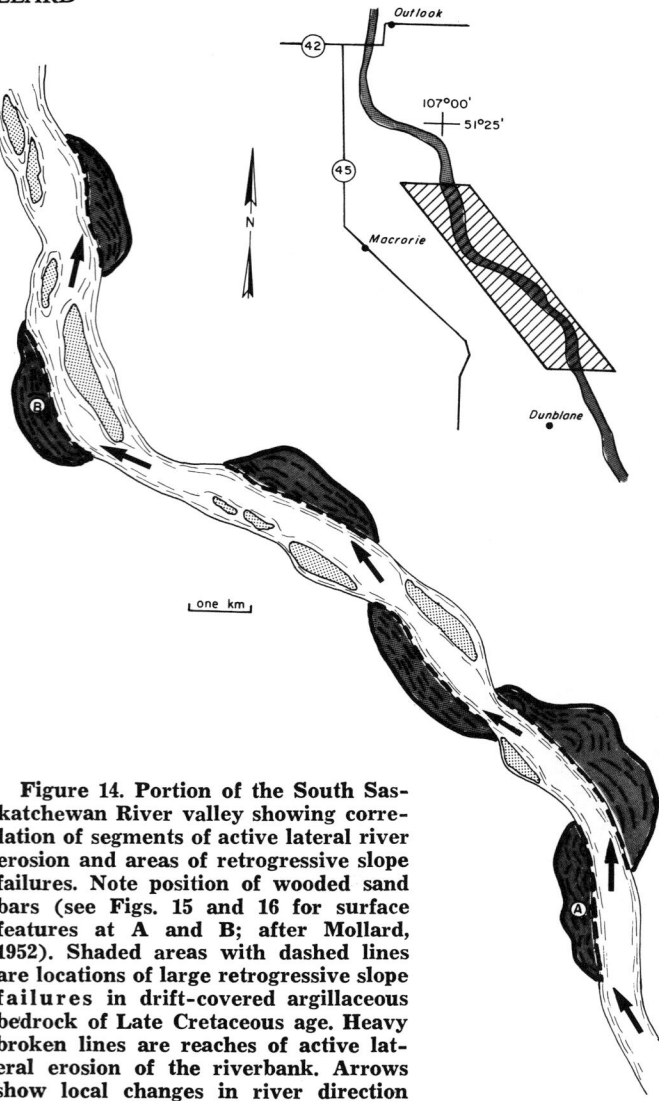

Figure 14. Portion of the South Saskatchewan River valley showing correlation of segments of active lateral river erosion and areas of retrogressive slope failures. Note position of wooded sand bars (see Figs. 15 and 16 for surface features at A and B; after Mollard, 1952). Shaded areas with dashed lines are locations of large retrogressive slope failures in drift-covered argillaceous bedrock of Late Cretaceous age. Heavy broken lines are reaches of active lateral erosion of the riverbank. Arrows show local changes in river direction related to slip-off and undercut slopes.

Figure 15. Oblique aerial view of landslide topography in the South Saskatchewan River valley, central Saskatchewan. Photograph is of locality A in Figure 14. Arrows point to concentric arcuate ridges and depressions (Mollard, 1952).

Figure 16. Oblique aerial photograph of landslide topography in the South Saskatchewan River valley, central Saskatchewan. Note that point A is located on the upslope scarp of a block that has moved toward the river; point B is a pond in a sinking "graben" block; and point C shows the end of the youngest, still-developing headscarp. Row of arrows indicates a small drainageway, the initial development of which predates the failure scarp it crosses (Mollard, 1952).

The occurrence of multiple, subparallel, elongate ridges resulting from "out" or "down-and-out" translated slide blocks is characteristic. Dominantly translational movements are usually accompanied by some rotational movement and by sinking "graben" movements on blocks next to the tops of valley walls (Figs. 16, B; 18). Poorly vegetated failure surfaces on blocks adjoining the upland look freshest because they are youngest (Fig. 16). As can be seen in Figures 15 and 16, ponds, swamps, and dry, closed depressions are common; trees and scrub brush grow in wet hollows, whereas the drier ridges are generally covered by grass.

Landslide locations generally correspond to reaches of active lateral river erosion (Fig. 14) and to deeply incised tributaries in which active downcutting is still taking place (Figs. 17, 19, 20). High stable banks, on the other hand, can be found where aquifers in buried bedrock valleys act as underdrains, thereby depressing the water table. Although major slope failures are common wherever major valleys have been incised into Upper Cretaceous bentonitic marine shales, they can also be seen on the flanks of such mesalike uplands as the Horn Plateau, the Birch and Caribou Mountains, and the Swan Hills in the Interior Plains region.

Some fine-grained clastic rocks show a greater propensity to fail than others. The more landslide-prone argillaceous rocks tend to be heavily overconsolidated, highly plastic, bentonitic, and marine in origin (Figs. 15 to 20). Several generations of slope failures are usually detectable in the topography, which suggests that they are retrogressive in space and progressive in time (Fig. 20). Clearly translational and rotational block movements may be apparent from the position, elevation, and slope of level and backward-tilted horst surfaces. Features indicative of viscous flow are rarely associated with these movements; where they do occur (Fig. 20), the failure debris is probably composed of surficial lacustrine sediments overlying the shale rather than shale itself.

Stratigraphy and Lithology

The stratigraphy, lithology, and geotechnical properties of Upper Cretaceous argillaceous bedrock in the Interior Plains have been widely reported (Bjerrum, 1967; Caldwell, 1968; Locker, 1973; Mollard, 1952; Morgenstern and Eigenbrod, 1974; Ringheim, 1964; Scott and Brooker, 1968; Thomson and Morgenstern, 1974). Geotechnical engineers have referred to these argillaceous bedrock materials as preconsolidated clay shales (Hardy, 1957), overconsolidated clays (Hardy and others, 1962), and overconsolidated plastic clay and clay shales (Bjerrum, 1967). Geologists refer to them as poorly indurated argillaceous bedrock, but mostly as shale and silty shale. Scott and Brooker (1968) reported that most shale in the marine Bearpaw Formation, which is responsible for a large proportion of the landslides in the Interior Plains, contains 30% to 60% clay and 40% to 70% silt, has a plastic index between 40% and 80%, and a liquid limit between 65% and 100%. These figures compare with averages of 31% clay, 55% silt, and 14% sand in central Alberta (Locker, 1973). At the Gardiner Dam in Saskatchewan, where detailed studies were carried out over several decades, the average plastic limit is 23%; the liquid limit, 115%; and the plastic index, 92%. Here, the average clay content is 50%, and the activity is 1.8% (Bjerrum, 1967). In the South Saskatchewan River valley, the 345 m Upper Cretaceous Bearpaw Formation, estimated to have accumulated 74.5 to 70 m.y. ago, is divisible into 11 members, 6 of silty clay alternating with 5 of

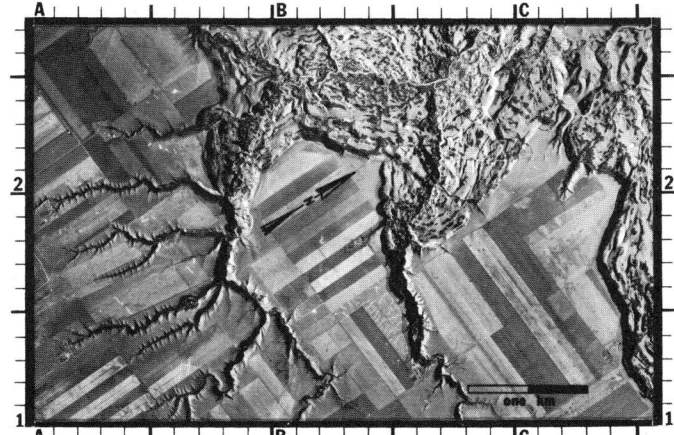

Figure 17. Topography developed on retrogressively failed Upper Cretaceous, argillaceous bedrock in western Canada. Note the subparallel ridges and depressions characteristic of landslide topography and the aligned gullies associated with rapid postglacial valley erosion and resulting rebound (A.0 to A.7, 1.0 to 2.1). They occur on a nearly level upland underlain by glaciolacustrine clay over till over bentonitic marine shales of the Bearpaw Formation. Along the South Saskatchewan River valley, north of Swift Current, Saskatchewan. Photo A 16372–73, courtesy of the National Air Photo Library, Ottawa (Mollard, 1976).

Figure 18. "Sinking Hill," west of Killdeer, Saskatchewan. At location "A", a single block is translating subhorizontally toward the creek, causing the successive sinking movements of a "graben" block behind. The "graben" started moving in 1932, and the trough is now 10 m deep. Large ovoid sandstone concretions, in exposed, vertical bedrock walls above the highly fissured graben surface, appear to have been "sliced through," and they resemble half of giant hamburgers. The vertical faces, however, possibly follow pre-existing joint surfaces. J. G. Locker (1976, oral commun.) reported that during test drilling in the graben, it was impossible to maintain mud circulation below about 24 m. Movement is associated with active toe erosion of a spur containing a horizontal bentonite bed near creek level. Similar former block movements have taken place at "B" (B.6, 1.9), which adjoins "A" (B.8, 2.0), and at "C" and "D" (C.1, 2.2 and A.5, 1.4). All movements have occurred where thinly drift-covered Upper Cretaceous Frenchman and Eastend Formations (Kfe) overlie landslide-prone, bentonite-rich, marine shale of the Upper Cretaceous Bearpaw Formation (Mollard, 1976). Air photo A 15115–139, courtesy of the National Air Photo Library, Ottawa.

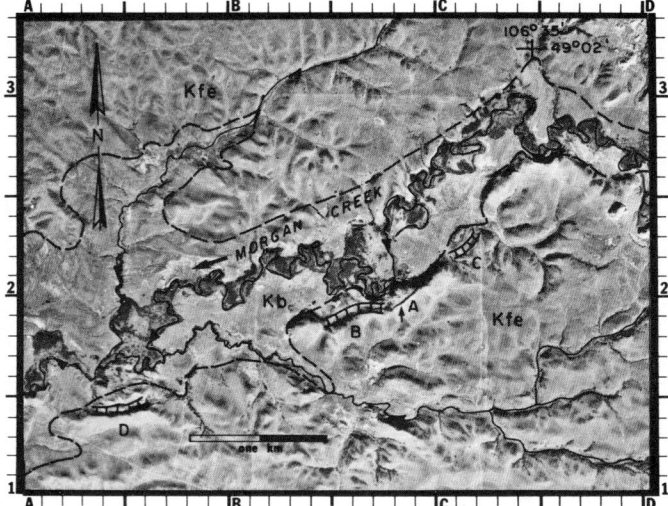

silty sand (Caldwell, 1968). Bentonite beds and bentonite-rich shale are common, and the dominant clay mineral is montmorillonite. The liquid limit in bentonite-rich shale may be as high as 265%. Where the bentonite content is very high, the weathered surface of the shale is commonly a dirty grayish white; has a "popcornlike" appearance; shows a finely cracked or checked surface; exhibits a nuggetlike structure, iron stains, and salt coatings; and is strewn with selenite crystals (Mollard, 1952).

Thomson and Morgenstern (1974) stated that the presence of bentonite seams and admixtures of bentonite is the most important single factor affecting the shearing resistance of Upper Cretaceous argillaceous rocks in Alberta. Pure bentonite beds, however, are rarely reported in the logs of test holes drilled during geotechnical investigations. The infrequent detection of pure bentonite during test hole drilling is commonly due to poor recovery; in some places, however, such lack of detection may also be related to its depositional environment, when the soft sediments were highly mobile, or possibly the result of bentonite beds having been "squeezed out" during glacial loading or postglacial translational failures associated with valley erosion. In any event, test drilling in the valley of Twelve Mile Lake, some 16 km south of Assiniboia, Saskatchewan, suggests that one or more of these processes must have taken place, because long pods and wedges of exposed bentonite several metres thick in silt and fine sand of the nonmarine Frenchman Formation are either very thin or absent a few metres away from the outcrops, which occur on seemingly unfailed valley walls (R. Gotts, 1976, oral commun.).

Thomson and Morgenstern (1974) stated that marine argillaceous bedrock is more prone to develop landslides than deltaic and fresh-water (nonmarine) sediments of equivalent age. Their finding agrees with my field observations in many parts of the Interior Plains region, where bentonite-rich strata in nonmarine argillaceous rocks are less prone to fail than

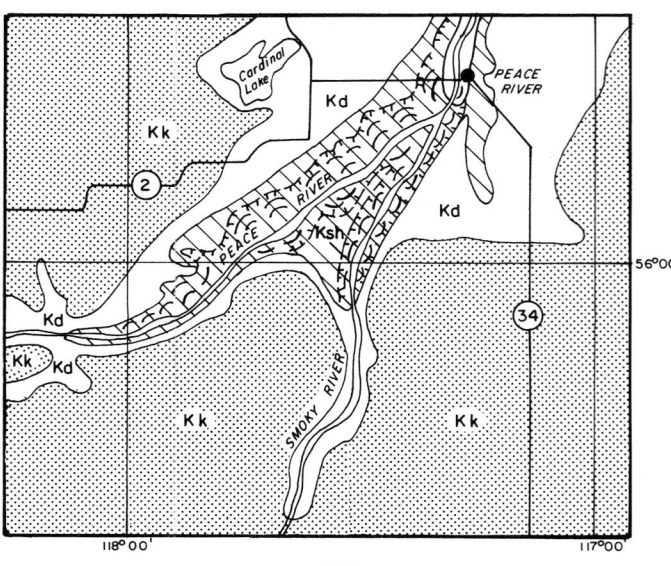

Figure 19. River erosion and size of area affected by slope failures near Peace River, Alberta. Map shows a correlation between the depth of river erosion in landslide-prone Upper Cretaceous bentonitic argillaceous bedrock (Ksh) and increasing area of retrogressive slope failure per unit length of river valley (see Fig. 20). Above the town of Peace River, Alberta. Bedrock formations after Research Council of Alberta (1972). Upper Cretaceous is indicated by Kk and Kd. Kk represents the Kaskapau Formation, which is made up of dark-gray silty shale and, thin concretionary ironstone beds, which are interbedded in the lower part with fine-grained quartzose sandstone and thin beds of ferruginous oolitic mudstone; marine. Kd represents the Dunvegan Formation, which is made up of gray, fine-grained, feldspathic sandstone with hard calcareous beds, laminated siltstone, and gray silty shale; deltaic to marine. Upper and Lower Cretaceous are indicated by Ksh in diagonal lines. Ksh represents the Shaftesbury Formation, which is made up of dark gray, fish-scale–bearing shale (silty in upper part), numerous nodules and thin beds of concretionary ironstone, and bentonite partings; the lower part has thin silty and sandy intervals; marine. Dashed lines with ticks and arcuate dashed lines indicate the extent of valley sides affected by massive, postglacial, retrogressive slope failures.

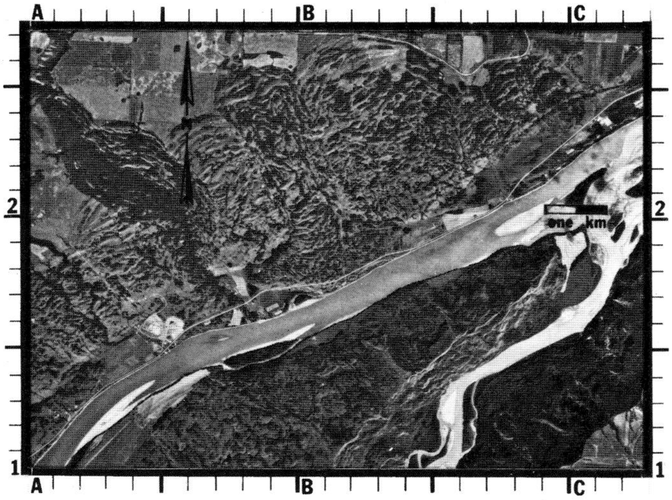

Figure 20. Retrogressive slope failures west of the town of Peace River, Alberta. Successive generations of retrogressive slope failure are indicated by a crudely arcuate pattern of subparallel ridges and intervening hollows. A lobate debris apron exhibiting the morphologic characteristics of a large viscous flow, probably involving failed surficial lacustrine clay, has moved over and locally buried the former flood plain of the Peace River in the area south of the north point (A.5 to A.8, 1.5 to 1.7). Air photo A 21819–41, courtesy of the National Air Photo Library, Ottawa. See also Figure 19 (after Mollard, 1976).

bentonite-rich marine beds. For example, along the south wall of Morgan Creek valley, west of Killdeer, Saskatchewan, nearly horizontal translational movements of single blocks took place and are still taking place where the creek has undercut the toe of slopes underlain by "soapy," bentonite-rich marine shale beds of the Bearpaw Formation (Fig. 18). Other Upper Cretaceous strata nearby have not failed, because the immediately overlying beds are nonbentonitic marine silt and fine sand and the younger bedrock strata above them are bentonitic but consist of nonmarine clay, silt, and fine sand (Fig. 18).

In the Interior Plains region, the five Upper Cretaceous formations exhibiting the greatest tendency to fail retrogressively are the Shaftesbury, Lea Park, Judith River, Bearpaw, and Riding Mountain. All but the Judith River Formation consist of dark gray marine shale and silty shale, with silt and sand intervals and concretionary ironstone beds. The Bearpaw, Shaftesbury, and Judith River Formations contain pure bentonite beds, bentonite-rich shale, or both. The Judith River is, however, dominantly nonmarine.

Geologic Conditions Affecting Failure

I made a series of studies of the geologic characteristics of large retrogressive slope failures in drift-covered Upper Cretaceous argillaceous bedrock between 1947 and 1955 (Mollard, 1952, 1955). These studies entailed air photo and field examination of more than 300 km of failed and unfailed valley slopes along the South Saskatchewan River valley and adjoining Qu'Appelle Valley. Detailed observations were made of 20 major slope failures, several of them more than 3 km long and extending 1.5 km or more back from the river's edge. In the late 1940s, unanswered questions centered around whether the extensively drift-covered, hummocky terrain between the prairie upland and the river's edge was morainal (glacial) or landslide (colluvial) in origin. Core samples obtained during borehole drilling and tested in the laboratory revealed that the overburden and underlying shale possessed a high strength when subjected to conventional short-term shear tests. The indicated high laboratory shear strength made it difficult to understand why slopes underlain by the marine Bearpaw Shale had failed so extensively and had done so on such gentle slopes (Mollard, 1952, 1955; Mollard and Pollock, 1955; Peterson, 1954; Peterson and others, 1960; Pollock, 1962; Ringheim, 1964). Whereas slip surfaces were rarely observed during this early period of investigation, the studies revealed that the failures were developed in postglacial time and were caused by rapid downward erosion of streams into drift-covered bentonitic marine shale of Late Cretaceous age. The fluvial erosion commonly extended 100 m below the upland and 30 m below river level. Slow retrogressive movements were thought to be largely horizontal and associated with shale rebound following glacial melting and rapid downward and lateral river erosion (Fig. 21). In certain locations, younger slope failures seemed to develop where previous movements in the shale could have taken place in preglacial, interglacial, glacial, or early postglacial time (Mollard, 1955). A significant finding was that the weathered and softened bentonite-rich shale zone just beneath the drift failed on slopes as low as 4° (1 vertical on 15 horizontal) — much flatter indeed than conventional laboratory tests of shear strength would suggest.

Many detailed geologic and geotechnical studies pertaining to factors affecting the strength of shale below valley walls have been made in the past 20 yr. A sequence of diagrams attempting to portray the geologic conditions affecting slope stability is given in Figure 21. The postulated sequence of events that follows is based on a synthesis of data accumulated and reported from extensive investigations carried out over a 30-yr period. Diagram 6 in Figure 21 is modified after Figure 7 in Peterson and others (1960). Of course, not all events in the postulated sequence may have occurred at any one landslide locality.

Event 1. Development of a badland topography on the preglacial surface. Deep weathering, erosion, and slumping, lasting more than 70 m.y. Estimated depth of preglacial subaerial erosion is 450 to 750 m, with an average depth of removal of about 600 m (Hardy, 1957; Mollard, 1952, 1955; Peterson, 1958; Ringheim, 1964; Scott and Brooker, 1968; Thomson and Morgenstern, 1974). Maximum past effective

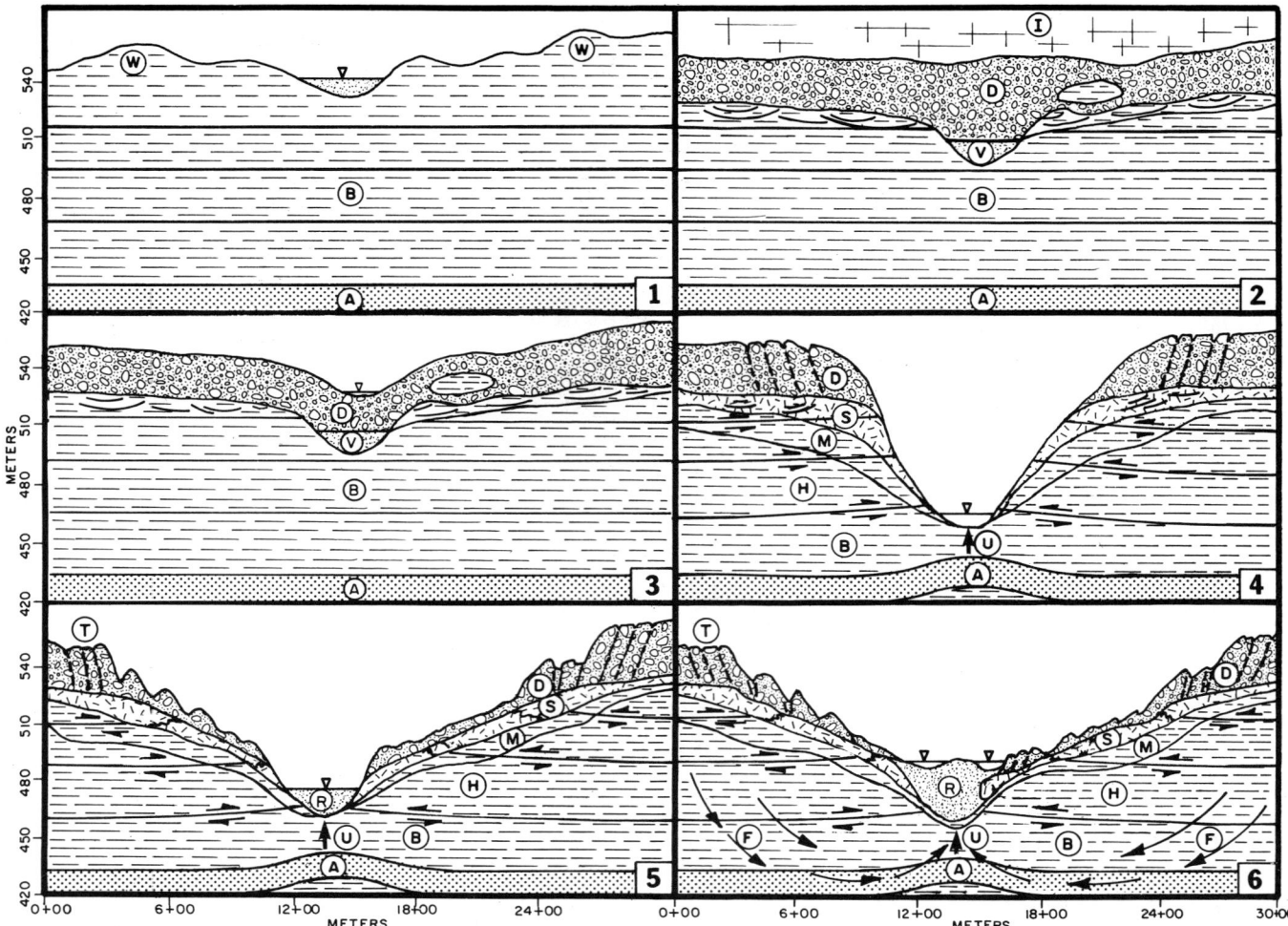

Figure 21. Conceptual diagrams of events determining the characteristics of retrogressive slope failures in argillaceous bedrock of the Canadian Western Interior. For explanation, see text. A, Ardkenneth Sand Member in the Bearpaw Formation. B, Bentonitic, marine, argillaceous strata of the Bearpaw Formation with bentonite seams (darker lines). D, Drift, mostly till, containing blocks of ice-thrust shale in the lower part; shear planes of ice-thrusting in the upper part of the shale. F, Direction of regional ground-water flow. H, Hard shale zone. I, Ice ($\pm$ 1,200 m thick). M, Medium shale zone. R, River alluvium, mostly fine to medium sand. S, Soft shale zone containing closely spaced joints, numerous slickensides, and deformed bentonite seams. T, Tension cracks in stiff glacial overburden on the upland near the top of the valley wall. U, Uplift of valley floor, producing a gentle anticlinal structure and upwarping of beds in valley side (exaggerated in these sections), associated with rapid postglacial valley erosion. V, Valley in bedrock, partly filled with stratified sand and gravel. W, Weathering and erosion.

pressure (that is, preconsolidation pressure) derived from curves expressing the void ratio versus the effective pressure is on the order of 100 to 150 kg/cm², or 9,800 to 14,700 kN/m² (Bjerrum, 1967; Peterson, 1958).

Event 2. Multiple glaciation, ice erosion, and deposition. Pre-existing bedrock valleys partly filled with sand and gravel. Formation and burial of interglacial valleys, with disruption of pre-existing surface drainage courses and glacial sand and gravel aquifers. Glacier erosion and ice-thrusting of the upper shale zone; detachment and transport of small and large blocks of bedrock, many of which are enclosed in drift (Christiansen and Whitaker, 1976); local brecciation, shearing, and drag folding of the upper 15 m or so of bedrock strata underlying the drift. In central Saskatchewan, an estimated 1,000 m of ice load may have resulted in the development of high pore pressures in the shale and in sand interbeds within the shale (Scott and Brooker, 1968).

Event 3. Stagnation and melting of the Wisconsinan glacier. Postglacial lakes formed and drained. Beginning of fluvial erosion in the drift.

Event 4. Unloading caused by rapid postglacial stream erosion produced raised valley rims, upward flexure of valley sides, and development of gentle anticlinal structures below the valley floor (magnitude exaggerated in Fig. 21). These developments are associated with the elastic rebound of argillaceous bedrock due to vertical stress relief. Rebound may have been as much as 10% of the valley depth, and typically is 1 to 3 m (Matheson and Thomson, 1973). Elastic rebound resulting from excavations into shale during construction of the Gardiner Dam, Saskatchewan, equals about 0.7% of the excavated depth at shallow depths and 0.4% at greater depths in the hard shale zone (Ringheim, 1964). Peterson (1958) stated that, for the purpose of geotechnical investigations, rebound can be divided into two phases: (1) elastic rebound, which occurs immediately upon release of load and varies directly with the load removed, and (2) "time rebound," which takes place over many thousands of years.

Gentle upwarping of shale strata induced slip between beds, along shears developed during former ice-thrusting, in bentonite seams and partings, in bentonite-rich shale zones, and, it is thought, mostly in the weathered soft shale zone. It is believed that this movement caused enough slip to reduce the angle of shearing resistance from peak value to some lesser value approaching that of the residual angle (Eigenbrod and Morgenstern, 1972; Thomson and Morgenstern, 1974). For example, a total inward movement of 50 cm in the weathered soft shale zone but only 5 to 7.5 cm in the underlying unweathered hard shale zone was measured by Peterson (1958) in a 90-m-long by 1.8-m-wide horizontal test tunnel excavated by hand. Typical consolidation curves (void ratio versus effective pressure) projected to a maximum past effective normal pressure of 150 kg/cm^2 (14,700 kN/m^2) indicate that the moisture content of the consolidated shale was then about 14%. Assuming that all rebound took place in a vertical direction and that the moisture content increased from 14% to the present values in the hard, medium, and soft shale zones, Peterson estimated that the top of the valley wall would heave 17 m and that the bottom of the valley side, approximately 900 m away, would rise approximately 11 m. Moreover, assuming that all of the rebound occurred in the horizontal direction, the toe of the slope would move 180 m toward the river over the width of the valley side, that is, from the upland to the base of the slope (Peterson, 1958).

Brecciated zones developed in bedrock near the base of the valley wall, and transverse tension cracks formed in drift on the upland near the top of the valley wall. Selective weathering and erosion accentuated these subparallel, subequally spaced tension cracks, or "fracture traces," in the landscape. Eventually they developed into small knicklike gullies arrayed on opposite sides of larger ravines — in general, oriented parallel to the top of the valley wall (Fig. 17). In fact, the style of landslide topography developed during retrogressive slope failure appears to have been influenced to some extent by the directions of regional jointing in the underlying Cretaceous bedrock (Mollard, 1952). Recent regional mapping of fracture traces and joint patterns (Babcock, 1973, 1974; Mollard, 1957, 1958; Westgate, 1976) reveals that two orthogonal joint systems characterize regional jointing over much of the plains area of Alberta, Saskatchewan, and Manitoba. In southern and central Alberta these are made up of sets striking approximately 55° and 140° (roughly normal and parallel to the Rocky Mountains) and approximately 5° and 95°. Moreover, the orientations of the sets persist across lithologic and stratigraphic boundaries. Although local deviations from these patterns exist, the two orthogonal systems extend over 200,000 km^2 (Babcock, 1974). Westgate (1976) suggested that the coincidence in orientation of joints in Cretaceous bedrock and basal till in the Athabasca Tar Sands region may be produced by slight adjustments along bedrock joints after deposition of the overlying glacial drift. Upward propagation of joints into the till probably occurred during deglaciation as a consequence of glacio-isostatic crustal rebound and unloading of the basal till (Westgate, 1976). Where the strike of the valley wall parallels the trend of a prominent joint set, the tendency toward failure seems to be accelerated, possibly owing to reduction in shearing resistance in wedge failures developed in the shale and overlying basal till.

The large number of closely spaced joints in the soft shale zone appears to be related to horizontal and vertical rebound, an increase in moisture content in the shale, swelling, and the development of numerous slickensides. There is a noteworthy decrease in joints and slickensides in the medium shale zone, and they are rarely detectable in core samples from the hard zone (Fig. 21). The thickness and characteristics of these three zones are well documented at the site of the Gardiner Dam in central Saskatchewan (Bjerrum, 1967; Peterson, 1954, 1958; Ringheim, 1964).

Event 5. During and following the events outlined above, progressive slip movements took place along pre-existing planes of weakness. Such movements reduced the angle of shearing resistance from peak value to near the residual angle. Landslide topography developed as a result, much of it originating from the slow progressive creep movements of valley walls toward the river channel. Horst and graben structures developed in the drift mantle. Sharp and angular at first, eventually these slide blocks became smoothly rounded by long-continued mass wasting (Figs. 15, 16). Complex systems of retrogressive movements are thought to result from slip on several surfaces (particularly in the bentonite seams) at different elevations, but mostly within the soft shale zone. Formation and enlargement of fractures in the relatively stiff glacial overburden on the upland was followed by rill and then gully erosion along the fracture lines (Fig. 17).

The role of artesian ground water is not particularly well understood owing to the extremely low permeability of the shale and the common difficulty of determining accurate piezometric water levels in the shale (Ringheim, 1964; Thomson and Morgenstern, 1974). That significant movement of ground water does, however, occur through vertical joints in the shale and in silty sand interbeds is suggested by four recent studies carried out by J. D. Mollard and Associates along the South Saskatchewan River, all of them concerned with induced infiltration of river water into deep alluvial sand and gravel aquifers containing brackish water. The chemistry of the ground water in alluvial sediments reveals a high content of total dissolved solids caused by the subsurface water having moved through marine strata and overlying drift sediments derived from these marine strata (Fig. 22; Table 2). Van Everdingen (1972) showed the ground-water flow regime adjoining and below the valley; he also showed that the piezometric surface of pore water in silty sand beds within the Bearpaw Formation is above the river level near Riverhurst, Saskatchewan (Fig. 22).

Event 6. Aggradation of the lower part of the stream-eroded valley with 20 to 40 m of postglacial alluvium, consisting mostly of medium-grained sand. Present-day lateral erosion

Figure 22. Location map and geologic cross sections showing the influence of ground-water regimes on slope stability. The two cross sections (A'–A' after Whitaker, 1970; B'–B' after Van Everdingen, 1972) show the inferred direction of flow (arrows) of ground water in bedrock strata beneath drift and alluvium. Small black squares in the location map indicate locations where river infiltration was induced into deep alluvial sand and gravel aquifers during studies of municipal and industrial ground-water supplies (see Table 2). Note artesian ground-water conditions (piezometric surface above the river flood plain) in B'–B'. Pattern 1 is postglacial alluvium (mostly fine to medium sand). Pattern 2 is glacial drift (mostly glaciolacustrine clay over till). Pattern 3 is Bearpaw Formation (bentonitic clay and silty shale with pure bentonite partings). Pattern 4 is sandstone beds (the lowest, thickest unit in A'–A' is the Judith River Formation). Note also that thin sand members occur in shale of the Bearpaw Formation, higher in the section.

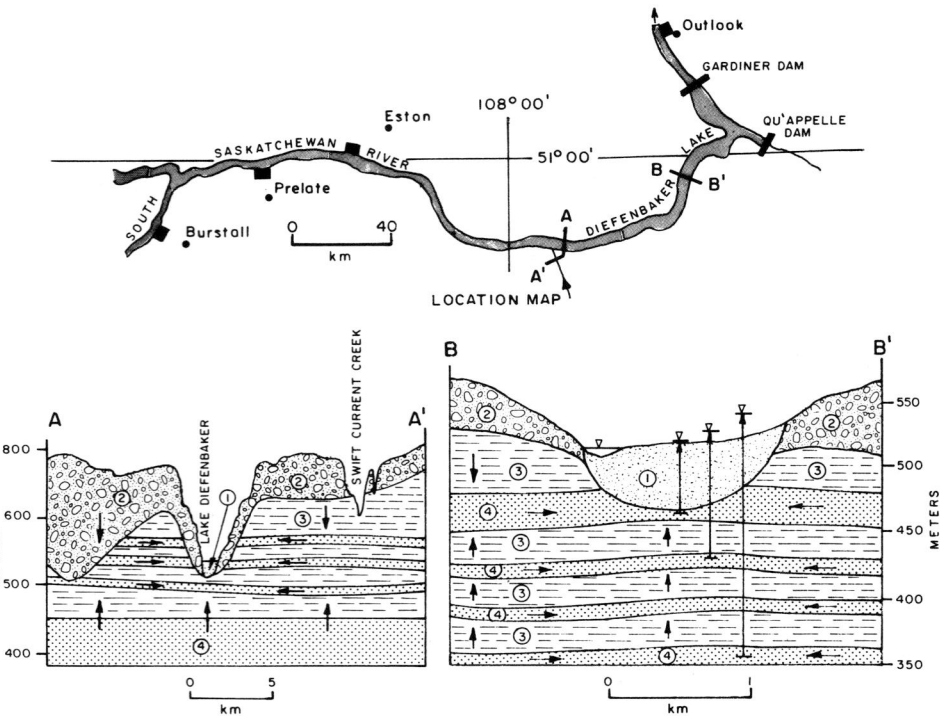

by the river and associated active downcutting by its tributaries in ravines, as well as man-made cuts, induce renewed movements of valley sides (Fig. 14) by gravity creep on pre-existing slip surfaces, where the residual angle of shearing resistance (Fig. 23) varies somewhere between 5° and 14° and is commonly between 8° and 10° (Bjerrum, 1967; Eigenbrod and Morgenstern, 1972; Insley and others, 1976; Ringheim, 1964; Thomson, 1971; Thomson and Hayley, 1975; Thomson and Morgenstern, 1974). Because in the 1950s and 1960s short-term laboratory tests of shear strength appeared to be unreliable indicators of long-term creep strength in nature, measurements were made to determine the safe angle of excavation slopes based on plots of cotangent slope versus slope height for unfailed and failed slopes underlain by Upper Cretaceous argillaceous rocks, predominantly shale (Krinitsky and Kolb, 1969; Lane, 1967; Peterson, 1954; Peterson and others, 1960; Ringheim, 1964).

Character and Rate of Movement

The character of slope movements in argillaceous bedrock of the Interior Plains has been described as slow progressive failure (Thomson and Hayley, 1975), large-scale retrogressive slope failure (Mollard, 1952), and massive gravity creep (Terzaghi, 1955). Hardy (1957) stated that failures along the Alaska Highway developed 10 or 12 yr after construction cuts were made. He added that the Grierson Hill shale landslide at Edmonton, Alberta, moved approximately 120 m toward the river between 1893 and 1957, or about 2 m/yr. Such movements tend to be irregular. For example, carefully measured movements at St. Michael's Retreat, located on an old landslide in bentonitic marine shale at Lumsden, Saskatchewan, amounted to 8 cm in 1974 but only 0.6 cm in 1975 (V. G. Beckie, 1976, oral commun.). At Gardiner Dam, on the South Saskatchewan River, creep movements of as much as 15 cm in soft shale were observed during dam construction. One particular failure involved 2.3×10^6 m^3 of shale and overburden material. However, all of the recorded movements stabilized after a short time (Ringheim, 1964). Data that I have obtained reveal average annual vertical movements of 6 cm along the Canadian Pacific Railway just west of Westerham, Saskatchewan, and of 45 cm along the Canadian Pacific Railway in the Red Deer River valley just west of that river's confluence with the South Saskatchewan River. Also, following rail-line abandonment, between April 1974 and April 1976, the railway grade below the Canadian Pacific tracks at Regina Beach dropped between 1 and 2 m, or 0.5 to 1 m/yr, depending on the location (Fig. 24). At the Morgan Creek block slide, the rate of downward vertical movement on a graben block has averaged approximately 30 cm/yr over a 44-yr period (Fig. 18). In exceptional years, however, the rate of movement was said to be several metres (B. Anderson, oral. commun.). In any event, the rate of movement of landslides in argillaceous bedrock in the Interior Plains is very slow, which is expressed by Terzaghi's term "gravity creep" (Terzaghi, 1955).

Engineering and Environmental Implications

That the existing factor of safety of failed valley walls in Bearpaw Formation bentonitic marine shale and similar argillaceous rocks is near unity is indicated by the reports of many failures that occurred shortly after man-made excavations were made in or near the toe of a slope. These creep move-

TABLE 2. QUALITY OF GROUND WATER FROM DEEP AND SHALLOW AQUIFERS IN ALLUVIUM INDICATING GROUND-WATER FLOW AND ARTESIAN PRESSURES IN BEDROCK BENEATH DRIFT

Location	Depth of alluvial sediments (m)	Total dissolved solids (original quality) of water in deep alluvial aquifer (ppm)	Total dissolved solids of South Saskatchewan River water (ppm)	Total dissolved solids (modified quality) of deep well water after pumping (ppm)	Time after pumping began (months)
West of Burstall	21	2,700	300	600	6
North of Prelate	39	2,700	300	400	S*
South of Eston	30	3,700	300	850	6
West of Outlook	24	4,000	350	400	S*

*S, Sandpoint installed at shallow depth in sand of riverbed after pumping poor-quality ground water from deep wells and abandonment of wells.

ments, which usually stop after a short time, are, however, easily reactivated, not only by unloading of lower slopes but also by the weight of heavy earthwork and building structures placed on upper slopes (Ringheim, 1964). Slope movements may also be reactivated by the introduction of seepage water and increased pore pressure associated with subsurface drainage from sewage lagoons and cracked concrete swimming pools, broken water and sewer mains, improper hillside drainage installations, and excessive watering of lawns. The areal extent of old landslides should be mapped and excavation in them avoided if at all feasible. Where retrogressive slope movements are likely to cause serious damage, zoning regulations should call for the landslide area to be maintained as pasture, park, or some similar low-intensity use. Failed shale slopes should be investigated thoroughly before engineering structures or heavy buildings are erected on them. Landslide terrain can be classified and mapped in terms of various sizes and ages of past slope failure. Areas having different degrees of potential instability — such as stable, unstable, and marginally stable or quasi-stable — can usually be recognized and mapped. Such potential instability ratings should be clearly indicated on maps. Areas that are likely to experience rapid drawdown of water levels after flooding and bank saturation should be avoided; so should places where accelerated wave or stream erosion opposite wooded islands or at undercut banks may trigger movement. Excavation depths should be kept to a minimum in order to limit elastic rebound, softening and swelling of the shale, and loss in shear strength. Sauer (1975) emphasized that many old failures are marginally stable and may be moving so slowly that the movements go unnoticed unless they happen to be carefully monitored or observed. Where movements are reactivated, they can cause costly damage to property, especially to masonry structures.

Case Histories

Morgan Creek Slide (Fig. 18). The case history of the Morgan Creek translational block slide, locally called the "Sinking Hill," was selected for several reasons. First of all, the beginning of movement of the most recent valley-wall failure (about 1932) is documented, and the progress of movement over a span of 40 yr has been watched by the local ranchers. Whereas one block slide is presently active, three other almost identical old slides occur within 1.5 km (Fig. 18). All four failures are in the same geologic setting. They are located where Morgan Creek previously undercut, or is now undercutting, the toe of the valley slope underlain by the landslide-prone bentonite-rich Bearpaw Formation. The character of movement at all four slides has been similar: the slow, subhorizontal translation movement of a single block adjacent to the undercut creek bank, with sinking "graben" movements of a slice of ground behind. The drift cover here is only about 1 m thick, and it overlies the predominantly nonmarine Frenchman Formation, which is also bentonitic but has not failed, probably because it is not only nonmarine but also contains abundant silt- and sand-sized particles. The Eastend Formation, below the Frenchman Formation and above the Bearpaw Formation, is marine but not bentonitic. It also contains much fine-grained sand and silt, whereas the Bearpaw Formation consists almost entirely of silt and clay particles. The amount of "graben" sinking since 1932 has been about 10 m; it has occurred in a spur whose original elevation was some 42 m above the level of Morgan Creek. Close by, a bentonite bed can be seen to crop out in the valley

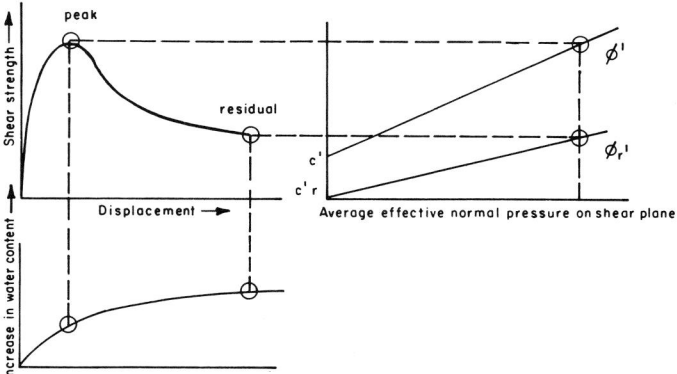

Figure 23. Shear-strain characteristics of Upper Cretaceous argillaceous bedrock in the Interior Plains region of western Canada. Note that as valley erosion (unloading and stress relief) occurs and interbed slip displacement takes place, the shearing resistance is reduced from peak to residual value, and there is a corresponding increase in moisture content (lower diagram) and swelling of the overconsolidated clay shale (after Scott and Brooker, 1968).

Figure 24. Condition of the Canadian Pacific Railroad track at Regina Beach, Saskatchewan, 2 yr after abandonment. The railway grade below the tracks dropped between 1 and 2 m in the 2-yr period between April 15, 1974, and April 15, 1976. A row of deep wood-pole piles (right) have failed owing to slow slippage toward an undercut slope at the shoreline of Last Mountain Lake. White arrow points to head and shoulders of boy standing between ties on the railway grade that has dropped.

wall a short distance above the elevation of Morgan Creek. The position of the bentonite seam is thought to be responsible for the subhorizontal translation of three former and stabilized and one active block slides and related sinking "graben" movements (Fig. 18).

EARTHFLOWS AND FLOW SLIDES IN SENSITIVE CLAYS IN EASTERN CANADA

Sensitive clays, also called "quick clays," are defined as having a remolded strength of 25% or less of their undisturbed strength — that is, 1 to 4 or less. However, some marine clays are even more sensitive, having a remolded to undisturbed strength ratio of 1 to 20, or even 1 to 100 or less (Peck and others, 1951). That is to say, they will transform from a brittle material to a liquid mass when sufficiently disturbed (Crawford, 1968). Typically, quick clays have a natural moisture content equal to or greater than their liquid limits, and determination of this condition may reveal that a sediment is sensitive. Extensive deposits of quick clays occur in the Ottawa River and St. Lawrence River valleys, along the eastern coasts of James Bay and the lower part of Hudson Bay, and along the lower Hamilton River valley west of Goose Bay, Labrador (Figs. 3 and 25). Such clays may have been preconsolidated by partial desiccation or by the weight of previously existing but subsequently eroded materials.

Discussing the character of these sediments along the lower La Grande River, east of James Bay, Skinner (1974) mentioned that below surficial sand and a stiff, coherent stratum of clay, the marine silty clay is in a sensitive state. The open, flocculated structure of this sensitive material can be likened to that of a "house of cards." If the pore pressure is increased, the "cards" (the loosely bonded, platy fragments of silty clay) separate, and the mass behaves like a liquid. When this happens, long, narrow, transverse blocks of overlying, less sensitive strata subside, are stretched, and are rafted in the

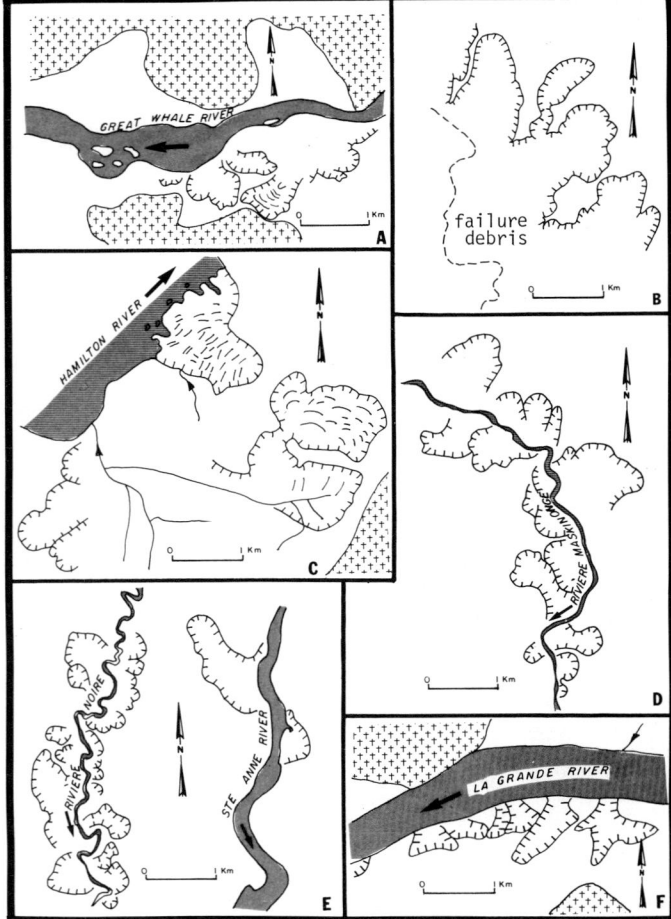

Figure 25. Major retrogressive flow slides and earthflows in eastern Canada. Areas A (lat 55°17'N, long 77°37'W) and F (lat 53°44'N, long 78°13'W) are located near the east coast of James Bay, Quebec. Area B (lat 45°31'N, long 75°11'W) is southeast of Ottawa, Ontario. Area C (lat 53°03'N, long 61°11'W) is west of Goose Bay, Labrador. Areas D (lat 46°15'N, long 73°03'W) and E (lat 46°44'N, long 72°05'W) are in the central St. Lawrence Lowland region of Quebec. Note the exposed Precambrian bedrock (cross pattern) in valley walls in A, C, and F. Generally, crater areas appear larger and crater depths appear shallower on old failures, many of which likely occurred during the early stages of stream downcutting. The reason old failures are usually larger and shallower than historic ones is that, at the time of high failure incidence, the freshly deposited sediments had neither drained nor consolidated to any significant degree. Accordingly, as a general condition, natural water content tended to be much higher and soil strengths much lower than they are today along the margins of abandoned terraces and river banks.

fluid mud. The bank commonly retrogresses in successive slices in a more or less concentric pattern until stability is attained. These large failures have been referred to in the literature as major retrogressive flow slides, earthflows, lateral spreads, and quick-clay slides.

Quick-clay failures are numerous in the lowlands of eastern Canada, where more than 750 of them have been counted. Published data on 50 large failures indicate that loss of life exceeds 100 persons, and loss of upland soil exceeds 49,000 ha (Mitchell and Markell, 1974). They are conspicuous features along the lower Gatineau and South Nation Rivers near Ottawa; along the lower Yamaska, Nicolet, Maskinonge, and Bastican Rivers between Ottawa and Quebec City in Quebec; along the lower Little Whale, Great Whale, La Grande, and Rupert Rivers on the lower east coast of Hudson Bay and the adjoining coast of James Bay; and along the lower Hamilton River in Labrador (Chagnon, 1970; Conlon, 1966; Eden and Mitchell, 1973; Hurtubise and others, 1957; Mollard, 1976).

Morphologic Features

The diagnostic identifying features of failures in sensitive marine deposits are so distinctive that for many years I have used them as an aid in distinguishing marine deposits from fine-grained lacustrine and fluvial sediments when carrying out air-photo studies for site and route investigations in eastern Canada. Not all fine-grained marine deposits in eastern Canada are sensitive, however.

Characteristically, failure craters are bowl-shaped, have steep head scarps and lateral margins, and have floors that are flat or have only a gentle gradient. Where the periphery of a crater is not distorted by nearby cliffs of hard bedrock or by pre-existing, deeply incised ravines and creek valleys that limit the growth of a crater, crudely semicircular, horseshoe-shaped, oblong, spatula-shaped, and pear-shaped outlines predominate. In detail these outlines may be either regular or irregular (Fig. 25). A distinct bottleneck may also be present, commonly where a river has incised deeply into thick, sand-veneered marine silt and clay in former deep estuaries bounded by high Precambrian rock walls (Fig. 25, A and C). Failure craters with nearly level to gently sloping floors in some places coalesce, and adjoining crater floors may be stepped as a result of failures having occurred at different times and at different elevations.

The ratio of the length of retrogression to original slope height, R/H, is usually more than 10 and is commonly between 20 to 80 (Mitchell and Markell, 1974; Mitchell, 1976). Crater lengths generally vary from 100 to 1,000 m. Twenty-two documented major retrogressive flow slide/earthflow failures reveal the following geometric parameters (Mitchell, 1971): (1) minor retrogressions in homogeneous clay without weak layers have a maximum retrogression length of 0.4 H, suggesting partly drained failure, and (2) major retrogressions in sand-veneered, stiff clay over soft, banded clayey silt and silty clay with silt and fine sand seams have retrogression distances of 4 R to 80 H, but mostly less than 20 H, with outlines of failure craters that vary from crudely pear-shaped to crudely bowl-shaped. Where not removed by river erosion, the spoil from recent earthflows may still show a "catspaw" terminal area, with splaying, ragged, longitudinal ridges and intervening grooves (Fig. 25, C). Many original backscarp slopes are steep (30° to 50°), and their intersection with the unfailed upland can be angular (Fig. 26). With time, backscarps become flatter owing to weathering and erosion.

Sand-veneered transverse ridges, or ribs, on the crater floor initially may appear flat-topped and steep-sided in cross section; eventually, however, they become smoothly rounded, because in most places the transverse slices, or "ribs," have been subjected to mass wasting over a long time. Indeed, they can be very subtle microrelief features if they are old and have been cultivated for many years. A highly characteristic thumbprintlike pattern (Mollard and Hughes, 1973) is formed by the crudely concentric, transverse, sandy ribs on the flattish floors of some bowls (Fig. 27). The low gradient and the ribbed pattern on the floor of the bowl seem to provide clues to the near-surface stratigraphy with which they are associated. The ribbed pattern is better expressed where the

Figure 26. Scarps of South Nation River earthflow. View showing portions of the steep lateral margin and headscarp of the main failure crater. Note the nearly horizontal tops (for example, A.5 to C.8, 4.0) and steep bounding scarps on transverse blocks that have been stretched and rafted, here only a short distance on remolded sensitive marine clay. Location: South Nation River, 6.5 km north of Casselman, Ontario. Photo courtesy of the Geotechnical Division, National Research Council of Canada, Ottawa.

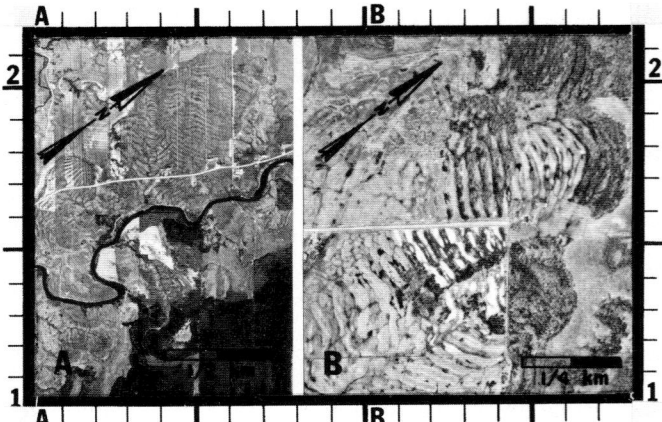

Figure 27. Thumbprint patterns on crater floors (A.1, 1.7; A.5, 1.9; also B.2 to B.7, 1.6 to 1.9). A, St. Marc des Carrières, Quebec; photo 904A–26, courtesy of Ministry of Lands and Forests, Quebec. B, Bourget, Ontario; photo A 19866–41, courtesy of the National Air Photo Library, Ottawa. The pattern of subequally spaced transverse ribs seems to originate wherever a surficial cover of fluvial and deltaic, fine to medium sand (locally with some gravel) overlies soft, laminated, sensitive marine clayey silt and silty clay having a high proportion of silt and fine-sand partings. The ribs (to the right of the north arrows) are usually less evident or absent where a desiccated, cohesive clay crust overlies a soft, laminated silt and silty clay substratum.

Figure 28. Rafted blocks, South Nation River, Ontario (A.0, 2.5 to B.6, 1.0.). Trees can be seen standing upright in sand that forms the top layer of blocks composed of stiff cohesive clay; these blocks have been rafted into the South Nation River on a liquid mud composed of remolded marine clay. Photo courtesy of Geotechnical Division, National Research Council of Canada, Ottawa.

surficial sediments are sand rather than desiccated clay. Yet not all bowl floors show a classic thumbprint pattern, and the bottoms of some large craters have been dissected by surface runoff. Many of the characteristic physical features of failures in sensitive marine deposits are discussed in the literature (Chagnon, 1968, 1970; Conlon, 1966; Crawford and Eden, 1967; Eden and Mitchell, 1970; Karrow, 1972; LaRochelle and others, 1970; LaSalle and Chagnon, 1968; Mollard and Hughes, 1973; Mollard, 1976).

Although they appear to be relatively rare, in places large, pancake-shaped accumulations of failure debris, called "spoil aprons" or "failure debris aprons," are present. Probably one reason the spoil aprons are not more common is that the failure detritus from the crater transforms into liquid mud, which is usually carried into rivers, where in time the spoil is removed by erosion. Spoil aprons are best expressed where the spoil has flowed from the rim of a terraced sand plain onto a level plain fronting the base of the terrace scarp. Many examples of this can be seen between Ottawa and Casselman, Ontario (Fig. 27, B). In places, spoil in the failure apron shows subparallel, crescent-shaped ridges that are bowed downslope. Trees (Fig. 28), as well as buildings and construction equipment (Fig. 30), rafted on slices of stiff clay commonly stand upright after having travelled appreciable distances (Hurtubise and Rochette, 1956; Mitchell, 1976).

Stratigraphy and Lithology

Major retrogressive flow slides and associated earthflows are characteristic features where toe erosion has occurred at the base of terraces composed of sand-veneered, laminated, fine-grained marine sediments. Gadd (1957) noted that in the glaciomarine environment of the Champlain Sea, water-laid silt and clayey silt, with glacial pebbles and boulders, were deposited in thicknesses of as much as 60 m. Although these marine sediments act and feel much like clay, they usually contain minor clay sizes and commonly lack cohesion and plasticity.

The sequence and lithology of stratified materials and characteristics of their shear strength determine the repetitive features one observes on slope failures. Generally, a surficial layer of fluvial and deltaic fine- to medium-grained sand with some gravel overlies a stiff, cohesive silty clay stratum. These units in turn commonly overlie massive to laminated gray clay, generally with a high silt content and fine sand partings. The consistency of this unit varies from very soft, usually just below or near the overlying stiff layer, to medium stiff at greater depths. This structurally weak unit in turn commonly overlies a subhorizontal stratum consisting of stiff silt, till, or bedrock (Fig. 29). Characteristically, the total salt content and the ratio of Na to Ca increases with depth in the marine sediments. With minor variations this succession of units has been noted in several different localities (Ballivy and others, 1971; Bilodeau, 1957; Conlon, 1966; Eden and Mitchell, 1973; Fransham and others, 1976; Hillaire-Marcel and de Boutray, 1975; Mitchell, 1971, 1976; Mollard and Hughes, 1973; Skinner, 1974).

The fine-grained marine sediment is usually highly sensitive (mostly 10 to 100) and consists of nonplastic to slightly plastic clayey silt and silty clay; typically the laminated facies contains much silt and fine-sand partings between finer grained bands. Peck and others (1951) reported a sensitivity of around 150, a clay-size fraction of 36%, and a plasticity index of 12 for sediments in the slope failures at St. Thuribe, Quebec.

The upper units of saturated sand and stiff clay not only add to the weight of the overburden, γH, but the surficial sand layer also acts as a ground-water reservoir that contributes water and high pore pressures to the underlying laminated silt and clay unit. Around Ottawa, however, a stiff, fissured "crust" without the cap of surficial sand overlies softer and weaker silty clay in some places.

Discussing changes in regional lithology and geotechnical parameters of the marine sediments, Eden and Mitchell (1973) stated that in the Lake St. Jean and the lower St. Lawrence Valley the clays tend to be heavily overconsolidated, very strongly bonded (that is, cemented at particle contacts), or both. This includes sediments involved in the catastrophic landslides at St. Jean Vianney and Toulnustouc, which, however, Eden and Mitchell did not regard as true earthflows because they have ratios of overburden pressure to undrained strength of less than 4 and because the ratio of length of retrogression to slope height is less than 20. The central St. Lawrence Valley (including the area between Bourget and Casselman, Ontario), the lower reaches of major rivers flowing into the southeastern shore of Hudson Bay and eastern James Bay, and the Goose Bay estuary are all characterized by major retrogressive flow slides, huge earthflows, or both. Around Ottawa, on the other hand, minor retrogressive slides and infinite slope failures are thought to occur in a stiff, fissured crust of reworked, fine-grained marine or, what is more likely, fluviodeltaic clay deposited in fresh water following regression of the Champlain Sea (Fransham and others, 1976).

Sensitive marine silt and clay deposits generally have low strengths and have been involved in large flow-type landslides, whereas the freshwater fluvial and deltaic silt and clay deposits have relatively higher strength and have produced mainly small bank failures (Fransham and others, 1976; Gadd, 1976). Major retrogressive flow slides and liquefaction are restricted to the fine-grained marine sediments. Where the valley intersects marine deposits along the South Nation River, retrogressive flow slides are common, and where the valley intersects freshwater sediments, small bank failures are the characteristic landslide form. To a significant degree, the contrast in landslide frequency and type between marine and freshwater sediments is probably due to fundamental physical and chemical differences in these sediments (Gadd, 1976).

Geological Conditions Affecting Failure

In most places where there are slope failures, clay and silt deposited during transgression and regression of the Champlain Sea are veneered by fluvial and deltaic, fine- to medium-grained sand with some gravel. Upon postglacial uplift, streams cut into these sediments, forming sand terraces at different elevations. Locally, the terraced sands show features of abandoned shorelines, and dunes and "blowouts" may be seen where fine and medium sands have been reworked by the wind.

Some investigators have believed that salt was leached from the upper zone of marine sediments during postglacial time (Chagnon, 1968, 1970; LaRochelle and others, 1970). However, this view was not shared by Eden and Mitchell (1973),

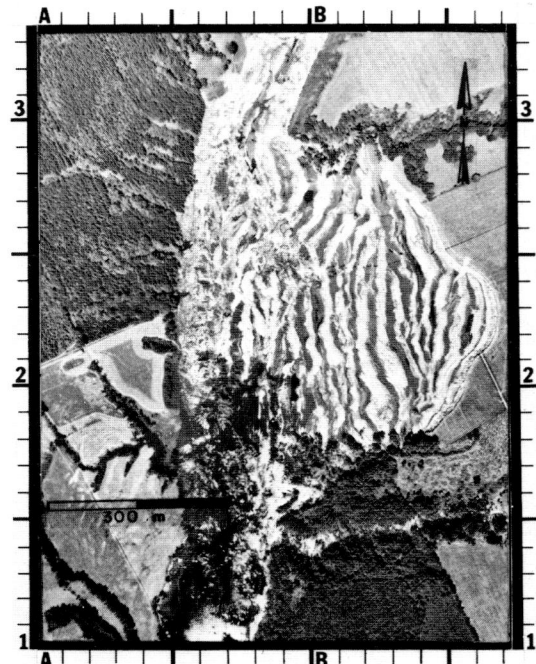

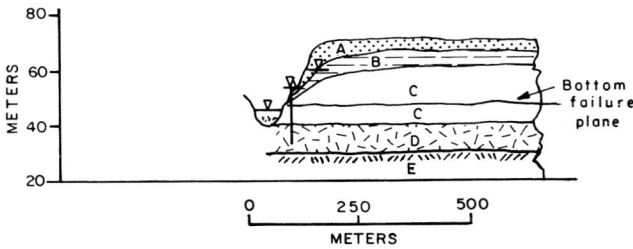

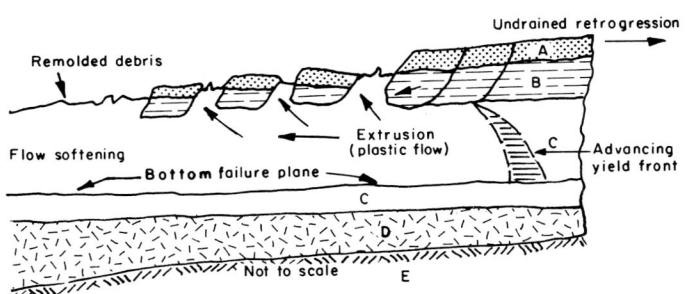

Figure 29. Characteristics of the South Nation River earthflow. Top: Black-and-white print made from a color air photo taken shortly after the South Nation River earthflow occurred. Note the dark-toned, grass-covered, transverse sand ridges and the light-toned, intruded, remolded silt and clay in the fresh failure (A.8 to B.6, 1.8 to 2.8). A similar pattern shows on the bowl of an older crater west of the river. Middle: Stratigraphy of the failed river bank. Bottom: Inferred movements during undrained condition. Letters in the middle and lower diagrams indicate the following materials: A, medium-dense, fine to medium sand; B, stiff clay and silt; C, very soft to medium stiff, silty clay and clayey silt, with silt and fine-sand partings; D, dense silty till; E, bedrock. Photo A 30362–151, courtesy of the National Air Photo Library, Ottawa.

Sangrey and Paul (1971), or Fransham and others (1976) for slope failures near Ottawa. Sangrey and Paul (1971) determined Na/Ca ratios of 30 to 100 in marine sediments but ratios of only 1.7 to 5.5 (mostly 1.7 to 3.5) in fine-grained fluviodeltaic deposits.

In an attempt to explain the style of failure and repetitive pattern of transverse ribs in craters in sensitive marine deposits (Fig. 27), Mollard and Hughes (1973) postulated a failure mechanism in which plastic flow and extrusion takes place in a soft layer underlying a substantially stiffer layer that underlies surficial saturated sand. I believe that extrusion of the underlying, remolded soft clay very rapidly transported successive and intact slices of the stiff layer away from the retreating backscarp. Elongate, crescent-shaped slices appear to subside, more or less intact, as the softened clay is extruded into the river, with ridges and pinnacles of the remolded clay being squeezed upward locally between successive blocks. Such a mechanism, I felt, would explain the thumbprintlike pattern, the steep amphitheatrelike head scarp, the very gentle slope on old earthflow craters, and the common absence of failure spoil in a downslope direction.

In the Bourget and Casselman area of eastern Ontario, almost all of the landslide terrain is associated with fault-controlled abandoned channel systems, rather than with modern streams occupying narrower channels. The extent of retrogression is much greater in old failures along abandoned channels than along modern channels, even though the initial bank heights were much lower in the older slides. Generally, modern stream banks are stable at much greater heights. With the drop in sea level and lowering of the water table that followed, marine silt and clay deposits have become better drained, stronger, and less susceptible to liquefaction. As a result, modern landslides are generally deeper and narrower than earlier ones. The size-shape parameters of landslides have changed over the last ten thousand years, mainly due to dewatering of the sediments (Gadd, 1976). Even so, present-day saturation and high pore-water pressures in marine silt and clay is a principal factor in the landslide mechanism, and the ground-water regime is very much affected by the presence or absence of a surficial sand cover, which results in low runoff and a high level of recharge on a year-round basis.

Antecedent high rains and heavy runoff, particularly from snowmelt, are believed to create high pore pressures, which can initiate slope failure. Flooding accompanied by rising water levels in river banks and subsequent river drawdown — with consequent build-up of high pore-water pressures in the lower part of the slope — is also an initiating mechanism (Mitchell, 1976). Tension cracks have been observed prior to failure, and high pore pressures may be due in large measure to surface water infiltrating these fissure openings (Eden and Mitchell, 1973). A typical ground-water regime in stratified marine deposits was illustrated by Skinner (1974), who also suggested that severe niching can occur where ground water emerges from these slopes. In addition, small, slightly incised tributary creeks may flow at higher elevations than and parallel to the trend of larger river valleys, thereby contributing excess ground water to permeable layers within the marine sediments and promoting slope failure in the larger valleys (Conlon, 1966). Active stream erosion at the toe of slopes and vibration (dynamic stresses from blasting, heavy traffic, or earthquakes) may also precipitate failure. Howell (1973), in fact, listed eight earthquakes with intensity greater than 8 in Quebec for the period 1622 to 1971. He also stated that the St. Lawrence Valley has experienced one or more earthquakes of intensity 7 or greater in each 35-yr period since 1622, except 1692 to 1726 (see also E. A. Hodgson, 1927, and J. H. Hodgson, 1965).

Character and Rate of Movement

Earthflows in quick clay exhibit many of the characteristics of lateral spreads, where a stiff upper layer is broken into strips, stretched, and rafted by plastic flow and (or) liquefaction in the underlying remolded clay and silt. The strips translate and subside but rarely liquefy. Blocks at the headscarp sink as "grabens." The failure mechanism involves the dominant elements of upper-layer translation — rather than rotation — and subjacent layer flow, with relatively little rotational movement of the blocks after failure has proceeded to undrained retrogression (Mitchell, 1976).

Eden and Mitchell (1973), Mitchell and Markell (1974), and Mitchell (1971, 1976), who carried out detailed studies on shear-strength characteristics and the failure mechanism of sensitive, fine-grained marine deposits in eastern Canada, believed that three phases of retrogression take place. Retrogression spreads rapidly after an initial rotational slip failure, quickly developing into a fast-retreating scarp by the mechanism of major retrogressive flow sliding. Finally, failure involves massive plastic extrusion (Fig. 29). Mitchell (1976) stated that a ratio of overburden pressure (γH) to undrained strength (Cu) exceeding 6 must exist within the depth of potential flow for continuous plastic flow to develop earthflows. Moreover, he stated that this ratio serves to distinguish earthflows from retrogressive circular arc sliding, called "flow sliding." Because the very soft to medium stiff zone is commonly thicker than the stiff upper layer, considerable plastic extrusion at the toe can take place (as much as two-thirds of the total volume of failed material).

Earthflows have been estimated to move at rates of 26 km/h (Tavenas and others, 1971). Mitchell (1976) noted that movements of 300 to 450 m may occur in a matter of minutes.

Engineering and Environmental Implications

It is difficult to predict the precise location of future earthflows and to estimate in advance their area and volume. Nevertheless, high-intensity land use should be avoided in "suspect" areas, because catastrophic failures can occur suddenly. Thorough geological and geotechnical investigations should be made. Building in areas of past failure and in areas having a similar geologic history, material composition, structure, and physical setting should also be avoided. Failures are most likely to occur at special geomorphic settings: (1) the banks of actively eroding streams, (2) high, steep bluffs along abandoned terraces, and (3) sites where ground-water flow is concentrated on account of buried bedrock valleys or a position between deep gullies (Fransham and others, 1976).

The construction of buildings should not be undertaken at the top, on, or below high (10 to 80 m) or steep (10° to 50°) valley walls and terrace faces that have had a history of past failure. Several historic failures have been highly destructive, as revealed by eyewitness accounts and observations made shortly after failure (Eden and others, 1971; Tavenas and others, 1971).

Parkes and Day (1975) attempted to devise a hazard rating for potential quick-clay slope failures, as shown in Table 3. A high hazard is indicated by 30 points or more. Most bank heights reported by Mitchell (1971, 1976) fall in the range of 15 to 75 m. He also noted that for slopes of more than 30°, the shear stresses at the base of a slope typically exceed the undrained strength. Mitchell (1976) has proposed the following simple criteria for evaluating the maximum length of retrogression, R, of earthflows based on the stability number, N_s, or $\gamma H/C_u$:

$$R < 100 \text{ m for } N_s < 5$$
$$R = 100(N_s - 4) \text{ for } 5 \leq N_s \leq 12$$
$$R = 100 H \text{ (slope height) for } N_s > 12.$$

Case Histories

South Nation River Earthflow (Figs. 26, 28, 29). On May 16, 1971, a large earthflow occurred on the South Nation River about 6.5 km north of Casselman, Ontario. Mitchell's (1976) data for the stratigraphy, as revealed in the backscarp, which is 9 m high and originally had a slope of 50° but now has a slope of 35° to 45°, is shown in Table 4.

The upper part of the failure in the crater consists of intact blocks that dropped 8 to 10 m vertically in remolded material. Intact grass-covered blocks have a total length of 200 m, or about one-half the total length of the retrogression (Fig. 29). Generally, block surfaces show little forward or backward tilting, indicating subhorizontal translation with only minor rotation (Fig. 26). Volume of spoil in the river is estimated to consist of 0.8×10^6 m³ of intact, stiff block material and 1.6×10^6 m³ of remolded clayey silt and silty clay, totaling 2.4×10^6 m³ (Mitchell, 1976).

Large blocks of cohesionless silty and sandy material resting on stiff clay and rafted in remolded silty clay filled the river channel to a depth of 8 to 10 m. Trees stood upright in this spoil, rooted in the sandy topsoil (Fig. 28), again suggesting minor rotational movement and a pulling apart and rafting of the stiff layer. The crudely semicircular crater that formed is 650 m wide and 400 m long (Fig. 29). An old crater scar with light-toned (sandy) transverse ribs can be seen on the opposite bank of the river (Fig. 29, top).

Factors that promoted failure, according to Mitchell (1971, 1976), were as follows: (1) low undrained strength of a layer within the soft silty clay; (2) high unit weight of the saturated sand and stiff clay and silt layers below the sand; (3) no cohesive strength in the saturated sand layer; (4) rapid drawdown conditions at the toe due to flood levels in the river, and high ground-water pressures caused by snowmelt and rainfall; (5) a stiff horizontal layer at the base of failure, which limited the depth of failure and thereby helped to direct the failure energy in a horizontal direction, away from the retrogressing backscarp; and (6) unusually soft and highly sensitive clay layers, which are subject to plastic flow, and uniform silt and fine sand layers that may be subject to liquefaction, both of which promote long distances of retrogression.

Mitchell (1976) thought that this failure in soft sensitive clay, with seams of silt and fine sand, developed in three distinct stages: (1) an initial rotational slip on the valley bank; (2) major retrogressive sliding, termed "flow sliding" (that is, retrogressive circular arc sliding); and (3) plastic extrusion of soft remolded silt and clay, termed "earthflow," representing about two-thirds of the total volume of failed debris.

Mitchell (1976) thought that an initial minor rotational slip occurred on the river bank and that this movement was followed by a 150-m retrogression, occurring as circular arc slips and progressing from the slower initial slide to more rapid undrained rotational slips. This failure movement was followed in turn by a 250-m retrogression that occurred as a continuous subhorizontal translation without rotational slips, with the soft remolded sediments being extruded under the weight of fine sand and stiff silty clay — their weight driving the spoil rapidly downslope. A transverse pattern of ridges formed on the spoil debris, but it was seen for a short time only, because it was quickly removed by river erosion.

Nicolet Slide (Fig. 30). A 2.5×10^5 m³ volume flowslide occurred at Nicolet, Quebec, on the right bank of the Nicolet River at 1145 h on November 12, 1955 (Béland, 1956; Crawford and Eden, 1957). It killed three people and caused $5 million in property damage. The front part of a large cathedral

TABLE 3. HAZARD RATINGS FOR POTENTIAL QUICK-CLAY SLOPE FAILURES

Item	Rating
Bank height more than 7.5 m	10
Bank slope more than 20.6°	10
Located in an area of previous slope failure	8
Active toe erosion	8
Presence of strong vibrations from heavy traffic, blasting, earthquakes, and so forth	2

Note: From Parkes and Day (1975).

TABLE 4. STRATIGRAPHY OF THE SOUTH NATION RIVER EARTHFLOW

Stratigraphic unit	Thickness (m)	Salts (ppm Na)	Na/Ca ratio
Medium dense sand	4	<100	<1.5
Stiff clay and silt	3.5	<100	<1.5
Soft to very soft clayey silt and silty clay	6.5	100 to 125	1.5 to 2.5
Medium stiff to very soft silt and clay	10.5	250 to 500	12 to 25
Stiff silt	4.5		
Dense till	10		
Bedrock			

Note: Data from Mitchell (1976).

Figure 30. Nicolet slide. This vertical air photo shows the pile of spoil from a flow slide at Nicolet, Quebec. Note that part of a large cathedral (B.4, 2.4) and a house (B.5, 2.2) have been rafted in the remolded, sensitive marine clay and that many trees are still standing upright. Air photo 377–43, courtesy of the Ministry of Lands and Forests, Quebec.

and a bulldozer were carried on the failure spoil; also, one house on the flow slide remained intact, whereas another was badly damaged (Fig. 30). The retrogressive flow slide failed in a series of slices, the first two slices failing in rapid succession (Hurtubise and Rochette, 1956). At the failed slope, 2 to 4.5 m of sand overlies 27 m of sensitive marine clay, which in turn overlies a thin, dense, bouldery till on bedrock. The marine sediments consisted, on the average, of 60% clay, 30% silt, and 10% fine sand; the moisture content of these sediments was above the liquid limit of 40% to 45%. Both water content and plasticity decreased with depth. A salt content that increased with depth indicated a change from a brackish to a marine depositional environment. The upper clay, light gray near the surface, changed to dark gray marine clay at depth. Hurtubise and Rochette (1956) stated that the stability of the marine clay decreases with time on a geologic time scale, but that failure may be accelerated by accidental factors, such as the adverse actions of man.

Bilodeau (1957) gave the following factors that could have contributed to slope failure: (1) the heavy weight of a large building located near the crest of the bank that failed; (2) a bank having a height of 17 m and a slope of 26°, or approximately 1 on 2; (3) water that accumulated in the surface sand, adding overburden weight and feeding water to the underlying clay as the bank moved out toward the river; and (4) vibrations resulting from the detonations of a cannon and from heavy traffic associated with a road detour.

SLOPE FAILURES IN PERMAFROST IN NORTHERN CANADA

A wide variety of mass movements in permafrost regions affect engineering works and present environmental concerns. Of the several types of mass movements that can occur, attention here is directed to two common types: skin flows and bimodal flows. Each is found in the Lower Mackenzie Valley region (Fig. 4). Although mass movements in permafrost regions have received only sporadic attention for a long time on account of their remote location, the small population and scarcity of settlements in the area, and the relatively small areas affected by the movements, they have recently assumed major significance because of accelerated exploration for oil and gas and proposals for the construction of pipelines to transport these petroleum products.

Skin Flows

Skin flow is defined as a failure in which the active layer of soil and covering vegetation detaches and moves rapidly downslope on a planar, inclined surface (McRoberts and Morgenstern, 1973, 1974). Commonly, individual ribbonlike features coalesce into broad sheets. These shallow, elongate, planar, fingerlike landforms develop most conspicuously on lightly wooded and burned-over steep hillsides and valley walls (Fig. 31). Their movement is dominantly translational, and they are confined to the active layer. As a result, they are variously called "infinite slope failures" (Isaacs and Code, 1972), "active layer glides" (Mackay and Mathews, 1973), and "detachment failures" (Hughes, 1972). Mostly they occur on south- and west-facing slopes on shallow, silty colluvial soil material overlying weathered shale in the continuous and the widespread discontinuous permafrost zones. They are also common in valley slopes that have been deeply incised into glacial lake basins. In the Lower Mackenzie Valley, they can be seen locally on escarpments and valley sides having slopes of more than about 15%, and they are especially well ex-

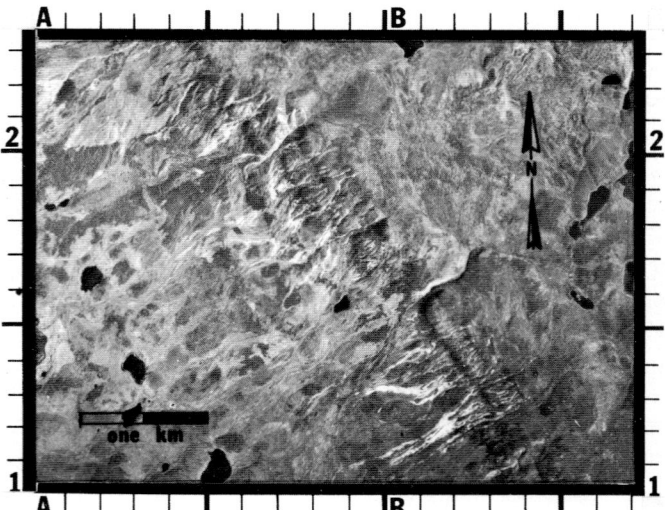

Figure 31. Skin flows east of Little Chicago, Northwest Territories. The single and coalescing, whitish, streamerlike skin flows occur in silty colluvium over weathered, interbedded shale and siltstone of Middle Devonian age. They occur along an escarpment that extends from B.4, 1.0 to A.3, 2.3. Failure may occur in a catastrophic manner. Air photo A 22418–208, courtesy of the National Air Photo Library, Ottawa.

pressed on the east side of Yeltea Lake, east of Inuvik, and along the walls of a former meltwater channel now occupied by Thunder Creek (Fig. 4).

Bimodal Flows

Bimodal flows have biangular profiles consisting of a steeply sloping, semicircular to spatulate ablating headscarp and a low-angle, flattish tongue or lobe (McRoberts and Morgenstern, 1973, 1974; Brown and Kupsch, 1974). Thin veneers of icy soil melt and flow down the headscarp. Overhanging moss and other tundra plants may protect the headscarp from rapid ablation and washing. Many bimodal flows show a bulging terminal area consisting of mud, with multiple transverse ridges. Bimodal flows are conspicuous features on the upper Peel plateau southwest of Fort McPherson, where they are horseshoe-shaped in outline. I have mapped many of them in ice-rich silty sediments in the tundra, originally calling them "mudflows" (Mollard, 1972, 1977). The slope of the headscarp typically varies from 30° to 50°, and the slope on the tongues commonly ranges from 3° to 14°.

Many bimodal flows are active in late fall, appearing as streams of liquid mud. The oozy mud thrusts into the sides of streams and thaw depressions, where the frontal terminus is removed by erosion. They are a common feature in fine-grained lacustrine, marine, and deltaic deposits. They are particularly numerous in pre–classical Wisconsinan water-laid deposits located on the thermokarst-dotted tundra of the Lower Arctic Coastal Plain, both east and west of the Mackenzie Delta. On Richards Island, in the Mackenzie Delta, where great numbers of them have been mapped, they are found along the margins of actively thawing lake and pond shores (Fig. 32).

Bimodal flows have been called "retrogressive thaw flow slides" by Hughes (1972). Kerfoot (1969) called them "mudflows" if there is considerable mass movement in the extended tongue-shaped terminus and "mudslumps" if there is little or no transport of soil particles away from the ablating scarp. A thermokarst depression eventually develops from "mudslumps."

SUMMARY

Although a wide range of landslide types occurs in Canada, there are certain ones that are confined to specific geographic regions and tend to be concentrated in specific geologic settings within these regions. Examples include rapid rock avalanches, which are more spectacular but much less common than deep-seated rotational slumps in the mountains of western Canada. Slow retrogressive slope failures in Upper Cretaceous, bentonitic, marine, argillaceous bedrock are common along valley walls in the Interior Plains of western Canada. Much of this old failure terrain is marginally stable at the present time. More than 750 retrogressive flow slides and earthflows have been identified in sensitive marine clays of eastern Canada. The stability of high, steep slopes underlain by these sediments is related to fluvial erosion at the toe of the slope, to the development of high pore pressures follow-

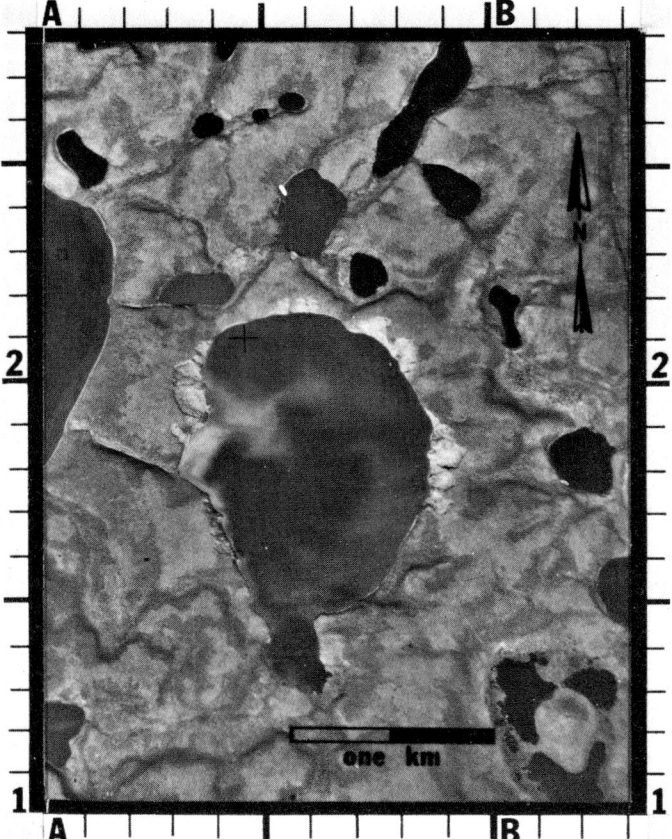

Figure 32. Bimodal failures, Richards Island, Northwest Territories; around the margin of a large thaw lake (A.3, 2.0 and A.9, 1.8). Note that failure debris in the tongue is absent in the large lake because, when ice in the supersaturated fine-grained sediment thaws, the mass at the toe of the slope has much less volume than the cavity from which it originated. Retrogressive melting and failure of these icy soils take place as freshly exposed subsoils are subjected to solar radiation during the warm summer months. Note the high, steep headscarp, which in these ice-rich sediments is actively ablating. Such ablation-dominated bimodal (two-sloped, planar) flows are also called retrogressive thaw flow slides, active-layer glides, and mudflows. They occur on slopes of 1° to 15°, but mostly on slopes of 3° to 9°. Air photo A 12918–193, courtesy of the National Air Photo Library, Ottawa.

ing heavy precipitation, and to vibrations from earthquakes, traffic, and other dynamic stresses. Active-layer skin flows and bimodal flows occur locally on steep slopes in the subarctic and arctic regions, where the terrain is affected by permafrost.

Investigators of landslides should be able to recognize the identifying morphologic features of these important slope failures and understand their geologic history, structure, and stratigraphy, particularly the geologic processes affecting the character and rate of their movement. Destructive landslides can have far-reaching economic, engineering, environmental, and land-use implications, and catastrophic failures in Canada have already had a tremendous impact in terms of lost life and property damage. Although a good deal of progress has been made, much remains to be learned about the behavior of the regional landslide types discussed.

ACKNOWLEDGMENTS

I thank many people with whom I discussed different pieces of information presented in this paper, who sent me copies of their published and unpublished papers, or who provided ground and oblique aerial photographs showing different types of slope failure. These people include L. A. Bayrock, V. G. Beckie, C. O. Brawner, D. M. Cruden, W. J. Eden, G. H. Eisbacher, R. J. Gotts, D. W. Hayley, A. E. Insley, J. Krahn, J. G. Locker, W. H. Mathews, A. M. McCann, E. A. McRoberts, R. J. Mitchell, N. R. Morgenstern, F. D. Patton, D. R. Piteau, D. H. Pollock, M. A. Roed, J. S. Scott, A. Stende, and D. F. VanDine. In addition, D. R. Coates, W. O. Kupsch, R. J. Mitchell, and N. R. Morgenstern read the manuscript and made many helpful suggestions.

REFERENCES CITED

Armstrong, J. E., 1976, Geomorphology of the Canadian Cordillera and its bearing on mineral deposits: Geoscience Canada, v. 3, p. 111–112.

Babcock, E. A., 1973, Regional jointing in southern Alberta: Canadian Jour. Earth Sci., v. 10, p. 1769–1781.

———1974, Jointing in central Alberta: Canadian Jour. Earth Sci., v. 11, p. 1181–1186.

Ballivy, G., Pouliot, G., and Loiselle, A., 1971, Quelques charactéristiques géologiques et minéralogiques des dépôts d'argile du Nord-Quest du Québec: Canadian Jour. Earth Sci., v. 8, p. 1525–1541.

Béland, J., 1956, Nicolet landslide, November 1955: Geol. Assoc. Canada Proc., v. 8, p. 143–156.

Bilodeau, P. M., 1957, The Nicolet landslide: Natl. Research Council Canada Associate Comm. Soil and Snow Mechanics Tech. Memo. 46, p. 11–13a.

Bjerrum, L., 1967, Progressive failure in slopes of over-consolidated plastic clay and clay shales: Am. Soc. Civil Engineers Proc., Jour. Soil Mechanics and Found. Div., v. 93, no. SM5, p. 1–50.

Brawner, C. O., 1964, The use of aerial photography in the study of landslides: British Columbia Prof. Engineer, April, 5 p.

———1975, Case examples of instability of rock slopes: British Columbia Prof. Engineer, February, 8 p.

Brown, R.J.E., and Kupsch, W. O., 1974, Permafrost terminology: Natl. Research Council Canada Associate Comm. Geotech. Research Tech. Memo. 111 (NRC 14274), 62 p.

Caldwell, W.G.E., 1968, The Late Cretaceous Bearpaw Formation in the south Saskatchewan River valley: Saskatchewan Research Council Geology Div. Rept. 8, 86 p.

Chagnon, Jean-Y., 1968, Les coulées d'argile dans la Province de Québec: Naturaliste Canadien, v. 95, p. 1327–1343.

———1970, Geotechnique—A regional approach applied to landslide investigation: Geol. Assoc. Canada Proc., v. 22, p. 37–43.

Christiansen, E. A., and Whitaker, S. H., 1976, Glacial thrusting of drift and bedrock: Royal Soc. Canada Spec. Pub. 12, p. 121–130.

Conlon, R. J., 1966, Landslide on the Toulnustouc River, Quebec: Canadian Geotech. Jour., v. 3, p. 113–144.

Crawford, C. B., 1968, Quick clays of eastern Canada: Eng. Geology—Internat. Jour., v. 2, p. 239–265.

Crawford, C. B., and Eden, W. J., 1957, Report on the Nicolet landslide of November 1955: Natl. Research Council Canada Div. Bldg. Research Internat. Rept. 128.

———1967, Stability of natural slopes in sensitive clay: Am. Soc. Civil Engineers Proc., Jour. Soil Mechanics and Found. Div., v. 93, no. SM4, p. 419–436.

Cruden, D. M., 1976, Major rock slides in the Rockies: Canadian Geotech. Jour., v. 13, p. 8–20.

Cruden, D. M., and Krahn, J., 1973, A re-examination of the geology of the Frank slide: Canadian Geotech. Jour., v. 10, p. 581–591.

Dishaw, H. E., 1967, Massive landslides: Photogramm. Eng., v. 33, p. 603–608.

Eden, W. J., and Mitchell, R. J., 1970, The mechanics of landslides in Leda Clay: Canadian Geotech. Jour., v. 7, p. 285–296.

———1973, Landslides in sensitive marine clay in eastern Canada: Highway Research Rec. 463, p. 18–27.

Eden, W. J., Fletcher, E. B., and Mitchell, R. J., 1971, South Nation River landslide, 16 May 1971: Canadian Geotech. Jour., v. 8, p. 446–451.

Eigenbrod, K. D., and Morgenstern, N. R., 1972, A slide in Cretaceous bedrock, Devon, Alberta, in Brawner, C. O., and Milligan, V., eds., Geotechnical practice in open pit mining: New York, Am. Inst. Mining, Metallurgy and Petroleum Engineers, p. 223–238.

Eisbacher, G. H., 1971, Natural slope failure, northeastern Skeena Mountains: Canadian Geotech. Jour., v. 8, p. 384–390.

Fransham, P. B., Gadd, N. R., and Carr, P. A., 1976, Geological variability of marine deposits, Ottawa–St. Lawrence lowlands: Canada Geol. Survey Paper 76-1A, p. 37–41.

Gadd, N. R., 1957, Geological aspects of eastern Canadian flow slides: Natl. Research Council Canada Associate Comm. Soil and Snow Mechanics Tech. Memo. 46, p. 2–8.

———1976, Surficial geology and landslides of Thurso–Russell map-area, Ontario: Canada Geol. Survey Paper 75-35, 7 p.

Gary, M., McAfee, R., Jr., and Wolf, C. L., eds., 1972, Glossary of geology: Washington, D.C., Am. Geol. Inst., 805 p.

Gogeul, J., and Pachoud, A., 1972, Geology and dynamics of the collapse of Mount Granier in the Chartreuse Massif in November 1948: Bureau Récherches Géologie et Minéralogie, v. 3, p. 29–38.

Hardy, R. M., 1957, Engineering problems involving pre-consolidated clay shales: Eng. Inst. Canada Trans., no. 1, p. 5–14.

Hardy, R. M., Brooker, E. W., and Curtis, W. E., 1962, Landslides in overconsolidated clays: Eng. Jour. Canada, v. 45, p. 81–89.

Hillaire-Marcel, C., and de Boutray, B., 1975, Les dépôts meubles Holocènes de Poste-de-la-Baleine, Nouveau-Québec: Québec City, Univ. Laval Centre d'Etudes Nordiques, Coll. Nordicana 38, 47 p.

Hodgson, E. A., 1927, The marine clays of eastern Canada and their relation to earthquake hazards: Royal Astron. Soc. Canada Jour., v. 21, p. 257–264.

Hodgson, J. H., 1965, Canadian earthquakes: Canadian Geog. Jour., v. 71, no. 1, p. 30–39.

Howell, B. F., 1973, Earthquake hazard in the eastern United States: Earth and Mineral Sci., v. 42, p. 41–45.

Hughes, O. L., 1972, Surficial geology and land classification, Mackenzie Valley transportation corridor: Natl. Research Council Canada Div. Bldg. Research Tech. Memo. 104, p. 17–24.

Hurtubise, J. E., and Rochette, P. A., 1956, The Nicolet slide: Canadian Good Roads Assoc., 37th conv., Proc., p. 143–155.

Hurtubise, J. E., Gadd, N. R., and Meyerhof, G. G., 1957, Les éboulements de terrain dans l'est du Canada: Internat. Conf.

Soil Mechanics and Found. Eng., 4th, London, Proc., v. 2, p. 325–329.

Hsü, K. J., 1975, Catastrophic debris streams (Sturzstroms) generated by rockfalls: Geol. Soc. America Bull., v. 86, p. 129–140.

Insley, A. E., Chatterji, P. K., and Smith, L. B., 1976, Rational slope stability analysis using residual strengths: Canadian Geotech. Conf., 29th, Vancouver 1976, Proc. (in press).

Isaacs, R. M., and Code, J. A., 1972, Problems in engineering geology related to pipeline construction: Natl. Research Council Canada Div. Bldg. Research Tech. Memo. 104, p. 147–178.

Karrow, P. F., 1972, Earthflows in the Grondines and Trois Rivières areas, Québec: Canadian Jour. Earth Sci., v. 9, p. 561–573.

Kerfoot, D. E., 1969, The geomorphology and permafrost conditions of Garry Island, N.W.T. [Ph.D. thesis]: Vancouver, Univ. British Columbia.

Krahn, J., and Morgenstern, N. R., 1976, The mechanics of the Frank slide: Am. Soc. Civil Engineers Conf. on Rock Engineering, Edmonton 1976 (in press).

Krinitzky, E. L., and Kolb, C. R., 1969, Geological influences on the stability of clay shale slopes, in Engineering geology and soils engineering symp., 7th, Proc.: Boise, Idaho Dept. Highways, p. 160–175.

Lane, K. S., 1967, Stability of reservoir slopes, in Failure and breakage of rock—Symposium on rock mechanics, 8th, Proc.: New York, Am. Inst. Mining, Metallurgy and Petroleum Engineers, p. 321–336.

LaRochelle, P., Chagnon, J.-Y., and LeFebvre, G., 1970, Regional geology and landslides in the marine clay deposits of eastern Canada: Canadian Geotech. Jour., v. 7, p. 145–156.

LaSalle, P., and Chagnon, J.-Y., 1968, An ancient landslide along the Saguenay River, Quebec: Canadian Jour. Earth Sci., v. 5, p. 548–549.

Locker, J. G., 1973, Petrographic and engineering properties of fine-grained rocks in central Alberta: Research Council Alberta Bull. 30, 144 p.

Mackay, J. R., and Mathews, W. H., 1973, Geomorphology and Quaternary history of the Mackenzie River valley near Fort Good Hope, N.W.T., Canada: Canadian Jour. Earth Sci., v. 10, p. 26–41.

Matheson, D. S., and Thomson, S., 1973, Geological implications of valley rebound: Canadian Jour. Earth Sci., v. 10, p. 961–978.

Mathews, W. H., and McTaggart, K. C., 1969, The Hope landslide, British Columbia: Geol. Assoc. Canada Proc., v. 20, p. 65–75.

McConnell, R. G., and Brock, R. W., 1904, Report on the great landslide at Frank, Alberta: Canada Dept. Interior Ann. Rept., pt. 8, 1902–1903, 17 p.

McRoberts, E. C., and Morgenstern, N. R., 1973, A study of landslides in the vicinity of the Mackenzie River, mile 205 to 660: Environmental-Social Comm., Northern Pipelines, Task Force on Northern Oil Development Rept. 73-35, 96 p.

——1974, The stability of thawing slopes: Canadian Geotech. Jour., v. 11, p. 447–469.

Mitchell, R. J., 1971, The South Nation River landslide, 16 May 1971, Russell County, Ontario: Kingston, Ontario, Queen's Univ. Dept. Civil Engineering Rept., 14 p.

——1976, Earthflow terrain evaluation in Ontario: Kingston, Ontario, Queen's Univ. Dept. Civil Engineering Final Rept., Project Q-53, 23 p.

Mitchell, R. J., and Markell, A. R., 1974, Flowsliding in sensitive soils: Canadian Geotech. Jour., v. 11, p. 11–31.

Mollard, J. D., 1952, Aerial photographic studies in the Central Saskatchewan Irrigation Project, Pt. III, Landslides in Bearpaw Shale [Ph.D. thesis]: Ithaca, New York, Cornell Univ., 303 p.

——1955, Summary report on the geology of the South Saskatchewan River damsite: Regina, Saskatchewan, Canada Dept. Agriculture Prairie Farm Rehabilitation Admin. Air Photo Analysis and Eng. Geology Div., Eng. Services, 38 p.

——1957, Aerial mosaics reveal fracture patterns on surface materials in southern Saskatchewan and Manitoba: Oil in Canada, v. 9, no. 40, p. 26–50 (p. 18140–18164).

——1958, Photogeophysics—Its application in petroleum exploration over the glaciated plains of western Canada, in Second Williston Basin symposium: Bismarck, N. D., Conrad Pub. Co., p. 109–117.

——1962, Photo analysis and interpretation in engineering-geology investigations: A review, in Fluhr, T., and Legget, R. F., eds., Reviews in engineering geology: Geol. Soc. America Rev. Eng. Geology, v. 1, p. 105–128.

——1972, Terrain classification and mapping for northern pipeline studies: Natl. Research Council Canada Div. Bldg. Research Tech. Memo. 104, p. 105–127.

——1976, Landforms and surface materials of Canada. A stereoscopic atlas and glossary (5th ed.): Regina, Saskatchewan, J. D. Mollard and Assoc., Ltd., 366 p.

——1977, Site and route investigations in permafrost regions: Natl. Research Council Canada Div. Bldg. Research Permafrost Eng. Manual (in press).

Mollard, J. D., and Hughes, G. T., 1973, Earthflows in the Grondines and Trois Rivières areas, Québec—Discussion: Canadian Jour. Earth Sci., v. 10, p. 324–326.

Mollard, J. D., and Pollock, D. H., 1955, Discussion—Studies of Bearpaw Shale at a damsite in Saskatchewan by R. Peterson: Am. Soc. Civil Engineers Proc., v. 81, separate 656, p. 17–23.

Morgenstern, N. R., and Eigenbrod, K. D., 1974, Classification of argillaceous soils and rocks: Am. Soc. Civil Engineers Proc., Jour. Soil Mechanics and Found. Div., v. 100, no. GT10, p. 1137–1156.

Parkes, J.G.M., and Day, J. C., 1975, The hazard of sensitive clays—A case study of the Ottawa-Hull area: Geog. Rev., April, p. 198–213.

Patton, F. D., and Hendron, A. J., Jr., 1974, General report on "mass movements": Internat. Cong. Internat. Assoc. Engineering Geology, 2nd, São Paulo, Brazil, v-GR1-57.

Peck, R. B., Ireland, H. O., and Fry, T. S., 1951, Studies of soil characteristics in earthflows of St. Thuribe, Quebec: Illinois Univ. Soil Mechanics Ser. 1.

Peterson, R., 1954, Studies of Bearpaw Shale at a damsite in Saskatchewan: Am. Soc. Civil Engineers Proc., v. 80, separate 476, 28 p.

——1958, Rebound in the Bearpaw Shale, western Canada: Geol. Soc. America Bull., v. 69, p. 1113–1124.

Peterson, R., Jaspar, J. L., Rivard, P. J., and Iverson, N. L., 1960, Limitations of laboratory shear strength in evaluating stability of highly plastic clays: Am. Soc. Civil Engineers Conf. Shear Strength Cohesive Soils, Boulder, Colo., p. 765–791.

Piteau, D. R., Mylrea, F. H., and Blown, I. G., 1976, The Downie slide, Columbia River, British Columbia, in Voight, B., ed., Landslides: Amsterdam, Elsevier Pub. Co. (in press).

Pollock, D. H., 1962, Geology of the South Saskatchewan River Project: Eng. Jour. Canada, April, p. 3–12.

Research Council of Alberta, 1972, Geological map of Alberta: Research Council Alberta Map 35.

Ringheim, A. S., 1964, Experiences with the Bearpaw Shale at the South Saskatchewan River Dam: Internat. Cong. on Large Dams, 8th, Edinburgh, v. 1, p. 529–550.

Sangrey, D. A., and Paul M. J., 1971, A regional study of landsliding near Ottawa: Canadian Geotech. Jour., v. 8, p. 315–335.

Sauer, E. K., 1975, Urban fringe development and slope instability in southern Saskatchewan: Canadian Geotech. Jour., v. 12, p. 106–118.

Scott, J. S., and Brooker, E. W., 1968, Geological and engineering aspects of Upper Cretaceous shales in western Canada: Canada Geol. Survey Paper 66-37, 75 p.

Shreve, R. L., 1966, Sherman landslide, Alaska: Science, v. 154, p. 1639–1643.

——1968, The Blackhawk landslide: Geol. Soc. America Spec. Paper 108, 47 p.

Skinner, R. G., 1974, Terrain studies in the James Bay Development Area: Canada Geol. Survey Open-File Rept. 219, p. 6–8.

Tavenas, F., Chagnon, J.-Y., and LaRochelle, P., 1971, The Saint-Jean–Vianney landslide; observations and eyewitness accounts: Canadian Geotech. Jour., v. 8, p. 463–478.

Terzaghi, K., 1955, Influence of geological factors on the engineering properties of sediment: Harvard Soil Mechanics Ser. 50, 61 p.

Thomson, S., 1971, Analysis of a failed slope: Canadian Geotech. Jour., v. 8, p. 596–599.

Thomson, S., and Hayley, D. W., 1975, The Little Smoky landslide: Canadian Geotech. Jour., v. 12, p. 379–392.

Thomson, S., and Morgenstern, N. R., 1974, Landslides in argillaceous bedrock, Alberta, Canada, in Voight, B., ed., Landslides: Amsterdam, Elsevier Pub. Co., (in press).

Van Everdingen, R. O., 1972, Observed changes in groundwater regime caused by the creation of Lake Diefenbaker, Saskatchewan: Canada Dept. Environment Inland Waters Branch Tech. Bull. 59, 65 p.

Westgate, J. A., 1976, Discussion of hydrogeologic and hydrochemical properties of fractured till in the Interior Plains region by J. A. Grisek et al.: Royal Soc. Canada Spec. Pub. 12, p. 334.

Whitaker, S. H., 1970, Geology and groundwater resources of the Swift Current area (72J): Saskatchewan Research Council Geology Div. Map 11, scale 1:250,000.

Manuscript Received by the Society September 7, 1976
Manuscript Accepted September 17, 1976

PART 2
Regional Studies

3

Causes of rock-slope failure in a cold area: Labrador-Ungava

KARL-HEINZ WYRWOLL
Department of Geography, University of Western Australia, Nedlands, Western Australia 6009

ABSTRACT

The failure of high-angle rock slopes was studied in the Schefferville area of Labrador-Ungava. From this it is concluded that the failure of rock slopes took place initially as the result of the external factors — glacial and glaciofluvial action — that created the slope. Through time, residual stress release, joint water pressure, and weathering have perpetuated failure, although it continued on a lesser scale. The traditional explanation — that rock-slope failure in this area is the result of frost action — is not supported by this study. Furthermore, it is likely that this conclusion can be extrapolated to other cold areas. The general model of hillslope development in cold areas, which emphasizes the role of frost action in rock-slope failure, may need to be reconsidered.

INTRODUCTION

The failure of high-angle rock slopes and associated talus development have been widely recognized as characteristic features of cold areas and generally have been attributed to the action of frost shattering. Previous work dealing with this topic has been largely descriptive and has not considered the mechanics of rock-slope failure under these conditions. In fact, hillslopes in cold areas have generally been neglected by studies that attempt to understand the mechanics of slope processes. This may account for the wide acceptance of Peltier's (1950) conceptual scheme of hillslope evolution; similar conclusions are mirrored in works dealing specifically with cold-area processes (Bird, 1967; Embleton and King, 1968; Tricart, 1970).

The importance attributed to the role of frost action in rock-slope failure does not seem to be supported by the few measurements of this process that are available. When actual rates of free-face retreat, which is attributed to frost action (Table 1), are compared with the large volumes of Holocene talus accumulations that are so widespread in cold areas, it seems questionable that such low rates of supply could account for such substantial volumes of material. Clearly, it could be argued that rates of frost shattering may have been formerly greater, during, for example, a period of more severe climate (Andrews, 1961). However, since so little is known of rock-slope failure, some caution should be exercized before adopting such an explanation.

The purpose of this paper is to examine the problem of rock-slope failure from the point of view of the mechanics of rock slopes. Through adopting this general approach it is hoped that (1) some understanding of the nature and causes of rock-slope failure, in an area where failure has traditionally been explained by frost action, can be obtained and (2) a more satisfactory explanation of the factors responsible for rock-slope failure in cold areas can be outlined.

REGIONAL SETTING

The study area is located near Schefferville in central Quebec–Labrador (lat 54°54′N, long 66°45′W), within the zone of transition between boreal spruce forest and tundra. In the Köppen–Geiger climatic classification it falls into the division "Dfc" (subarctic). It has a mean annual tempera-

TABLE 1. RATES OF FREE-FACE RETREAT

Rock unit	Rate (mm/yr)
Mica schist, Kärkevagge, northern Sweden (Rapp, 1960a)	0.04–0.15
Limestone-sandstone, Templefjorden Spitzbergen (Rapp, 1960b)	0.02–0.2
Quartzite, dolomite, slate, Yukon Territory (Gray, 1971)	0.02–0.17
Syenite, diabase, Yukon Territory (Gray, 1971)	0.007–0.03

ture of −4.4 °C and receives a mean annual precipitation of approximately 722 mm, of which 401 mm is rain and 321 mm (water equivalent) is snow.

The area lies within the Labrador "trough," which is the preserved part of an early Proterozoic (Alphebian) geosyncline (Dimroth, 1970). The Aphebian rocks compose, in the west, a miogeosynclinal sequence of shale, dolomite, arkose, and chert (Table 2). These are the rocks with which we are concerned. The region is folded into a series of northwest-trending synclines and anticlines. In places, this fold pattern is tight and further complicated by prominent strike faults with reverse dip-slip movement. Later cross-faults further disrupt the earlier structures. A "sample area" showing the general geologic features mentioned is given in Figure 1.

The major physiographical elements of the area are controlled by the underlying structure and bedrock. The structural geology gives rise to the pronounced northwest-southeast grain of the country, which is further accentuated by differential denudation of the various lithologies. These factors combine to produce a ridge and valley topography; along the sides of these ridges, as well as within the major meltwater channels, large-scale rock-slope failure and talus formation have taken place. The area was probably deglaciated 6,000 ± 1,000 yr ago (Bryson and others, 1969). Rock-slope failure and talus deposition have taken place throughout the postglacial period. There is no evidence to suggest that talus deposits formed at an earlier period and survived glaciation.

TABLE 2. GEOLOGIC SUCCESSION IN THE SCHEFFERVILLE AREA

Group	Formation	Lithology
Cretaceous	Redmond	Clay, argillite, ferruginous talus, and rubble
Proterozoic	Menihek 300 m	Gray to black carbonaceous slate; pyritic slate; impure dolomite; graywacke; minor chert
	Sokoman ± 200 m	Iron-formation: banded silicate, banded jasper, banded chert, cherty iron carbonate, massive chert, lean chert, and slaty members
	Ruth 20 m	Black to greenish ferruginous and carbonaceous slate; local chert; black chert at base
	Wishart 30 m	Quartzite, arkose; minor chert at base; minor slate and carbonate at top
	Fleming 70 m	Massive chert; chert breccia; quartzite with chert cement; cherty slate; chert-pebble conglomerate
	Denault 200 m	Dense dolomite; arenaceous dolomite; dolomite breccia cemented by dolomite and (or) chert; minor chert, slate, and quartzite interbeds
	Attikamagen	Varicolored slates; local interbeds of dolomite and granual chert near base
"Archean" basement	Laporte Ashuanipi	Biotite and hornblende schists; biotite, hornblende, garnet gneiss; amphibolites; granitic intrusions

ROCK-SLOPE FAILURE IN THE STUDY AREA

General Characteristics

Much of the previous work in this area has stressed the importance of frost shattering as a contemporary process (Twidale, 1957; Derbyshire, 1962). Twidale, in fact, suggested that the most important process accomplishing rock breakdown is congelifraction. Later work by Andrews (1961) questioned this view and suggested that frost shattering is only active to a limited extent under present climatic conditions.

The evidence on which this earlier work based its conclusions is the large-scale occurrence, throughout the field area, of extensive talus deposits found at the foot of failed high-angle rock slopes. Such talus accumulations are especially characteristic of the more resistant ridge-forming lithologies, such as the Fleming Chert Breccia, Denault Dolomite, and Wishart Quartzite (Fig. 2).

Apart from the talus accumulations, which result from the failure of high-angle rock slopes, rock types other than shale show little sign of frost action. Some parts of the outcrops of the Attikamagen Shale are extensively covered by shattered material, and accumulations as thick as 0.5 m mantle some of the footslopes on this unit. Outcrops of the other lithologies can show an almost complete absence of frost-shattered material, especially where the rock units are massive. Despite this, however, the talus slopes found at the foot of many of the steep rock slopes on these massive lithologies have led some workers to suggest that, even on these rock types, frost action plays a major role in rock-slope failure.

Field Technique

In order to assess the stability of some of the rock slopes and the reason for their widespread failure, the geometry of the rock masses on which these slopes had developed was established. This was done primarily through the measurement of joint orientation and joint spacing, in conjunction with the surveying of the profiles of the rock slopes studied, as well as through observations of the general nature of the rock mass. Thus, it is possible to assess the stability of the slope and to establish the type and cause of rock-slope failure.

The profile of the rock slope was surveyed in those areas where rock-slope failure had taken place, and a joint survey was also undertaken in its immediate vicinity. Available joint blocks were used in this survey. This prevented the creation of a "blind zone," which occurs when a line-sampling procedure is adopted (Terzaghi, 1965). A line-sampling scheme would also have been impossible because the physical constraints imposed by rock slopes often prevent any attempt at a rigorous sampling design. Similarly, the spacing of various joint planes was often impossible to obtain because adjacent joint planes could not be reached. Wherever possible, additional measures of spacing were taken from photographs. Altogether, 19 independent failed slopes were profiled, and joint surveys were undertaken in their immediate vicinity, with nearly 2,000 joints recorded in the study.

The results of the joint surveys are presented by means of

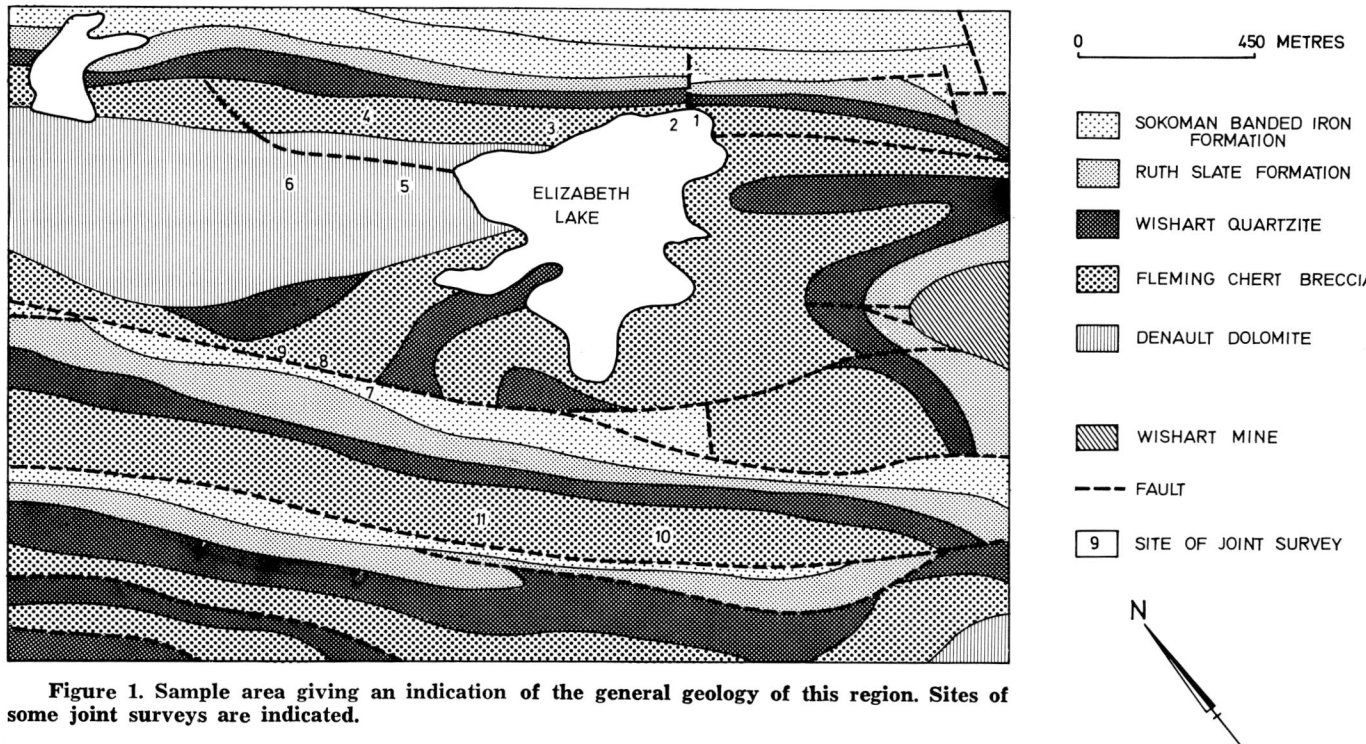

Figure 1. Sample area giving an indication of the general geology of this region. Sites of some joint surveys are indicated.

polar stereographic projections using the equal-area or Schmidt grid. Through plotting the normal to the joints, a point distribution was obtained, and a 1% counting circle was used to construct a contour diagram showing the concentration of poles obtained. The number of poles constituting the contour diagram is indicated in the figures.

MECHANICS OF FAILURE

Following Terzaghi (1950), the mechanisms leading to slope failure can be divided into external and internal factors. The external factors cause an increase in the shearing stresses within the slope. These result from any heightening or steepening of the slope through the processes of slope creation, such as glacial or river erosion. The internal factors that can lead to failure bring about a reduction in the shear strength of the material making up the slope. In rock slopes the most important of these are joint water pressure, residual stress, and weathering.

EXTERNAL FACTORS

Three major modes of rock-slope failure were recognized in this area: (1) sliding, (2) toppling, and (3) wedge separation. Some minor modes, which were modified forms of these major categories, were also noted, but they are only of limited occurrence. Although in the following examples two lithologies are highlighted, this is only because they were the most intensively studied. The conclusions drawn from these also generally apply to the other rock units in this area.

Sliding. The failure of high-angle rock slopes on the Fleming Chert Breccia has been extensive and, at times, spectacular. The most common mode of failure on this rock unit is that in which a major joint set dips out of the slope so that, after failure, the failure plane constitutes the new slope surface. Two examples are shown in Figure 3. In both cases the major joint set dips out of the rock mass and constitutes the slope face.

In many cases, slope failure has occurred through the separation of large, individual joint blocks from the main rock mass. However, in other instances, the entire slope face has separated from the rock mass. This occurs where large tension joints, running parallel to the slope, have developed; they

Figure 2. Massive failure of some Fleming Chert Breccia slopes. Little of the initial rock slope now remains.

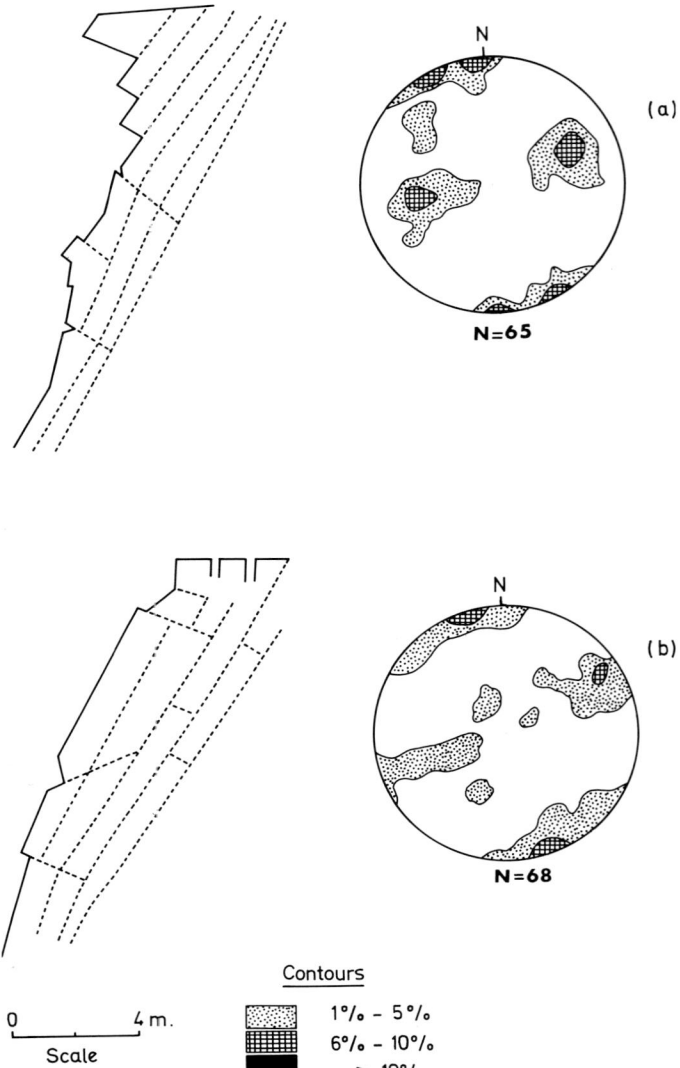

Figure 3. General rock-mass geometry and associated slope forms developed along a ridge of Fleming Chert Breccia. Before failure, these slopes had a height of about 50 m. (N indicates the number of joint poles making up the contour diagram.)

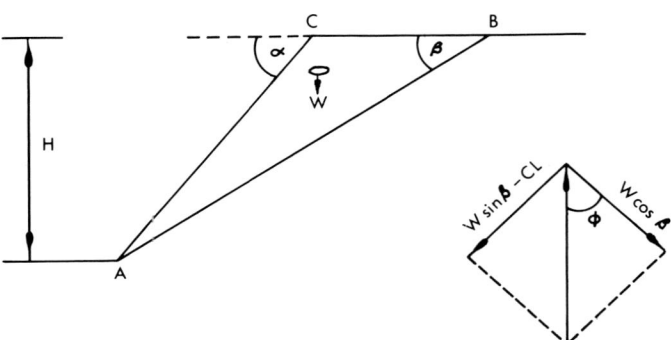

Figure 4. Geometry of a rock slope divided into a wedge by a planar joint and the forces acting on this joint as the result of self-weight–induced stresses (adapted from Barton, 1972).

were observed as much as 10 m behind the slope and were often wide enough to climb into.

To obtain some understanding of why failure of the rock slopes has taken place, the problem has to be considerably simplified. It is assumed that the rock masses of the Fleming Chert Breccia, before glaciation, stood at an angle below that of limiting stability and that only with the advent of glacial action did these rock masses develop a steeper angle. When such assumptions are made, the nature of rock-slope failure can be demonstrated.

If, as the result of glacial action, the particular rock mass stood at an angle higher than the angle of inclination of the joint set dipping steeply to the southwest (Fig. 3), the rock mass immediately adjacent to the slope would be divided into a number of wedges. The size of these wedges would depend upon the spacing of the joints. An ideal situation is assumed, where the dominant failure plane passes through the toe of the slope (illustrated in Fig. 4). It is emphasized that this is only envisaged as an ideal state immediately after the glacial "support" was removed.

The joint planes dip out of the slope at an angle β, and the slope stands at an angle α. Assuming only "self-weight" stresses and that failure will take place along plane AB, the stability of the wedge will depend upon the forces mobilized along this plane and the strength that the plane possesses.

Following Barton (1972), from the geometry of the figure,

$$CB = H (\cot \beta - \cot \alpha) \tag{1}$$

$$W = \gamma \frac{H^2}{2} (\cot \beta - \cot \alpha), \tag{2}$$

where W = weight of the wedge, H = height of the slope, γ = unit weight of the rock mass, $\beta°$ = inclination of the assumed failure plane, and $\alpha°$ = inclination of the overall slope. At limiting equilibrium,

$$W \sin \beta = cL + W \cos \beta \tan \phi, \tag{3}$$

where L is the length of the failure plane $AB = H/\sin \beta$. The stability of the slope thus depends upon the inclination of the failure plane and the angle of shearing resistance (ϕ) and upon the cohesion (c) of this plane. Under conditions of limiting equilibrium, the value of ϕ necessary for stability to continue can be found by substituting for W and L in equation 3, so that

$$\tan \phi = \tan \beta \left[1 - 2c/\gamma H \frac{1}{(\cot \beta - \cot \alpha) \sin^2 \beta}\right]. \tag{4}$$

If the value of ϕ is not sufficient, failure occurs.

Unfortunately, equation 4 cannot be accurately solved, because the absolute values of its parameters are not known. Furthermore, assumptions underlying the equation are not strictly valid. There is no justification for the assumed uniform stress distribution along the potential failure plane. In fact, the wedge should be divided into a number of slices, and the stress distribution beneath each slice needs to be calculated (Barton, 1972). In spite of all these limitations, use of this general and much simplified approach still gives some insight into the general mode of, and reasons for, failure.

If the stresses are assumed to be only self-weight stresses, they can be calculated from the initial height of the slope and the geometry of the rock mass. Using the slopes shown in Figure 3 as examples, the basic mechanics of failure can be demonstrated. From contour maps it seems likely that the initial height of these slopes before failure could have been, at most, about 51 m. It will be assumed that owing to glacial erosion this initial slope had an inclination of 80°. Such an assumption is not unreasonable, because the joint sets that now constitute the slope have an inclination of 70°; therefore, the slope would only have to be oversteepened by 10°, which is considered to be quite a conservative assumption. The angle of shearing resistance of the joint surface necessary to sustain a slope 51 m high, standing at an angle of 80°, is obtained from equation 4. Assuming the extreme case of a bulk density (γ) of 3,078 kgf/m³ and a cohesion (c) along the joint surface of 1,944 kgf/m², the angle of shearing resistance necessary to retain stability would have to be about 65°. If a more stable situation is assumed by reducing the value of the bulk density (γ) to 2,268 kgf/m³ and increasing the cohesion (c) to 3,888 kgf/m², an angle of shearing resistance of 58° would have been necessary to retain stability. Thus, failure was quite likely under both conditions.

It has to be borne in mind that the "most stable" situation has been assumed. The effect of, for instance, tension joints, which are clearly present, has not been considered. Furthermore, from the nature of the immediate surface zone of the rock mass and the joint surfaces, it seems unlikely that cohesion is of any great importance in the shear strength of the joint planes in this zone. So again, the potential stability of these slopes has been overestimated.

Toppling. Sliding failure is by far the most common mode of failure of the chert breccia slopes, but there are a number of cases where the slope has failed by toppling. In Figure 5, a large rock block with an approximate volume of 150 m³ has lead to the failure of the slope. From the regional joint sets that were mapped, the rock slope should have failed in the same manner as the slopes shown in Figure 3; however, this did not occur because the block, when still enclosed in the rock mass, rested upon a joint plane that was not in accord with the regional sets. This joint plane was inclined in such a way with respect to the regional joint set that once the rock mass had failed and the block was no longer enclosed, toppling of the block was inevitable. Under these conditions the center of gravity lay beyond the pivot point of the block, and hence the block had to rotate about its base.

The explanation for this mode of failure is quite simple and can be demonstrated by using a single rectangular block standing on an inclined plane (Fig. 6). The breadth/height (b/h) ratio and the inclination (α) of the block will determine whether it will topple. When $b/h < \tan \alpha$, the center of gravity of the block will lie beyond the pivot point, and toppling about this point will occur (Ashby, 1971). From the diagram it is clear that under certain inclinations (α) a combination of toppling and sliding occurs. This explanation really only treats the simplest and ideal case. In the given example, for instance, the likelihood of toppling is further accentuated by the "rounded" geometry of the block. Unfortunately, this mode of slope failure, though described by a number of geomorphologists, has never been satisfactorily explained, and Ashby's (1971) terminology and explanation should be adopted.

Wedge Separation. The mode of failure in those rock masses where joint planes dip out of the slope can be quite easily explained, but it is more difficult when the joint planes are horizontal. Examples of this are the slopes shown in Figure 7. These slopes have developed on the Denault Dolomite and

Figure 5. An example of toppling failure on the Fleming Chert Breccia, where a whole block separates from the slope.

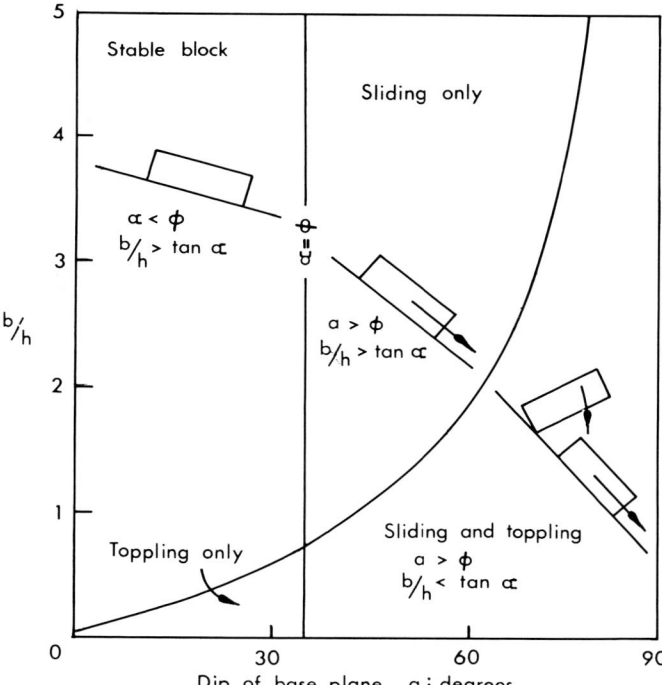

Figure 6. Conditions for sliding and toppling on an inclined surface having an angle of friction of 35° (after Ashby, 1971).

constitute part of a ridge that has failed in those sections opened by glacial action. From the inclination and length of the talus slopes and the height of the remaining rock slopes, it is likely that the rock slopes before failure could have been as high as 50 m.

The rock mass is divided into regular blocks by three joint sets. A bedding set dips slightly to the north at an angle of about 5°, with a spacing of 0.76 m. The second set has an approximate eastward strike and dips almost vertically either north or south, with a mean spacing of about 0.6 m. The third set has a north-south strike and is more or less vertical, with a spacing of about 0.5 m. The rock mass is thus little more than an amalgam of medium-sized joint cubes. The slopes that have developed in this rock mass have failed extensively, so that the remaining rock slopes are seldom more than 10 m high and in some places as low as 2.5 m.

A rock mass that has this structural geometry must fail if undercut by glacial action, and, even if the rock mass is only steepened and the slopes are created without undercutting, failure may still take place. If, initially and for the sake of simplicity, it is assumed that the rock mass behaves as a continuous cohesive body, the Culman wedge type of analysis would give some impression of the limiting height of the slope and the mode of failure. Although this method of stability analysis is not applicable, some of its general consequences are still valid.

A rock slope is able to support a load in unconfined compression, so that the rock mass, as a whole, does in part behave like a cohesive body. Because of the discontinuous nature of a rock mass, however, the shear plane, which will limit the vertical height of a slope, will be modified to form a "zigzag" zone through the rock mass. At the same time, as the slope was created and perpetuated through the removal of the outermost part of the rock mass, active earth-pressure conditions have arisen, resulting in a zone of tension in the upper part of the slope. This is an important feature, because, from the geometry of the rock mass, it is evident that the whole rock mass is extremely weak in tension. The two factors combined may clearly result in the failure of an outer wedge of the rock mass.

The examples that have been discussed make it clear that many of the steep rock slopes, once created, had to fail because of their rock-mass geometry. Joint and slope surveys demonstrated that this was widely the case and was not a feature confined to any one rock type. When a steep slope had been created, regardless of lithology, the geometry of the rock mass would often make failure inevitable.

Internal Factors

Although rock-slope failure is likely to have occurred on a greater scale during the immediate postglacial period (which would concur with Church and Ryder's (1972) ideas on paraglacial sedimentation), failure was not limited to this period. From the degree of lichen development on talus deposits and the presence of fresh failure scars on rock slopes, it is evident that the process has continued through time and is still quite active today. The factors that have perpetuated failure through time are the internal factors active within a rock mass; these are (1) residual stresses, (2) joint water pressure, and (3) weathering processes.

Residual Stress. The discussion so far has only emphasized that failure is likely because of the self-weight stresses that arise as the result of the geometry of the rock mass. It is quite likely that this assumption underestimates stresses in rock slopes. Large horizontal stresses may also be present. After all, this area was the center of the Laurentide Ice Sheet (Ives, 1968) and may have been covered by as much as 3 km of ice (Flint, 1971). The horizontal stresses resulting from such a load could be calculated if the particular rock mass is treated as an elastic continuum. In such a case the horizontal stresses would be

$$\sigma_H = \frac{v}{1-v} \sigma_v, \quad (5)$$

where σ_H is the horizontal stresses, σ_v the vertical stress resulting from the ice load, and v is the Poisson ratio.

Unfortunately, these assumptions are not valid for the situation in which the existing discontinuities within the rock mass force the rock mass to behave in an elastic-plastic manner (Müller, 1972). On subsequent unloading, this results in only part of the vertical deformation being regained; the rest will be retained by the rock mass as "pseudoplastic" deformation. Nevertheless, horizontal stresses will still arise, because with unloading, the stresses in the horizontal direction will be reduced by less stresses in the vertical direction (Herget, 1974). Hence,

$$\Delta\sigma_H = \frac{v}{1-v} \Delta\sigma_v, \quad (6)$$

where $\Delta\sigma_v$ is the reduction of vertical stress, $\Delta\sigma_H$ is the reduction of horizontal stress, and v is the Poisson ratio.

Although no actual measurements are as yet available for this area, horizontal stresses three times the magnitude of the vertical stress have been recorded in Sweden, a region with a similar geologic history (Haste, 1958). There is every

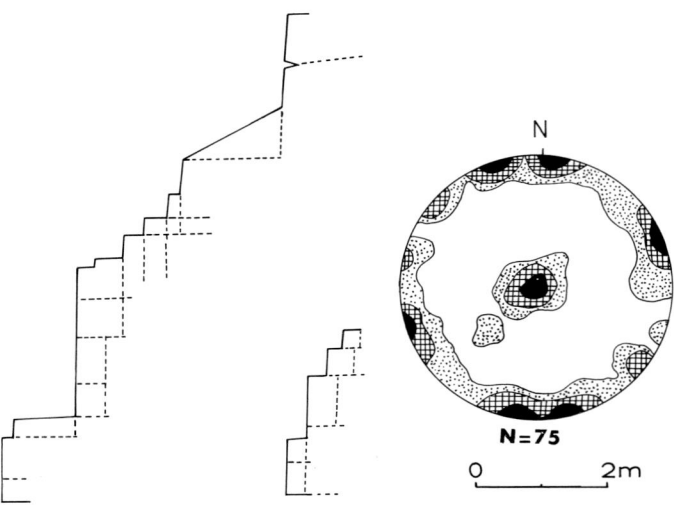

Figure 7. Rock-mass geometry and associated remaining rock slopes of parts of the Denault Dolomite. The initial rock slope before failure had a height of about 50 m.

reason to expect similar stress conditions to prevail in this area, a view indirectly substantiated by the widespread occurrence of postglacial sheeting structures on the chert breccia.

In a number of locations on the Fleming Chert Breccia, large-scale sheeting has developed. This is not related to the regional structural geology, but coincides with the topography of the outcrop. The sheets can vary in thickness from about 0.3 to 1 m. They usually drape the rock mass, and removal of part of a sheet makes the whole slope susceptible to failure. Figure 8 shows an example of how, through sheet structures, slabs of bedrock have separated from the slope face.

Extensive field work revealed sheeting only on the Fleming Chert Breccia, which is not surprising when the brittle nature of chert is considered. It would seem likely that these sheet structures result from glacial unloading. It must be remembered, however, that not all workers have considered such structures to be the result of the release of residual strain energy. Wolters (1969), for example, proposed that large variations in temperature give rise to such features, but the available literature on this topic would seem to suggest that this conclusion is highly unlikely.

So far as the field area generally is concerned, it seems valid to propose that part of the ice-load deformation was elastic and is "stored" as residual strain energy in most rock types. If high horizontal stresses are present, the stress distribution in the rock slope will be very different from what it would be under "self-weight"—induced stresses. Dodd and Anderson (1972) have demonstrated that if residual horizontal stresses are high enough, they can result in major shifts in the orientation of the principal stresses, with a significant region of tensional stresses developing behind the slope; this will further add to the instability of the rock slope.

Once the slope is "created" by the opening of the rock mass and failure of a surface zone, the region behind this zone can now "relax." This leads to rebound of the slope wall (Matheson and Thompson, 1973), which is predicted by finite-element methods when applied to an excavation made into a rock mass (Duncan and Goodman, 1968). The magnitude of the rebound depends upon the lateral stress relief and the modulus of elasticity of the rock. The new surface zone has to take on physical characteristics different from the more interior parts of the rock mass. It will have less strength, primarily because of a loss of cohesion across joint planes and the reduction in "locking" stresses, which formerly acted in the more confined state. The strength of the joint surface and the stability of the slope will now largely depend upon the angle of shearing resistance acting along the joint. It must be borne in mind that the angle of shearing resistance is also dependent upon the normal stresses, so that the joint surface in the surface zone will have a different value than those in the interior of the rock mass. This general relaxation of the surface zone may lead to immediate failure. Because of the time-bound nature of rock-mass relaxation, however, this may not take place for some time after the new slope has formed and will continue until the stresses are dissipated.

Joint Water Pressure. A further factor that has to be considered in rock-slope failure is joint water pressure. The importance of this is now generally accepted and is well demonstrated, for example, in the problems solved by Hoek and others (1973). From the nature of some of the rock masses and especially some of the deep tension joints already described, the development of joint water pressures would seem quite likely, especially since some joints were blocked in August by ice that had formed in the previous winter.

A feature possibly reflecting the importance of joint water pressure is the mechanism leading to the failure of the open-pit slopes in the field area (Rana and Bullock, 1969). There seems to be some relationship between failure and hydrologic conditions, with the distribution of the failure similar to that which Terzaghi (1965) showed for rockfalls and rockslides on natural slopes in Norway (Fig. 9).

Terzaghi attributed these failures to the occurrence of high joint water pressure during the thaw period, when joints are blocked with ice, and to the development of high joint water pressure at times of heavy precipitation in the autumn months. A similar explanation may account for the occurrence of slides in the open pits of the study area. However, Terzaghi's conclusions must be treated with some caution, because the same information was interpreted by Bjerrum and Jörstad (1968) as reflecting the importance of frost shattering.

Chemical Weathering. Although the role of physical weathering in the failure of rock slopes in cold areas has been emphasized, the potential of chemical weathering has never fully been explored. Chemical weathering acts on the face of a rock slope and, by utilizing joints and other discontinuities, can lead to the decay of the interior of the rock mass. Chemical weathering along a joint plane may lead to a reduction in the shear strength of the plane, which can result in failure of the slope.

As is to be expected from a consideration of carbonate weathering, chemical weathering is intense on the Denault Dolomite. When a joint surface in a dolomite rock mass is intensively weathered, individual joint blocks are often removed from the rock mass. Because of the nature of rock slopes, the failure of one individual block has considerable consequences for the whole slope. The stresses that were supported by the now-removed joint block are transferred to

Figure 8. Failure of slabs of rock from the slope face as the result of sheet structures.

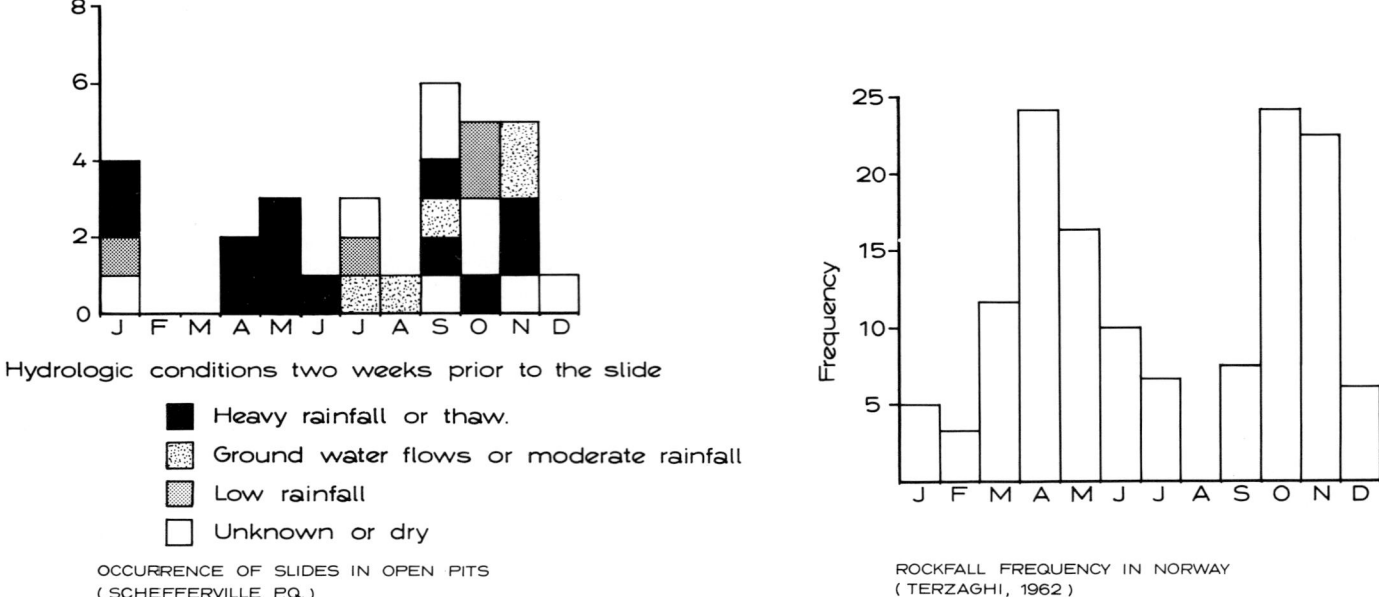

Figure 9. Frequency of occurrence of slides in the open-pit workings of the Schefferville area and rock-fall frequency on natural slopes in Norway.

the surrounding blocks. If the stress concentrations become large enough, failure of the surrounding material may result, which, in turn, causes further stress concentrations and so on until failure has spread through the entire slope. After this is complete, the slope can attain a new temporary stability.

Chemical weathering as a factor in slope failure is really only important in the Denault Dolomite. On the other rock units it is almost of no importance, although in two locations on the Wishart Quartzite chemical weathering was responsible for the removal of small-sized joint blocks in the rock slope.

Internal Factors Combined. The various internal processes and factors that have been mentioned as playing a role in rock-slope failure seldom act individually; most often they combine to decrease the shear strength of a rock mass through time. They may vary in intensity and importance, but they inevitably act in concert in such a way that there can be a progressive reduction in the shear strength of any potential joint surface. The general nature of a joint surface lends itself to this, especially as joint "roughness" (Patton, 1966) exerts a considerable influence on the shear strength (S) of a joint surface. Joint roughness leads to

$$S = c + \sigma \tan (\phi + \phi_r), \quad (7)$$

where σ_r is the increase in the angle of shearing resistance of the joint surface. This value is the average angle between undulations along the joint.

Variations in shearing stresses caused by changes in the elastic behavior of the rock mass and fluctuations in joint water pressure, and also possible reductions in the shear strength of the joint surface by chemical weathering, will decrease the resistance of this plane to failure. Such a decrease in shear strength may be partly accomplished by the grinding down of the asperities along the joint surface. In this way the peak shear strength of the plane will be overcome, and the residual strength will be mobilized. The reduction in the strength of the plane would take place in steps, where the more heavily stressed areas fail first, followed by the transfer of the stress burden to other positions along the joint surface. Once the shear strength of the joint has been sufficiently reduced, overall failure of a slope may take place.

CONCLUSIONS

When the failure of rock slopes in the Labrador-Ungava area is considered in detail, it becomes clear that a variety of factors play a role. The assumption of earlier workers that high-angle rock-slope failure in cold areas can be largely attributed to frost shattering does not seem to be an entirely satisfactory conclusion. This is not to say, however, that the process plays no role at all. While accepting frost shattering as one process of physical weathering, this process cannot play an equally significant role in the large-scale failure of rock slopes as it does in the limited disintegration of intact rocks in outcrops. From the rock-mass geometries, it is clear that the initial factors that led to the failure of these slopes were the external factors of slope formation, which in this case were glacial and glaciofluvial action. Failure was perpetuated by the internal factors of residual stress, the action of joint water pressure, and physical and chemical weathering.

It is possible that rock-slope failure in cold areas is often the direct consequence of the creation of the slope by external factors. Many cold regions are recently glaciated landscapes, and, in such settings, many have been oversteepened and undercut. It is therefore quite logical that rock-slope failure and talus-slope development should characterize such areas. This conclusion is by no means new. Young (1972, p. 229) noted that in cold areas the frequency of high-angle rock slopes and talus slopes is in part a relict from glaciation. Rapp (1960b), in his Kärkevagge study, drew attention to the fact that massive rock-slope failures, which cannot be attributed to frost action, have taken place.

Several authors have expressed general reservations about the importance of traditionally accepted cold-area slope processes. Certainly, as far as rock slopes in cold areas are concerned, more detailed work is needed on the factors responsible for failure. There seems to be some indication that the traditional emphasis on cold-area processes, when explaining rock-slope failure in such areas, may not be valid. The validity of this assumption could be ideally tested in some of the unglaciated areas of western Canada. Until such work has been done, though, caution should be exercised in adopting or proposing any general scheme of hillslope evolution in cold areas.

ACKNOWLEDGMENTS

I thank Bruce Thom for his help. I am grateful for the comments and criticisms of the two referees, J. T. Andrews and N. Caine. D. Haddow, M. A. Hirst, A.J. Conacher, and J. Gentilli were kind enough to comment on the manuscript.

REFERENCES CITED

Andrews, J. T., 1961, The development of scree slopes in the English Lake District and central Quebec–Labrador: Cahiers Géographie Québec, no. 10, p. 219–231.

Ashby, J. P., 1971, Sliding and toppling modes of failure in jointed rock slopes [M.Sc. thesis]: London, Univ. London, 125 p.

Barton, N., 1972, Progressive failure of excavated rock slopes: Stability of rock slopes: Am. Soc. Civil Engineers 13th Symp. on Rock Mechanics, Proc., p. 139–170.

Bird, J. B., 1967, The physiography of arctic Canada: Baltimore, N.J., Johns Hopkins Univ. Press, 336 p.

Bjerrum, L., and Jørstad, F., 1968, Stability of rock slopes in Norway: Nørges Geotek. Inst. Pub. 79, p. 1–11.

Bryson, R. A., Wenland, W. M., Ives, J. D., and Andrews, J. T., 1969, Radiocarbon isochrones on the disintegration of the Laurentide ice sheet: Arctic and Alpine Research, v. 1, p. 1–14.

Church, M., and Ryder, J. M., 1972, Paraglacial sedimentation: A consideration of fluvial processes conditioned by glaciers: Geol. Soc. America Bull., v. 83, p. 3059–3072.

Derbyshire, E., 1962, Fluvial glacial erosion near Knob Lake central Quebec–Labrador [M.Sc. thesis]: Montreal, McGill Univ., 326 p.

Dimroth, E., 1970, Evolution of the Labrador geosyncline: Geol. Soc. America Bull., v. 81, p. 2717–2742.

Dodd, J. S., and Anderson, W. W., 1972, Tectonic stresses and rock slope stability: Stability of rock slopes: Am. Soc. Civil Engineers 13th Symp. on Rock Mechanics, Proc., p. 171–182.

Duncan, J. M., and Goodman, R. E., 1968, Finite element analyses of slopes in jointed rocks: Berkeley, California Univ., Rept. S-68-3, 76 p.

Embleton, C., and King, C.A.M., 1968, Glacial and periglacial geomorphology: London, Edward Arnold Ltd., 608 p.

Flint, R. F., 1971, Glacial and Quaternary geology: New York, John Wiley & Sons, 892 p.

Gray, J. T., 1971, Processes and rates of development of talus slopes and protalus rock glaciers in the Ogilvie and Wernecke Mountains, Yukon Territory [Ph.D. thesis]; Montreal, McGill Univ., 220 p.

Haste, N., 1958, Measurement of rock pressure in mines: Sveriges Geol. Unders ökning Årsb., v. 52, p. 23–31.

Herget, G., 1974, Ground stress determination in Canada: Rock Mechanics, v. 6, p. 53–64.

Hoek, E., Bray, J. W., and Boyd, J. M., 1973, The stability of a rock slope containing a wedge resting on two intersecting discontinuities: Quart. Jour. Eng. Geology, v. 6, p. 1–55.

Ives, J. D., 1968, Late Wisconsin events in Labrador-Ungava: An interim commentary: Canadian Geographer, v. 12, p. 192–203.

Matheson, D. S., and Thompson, S., 1973, Geological implication of valley rebound: Canadian Jour. Earth Sci., v. 10, p. 961–978.

Müller, L., 1972, Geomechanische Auswirkungen von Abtragungsvorgängen: Geol. Rundschau, v. 59, p. 163–178.

Patton, F. D., 1966, Multiple modes of shear failure on rocks: Internatl. Soc. Rock Mechanics Proc., v. 3, p. 509–514.

Peltier, L. C., 1950, The geographical cycle in periglacial regions as it is related to climatic geomorphology: Ann. Assoc. Am. Geographers, v. 40, p. 214–236.

Rana, M. H., and Bullock, W. D., 1969, the design of open pit mine slopes: Canadian Mining Jour., v. 90, no. 8, p. 58–62.

Rapp, A., 1960a, Recent developments of mountain slopes in Kärkevagge and surroundings, northern Scandinavia: Geog. Annaler, v. 43, no. 58, p. 71–200.

——1960b, Talus slopes and mountain walls at Templefjorden, Spitzbergen: Norsk Polarinst. Skr. 119, 96 p.

Terzaghi, K., 1950, Mechanism of landslides, in Application of geology to engineering practice (Berkey volume): Boulder, Colo., Geol. Soc. America, p. 83–122.

——1962, Stability of steep slopes on hard unweathered rock: Géotechnique, v. 12, p. 251–270.

Terzaghi, R. D., 1965, Sources of error in joint surveys: Géotechnique, v. 15, p. 287–304.

Tricart, J., 1970, Geomorphology of cold environments: London, MacMillan, 418 p.

Twidale, C. R., 1957, Development of slopes in central New-Quebec–Labrador [Ph.D. thesis]: Montreal, McGill Univ., 269 p.

Wolters, R., 1969, Zur Ursache der Enstehung oberflächenparalleler Klüfte: Rock Mechanics, v. 1, p. 53–70.

Young, A., 1972 Slopes: Edinburgh, Oliver & Boyd, 288 p.

MANUSCRIPT RECEIVED BY THE SOCIETY SEPTEMBER 7, 1976
MANUSCRIPT ACCEPTED SEPTEMBER 17, 1976

Printed in U.S.A.

Geological Society of America
Reviews in Engineering Geology, Volume III
© 1977

4

Large landslides of the Columbia River Gorge, Oregon and Washington

LEONARD PALMER
Department of Earth Sciences, Portland State University, Portland, Oregon 97207

ABSTRACT

Large landslides from 10 to 50 km² occur in the Columbia River Gorge where construction is required for major transportation and hydrologic facilities. Geologic conditions conducive to landsliding include more than 1,200 m of topographic relief, high rainfall (250 cm/yr), erosional exposure of water-saturated plastic clayey layers under permeable rock masses, and a regional 5° to 30° dip into the gorge. In more than 130 km² of landslide deposits, 13 km² are known to be active with slow movement (less than 15 m/yr). Rapid motion (possibly more than 10 m/s) of 14 km² of the Bonneville landslide occurred about 700 B.P. and was the last landslide to dam the Columbia River. In evaluation of the landslides, geological techniques and analyses have been variable; this indicates a need for uniform procedures and criteria to define engineering constraints.

INTRODUCTION

More than 130 km² (50 mi²) of landslide deposits have been mapped in the Columbia River Gorge (Fig. 1). A combination of wet climate, weak bedrock units, and steep terrain have caused the sliding.

Most of the landslide deposits are not now in motion, but less than 13 km² (5 mi²) of actively sliding land have caused millions of dollars worth of damage or costs to prevent damage and still pose a great problem for geologists and railroads, and other structures in this major transportation engineers concerned with the dams, reservoirs, highways, railroads, and other structures in this major transportation corridor. Bonneville Dam impounds more than 20 km³ (500,000 acre-feet) of water. The dam is one of the major engineering projects that required special design to accommodate its foundations, which are located upon landslide deposits. Some landslide deposits appear stable, and some sliding areas have been stabilized. Others, however, are active and must be avoided or accommodated by frequent maintenance, repair, or innovative structures (Fig. 2).

This chapter describes the larger landslides of the Columbia River Gorge as summarized from various reports and investigations and briefly evaluates the engineering geology investigations for such geologic features.

Information on the landslides of the Columbia River Gorge has been culled primarily from unpublished governmental and consulting reports (see App. 1) describing the natural conditions for engineering structures at individual sites in the gorge. The reports were developed over more than 50 years and reveal varying objectives and investment of effort. Even the journals of the Lewis and Clark expedition of 1804–1806 refer to the presence of landslides (Lawrence and Lawrence 1958). Differences in description of stratigraphic units and terms, in suggested mechanisms of landsliding, and in the delineation of landslide deposits are almost as numerous as the number of reports. Only recently has a more unified and comprehensive description of the landslides been attempted by geologists of the U.S. Army Corps of Engineers.

GEOLOGIC SETTING

The Columbia River Gorge is a dramatic canyon cutting through the Cascade Range for about 120 km (75 mi). Cliffs of resistant Columbia River (Yakima) Basalt and younger volcanic rocks rise over 1,200 m (4,000 ft) above the river and expose a cross section of predominantly volcanic strata that overlie older incompetent volcaniclastic rocks. The younger, more resistant units have been undercut by canyon and flood-water erosion; this has allowed large landslides to move into the canyon (Fig. 2).

A columnar diagram summarizing the geologic history (Wise, 1970) is shown in Figure 3, and a diagrammatic north-south cross section of the Columbia River Gorge is shown in Figure 4.

Geologic events that occurred during the formation of the

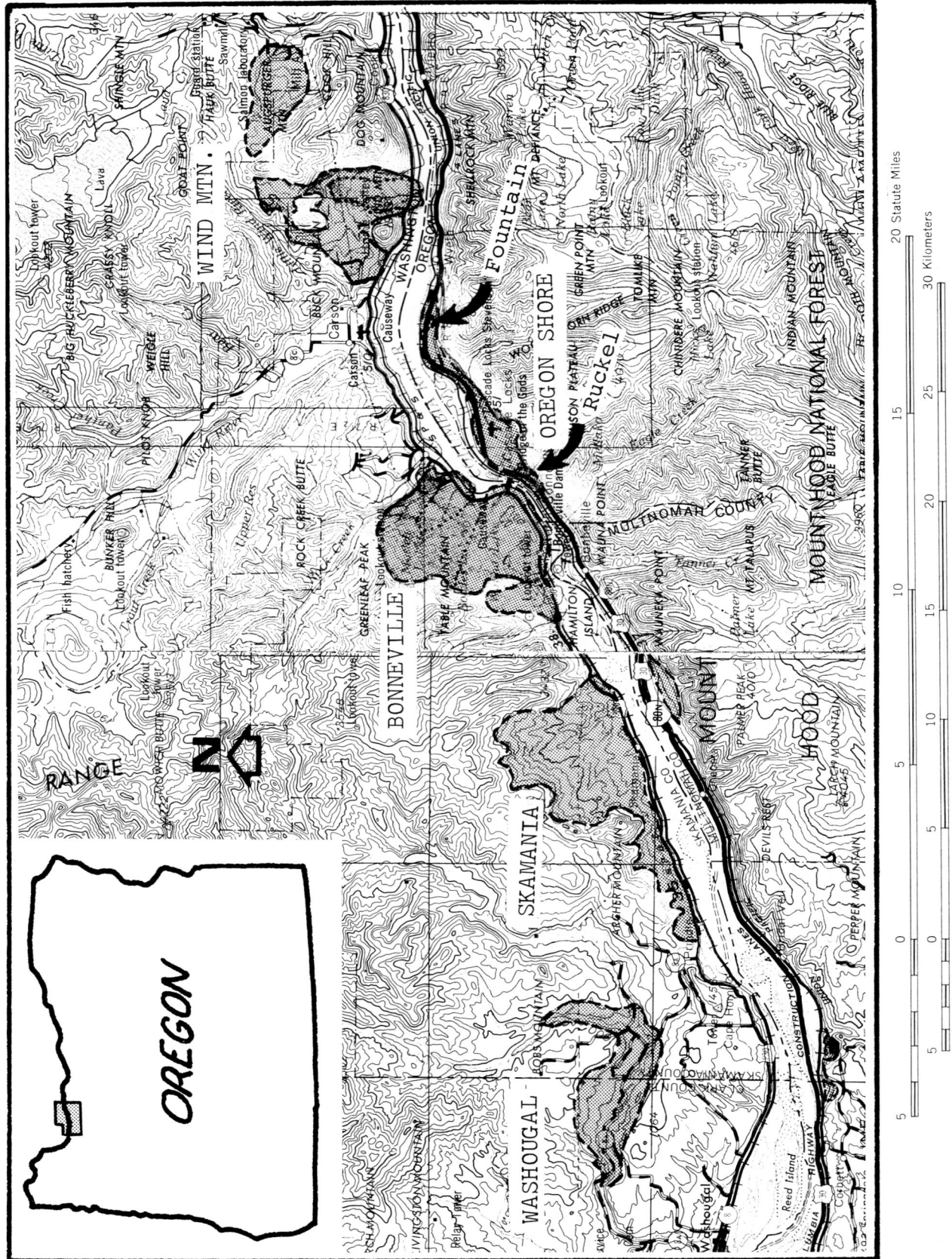

Figure 1. Location map showing major landslide deposits (shaded areas outlined by dashes) and some actively sliding areas (shaded areas outlined by dots).

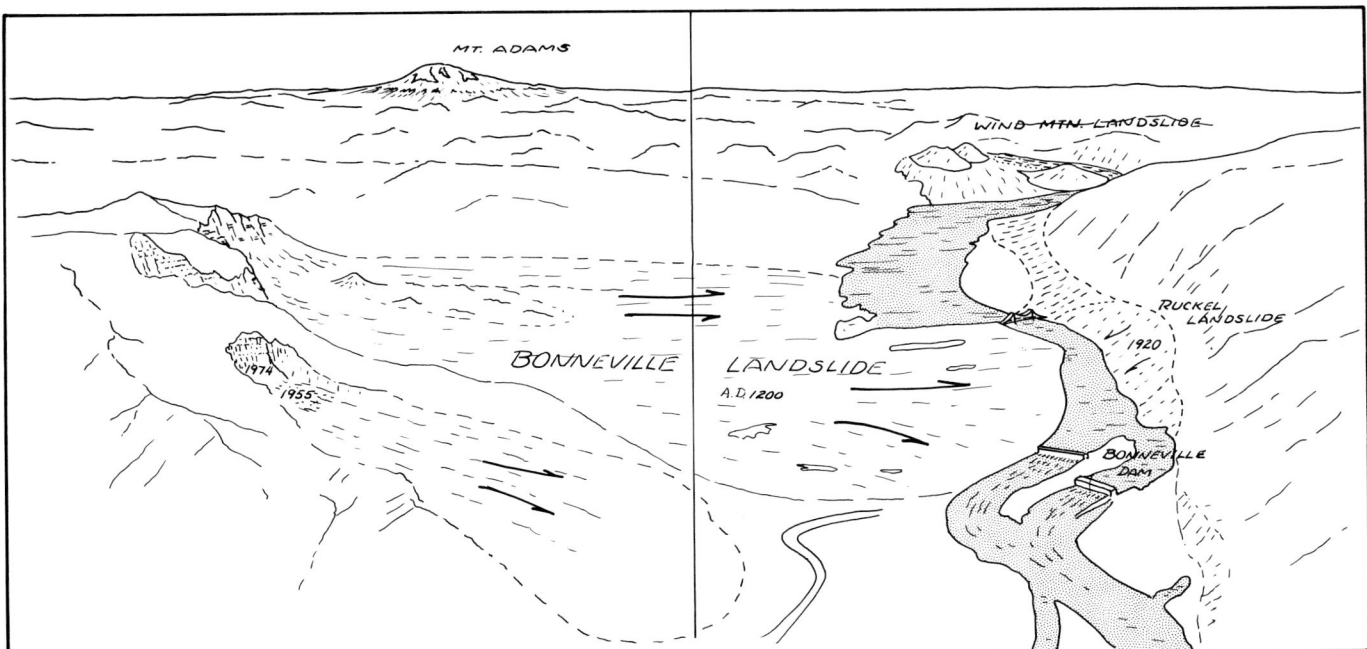

Figure 2. Two-photo aerial view of the Columbia River Gorge looking east from Bonneville Dam to Wind Mountain and diagrammatic sketch of photo area (photographs copyright Leonard H. Delano; used by permission).

Columbia River Gorge are summarized as follows (modified from unpublished data of D. A. Williamson and J. E. Allen, 1971):

1. Ohanapecosh and Eagle Creek volcaniclastic debris accumulated in a submerged trough. Because of their properties after weathering, these two units affect land stability.
   a. The Ohanapecosh Formation (Eocene to Oligocene) is a thick (more than 3,000 m [10,000 ft]) deposit ranging from varied claystone to pebble conglomerate composed of sedimentary and volcanic materials. The rocks have been weathered, probably by hydrothermal alteration, so that now-weak rock fragments commonly break with the matrix. Celadonite, zeolites (mostly laumontite), and montmorillonite are present owing to the devitrification of volcanic glass. Plagioclase phenocrysts are zeolitized and albitized. Miocene weathering and erosion resulted in a slight angular unconformity with the overlying Eagle Creek Formation. For 9 to 15 m (30 to 50 ft) below the unconformity, the Ohanapecosh rocks have weathered to a soft, clay-rich saprolite that retains the original textures. Generally overlying the saprolite is a red, more massive, 3-m-thick (10 ft) weathered zone, but it was removed locally before the Eagle Creek conglomerates were deposited (Wise, 1970, p. 8, 10). The rock unit as a whole is weak and contains beds of clay and clayey paleosoils. The beds dip 5° to 30° SE, toward the Columbia River Gorge.
   b. The Eagle Creek Formation is also volcaniclastic, but differs from the underlying Ohanapecosh Formation in being less weathered and containing larger rock fragments — commonly more than 15 m (50 ft) in diameter. The deposits originated as mudflows from volcanic eruptions a short distance to the north. The unit thins from 400 m (1,300 ft) to 150 m (500 ft) in 6.5 km (4 mi) from the head of the Bonneville landslide to the Columbia River (see Fig. 2).
2. Columbia River (Yakima) Basalt flooded and buried the Eagle Creek deposits and lapped around the volcanoes during the Miocene and Pliocene time.
3. Late Pliocene weathering, erosion, and folding of the Columbia River (Yakima) Basalt coincided with the beginning of the ancestral Columbia River.
4. Rapid erosion by glacial waters accompanied by volcanic intrusions and extrusions occurred in Pliocene-Pleistocene time. Intracanyon lava flows dammed the Columbia River Gorge on several occasions and at several locations.
5. Gigantic floods from glacial Lake Missoula poured through the gorge and scoured the channel to below sea level while discharging more than 2,000 km³ (500 mi³) of water over a few weeks time (Bretz, 1925; Baker, 1973).
6. Pleistocene and Holocene slope failures due to oversteepening resulted in large landslides. River erosion cut through Columbia River (Yakima) Basalt into the underlying, weak Eagle Creek and Ohanapecosh Formations; this triggered large-scale landsliding — the largest moving downdip from the north. The Columbia River was most recently dammed by a landslide at Bonneville about 700 yr ago.

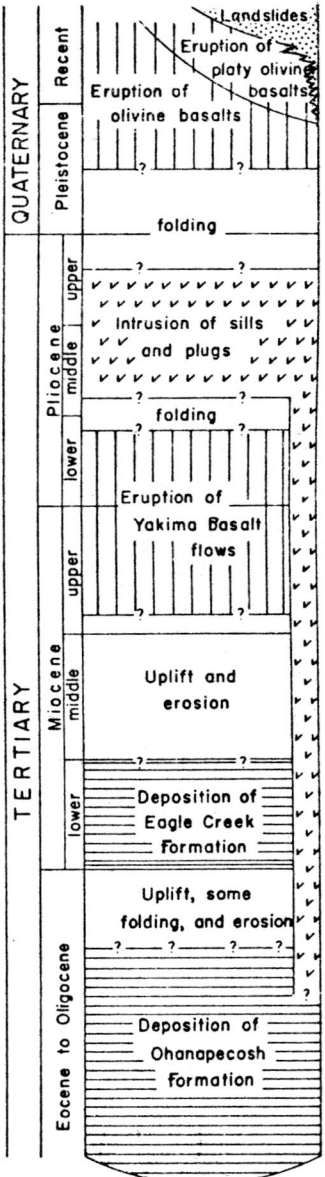

Figure 3. Columnar summary diagram of the geologic history of the Columbia River Gorge (modified from Wise, 1970, Fig. 14).

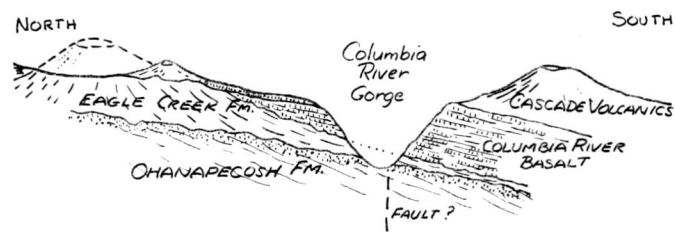

Figure 4. Diagrammatic cross section of the Columbia River Gorge illustrating the structural and stratigraphic relations (modified from Allen, 1958, Fig. 5).

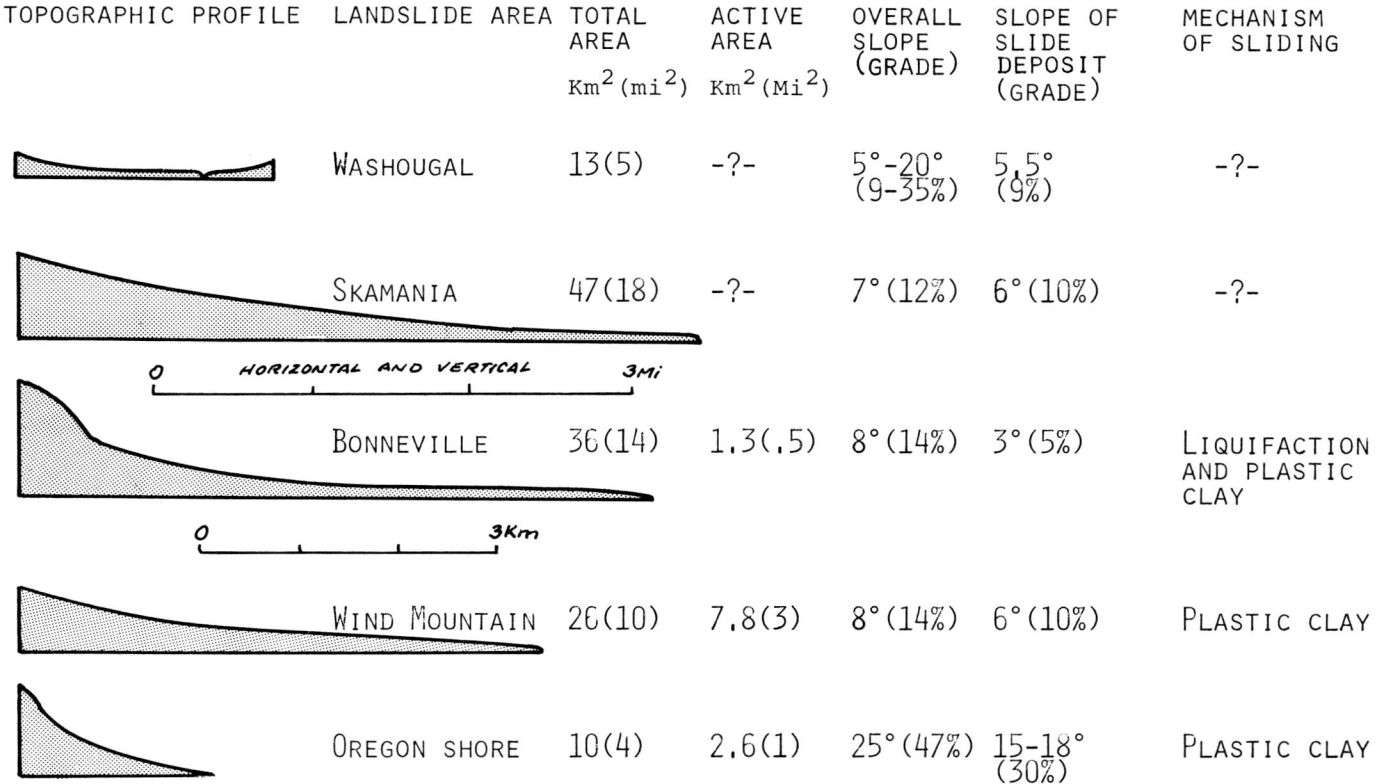

Figure 5. Comparative landslide data.

7. Instability and movement of some landslide deposits, and active raveling of head scarps continues at the present time. Slope failures are triggered or accelerated during periods of high rainfall over several weeks. Many landslide deposits are stabilized, but reactivation failures have been triggered by excavation and grading for highways and railroad routes. Some stabilization of actively moving landslide deposits has been accomplished by drainage structures. Removal of large amounts of slide debris has failed to stabilize one slide.

Information on some of the larger landslide areas as shown in Figure 1 is presented below. Eastward along the Columbia River to Walulla, unmapped landslides occupy more than 10% of the river banks (mostly north of the river) for 120 km (75 mi), but none is as large as those reported here.

LANDSLIDES

Comparative physical data are shown in Figure 5 for the landslide areas of the Columbia River Gorge (Fig. 1). Landslide characteristics in the gorge are little known. Many observations have been made since 1900. A general understanding of the area has been developed and improved, but most reports differ in delineation of any particular landslide and in its interpretation, as well as in the methods and terminology used. The various origins, mechanisms, and engineering problems of landslides in the gorge are shown both by their physical properties and by the variations in reports about the same slides by different agencies and authors!

Annual precipitation of about 250 cm (100 in.) provides ample saturation of the weak rock units described above. Undercut dip slopes with clay layers, volcanic loading of the heads of slopes, seismicity, tectonism, flooding and rapid drawdown, and reduction of shear strength by weathering are all present in the gorge. Internal and external agents that might cause the slides are easily found (Terzaghi, 1950). However, the specific history of a given slide area in the gorge appears to involve a combination of processes at different times and places.

In the descriptions of individual areas below, some landslides are not known to have been studied, whereas others have been studied well enough to provide evidence of an interesting and varied history of activity.

Washougal Landslide Area

Along the Washougal River, a landslide area of 13 km² (5 mi²) is shown on the Washington State geologic map as determined from unpublished data of A. C. Waters (see also Waters, 1973). The area may well be one in which the Eagle Creek Formation or the Ohanapecosh Formation is near the land surface, but specific information is lacking.

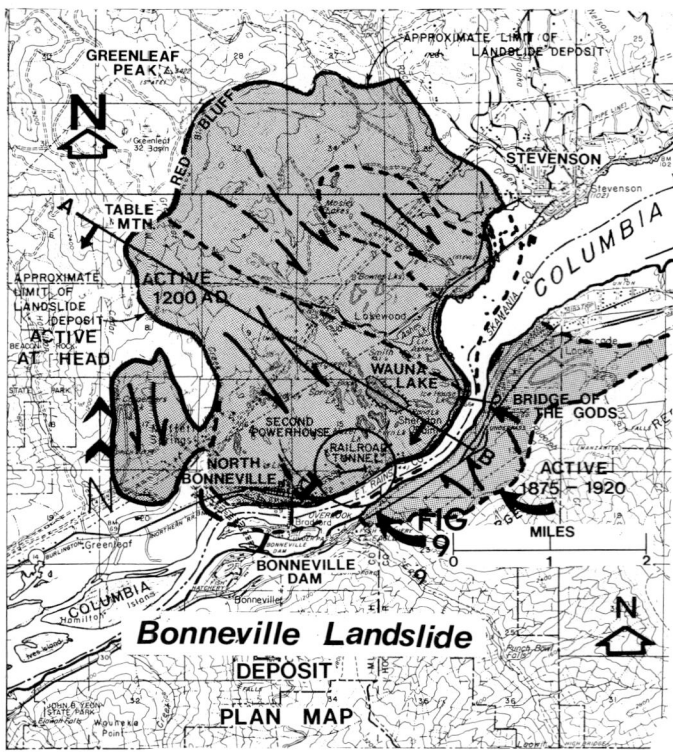

Figure 6. Map of the Bonneville landslide deposit area modified from Wise (1970) and USACE (1976).

Skamania Landslide Area

The distribution of landsliding near Skamania, Washington, as modified from A. C. Waters's unpublished data, is shown on the Washington State geologic map (Huntting and others, 1961). In this slide area are hummocky hills of broken and deformed Yakima Basalt, which is highly permeable and probably conducts water to saturated subsurface clayey layers at the top of the Eagle Creek or Ohanapecosh Formation. This is one of the largest areas of landslide deposits. It has a surficial gradient similar to that of the active Wind Mountain landslide, but its characteristics are as yet undermined. It is not known to be active.

Bonneville Landslide Area

Landslides in the vicinity of Bonneville Dam (Fig. 6) are better known than those described above, as shown by their multiple names — Cascade, Red Bluffs, or Bridge of the Gods landslide. However, from reports examined, it appears that the small, active parts of landslide deposits primarily on the Oregon side of the gorge have received most attention (C. P. Holdredge, 1937, unpub. data [see App. 1]). The relation of landslide deposits to the dam is shown in Figure 7.

The overall mass of the Bonneville landslide was mapped by Wise (1961, 1970), who first separated lobes of different ages. Physical data (Fig. 5) show the Bonneville landslide to have the lowest gradient of all slides in the gorge (3° compared to 6° to 20° for other slide areas). The surface form indicates that it is actually a composite feature resulting from a number of landslide events, which perhaps began in Pleistocene time. The total landslide area covers 30 to 36 km^2 (12 to 14 mi^2); individual slide deposits cover about 5 to 13 km^2 (2 to 5 mi^2).

The data presented here are largely from recent reports prepared in connection with expansion of generating capacity at the Bonneville Dam. Landslide stability was a concern in two places — where a new powerhouse was to be constructed in a channel excavated through the toe of the Bonneville landslide deposits, and where the highway and railroads might be endangered because of the effect of higher reservoir levels on the stability of the landslide deposits. After their studies, the U.S. Army Corps of Engineers concluded that reservoir fluctuations would have no significant effect on stability (USACE, 1971a, 1971b, 1975, 1976).

Wise and Waters have best mapped the landslides, even though they have been primarily concerned with the study of volcanic rocks. ("My work on Columbia River Gorge slides is not detailed — to me they have been a nuisance that hides the bedrock geology" [Waters, 1973].) Wise has hardly

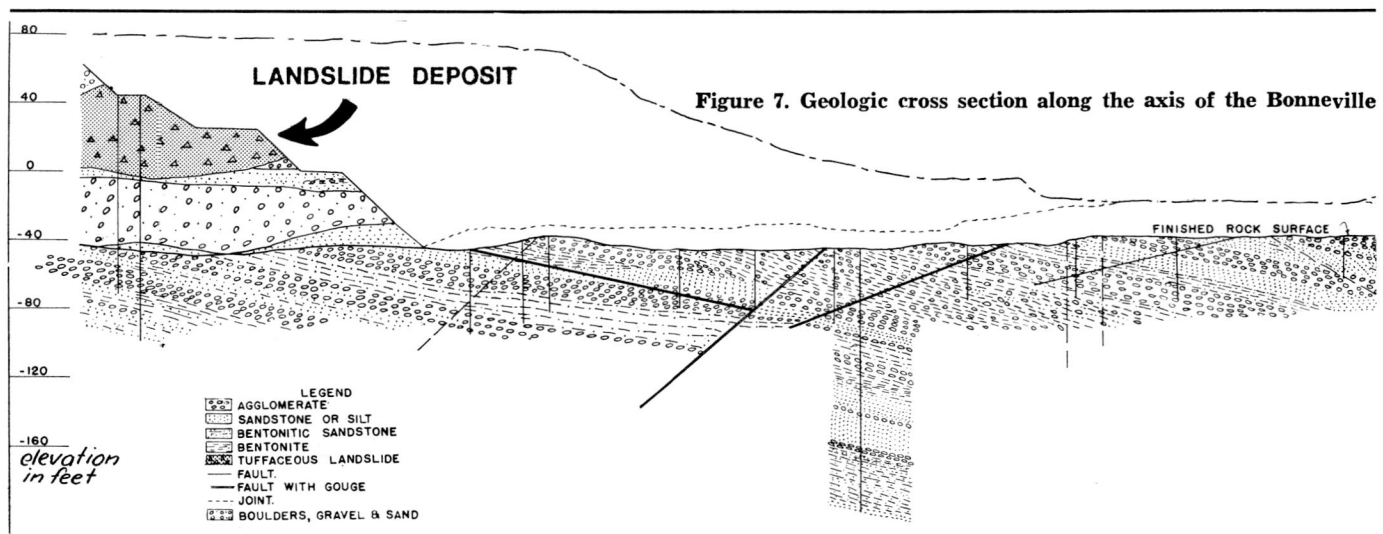

Figure 7. Geologic cross section along the axis of the Bonneville

mentioned the slides but has mapped them with best detail (Wise, 1961, 1970). The U.S. Army Corps of Engineers appears to have done the most in developing data on the slides, but their studies have focused on areas that relate to their special interest.

A summation of map data from Wise (1970), Waters (1955, 1973), U.S. Army Corps of Engineers (1976), and discussions with J. E. Allen and D. A. Williamson is shown in Figure 6.

Of the many exploratory drill holes for foundation studies at the Bonneville Dam area, no drilling is known to have been done for the purpose of evaluating the landslide area as a whole. The most-detailed study to date resulted in a 50 ft to 1 in. (600:1) scale cross section developed from a surface traverse of the slide (Fig. 8). The value of such surficial geological data, more than 6,000 m (20,000 ft) in length, is often underestimated when compared to drill hole information. Although the data are good, it would nevertheless be interesting to have some confirming borehole data.

At present, the lower part of the slide mass is settling probably because of reconsolidation and compaction. The head scarps are actively raveling, with large blocks more than 5 m (15 ft) in diameter crashing down from time to time, so that it is unsafe at the base of the cliffs. Rockfalls at the head of the western part of the slide area at Carpenters Lake were reported in 1955 and 1974, and the upper part of the slide is in motion.

A lobe of the Bonneville landslide with an area of about 14 km² (5.5 mi²) extends into the Columbia River, diverting the channel against the Oregon shore where it is today. Undissected topography and the 250-yr age limit to trees on the slide led Hodge (1938, p. 917) to suggest that the slide occurred not more than 500 yr ago. Carbon-fourteen measurements on wood in the slide date just over 700 yr B.P. and confirm the recency of that part of the slide (USACE, 1976; Lawrence and Lawrence, 1958). Legends of the native fishing tribes along the river include an account of a "bridge of the gods" over which the river could be crossed. The A.D. 1200 landslide event probably formed that bridge. Terminal lobes of some of the landslide units are found above the present river level, which suggests that some sliding occurred much earlier, possibly in Pleistocene time, perhaps before the erosion caused by the last giant flood (Bretz, 1925; D. A. Williamson, 1976, oral commun.).

The mechanism of sliding in the Bonneville area also appears to be composite, with different processes acting at different times and places. Rockfalls occur in oversteepened cliffs; plastic flow commonly occurs where permeable Columbia River (Yakima) Basalt and younger lava flows overlie impermeable and clayey rocks of the Eagle Creek and Ohanapecosh Formations. A third mechanism may also have operated at the toe of the A.D. 1200 slide lobe where it is composed of large landslide blocks more than 60 m (200 ft) across with a matrix of sandy and gravelly landslide debris (see Fig. 9). The mechanisms that could transport such large blocks in the absence of a clay slip plane suggest rapid original movement and probable stability of the resultant deposit (USACE, 1976).

Williamson (1976, oral commun.) interpreted the evidence to suggest that the landsliding began with slope failure but that when the slide moved into the river, liquifaction of micaceous river sand deposits resulted which allowed the slide mass to rise up the opposite (south) valley wall before coming to rest. Shear planes and sand dikes in the landslide toe suggest deceleration by retardation at the base, followed by dewatering and settling of the landslide mass. A detailed analysis of the entire Bonneville landslide, however, remains to be done.

Wind Mountain Landslide Area

The landslide at Wind Mountain (also called Collins Cove, Collins Point, and Girl Scout Camp landslide) contains the largest area of actively moving landslide deposits in the Columbia River Gorge. Almost 8 km² (3 mi²) of landslide deposits, 23 to 55 m thick (75 to 180 ft), is moving toward the Columbia River at up to 15 m (47 ft) per year. More than 190×10^6 m³ (350×10^6 yd³) may be in motion (USACE, 1971, Pl. 6).

Comparatively good data are available on the Wind Moun-

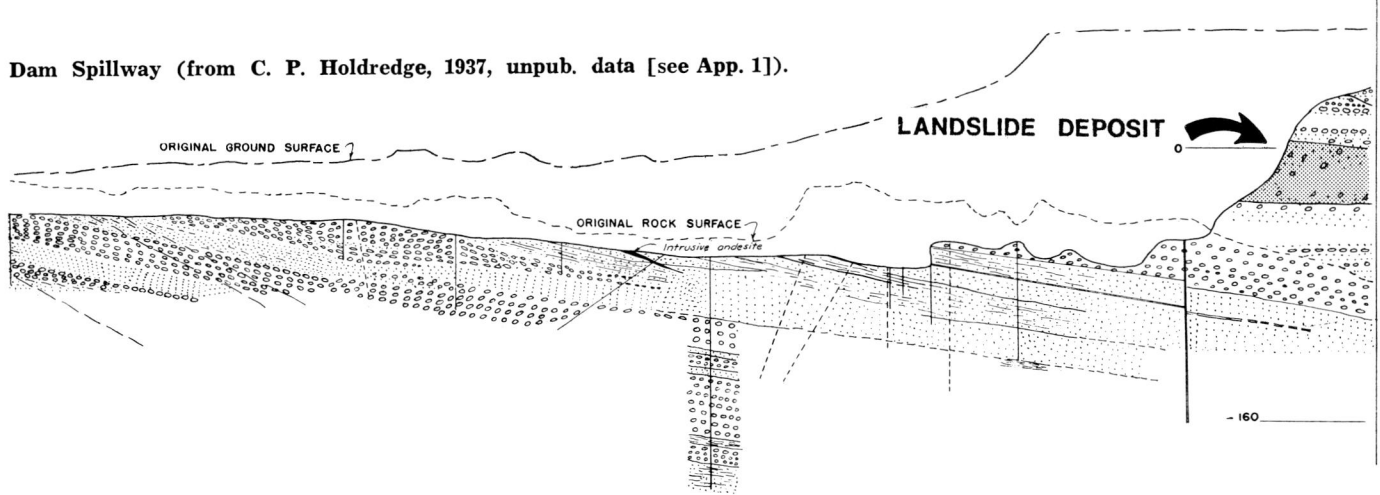

Dam Spillway (from C. P. Holdredge, 1937, unpub. data [see App. 1]).

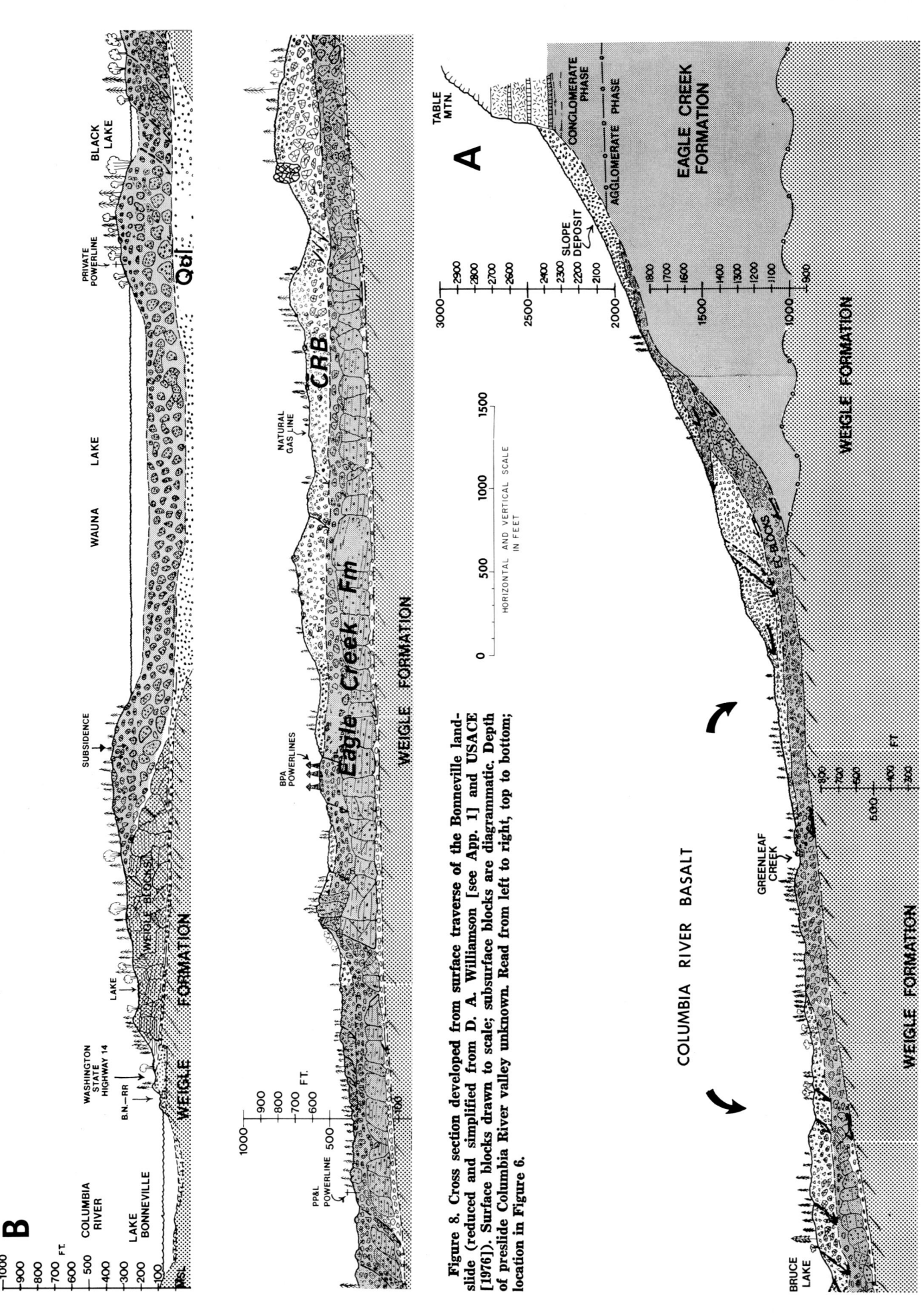

Figure 8. Cross section developed from surface traverse of the Bonneville landslide (reduced and simplified from D. A. Williamson [see App. 1] and USACE [1976]). Surface blocks drawn to scale; subsurface blocks are diagrammatic. Depth of preslide Columbia River valley unknown. Read from left to right, top to bottom; location in Figure 6.

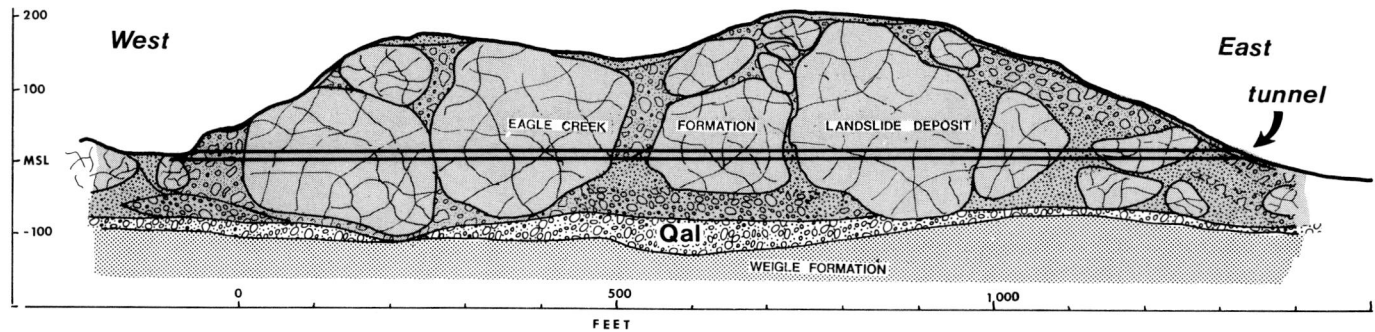

Figure 9. The toe of the Bonneville landslide at a site explored for a railroad tunnel contains large fractured slide blocks that range up to about 100-m (300-ft) diameter in a matrix of micaceous sand and coarser debris without significant clay.

tain landslide from a number of reports and surveys. These include four slope-indicator installations, six piezometer borings, and three survey lines (at the lower, middle, and upper part of the slide) and two geologic cross sections. The total area of the landslide deposits (26 km² [10 mi²]) is shown in Figure 10.

An approximate history of movement was collected from various records and reports beginning with railroad construction in 1907. The slide was recognized by Barnes and Butler (1930). The Corps of Engineers surveyed a power line in 1936 that was built by the Bonneville Power Administration in 1941 across the upper part of the active slide area. In 1936–1938, 0.6 m (2 ft) of movement was noted. By 1946, four transmission line towers were affected by movement of 0.3 to 0.6 m/yr (1 to 2 ft/yr). From 1952 until the power line was abandoned in the mid-1950s, as much as 10.5 m/yr (35 ft/yr) of movement occurred, and 16 towers were kept in line by mounting them on skids. Increased precipitation after 1945 had a marked effect on the landslide movement. An earthquake of intensity 8 in 1949 was also suggested as a trigger for slide activation (D. C. Birch, 1955, unpub. data [see App. 1]). However, this possibility may be disproved by the fact that without concomitant seismic activity, sliding greatly accelerated during the heavy rains of 1970. The rainfall and slide motion data shown in Figure 11 indicate that rainfall averaging more than 250 cm/yr (100 in./yr) for a 3-yr period is enough to accelerate motion of this slide. A map of the active part of the Wind Mountain landslide shows the distribution of deposits and movement (Fig. 12).

Coarse rock debris from the head scarp and flanking cliffs has spread over the slide surface that covers underlying clayey layers of reworked Ohanapecosh and Eagle Creek Formations. Water percolating through the coarse surface debris accumulates over the clayey layers; this maintains a saturated condition and several small lakes. Greater slide movement at the head of the slide than at the toe has been presented as evidence that rainfall infiltration at the head has more influence on sliding than the fluctuation of reservoir water levels does (USACE, 1971b).

A longitudinal geologic cross section developed from surface data by D. A. Williamson (Fig. 13) shows the character of the active part of the landslide. The overall structural form of the moving slide shows a variety of features not unlike those encountered in a glacier. Coarse rock debris forms a permeable mantle over much of the upper and middle part of the slide (Fig. 14a). Plastic flow in the saturated clayey layers causes extension, subsidence, and block faulting in the upper parts of the slide (Fig. 14b). Flow resistance of the soil strata on top of the slide mass causes slow overriding by clay at the surface (Fig. 14c). Burial of talus forms permeable subsurface layers through which lakes may intermittently drain (Fig. 14d).

Active movement of the slide is evident at the surface by offset of buildings, roadways, and pipelines along deep surface cracks. The boathouse at the Girl Scout camp has moved away from the lake. Trees have been torn in half along longitudinal cracks; others have been progressively swallowed by the slide in subsidence depressions.

Many structures have required changes because of the slide movement. Water supply pipes at the Girl Scout camp have been accommodated to slide motion by the installation of zig-zag courses, joints, and plastic piping. A stone lodge at the camp was destroyed by fire; it was replaced by a circle of covered wagons that could better accommodate the persistent movement. The power line crossing the upper part of the slide was moved off of the slide to avoid additional heavy maintenance costs. The highway, railroad, and Collins Point navigation light have required frequent maintenance at the toe even though the movement there is only 5 to 10 ft/yr — the slowest rate for any part of the slide.

Though the active part of the Wind Mountain landslide displays highly visible effects of movement, it is not expected to be a danger to life. Some structures have been abandoned or destroyed, but other recreational and agricultural land uses have been adopted that accommodate the movement.

Oregon Shore Landslides

A number of smaller landslide deposits along the Oregon shore of the Columbia River have received considerable attention because of their effects on the highway and railroad. The total area mapped as landslide deposits is only about 2.5 km² (0.8 mi²) — less than one-tenth of the total shown in Figure 6. These deposits occur between Bonneville Dam and the town of Cascade Locks (Fig. 1). Movement of the Ruckel landslide was a problem to railway construction as

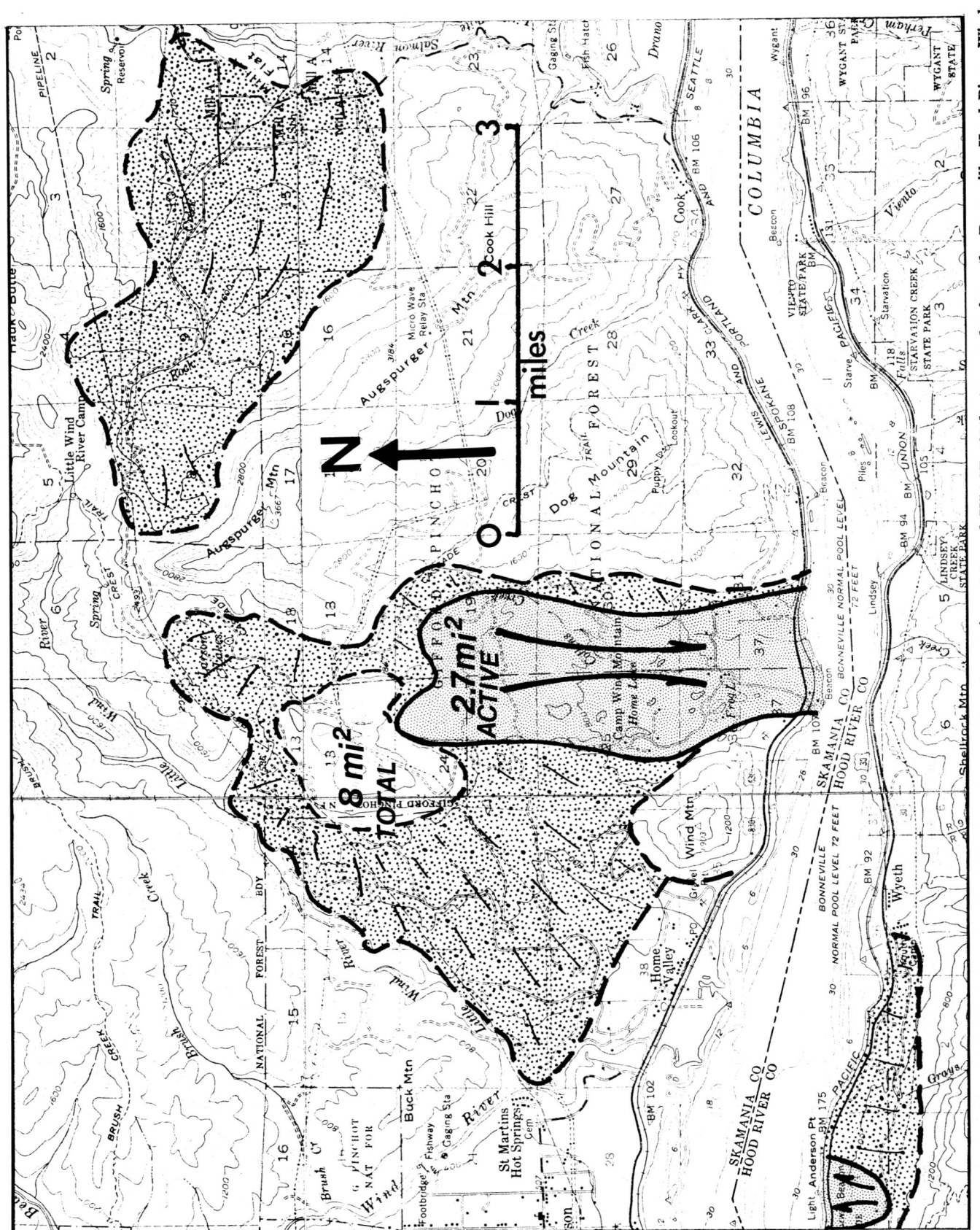

Figure 10. Map of the Wind Mountain landslide deposit area (data from Wise, 1970, and USACE, 1971b; topography from the Bonneville, Hood River, Wind River, and Willard 1:24,000 scale Geol. Survey topographic maps).

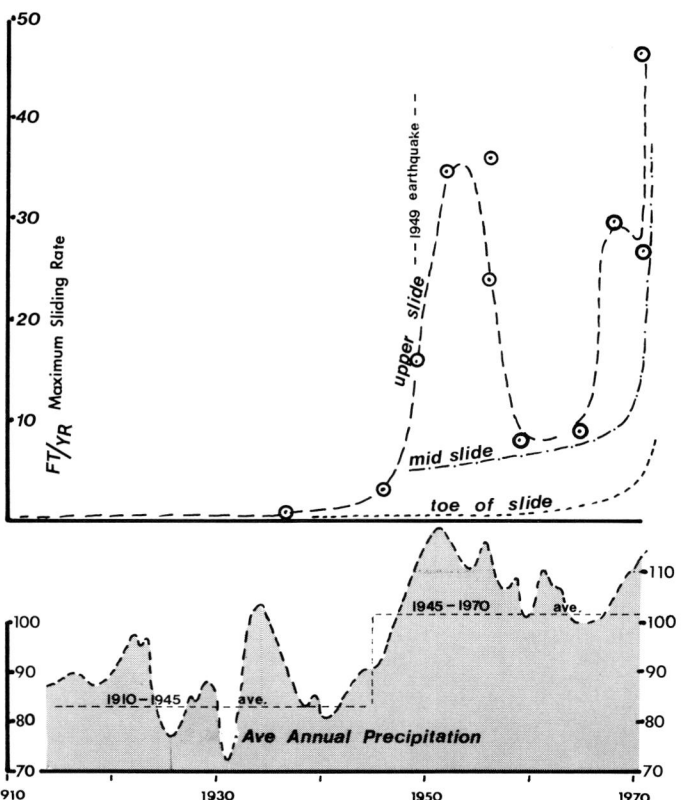

Figure 11. Rainfall and slide movement increases coincide after 1945. Upper part of the slide moves faster than the lower part (data from USACE, 1971b).

early as 1877 and intermittently continued with as much as 10 to 12 m (30 to 40 ft) of movement in 1913 (Murray, 1924). Horizontal drainage tunnels were installed beginning in 1918. These were so effective in reducing slide movement that horizontal drains have become one of the more popular landslide treatment techniques. Treatment by horizontal drains, however, has been ineffective for some small but very troublesome unstable areas in the eastern part of the Oregon shore landslides (near the Fountain landslide, Fig. 1).

At some points along the highway ... trouble with an unusual kind of slow earth movement is occurring. Over small areas, generally less than 500 feet long, the highway is heaving up in mounded masses. Some seem to be related to swelling of the toes of short landslide tongues, others are more enigmatic in form and movement. Unloading of thousands of cubic yards of loose talus from the slopes above one of them has not alleviated the problem. The saprolite, the Eagle Creek above it — which is thin in this area and absent from some parts of it — and the Yakima Basalt all dip 2° to 8° south in this area. The load of Yakima Basalt and overlying olivine basalt is great, 2,000 to 4,000 feet in most places. It seems likely that the great weight of overlying basalt is squeezing the saprolite mud up-dip toward the north, and forcing it to escape beneath the talus and landslides that spill down from the high cliffs. Here it may form a hidden wedge which is rafting the overlying talus downslope, or it may be heaving it up in broad, flat-topped pingo-like hills. Such movement would be rather similar to that of the Gatun slides in the Panama Canal which caused rise

of the floor of the canal without marked caving of the sides. [Waters, 1973, p. 151; see Fig. 15.]

A clear distinction between the act of sliding and the deposit making up a slide can help differentiate between active landsliding events and past landslide events that have stabilized but may be reactivated by grading, drainage diversion, or other construction practices. In many of the Oregon shore landslides, it would be appropriate to include excavation and grading among the landslide-activating processes. Road construction especially in the geologically youthful Columbia River Gorge has required undercutting of marginally stable deposits of talus and landslide debris. Such undercutting caused a small slide (230,000 m³ [300,000 yd³]) just east of Multnomah Falls in 1946 (Allen, 1958, p. 15).

APPLICATION OF ENGINEERING GEOLOGY

An engineering geologist studying the Columbia River Gorge is faced with providing a rational technique for evaluating natural conditions that can be applied to a proposed land use, grading plan or structural design. It would be preferable to have available some clear procedures for data collection, evaluation, and analysis, as well as for decision making. Ideally, the geologist should have good geologic comprehension and skill, occupy a position of influence in decision making, apply a uniform procedure in classification and description, and have specific criteria for decision making.

Engineering geology reports on the Columbia River Gorge provide a large sample of the practice of engineering geology in a critical area. Examples of knowledgeable and skilled geologists in a position of influence can be found, as with Hodge (1938) and C. P. Holdredge (1937, unpub. data [see App. 1]), but there are too many geologists with less influence. Furthermore, it is harder to find examples of uniform procedures in geologic classification and description, and I have not found specific criteria on which decisions have been made.

The engineering geologist is qualified to point out the presence of situations and processes that could lead to landslides (outlined by Terzaghi, 1950, Table 1) and can determine the weight, strength, and stability of rock and soil units involved and the slope stability (defined by the ratio of resisting forces to driving forces). The skills of the engineer are useful in helping determine some of these values, but to define slope stability by interpretation of geologic processes interacting through time requires the services of a capable engineering geologist.

If the engineering geology reports from the Columbia River Gorge area can be used as an example, there is a need to define the minimum data required to verify the suitability of a site for its intended use. The engineering geologist would then evaluate the site in relation to clear criteria and provide not just data, but *the data upon which a decision can be made*.

In the gorge area, past and present landsliding history provides evidence that major land instability problems will continue in the future. The engineering geology evaluation that will predict the relative stabilities of sites is not even

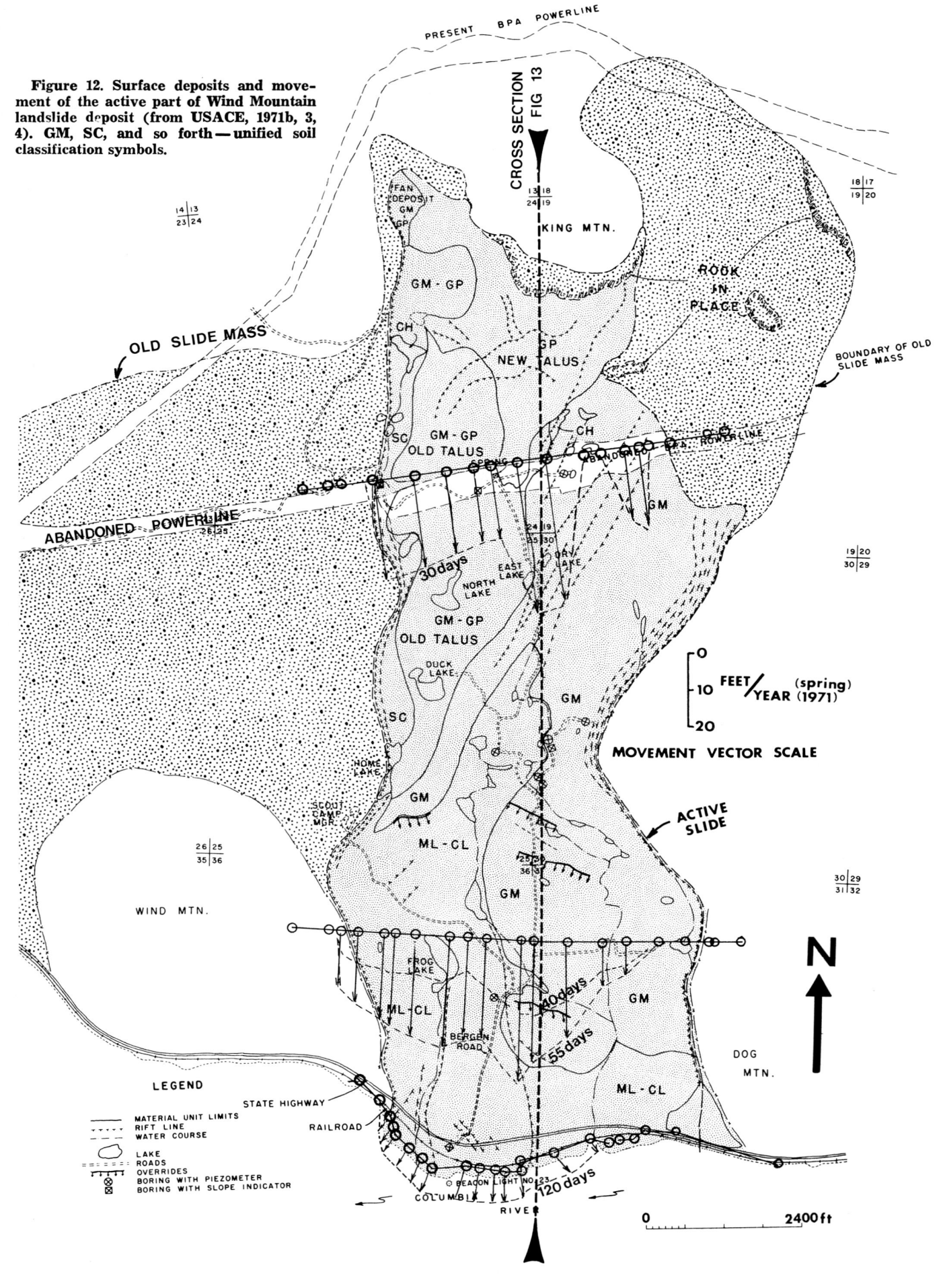

Figure 12. Surface deposits and movement of the active part of Wind Mountain landslide deposit (from USACE, 1971b, 3, 4). GM, SC, and so forth—unified soil classification symbols.

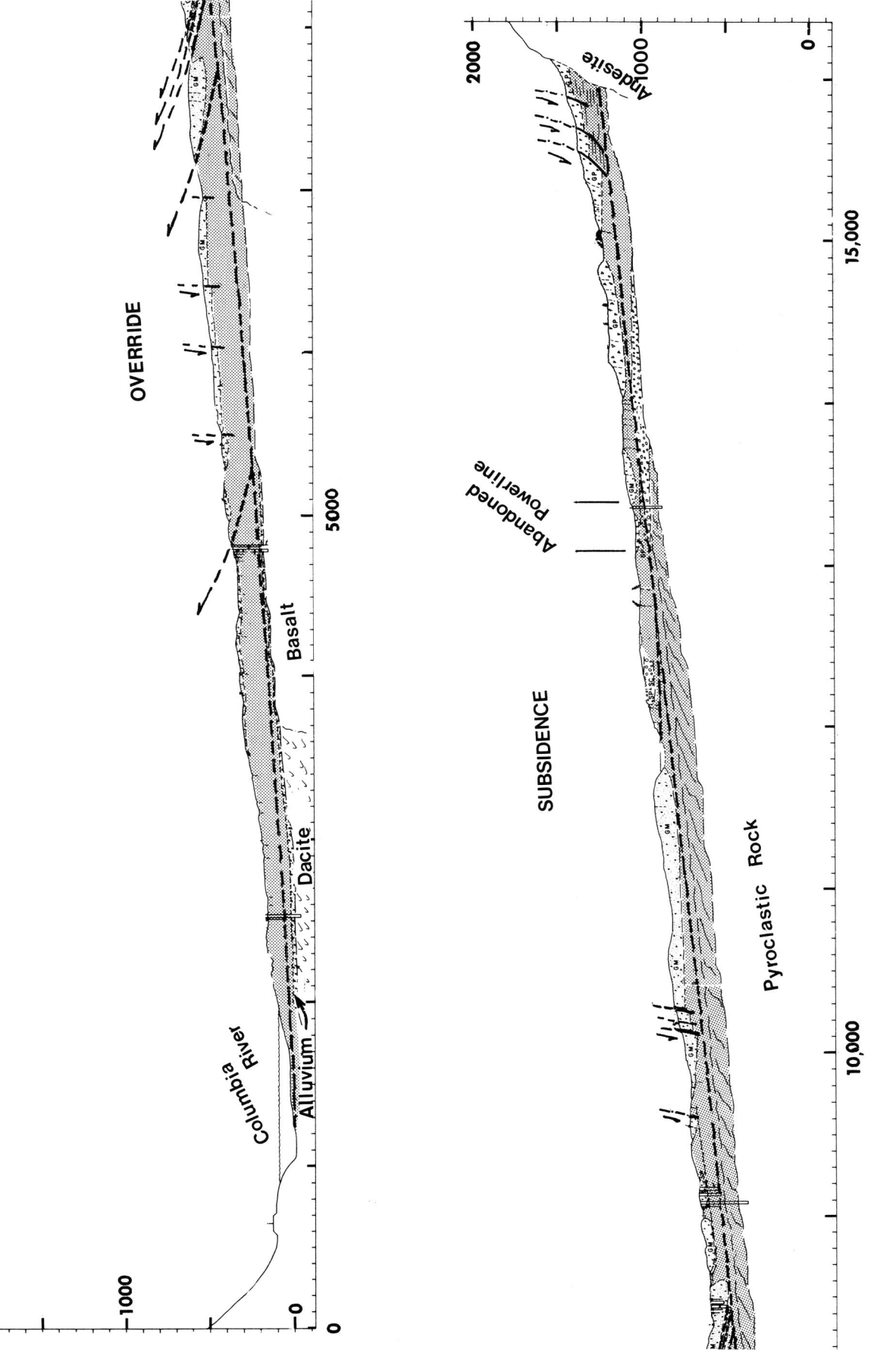

Figure 13. Cross section of the active part of Wind Mountain landslide deposit (from D. A. Williamson in USACE, 1971b, Pls. 6, 7). Horizontal distance in feet. Vertical distance in feet above mean sea level (approximate). GM, SC, and so forth — unified soil classification symbols.

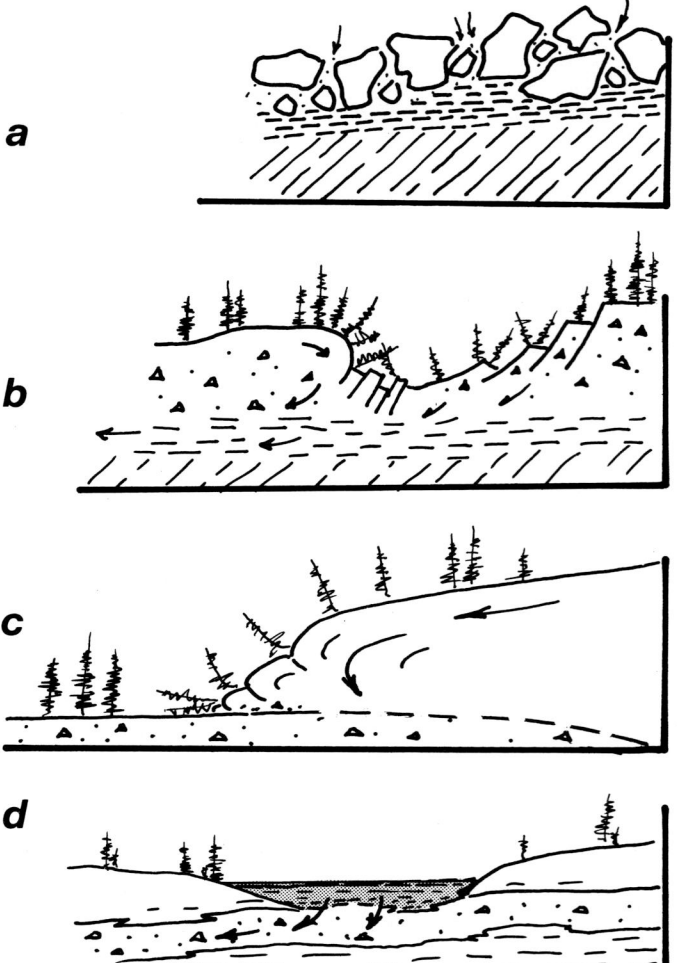

Figure 14. The active part of the Wind Mountain landslide displays a variety of drainage and deformation structures: (a) percolation through debris blocks to the clay underneath, (b) movement of clayey layer forms a subsidence depression, (c) mobile clay at the surface forms override structure, and (d) buried talus drains temporary lakes (after D. A. Williamson, 1971, unpub. data [see App. 1]).

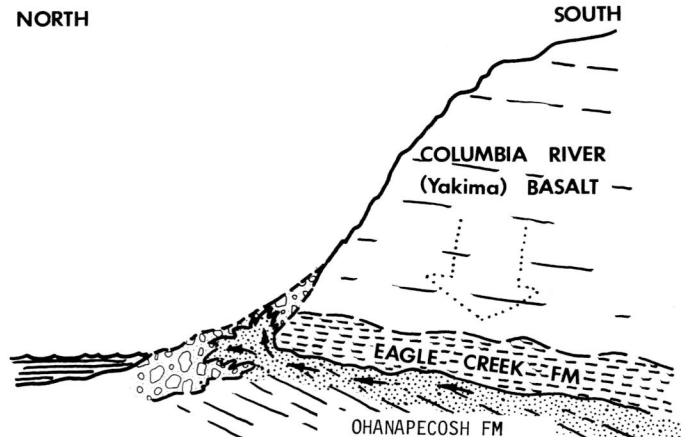

Figure 15. Diagrammatic representation of Waters's (1973) proposed "hydraulic ram" mechanism for some Oregon shore landslides.

defined, and, therefore, not yet done. Existing reports provide much of the necessary geologic data and are most useful, but the uniform procedure and specific criteria for decision making are not yet available to the extent proposed by Terzaghi (1950).

As the engineering geology profession develops procedures and criteria (with geologic comprehension and skill assumed to be present factors), influence should follow.

SUMMARY

The Columbia River Gorge contains more than 130 km² (50 mi²) of spectacular landslides of various types and in various stages of activity. Conditions leading to slope failure include oversteepening of slopes by erosion; undercutting of inclined bedding; volcanic loading of the head area; glacial erosion and debris deposition; tectonic uplift, tilting, and earthquake vibration; giant glacial floods that undercut slopes; high rainfall (250 cm/yr [100 in./yr]); hydrothermal alteration of underlying bedrock to montmorillonite-rich clay; heavy masses of permeable rock over weak impermeable units; and others.

Some areas are actively sliding, some are stabilized deposits after former landsliding, some are in the process of stabilizing by minor adjustments, and some areas are probably about to fail where no landslide now exists.

The known mechanisms and processes of landslide movement in the gorge vary in time and place. Slow (measured in centimetres per year) plastic deformation of water-saturated clayey layers in the Eagle creek and Ohanapecosh Formations allows gravitational movement of large masses into the gorge. Pressure of overlying rock masses may cause hydraulic rams — extrusion and even in some places upward motion of the clayey layers. Gravitational collapse of cliffs and landslide head scarps causes high-velocity (measured in metres per second) rockfalls. In the Bonneville slide, a large rapid failure or rockfall instigated liquefaction of river sediments over which the toe of the slide was able to move with considerable speed onto the opposite valley wall.

The style of sliding also varies from mass failures of variable velocity that displace whole slides as a unit to progressive failures that occur by increments of slope failure beginning at the toe and progressing headward. The Wind Mountain landslide, however, shows more accelerated movement at the top that is causing the upper part to override the slower-moving toe.

CONCLUSIONS

The Columbia River Gorge provides particularly varied and spectacular examples of landslides. The methods used to describe them by engineers and geologists are also varied.

Progress in understanding of the geologic conditions of the area has been made, but it is seldom possible to utilize the results of geologic studies in a way that can verify the slope stability (the ratio of resisting forces to driving forces). It would be desirable to have procedures and criteria by which planning, design, and construction decisions could be made. This could advance the practice of engineering geology in the gorge from authoritarian to authoritative.

ACKNOWLEDGMENTS

In this compilation data were derived primarily from the reports of others and from field trips I have taken with students. I am particularly thankful for the information and advice provided by John Eliot Allen and Douglas A. Williamson, both of whom have worn their boots thin in the gorge and greatly improved our comprehension of its geology. Phil Grubaugh of the Corps of Engineers has been most kind in helping me find the considerable amount of data developed by the agency.

APPENDIX 1. UNPUBLISHED REPORTS FOR COLUMBIA RIVER GORGE AREA

Birch, D. C., 1955, The Wind River slide: Bonneville Power Adm., 6 p.

Bush and Bartholomew, W. S. [untitled, Farley slide study]: Oregon State Highway Dept.

Deacon, Robert J. [undated], Preliminary report, ranger station slide: Oregon State Highway Div.

Elmer, W. W., and Torpen [date unknown], Probable seepage around north abutment and into cofferdam area, Bonneville Dam: U.S. Army Corps Engineers.

Ewen, Irv [about 1958], Girl Scout Camp landslide: Portland State Univ.

Gano, D., 1969, Progress report, Fountain slide, Columbia River Highway, Hood River County: Oregon State Highway Dept.

———1971, Fountain slide section, progress report, Columbia River Highway, Hood River County: Oregon State Highway Dept.

Hodge, E. T., 1932, Report of dam sites on the lower Columbia River: U.S. Army Corps Engineers [Portland District library].

———1934, Ruckel slide stability report: U.S. Army Corps Engineers, Portland District.

Holdredge, C. P. [date unknown], History of geological exploration, Booneville Dam project: U.S. Army Corps Engineers.

———1937, Final geological report on the Bonneville project: U.S. Army Corps Engineers, 39 p.

Meyers, J. D., 1953, A report on the geology and possible correction of landslide along the Columbia River highway between Cascade Locks and Wyeth: Oregon State Highway Div.

———1961 [untitled, Farley slide study]: Oregon State Highway Dept.

Murray, Samuel, 1971, Lake Bonneville-Ruckel slide tunnels, tunnel history and present conditions, *in* A review of Union Pacific Railroad maintenance and inspection records: U.S. Army Corps Engineers.

Scott and Deacon, R. J., 1951 [untitled, Farley slide study]: Oregon State Highway Dept.

Williams, Ira A., 1932, Preliminary geologic report on prospective dam sites, lower Columbia River: U.S. Army Corps Engineers.

———1937, Bonneville dam project, history of exploration: U.S. Army Corps Engineers.

Williamson, D. A., 1971, Wind Mountain landslide: Wind Mountain Girl Scout Camp, August, 21 p.

REFERENCES CITED

Allen, J. E., 1958, Columbia River Gorge, Portland to The Dalles, *in* Guidebook for field trip excursions, Cordilleran Section, Geol. Soc. America, University of Oregon, Eugene, Oregon, March 27, 1958: p. 4–23.

Baker, V. R., 1973, Paleohydrology and sedimentology of Lake Missoula flooding in eastern Washington: Geol. Soc. America Spec. Paper 144, 79 p.

Barnes, F. F., and Butler, J. W., 1930, The structure and stratigraphy of the Columbia River Gorge and Cascade Mountains in the vicinity of Mount Hood [M.S. thesis]: Corvallis, Oregon State Coll., 136 p.

Bretz, J., 1925, The Spokane flood beyond the channel scablands: Jour. Geology, v. 33, p. 97–115.

Hodge, E. T., 1938, Geology of the lower Columbia River: Geol. Soc. America Bull., v. 49, p. 831–930.

Huntting, H. T., Bennett, W.A.G., Livingston, V. E. Jr., and Moen, W. S., 1961, Geologic map of Washington, 1:500,000: Washington Dept. Conservation, Div. Mines and Geology.

Lawrence, D. B., and Lawrence, G., 1958, Bridge of the Gods legend, its origin, history and dating: Mazama, v. 40, no. 13, p. 33–41.

Murray, S., 1924, Drainage tunnels overcome great land slide: Railway Age, 29 Nov., p. 975–978 (duplicated in U.S. Army Corps of Engineers, 1971a; see below).

Terzaghi, C., 1950, Mechanics of landslides, *in* Paige, S., chairman, Application of geology to engineering practice [Berkey volume]: New York, Geol. Soc. America, p. 83–123.

U.S. Army Corps of Engineers, 1971a, Oregon shore slide study: U.S. Army Corps Engineers, Portland District, Design Memo. no. 1, Supplement no. 6, July, 19 p.

U.S. Army Corps of Engineers, 1971b, Collins Point, Washington, slide study: U.S. Army Corps Engineers, Portland District, Design Memo. no. 1, Supplement no. 7, Aug., 28 p.

U.S. Army Corps of Engineers, 1975, 2nd powerhouse cofferdam and seepage control: U.S. Army Corps Engineers, Portland District, Design Memo. no. 15, Jan., 41 p.

U.S. Army Corps of Engineers, 1976, 2nd powerhouse railroad tunnel: U.S. Army Corps Engineers, Portland District, Design Memo. no. 7, Supplement no. 4, May, 24 p.

Waters, A. C., 1955, Volcanic rocks and the tectonic cycle, *in* Poldervaart, A., Crust of the Earth: Geol. Soc. America Spec. Paper 62, p. 703–722.

———1973, The Columbia River Gorge: basalt stratigraphy, ancient lava dams and landslide dams, *in* Beaulieu, J. D., Geologic field trips in northern Oregon and southern Washington: Oregon Dept. Geology and Mineral Industries Bull. 77, p. 133–162.

Wise, W. S., 1961, The geology of the Wind River area, Washington, and the stability relations of celadonite [Ph.D. dissert.]: Johns Hopkins Univ., 258 p.

———1970, Cenozoic vulcanism in the Cascade mountains of southern Washington: Washington State Dept. Nat. Resources Bull. 60, 45 p.

MANUSCRIPT RECEIVED BY THE SOCIETY SEPTEMBER 7, 1976
MANUSCRIPT ACCEPTED SEPTEMBER 17, 1976

Printed in U.S.A.

5

Regional slope-stability controls and engineering geology of the Fraser Canyon, British Columbia

DOUGLAS R. PITEAU
D. R. Piteau and Associates Limited, Kapilano 100, Suite 708, 100 Park Royal South, West Vancouver, British Columbia V7T 1A2 Canada

ABSTRACT

A study of the 68-mi (109-km) section of the Fraser Canyon between Lytton and Hope, British Columbia, along the Canadian National Railway (CNR) was made to determine what factors controlled slope stability on a regional scale. Engineering geology aspects concerning regional faulting and related minor structure, lithology, drainage and hydrology, geomorphology, climate, river geometry, and effects of man were considered. The most significant cause of slope instability was deflecting of the Fraser River into its bank by the presence of either an alluvial fan at a tributary mouth or a river bend. This deflection allowed extensive lateral erosion and resulted in severe oversteepening, which undermined the toe of the slope. Rockfalls, rock and debris slides, and washouts have been recorded for more than 20 years by the CNR. These data indicate that about 66% of all such incidents occurred opposite alluvial fans or on outside curves of the river. The average numbers of incidents per mile occurring opposite alluvial fans and on the outside of river bends are 5.6 and 3.3 times greater, respectively, than the average number of incidents recorded for river stretches without these characteristics. Regional faulting, climatic conditions, and effects of man were also found to be important causes of slope instability.

INTRODUCTION

Alluvial fans and river bends proved to be significant factors in slope instability along a 68-mi section of the Canadian National Railway (CNR) in the Fraser Canyon between Lytton and Hope, British Columbia. Broad terrain-evaluation methods were employed to determine the relative significance of factors that influence the slope stability. Slope stability was evaluated quantitatively and comparatively on a regional basis. This paper describes the results of comparative analysis between locations where the river flow was deflected against the bank and locations of both recorded incidents of rockfalls, rock and debris slides, and washouts and large interglacial or postglacial (prehistoric) landslides. Also, fundamental considerations and an approach to regional slope-stability studies, as well as other factors involving the entire study — such as regional faulting, lithology, climate, and effects of man — are discussed.

FUNDAMENTAL CONSIDERATIONS IN REGIONAL SLOPE-STABILITY EVALUATION

Methods of assessing the stability of a local slope do not apply to slope-stability studies on a regional scale. To evaluate the stability of slopes on a regional basis, methods of quantitative and qualitative comparative analysis are used. A broad terrain analysis, therefore, deals with gross strength and degree of stability of entire rock sequences or formations, rather than the potential behavior of a single rock type, unless a certain type is widespread.

A broad overview lends itself to considering relative stability because only the gross slope-stability pattern is derived. The absolute stability conditions, order of priority, and any detailed slope remedial measures are considered only after the overview is complete. The problem resolves itself into one where engineering and geology complement one another, combining knowledge of precedent with the arts of estimation and judgment.

The collective effect of rock properties in the study is considered in terms of the average gross strength of the rock or as the ability of the rock to resist degradational processes. The regional and local effects of geologic structure and a comparison of the behavior of rock types present are integral parts of the analysis. Effects of man, such as railway and highway cuts, fires, logging activities, and reservoirs, are more difficult to assess; however, they will almost certainly change the existing environment.

In regional studies, the most significant evidence is prehistoric and recent rockfalls. From a regional perspective, the influence of local variations in rock strength and rock-mass behavior appears to average out. Hence, from identification and analysis of a large number of landslides in a particular area, a better appreciation of the regional controlling factors is achieved (Cleveland, 1971).

GENERAL APPROACH

Investigation Methods

Field work consisted of reconnaissance inspection of all slopes between Lytton and Hope. Both large- (1 in. = 1,000 ft) and small-scale (1 in. = 0.5 mi) photographs were required to delineate gross details. To determine where the major regional faults were located, the fault traces were identified on small-scale photos; then this data was transferred to the large-scale photos. Because of their magnitude, recognition of most major postglacial landslides was possible on the small-scale photos only.

Basic Aspects Considered in the Analysis

A brief description of the various factors and related aspects that were considered follows.

Major (Regional) Faults. Susceptibility of fault zones to weathering and deterioration; proximity of these features to the slopes; broad effects of faulting on topography, geomorphology, and slope and river geometry.

Minor (Local) Geologic Structure. Structural features induced by regional faults; local fault patterns; foliation (schistose and gneissic), joints, bedding, cleavage, sheet joints, and exfoliation features.

Lithology, Soil, and Related Weathering Characteristics. Slaking and spalling action and other forms of mechanical deterioration; removal of binding cement by dissolution processes; areal distribution of rock type, rock-sequence relationships, and vegetation cover.

Surface and Subsurface Water. Anomalous drainage patterns and seepage zones; correlation with freeze-thaw cycles; anomalous ground-water characteristics; linear shape of tributary streams, number of tributary streams per unit length of canyon.

Aggradational and Degradational Processes and Their Resultant Landforms. Alluvial fans, alluvial benches, and river bars; high-level alluvial-infilled river channels and related features; effects of relocation of the main river channel.

Valley Oversteepening and Deepening by Glacial Scouring. Variations in the degree of glacial scouring across and along the valley; slope-face directions that appear to be more susceptible to failure because of glacial action.

Climate. Precipitation and snowfall distribution; elevation; wetting and drying action; frost action and ice-jacking; subaerial weathering; thermal expansion; freeze-thaw cycles; type and extent of vegetation cover.

River Erosion. Variations of both lateral and vertical erosion by the river; effects of river bends on the degree of toe unloading and oversteepening of slopes in bedrock-controlled reaches; stream gradient.

River Configurations. Anomalous constrictions and widening of river bank; river fills, depressions, and other phenomena related to the river-channel geometry.

Mass Movement. Size, Type, and location of interglacial and postglacial rock and soil slides; demarcation of debris slides, snow avalanches, and any other mass-movement phenomenon.

Empirical Information. Date, location, type, and size of rock, soil, and snow slides, rockfalls, and washouts.

Effects of Man. Overloading of undisturbed soil, slide debris, and talus slopes by material cast aside from excavations or by roadfill material; oversteepening of cuts in unstable rock or soil; blasting damage in rock cuts; restriction of ground-water flow and other natural processes; dramatic changes in seepage conditions because of cuts and fills in slope; logging activities and fires.

PHYSIOGRAPHIC AND GEOLOGIC SETTING OF THE FRASER CANYON

The Fraser Canyon, which for purposes of definition and convenience is the area between Lytton and Hope, lies between the Coast and Cascade Mountains systems. The southern part of the canyon is typical of the Cascade system, with an axial core of folded, faulted, and slightly metamorphosed sedimentary and volcanic rocks (Monger, 1969). Toward the north the axial core broadens, and the Cascade system merges with the essentially granitic and high-grade metamorphic rocks of the Coast Mountains. The basic regional geology concerning the boundary between the Coast and Cascade Mountains is described by Cairnes (1944), Camsell (1920), Monger (1969), and Duffell and McTaggart (1952).

The northern 68 mi (109 km) of the canyon are discussed in this paper. The average gradient of the river in this section is about 5.5 ft/mi (1 m/km). The canyon exhibits features produced by destructive forces of erosion. This destructive erosional activity is particularly apparent in the section from North Bend to Yale.

The three physiographically distinct sections of the Fraser Canyon (described below) provide a natural subdivision for comparative analyses of the engineering geology data.

Lytton to Boston Bar. This 28-mi (45-km) stretch is characterized by a relatively broad valley with heavily timbered slopes, gravel terraces at the toes of the slopes, and few outcrops. Bands of relatively soft rocks (for example, at Jackass Mountain) and, in places, schists have in large part controlled the geomorphic expression of the valley. The river gradient is about 4 ft/mi (0.75 m/km).

Boston Bar to Yale. This 27-mi (43.5-km) stretch, which in part approximates a gorge, is the most constricted part of the entire canyon area. The river gradient approaches 8.0 ft/mi (1.5 m/km). The canyon walls are abnormally steep, and rock benches (instead of alluvial terraces) are common.

Yale to Hope. Except for the nature of the bedrock, the southernmost 13 mi (21 km) are somewhat similar to the area between Lytton and Boston Bar in that the valley quickly widens. Below Hope the mountains eventually merge into the deltaic region. The gradient of the river from Yale to

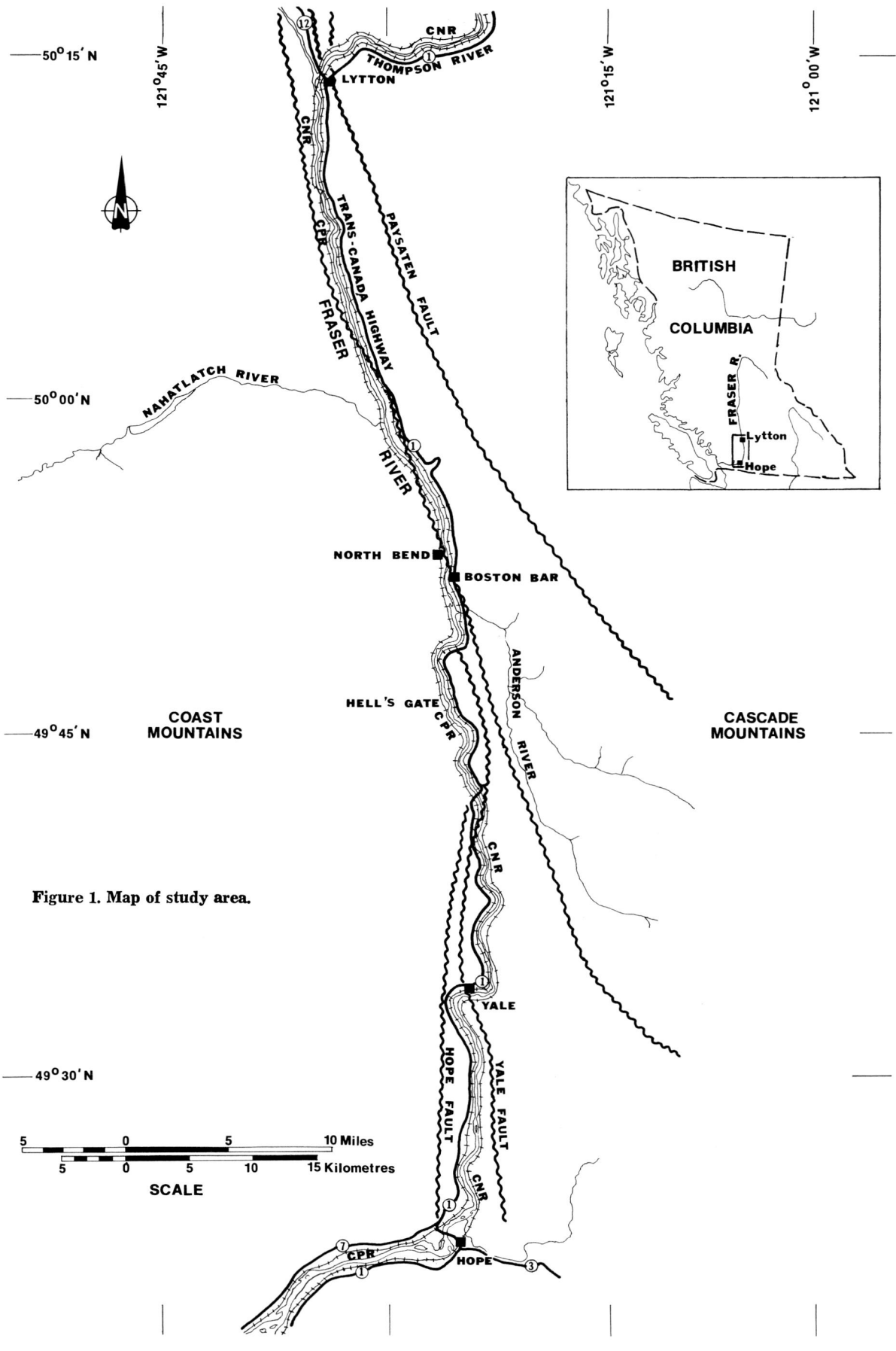

Figure 1. Map of study area.

Figure 2. Major postglacial slide immediately upstream of the town of Yale (1.5 mi wide × 1.5 mi from toe to crest). Note the well-defined linear furrows formed along the trace length of both the Yale fault and Hope fault. Also, note the large alluvial fan upon which the town of Yale is developed and the steep slopes opposite the fan. Explanation of symbols (on facing page) also applies to Figures 3 through 11.

Hope is only about 3.5 ft/mi (0.65 m/km). In this area the granitic rocks are somewhat friable and intensely sheared.

SIGNIFICANT GEOLOGIC ASPECTS

Regional Features

The Fraser Canyon is the site of a pronounced zone of faulting which trends roughly parallel to the river. The location of the faults in the area between Hope and Lytton is shown in Figure 1. The fault zones consist of severely broken rock. These broad zones of weak material have been the site of differential weathering, excessive glacial scouring, and river erosion. Both interglacial and postglacial erosion by the river has been particularly outstanding opposite alluvial fans and on the outside of river bends.

Slow regional uplift during Pliocene time and postglacial rebound appear to have accentuated the downcutting and

SYMBOLS

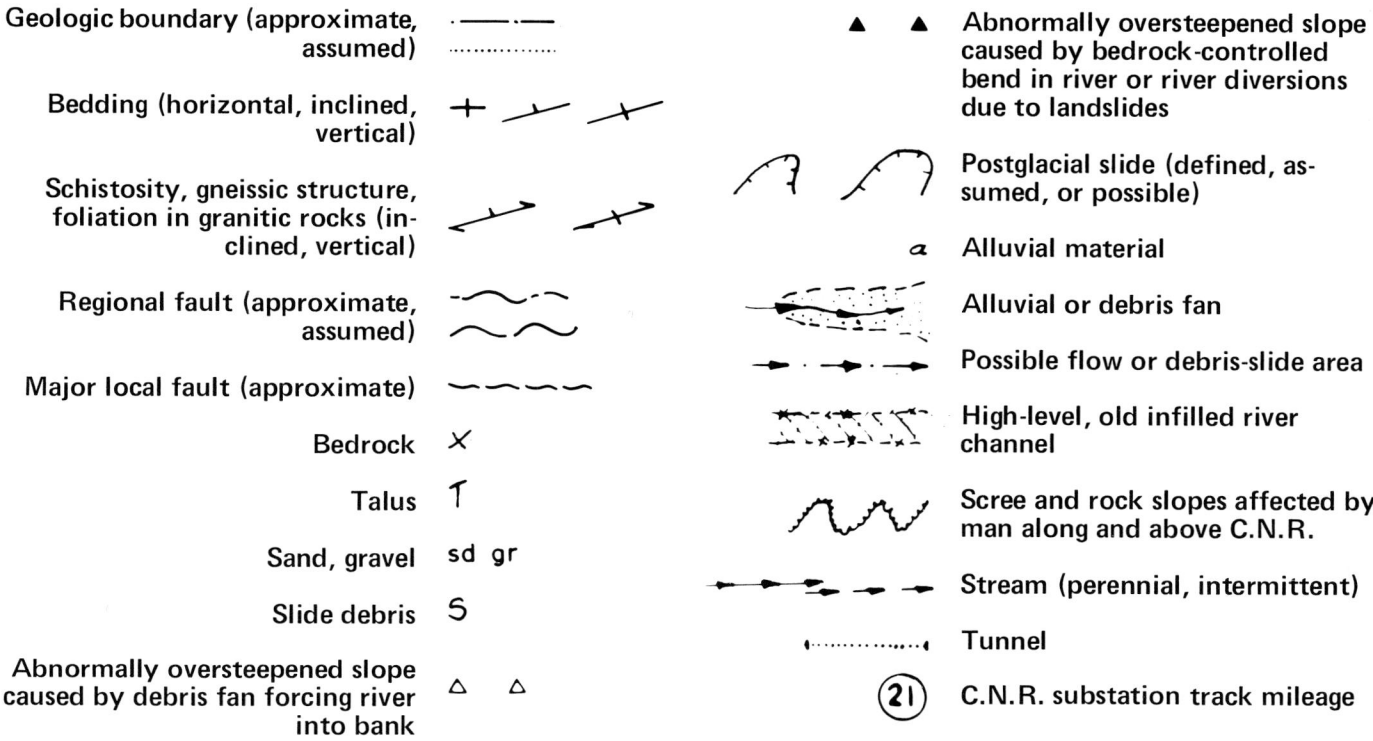

lateral erosion by the river. For example, an incised notch has developed on the floor of the present Fraser channel since glacial times. A well-defined topographic break consisting of spurs or rock benches about 1,200 to 1,500 ft (365 to 457 m) above the canyon floor in the Fraser and its tributary valleys is thought to represent the old pre-Pliocene valley floor. This indicates the amount of erosion since Pliocene time (Camsell, 1920).

In general, erosion appears to have taken place considerably faster than the canyon walls could adjust to the changing conditions. Evidence of multiple glaciations from the lower Fraser Valley is described by Armstrong and others (1965). The Fraser Canyon is typical of a glacially oversteepened valley. The result is the presence of numerous slopes of varying degrees of instability which are, geologically speaking, in an active state of failure.

With the exception of the outstanding influence of faulting and the lag between oversteepening and the natural tendency of the slopes to regain stability, the canyon in general has some characteristics of a typical glacial valley including an uneven longitudinal profile, steps, and steep valley sides. Tributary valleys, however, have all the characteristics normally ascribed to glacially carved valleys, such as U-shaped profiles, faceted spurs, and large radius of curvature on all bends. The glacially carved floors of the upper reaches of all these valleys are hanging, which indicates that downward erosion in the tributaries failed to keep pace with that in the main valley. At an archaeological site near river level near Yale, radiocarbon dating of materials indicates that the river had cut the canyon to its present depth 9,000 yr ago (Borden, 1965).

Anomalous bedrock constrictions, alluvial fan–controlled constrictions, and raised valley bottoms are probably related to large postglacial slides and (or) river aggradation in certain localities. Some of the isolated rock islands may have formed as a result of deflection of the river by landslides or may be massive remnants of landslide debris. Sudden widening (a "bulb") in the river is also a common feature and in many instances indicates the location of major postglacial or earlier landslides. These features and other evidence indicate that major landslides from the lower walls of the canyon have been a most significant geomorphic process in this area.

Steep scarplike rock faces form the dominant topographic features on the upper walls of the canyon. Scarp faces, some of which may be fault planes as well, are particularly predominant in the headwall areas of major postglacial slides. A number of these shear rock walls appear to have formed by breaking along the major regional north-trending faults that occur in the area. Also broad scree or talus slopes are formed in these areas because of the constant spalling of material from the fault-controlled scarp faces. In some instances, faults are the sites of long narrow slide chutes or gullies. As a result of differential weathering, the surface traces of most regional faults in the uplands appear as well-

Figure 3. Aerial photo mosaic showing an isolated, high-level, infilled river channel of the Fraser which appears to have been blocked off

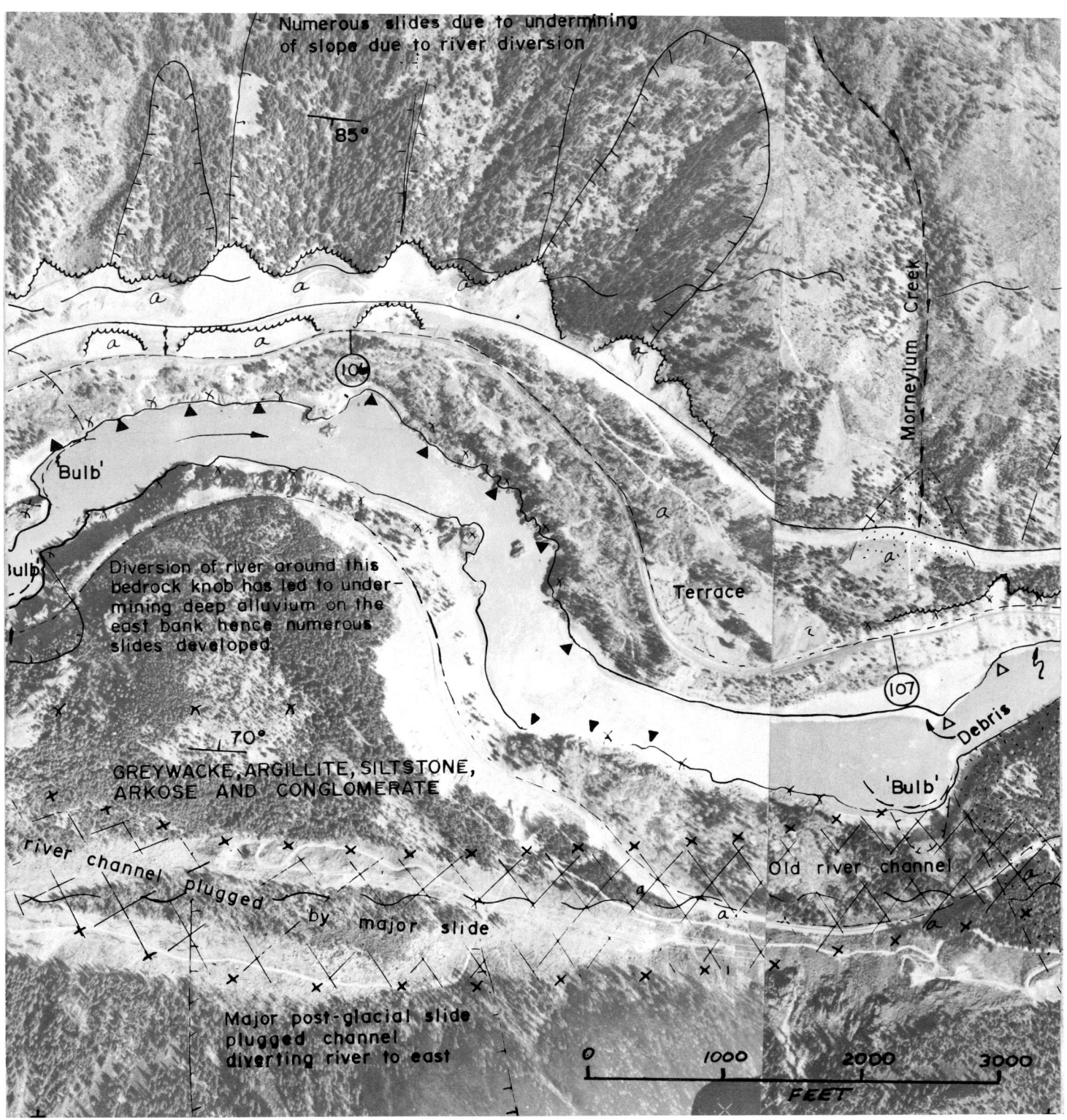

by a major slide. Note also the large failures that have developed where the river has been deflected around the large rock promontory.

defined linear furrows or troughs, which in places contain sag ponds (Fig. 2).

In a few localities, massive postglacial or interglacial slides appear to have completely blocked off the original Fraser River channel with the result that a major diversion in the course of the river has taken place. A good example of this is found about 6 mi (9.7 km) south of Lytton (Figs. 3, 4). In some instances, major landslides have, at least temporarily, forced the river against the opposite bank. The subsequent severe lateral erosion by the river, although probably of a relatively temporary nature, has caused significant oversteepening of the rock slopes opposite the slide areas.

Fraser River Fault Zone

Over most of the canyon length, the Fraser River follows the surface traces of two major, subparallel, steeply dipping faults — the Hope fault on the west and the Yale fault on the east (Read, 1960) (see Fig. 1). These two major faults, many other major cross faults that diverge into the canyon farther north, and their associated faults have been referred to together as the Fraser River fault zone. In some locations these two major faults seem to bound a graben. This fault zone has an overall trace length of 150 mi (240 km), extending northward from about lat 47°N in Washington State into British Columbia (McTaggart and Thompson, 1967).

The Fraser River fault zone — and, accordingly, a good part of the river — forms the boundary between the Cascade and Coast Mountains. A series of structural adjustments have taken place between the massive granitic intrusions to the west and the belt of interbedded volcanic and sedimentary rocks with related smaller granitic intrusions to the east. Serpentinized peridodite has been emplaced along or near the faults in many places.

Right-lateral movement is suggested by formation offsets but did not continue after intrusion of Cretaceous diorite rocks. The attitude of Eocene rocks along the fault zone suggests that they may have formed in the graben. South of Hope, Oligocene-Miocene plutons intrude the zone but are not themselves faulted. This indicates that, at least south of Hope, there has been no post-Miocene movement along the fault zone (Mathews, 1972). North of Hope, however,

Figure 4. Oblique aerial photo looking to the west showing the large landslide and old, high-level, infilled river channel which appears to have been blocked by a landslide from outlined area (see also Fig. 3).

Eocene beds are involved in the faulting (Duffell and McTaggart, 1952).

Rocks occurring within the range of influence of the Fraser River fault zone have been crushed, shattered, or intensely faulted and jointed. There is a definite change of trend of associated faults relative to their proximity to the Fraser River fault zone (Monger, 1969): they trend north near the fault zone, but farther away they trend northeast.

Details by Physiographic Area

From Lytton to Boston Bar the river occupies a somewhat broad flaring valley with a more typical glaciated appearance than in the gorge area south of Boston Bar. The change from the more open valley to the narrow rock canyon occurs at the bend of the river south of Boston Bar and is probably the result of a different developmental history for the two areas. However, varying resistence of the rocks to erosion appears to be significant as well. North of Boston Bar the river flows through a belt of easily eroded sedimentary and low-grade metamorphic rocks, whereas south of Boston Bar to Hope the canyon is cut mainly in resistant plutonic and moderately high grade metamorphic rocks.

For convenience in describing details in the canyon, specific mileages (shown by circled numbers on the aerial photos) are referred to along the CNR track. Lytton occurs at approximately mile 98 of the Ashcroft Subdivision; the Yale Subdivision starts at Boston Bar (that is, zero); the town of Yale is at about mile 26.

Lytton to Boston Bar. The trend of the canyon north of Boston Bar approximates the strike of the Cretaceous rocks. For most of the distance between Lytton and Boston Bar, the canyon has been cut in Cretaceous sedimentary rocks or along the contact between Cretaceous and older rocks. In this stretch the river also essentially follows the Fraser River fault zone.

Gravel terraces are conspicuous on the lower slopes except where lateral erosion by the river has caused the complete removal of alluvium and thus exposed bedrock. The Anderson and Nahatlatch Rivers merge with the Fraser approximately at grade, but through narrow rock notches several hundred feet deep. The upper reaches of these two rivers, as well as some of the creeks, occupy hanging valleys.

The confluence areas of the Fraser and most of the well-developed creeks and rivers are sites of very large alluvial fans, whose presence has forced the Fraser against the bank opposite them (see Fig. 5). Alluvial fans are well developed at the mouths of Nikaia (mile 99; Fig. 5), Kamiak (mile 101.3; Fig. 5), and Siska (mile 104.3) Creeks, a creek opposite the Jackass Mountain area (mile 109; Fig. 6), and Mowkokam (mile 112.8; Fig. 6), Ainslie (mile 120.1), Neopopulchin (mile 120.6), and Stoyama Creeks. In every case the alluvial fans have resulted in noticeable river deflection and significant lateral scouring.

Along Sawmill Creek (mile 100.2; Fig. 5) a large debris-slide fan deflects the river, which is now scouring the opposite bank. A very large debris slide or alluvial fan occurs at the mouth of Kwoiek Creek between miles 107 and 108 (see Fig. 4); here the Fraser River has been deflected 800 to 1,000 ft (240 to 306 m) toward the east bank, which has been significantly oversteepened.

A large rockslide on the west bank has blocked off the original Fraser channel between miles 105 and 107 (see Figs. 3, 4) with the result that a new channel has formed nearly 0.5 mi (0.8 km) to the east of the now-isolated higher-level infilled postglacial channel. This diversion caused considerable oversteepening of the east canyon wall; consequently, numerous postglacial slides have resulted. The Boston Bar airfield appears also to be on the site of a former Fraser River channel, which also may have been abandoned when the river was blocked by a landslide (see Fig. 7).

Nearly the entire, relatively straight 5-mi (8-km) section of the river north of Jackass Mountain appears to have developed along the trace of a regional fault zone (see Fig. 6 for extreme southern part of this section of the river).

Boston Bar to Yale. From Boston Bar to Yale the distance between the immediate canyon walls on opposite sides of the river is only about half as much as to the north and south of this reach. The gorge is essentially steeply U-shaped. The bottom of the gorge in turn has been notched to a depth of about 100 ft (30 m) by stream action with the result that a small V-shaped cut exists in the base of the glacial-scoured surface. Thus, the gorge in effect consists of a succession of short, narrow, rectangular box-shaped constrictions within a larger canyon and within which the water rushes with tremendous velocity. In contrast to the densely forested areas upstream of Boston Bar, the slopes in this area are generally steep, sparsely forested, and extremely irregular. Steep scarp faces, bare rock buttresses, and spurs are common features, as are numerous sharp bends and curves.

Lateral erosion by the river is as significant as downcutting in this area. Lateral erosion occurs where well-developed alluvial fans and sharp bends in the channel deflect the river flow into the bank on the outside of the curve. This has resulted in toe unloading and oversteepening of the outer-curve slopes. Lateral erosion is probably increased between Boston Bar and Yale because of the confined and narrow characteristics of this part of the canyon.

The upper reaches of the main tributaries — Skuzzy (mile 5.0) and Spuzzum Creeks (mile 16.4; Fig. 8) on the west side and Siwash Creek (mile 23.4) on the east side of the river — occupy pronounced hanging valleys whose floors are several hundred feet above the Fraser. These creeks flow in deeply incised bedrock notches in their lower reaches.

In this part of the Fraser River, with the exception of Anderson and Siwash Creeks, there is a conspicuous absence of alluvial fans along the east bank although there are several on the west side. A very large alluvial fan extending between miles 16 and 17 at the mouth of Spuzzum Creek appears to have caused a major deflection of the river (Fig. 8). The same can be said for the alluvial fan underlying the town of Yale (Fig. 2). This fan extends for about 0.75 mi (1.25 km) along the west bank at about mile 26.

The 3-mi (4.8-km) section of canyon in the Hell's Gate area between miles 6 and 9 is the site of at least six extremely large postglacial slides. All these slides could have caused a major, although probably relatively temporary, blockage of the Fraser River (Figs. 9, 10). The slopes between miles 23

Figure 5. Photo mosaic in the Lytton area showing large alluvial fans deflecting

and 24 also contain several massive postglacial slide scars.

A particularly large slide or series of slides occurred between miles 24 and 26 on the east bank immediately upstream of the town of Yale. This landslide area is so large that it is only recognizable on the small-scale aerial photo (Fig. 2). The slide area is about 1.5 mi (2.1 km) wide and 1.5 mi (2.1 km) from head to toe. The very steep slope on the west side of the river opposite the slide area may be attributable to this slide's forcing the river against the opposite bank. On the resulting oversteepened slope, several large slides are also evident.

Yale to Hope. From Yale to Hope the river gradient decreases considerably. Although the flow continues to be steady and strong, there are only a few moderately developed rapids. The river once again occupies a broad, flaring valley with densely forested slopes.

The rocks, being essentially granitic, are lithologically quite similar to the rocks upstream. However, closer examination reveals that these rocks have been highly sheared by large north-striking regional faults. These granitic rocks are whitish, highly shattered, and (or) jointed. In many localities they are somewhat crumbly and friable owing to bleaching and alteration. McTaggart and Thompson (1967) described these rocks in detail. Their nonresistant nature could explain why the valley south of Yale immediately opens and becomes relatively broad. These rocks were more easily scoured and excavated by glaciers than the steep slope-forming material immediately upstream.

An isolated, infilled former river channel containing Klahater Lake extends about 3 mi (4.2 km) on the west bank of the Fraser River between Stukawitis and American Creeks. This same channel swings across the river opposite American Creek and goes for about 2 mi (3.2 km) on the east bank (Fig. 11).

the river into the opposite bank. Note also the debris slide at about mile 100.

REGIONAL SLOPE-STABILITY FACTORS

Lithology

In most of the steepest parts of the canyon, such as between Boston Bar and Yale, the rocks are generally tough and of relatively high compressive strength and high shear strength. The granitic rocks south of Yale, although highly sheared, altered, and somewhat friable, are of little significance to the analysis since only a small percentage of that 13-mi length of track is exposed to bedrock slopes.

The relatively weak sedimentary rocks, particularly in the Jackass Mountain area as well as several other localities between Lytton and Boston Bar, have led to considerable slope problems. However, these are of only local significance because much of the railway line in this area occurs in alluvial-terrace material. Also, slopes in the lower reaches of the river are more gentle, as explained earlier, due to their greater susceptibility to erosion.

Geologic Structure

Effects of faults, bedding, joints, schistosity, gneissic structure, and contacts, although resulting in local slope instability either individually or in combination, cannot be extrapolated for significant distances along the canyon. Even though the weaker rocks and consequent flaring of the valley south of Yale can be attributed partly to regional faulting, slope-stability problems relative to faults are fairly insignificant in the study area.

Although the Fraser River fault zone is of a regional scale and the canyon is essentially parallel to it, there does not appear to be a relationship between the proximity of the individual faults and the slope failures recorded along the

Figure 6. Aerial photo showing extensive evidence of unstable slopes in Jackass Mountain rocks opposite a large alluvial fan that has deflected the Fraser into the east bank, causing severe lateral erosion.

CNR track. However, these faults provide broad scarplike faces on the canyon walls, generate broad talus and scree slopes, and form the headwalls of what appear to be large postglacial landslides in some localities. Hence, the importance of these faults in causing localized deep-seated failures and other mass movement cannot be underestimated.

Lateral Erosion

Study of the engineering geology of the canyon area shows that lateral erosion by the river of the bank opposite alluvial fans which have formed at the mouths of tributary rivers and creeks is the most important single factor that influences stability of the canyon walls. In addition, there is considerable lateral erosion and subsequent slope oversteepening on the outside of river bends.

For comparative purposes it is noteworthy that, of the 68 mi (109 km) between Lytton and Hope, 9.1 mi (1.45 km) of channel length have been affected by lateral erosion on the CNR (that is, east) side of the canyon, whereas only 5.1 mi (11.0 km) of channel length have been affected on the west side of the canyon. The number and size of alluvial fans and number of streams on both sides of the canyon between Lytton and Hope are given in Table 1.

Although there are only about 20% more streams draining into the Fraser on the west than on the east side of the canyon, there are about 65% more alluvial fans on the west side. Furthermore, twice the channel length [that is, 9.8 mi (16.0 km) compared to 4.9 mi (8 km)] is adversely affected by alluvial fans on the east bank. This is significant since 62 mi (98 km) of the total 68 mi (109 km) of CNR track between Lytton and Hope is on the east bank. The average fan width on the west side of the canyon is about 20% greater than that on the east bank.

The greater development of tributary streams (and, accordingly, of alluvial fans) on the west side of the canyon is probably related to climatic conditions. The Coast Mountains receive more precipitation than the Cascades, which are located farther inland. The greater number and size of streams in the Coast Mountains exist undoubtably in response to the higher precipitation. Regional geologic structural patterns may also be an important control in tributary-valley development in this area.

Ryder (1969, p. 386) noted that possibly in the upper Fraser Canyon "many of the Fraser fans were constructed during a relatively short span of time following deglaciation and that a constant state of equilibrium between fans and basins" was never attained "before lowering of the Fraser commenced." She also noted (p. 388) that possibly because of "climatic and topographic differences between the two

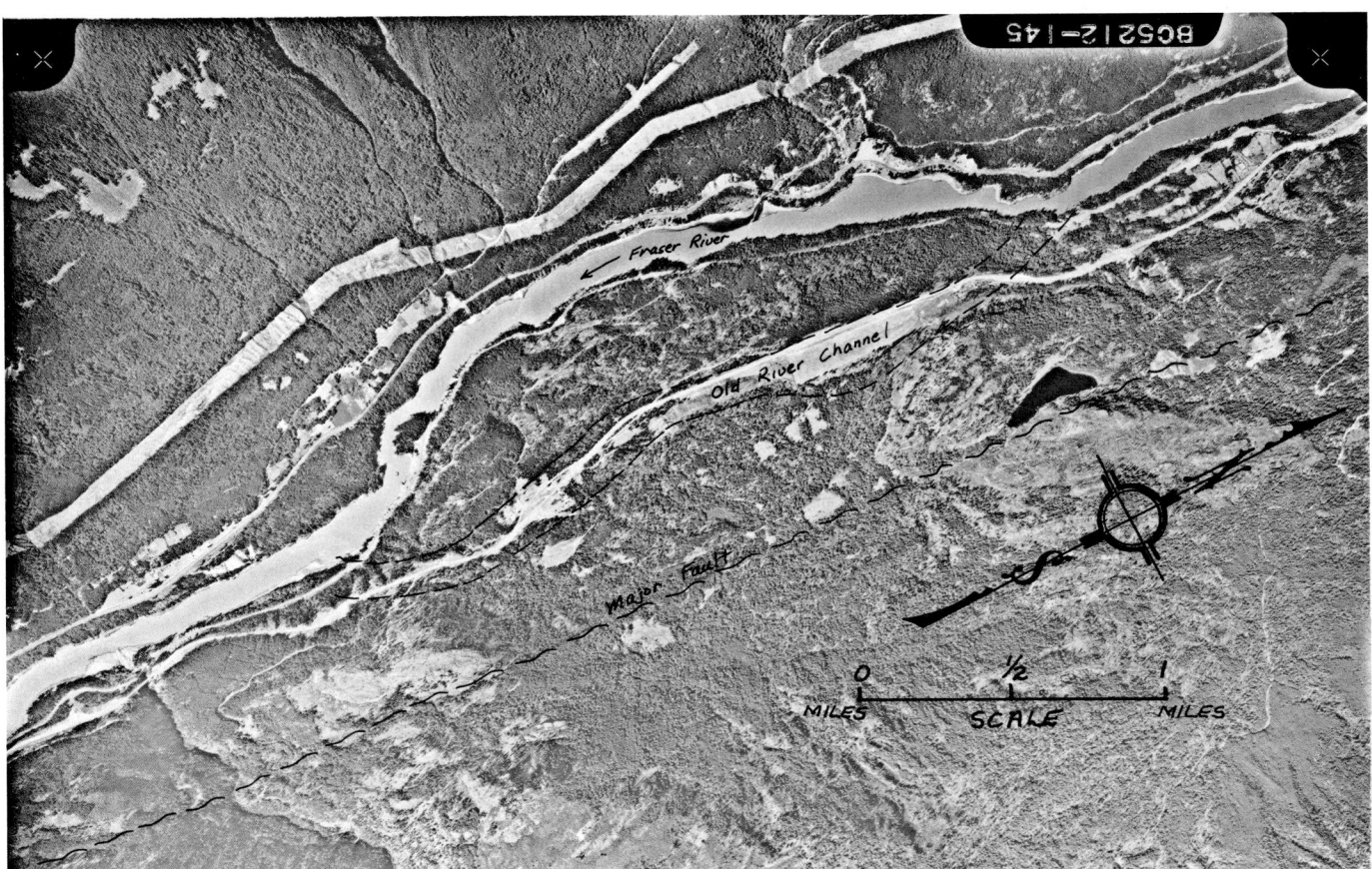

Figure 7. Aerial photo showing an old, isolated, high-level, infilled river channel, part of which is occupied by the Boston Bar airfield. This isolated channel may have developed because of a landslide blocking off the river in this locality.

Figure 8. Large but typical alluvial fan at the mouth of Spuzzum Creek. Fan causes a major deflection of the river into the east bank. Note the well-developed hanging valleys south of Spuzzum Creek.

banks and the prolongation of glacial effects in the Coast Mountains of the west bank, fan construction proceeded most rapidly (or for a longer time period) on the west bank."

Comparative Analyses According to Physical Circumstances at Various Locations

Locations of Recorded Incidents versus Locations Opposite Fans or on the Outside Bank of River Bends. Quantitative information concerning the main factors considered in this analysis is shown graphically in Figure 12. The locations of alluvial fans and large river bends are represented by heavy horizontal bars in Figure 12. Locations of slides, rockfalls, and other slope-instability incidents that were recorded by the CNR are shown as histograms. Data sets from three different sources are plotted in Figure 12. The upper set deals with incidents recorded along each 0.1 mi of track. The lower two sets are related to incidents recorded per mile. By counting the number of incidents that occur beneath the horizontal bars representing locations of alluvial fans and river bends, the quantitative significance of each can be evaluated. The total number of rockfall incidents considered in the analysis is about 700.

Results for (1) Lytton to Hope, (2) Lytton to Boston Bar, and (3) Boston Bar to Yale are summarized in Table 2. The average number of incidents per mile that occurred opposite alluvial fans, on the outside curves of river bends, and at areas unaffected by either of these factors is shown graphically in Figure 13 for the three reaches listed above and for the reach Lytton to Yale.

For the first three reaches mentioned above, the average number of incidents per mile of track opposite alluvial fans was calculated. These values are 5.65, 5.49, and 4.25 times greater, respectively, than the values for localities unaffected by river deflection (see Table 2). The average number of incidents per mile of track located on outside curves of river bends was calculated also. These values are somewhat less, although they are still significant, being 3.31, 2.81, and 2.82 times greater than the values for localities unaffected by river deflections.

From the data in Table 2 for the entire canyon from Lytton to Hope, simple calculations indicate that although

TABLE 1. DATA ON STREAMS AND ALLUVIAL FANS BETWEEN LYTTON AND HOPE

Side of the canyon	No. of intermittent streams	No. of perennial streams	No. of alluvial fans	Total width of all alluvial fans (mi)	Avg. width of all alluvial fans (mi)
West	28	15	23	9.8	0.425
East	23	12	14	4.9	0.350

localities affected by alluvial fans and river bends represent only 31.2% (13.4% and 17.8%, respectively) of the 68 mi (109 km) considered in the analysis, as much as 66.2% (37.2% and 29.0%, respectively) of all the recorded incidents occur in these localities.

Mapped Postglacial Landslide Locations versus Locations Opposite Fans and on the Outside Bank of River Bends. Analysis was made to determine the relationship of major postglacial (prehistoric) landslides and the effects of lateral erosion caused by alluvial fans and river bends. The results (Table 2, Fig. 14a) indicate that a strong correlation also exists between the occurrence of major landslides and the locations of fans and river bends.

Again, although locations affected by fans and river bends represent only 31.2% of the 68 mi (109 km) of track length considered in the analysis, calculations indicate that 59.7% of all major landslides (32.4% for locations opposite fans plus 27.3% for locations on outside curves of river bends) occur in these localities. As indicated in Table 2 and Figure 14a, the average number of landslides per mile that were opposite fans or on the outside curve of river bends is 4.15 and 2.64 times greater, respectively, than the average number of landslides that were in localities unaffected by either fans or undercutting at river bends.

The average plan area of the landslides occurring in localities affected by fans or bends is only slightly over 10% greater than localities which are not affected by either of these factors (Fig. 14b).

Locations of Incidents Recorded versus Postglacial Landslide Locations. The relationship between recorded rockfall incidents and mapped postglacial landslides in the reaches (1) Lytton to Hope, (2) Lytton to Boston Bar, and (3) Boston Bar to Yale is also given in Table 2. In the Lytton to Hope reach, for example, although the track length affected by postglacial landslides is only about 20% of the entire length considered, about 50% of the recorded incidents occurred in postglacial landslide areas. Also, for the three reaches considered, the number of recorded incidents that occurred in postglacial landslide areas are (1) 2.43, (2) 1.86, and (3) 2.23 times greater, respectively, than the average number of incidents per mile for the entire 68 mi of track considered (Table 2).

Climate

The proximity of the ocean, the prevailing westerly winds that move maritime air masses over the Fraser River basin, and topographic effects of the north-south mountain ranges are the important factors that affect the climate in the region. The dominant factors that influence the hydrology of the

TABLE 2. ROCKFALL AND LANDSLIDE OCCURRENCES BY GEOMORPHIC LOCATION AND STRETCH OF TRACK

Locations	Lytton to Hope					Lytton to Boston Bar			Boston Bar to Yale		
	T (mi)	R (mi^{-1})	R_D/R_U	L (mi^{-1})	L_D/L_U	T (mi)	R (mi^{-1})	R_D/R_U	T (mi)	R (mi^{-1})	R_D/R_U
Opposite alluvial and slide-debris fans	9.1	28.8	5.65	2.74	4.15	4.4	20.0	5.49	4.4	38.6	4.25
On outside curves of river bends	12.1	16.9	3.31	1.74	2.64	4.1	10.2	2.81	6.0	25.7	2.82
Unaffected by either fans or river bends	46.8	5.1		0.66		19.5	3.6		16.6	9.1	
All	68.0	10.3	2.03	1.13	1.71	28.0	7.2	1.97	27.0	17.6	1.93
			R_L/R_T					R_L/R_T			R_L/R_T
Postglacial landslide areas	13.5	26.05	2.43			6.5	13.38	1.86	6.30	39.20	2.23

Note: T = track length involved. R = average number of rockfalls recorded. R_D/R_U = ratio of R for locations of river deflections (opposite alluvial and slide-debris fans or on outside curves of river bends) to R for locations unaffected by river deflections. L = average number of postglacial (prehistoric) landslides mapped. L_D/L_U = ratio of L for locations of river deflection to L for locations unaffected by river deflection. R_L/R_T = ratio of R for locations in postglacial landslide areas to R for entire 68 mi of track in study area.

Figure 9. Aerial photo mosaic of the Hell's Gate Bluffs area showing the

numerous large postglacial landslides which have developed in this area.

Figure 10. Typical large (1,000 ft wide × 0.5 mi high) postglacial slide at Hell's Gate area. Note the extensive and deep slide debris overlying the granodioritic bedrock, well-developed flanks, and serrated nature of the crest of the slide.

Fraser River and tributaries are precipitation and temperature. The moist maritime air is forced up by the Coast Mountains, and the westward-facing slopes and central peaks are subjected to heavy precipitation in the process. After passing this range the air descends into the interior plateau where temperatures rise and precipitation diminishes.

Delineation of the meteorological conditions in the canyon area according to the Fraser River Board (1958) is given in Table 3. Although there is relatively little difference in average temperature between Hope, Hell's Gate, and Lytton, there is considerably less rainfall in Lytton than at the other two stations. Precipitation is greatest in the winter months, the maximum being around December (see Figs. 15, 16).

The meteorological map of the region, however, indicates

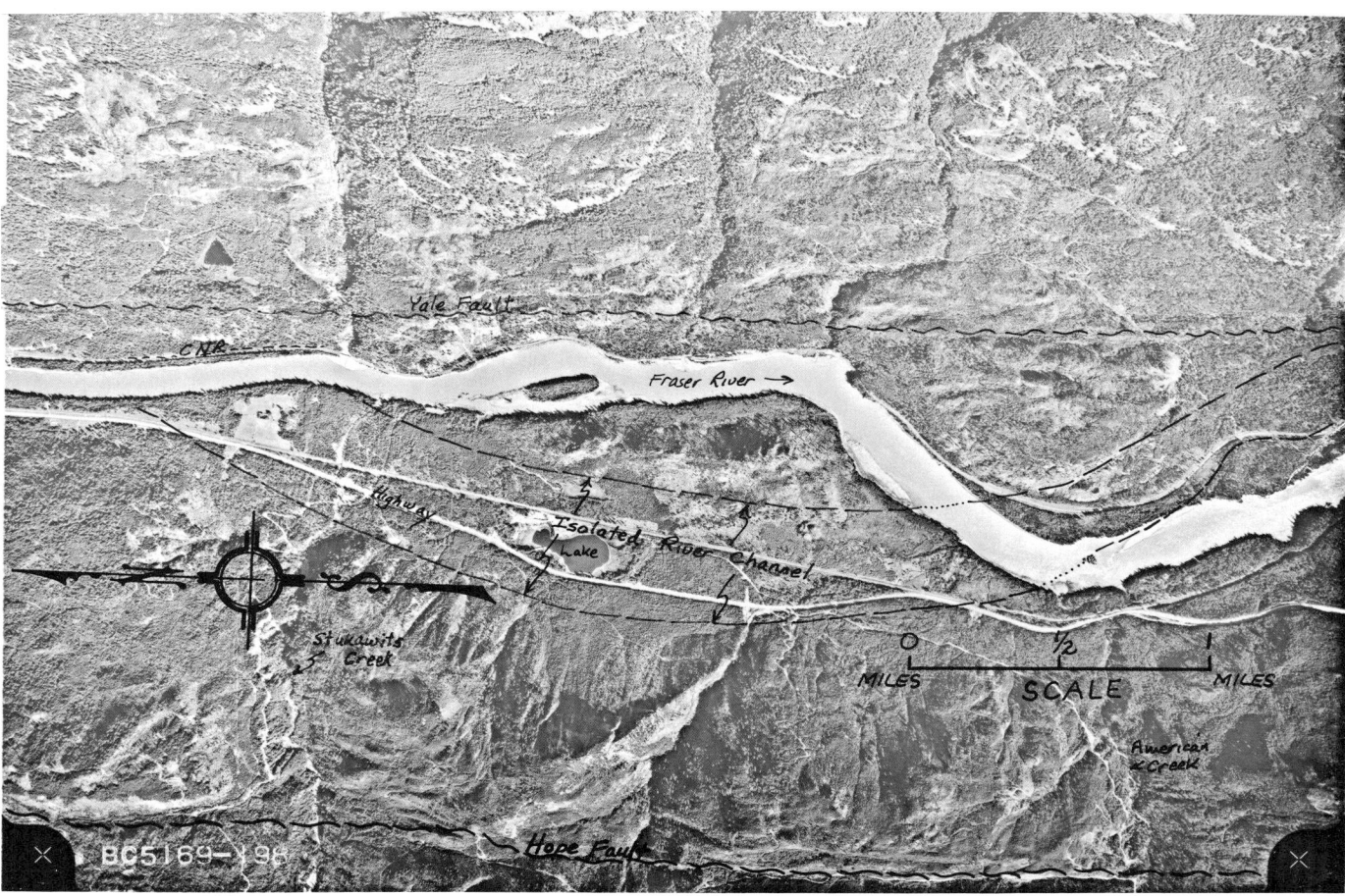

Figure 11. Aerial photo of the location of what appears to be an old, isolated, infilled river channel. This old channel extends for 3 and 2 mi on the east and west banks of the Fraser, respectively. Note the well-developed lineaments formed by the Yale and Hope faults.

a general west-to-east decrease in precipitation across southwestern British Columbia. The western slopes of the Coast Mountains receive between 60 to 100 in. (150 to 250 cm) and 40 to 60 in. (100 to 150 cm) of precipitation per year, whereas the Cascades fall in the 40 to 60 and 20 to 40 in. (100 to 150 and 50 to 100 cm) per year belt. This regional precipitation pattern could explain why there is greater development of tributary valleys and, accordingly, alluvial fans in the Coast Mountains compared with the Cascades.

Earlier studies by Piteau and others (1976) of climatic data of the Fraser Canyon indicate that the general area of the canyon between approximately Yale and Boston Bar is subjected to considerable ranges in daily temperatures, particularly in the winter months. As shown in Figure 15, it is not unusual to have as many as 15 freeze-thaw cycles per month during December, January, and February in this area (Piteau and others, 1976). The Fraser Canyon, particularly between Yale and Boston Bar, appears to be a transition zone between the coast climate and the interior climate. Depending on which of these two climates dominates at the time, temperatures can range from mild to severe. The distribution of average annual freeze-thaw frequencies in Canada presented by Fraser (1959) indicates that the middle Fraser Canyon lies in a zone having one of the highest freeze-thaw frequencies in the country.

A monthly correlation of incidents of slope instability in the canyon indicates that about 86% of all incidents occur in the late fall and winter months (October to March inclusive). These data also include snowslides, but there is little doubt of the significance of the effects of precipitation and frost action in this locality. The total number of rockfalls recorded for each month of the year in the canyon area south of Boston Bar is plotted against rainfall and temperature in Figure 16 from 1933 to 1970 data of Peckover (1975). This correlation clearly shows that the incidents are at a maximum during the winter months when precipitation and frequencies of freeze-thaw cycles are highest.

The disruptive process of freezing and thawing of water in cracks in the rock can be most outstanding. According to Reiche (1950), water experiences at 9.05% volume increase on freezing and thus exerts tremendous pressure when freezing in a confined space. He noted that "water-filled cracks or joints which terminate downward and which are narrow and perhaps irregular may be converted into essentially closed

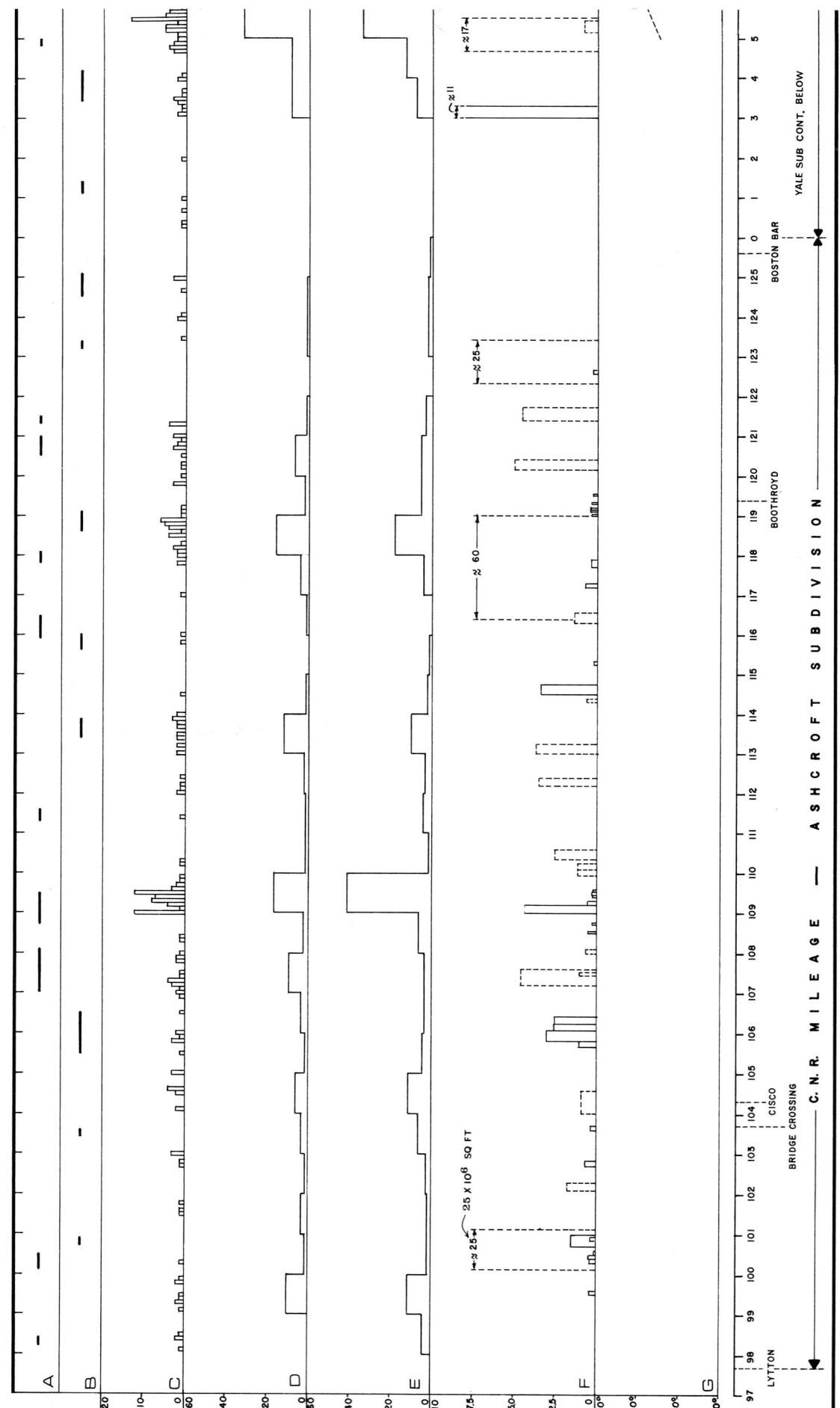

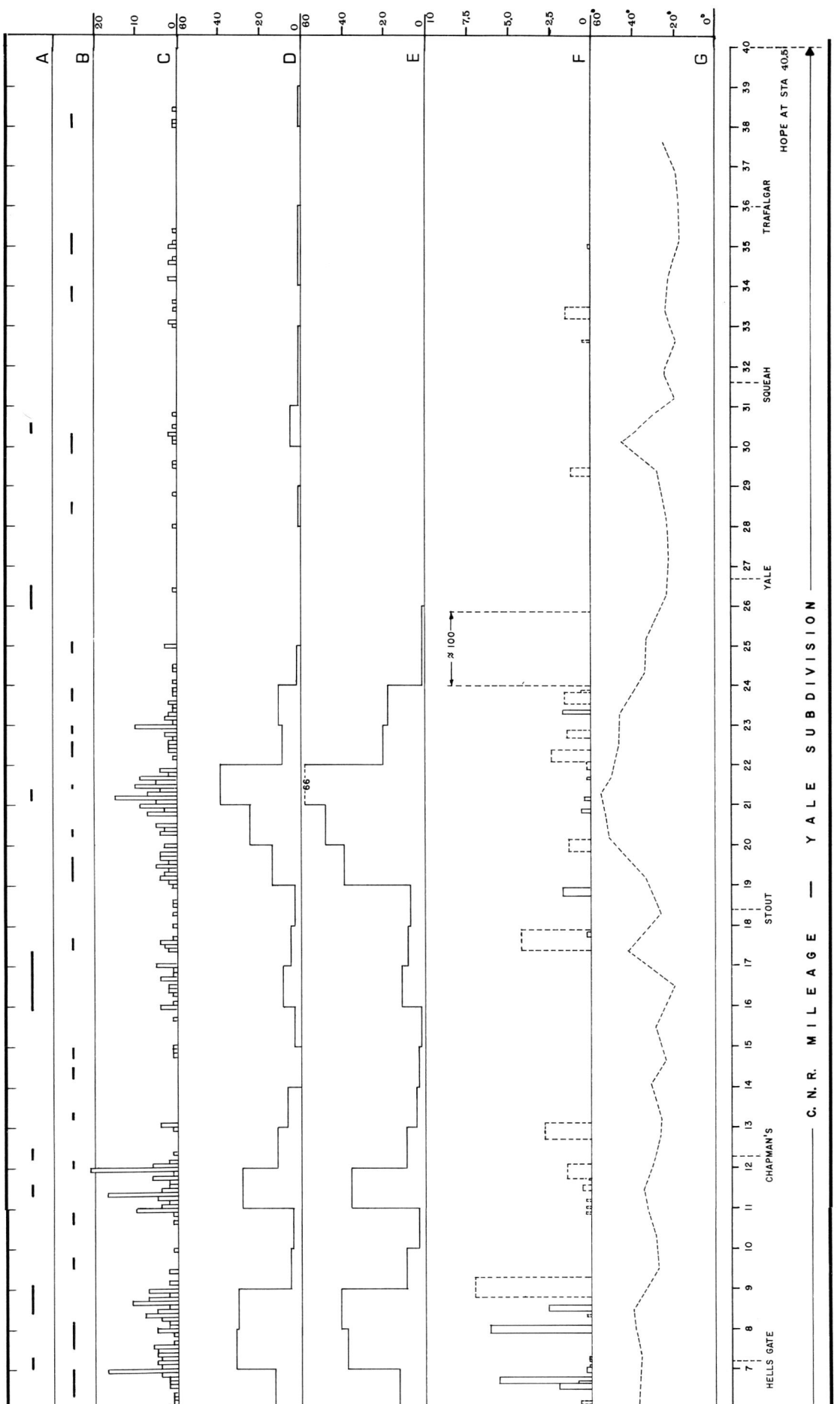

Figure 12. Plots along CNR showing the correlation of recorded and prehistoric slope failures with the presence of fans and river bends. Heavy horizontal bars in plots A and B represent stretches of track affected by oversteepened slopes caused by river deflection due to alluvial or debris fans (plot A) and river bends (plot B). Plot C shows histograms that indicate the number of slope failures (rock, debris, soil, and snow slides) per 0.1 mi of track, as recorded by the CNR from 1948 through 1968. Plots D and E show histograms that indicate the number of slope failures per 1 mi of track, as recorded by the CNR from 1952 through 1972 (in the case of D) and 1948 through 1968 (in the case of E). The data for plots C, D, and E were derived from three separate sources. Plot F shows the plan area ($\times 10^6$ f^2) for each mapped postglacial landslide (solid lines for well-defined slides; dashed lines for poorly defined slides). Plot G shows mean angle of slope of canyon walls along the track.

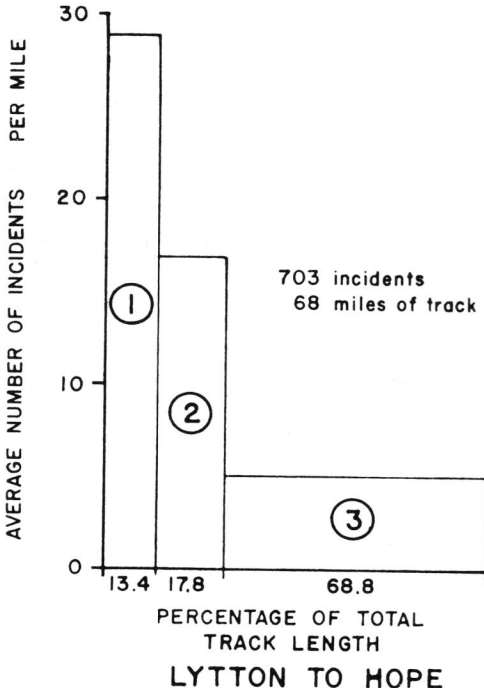
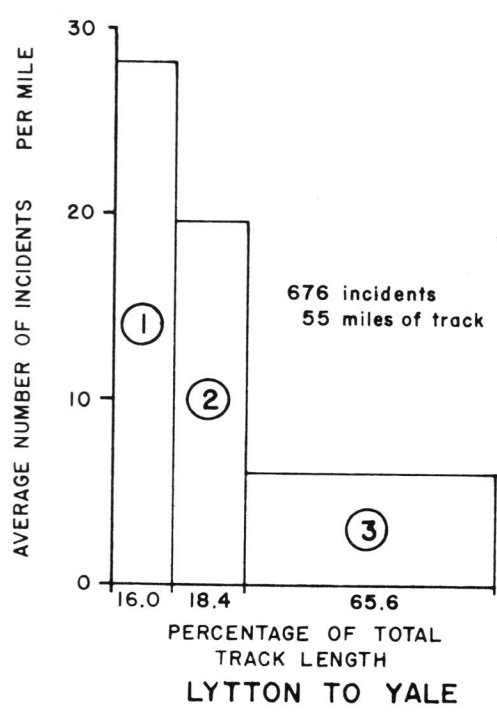
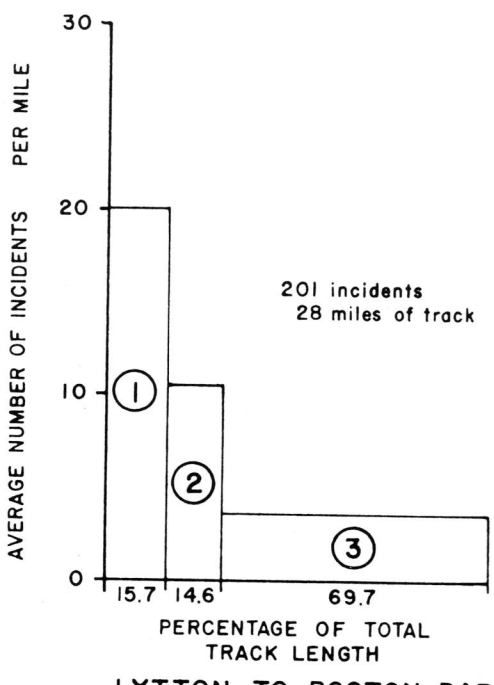
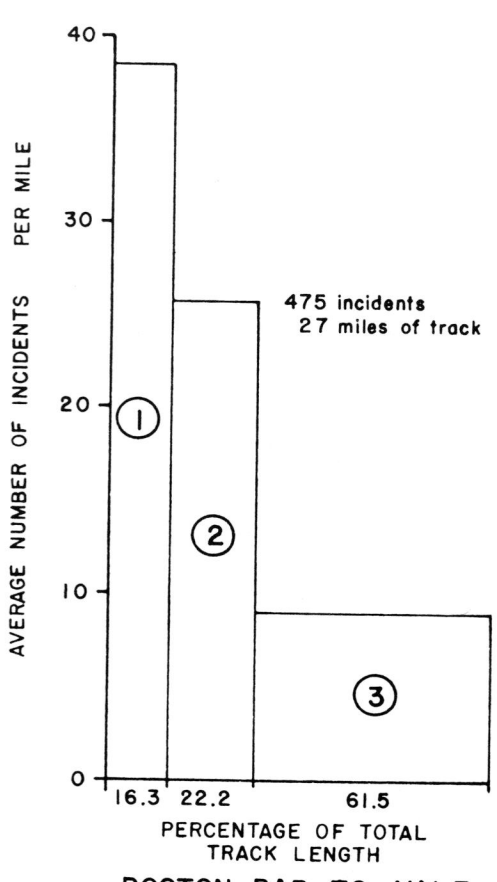

Figure 13. Percentage of incidents of slope failure occurring in (1) localities opposite fans, (2) localities on outside curves of river bends, and (3) localities not affected by either of these factors.

systems by preliminary freezing of the water in the superficial parts. In such cases the combination of expansion and low compressibility may exert a disruptive force which, if the temperature continued to fall and rock pressures permitted, would approach 30,000 pounds to the square inch at minus 22° C." Temperatures of approximately minus 22° C (minus 7.6° F) are not abnormal in temperate regions such as the Fraser Canyon.

The frequency of rockfalls originating on the steep rock faces is probably increased because of the absence of snow, vegetation, and soil that would otherwise insulate the rock from unusually low temperatures and from changes in temperature. Earlier studies by Piteau and others (1976) showed, for example, that snow insulation in the canyon area greatly retards the buildup in ground-water pressure as well as reducing freeze-thaw effects.

Effects of Man

The excavation and related work required to build both the highway and railway, particularly where the two routes are side by side, have affected the equilibrium of the slopes on the east side of the canyon, as discussed below. In many localities the highway acts as a catchment bench analogous to a berm. Although this situation is dangerous for highway users, the railway is sometimes protected from rockfalls and slides that originate upslope and that otherwise might reach the track.

Slopes in Postglacial Slide Debris and Talus Materials. There are extensive highway excavations across postglacial slide areas and deep talus slopes. In some of these instances, the excavated debris has been cast aside onto the immediate downslope side of the highway cut. Such a procedure results in general steepening of the slope between the highway and the CNR track. In some cases, the slope involved was already slightly oversteepened and out of equilibrium. Constant and possibly significant readjustment of the unconsolidated materials may take place on such slopes. Also, where the highway required massive excavations in unconsolidated materials, the CNR has commonly had to make such excavations lower on the slope. In these areas the slope has been surcharged with excavated material at highway level and unloaded at the toe by the CNR; of course, this aggravates the slope instability further.

These excavation procedures often result in sloughing, raveling, and imperceptible creep of the postglacial slide debris or talus material. Such slope readjustments will continue until the natural angle of slope is achieved. Although the moving debris is generally sand, gravel, and cobbles, slow to rapid translation of larger blocks is also involved, particularly in postglacial slide debris.

Highway construction upslope from the CNR has significantly affected the hydrologic regime — most particularly, the surface drainage and runoff conditions. These effects, however, are difficult to evaluate because they vary from area to area subject to variations in the local terrain conditions. Several washouts, debris slides, and mudflows were experienced by the CNR soon after the new highway was constructed and thus were attributed to highway construction.

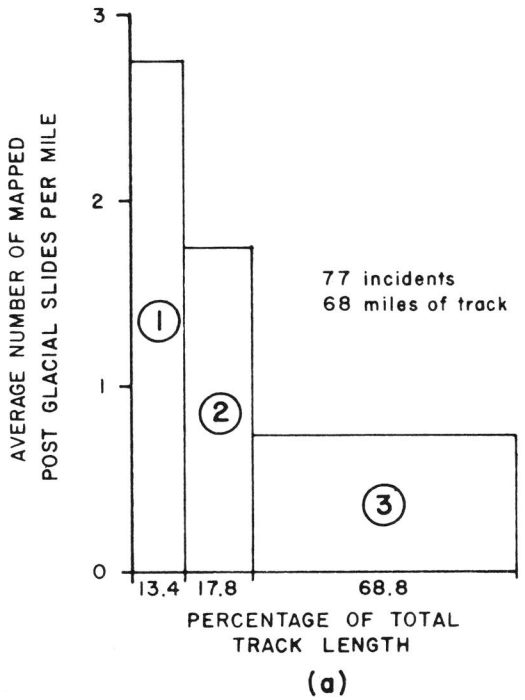

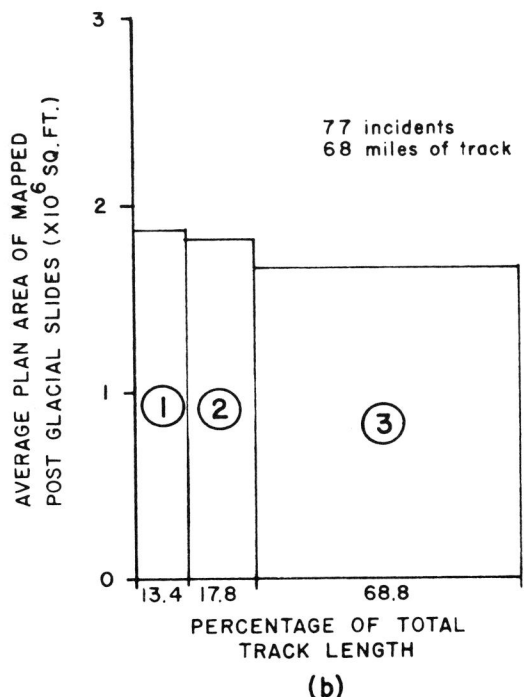

Figure 14. Abundance (a) and size (b) of mapped postglacial slides in (1) localities opposite fans, (2) localities on outside curves of river bends, and (3) localities not affected by either of these factors.

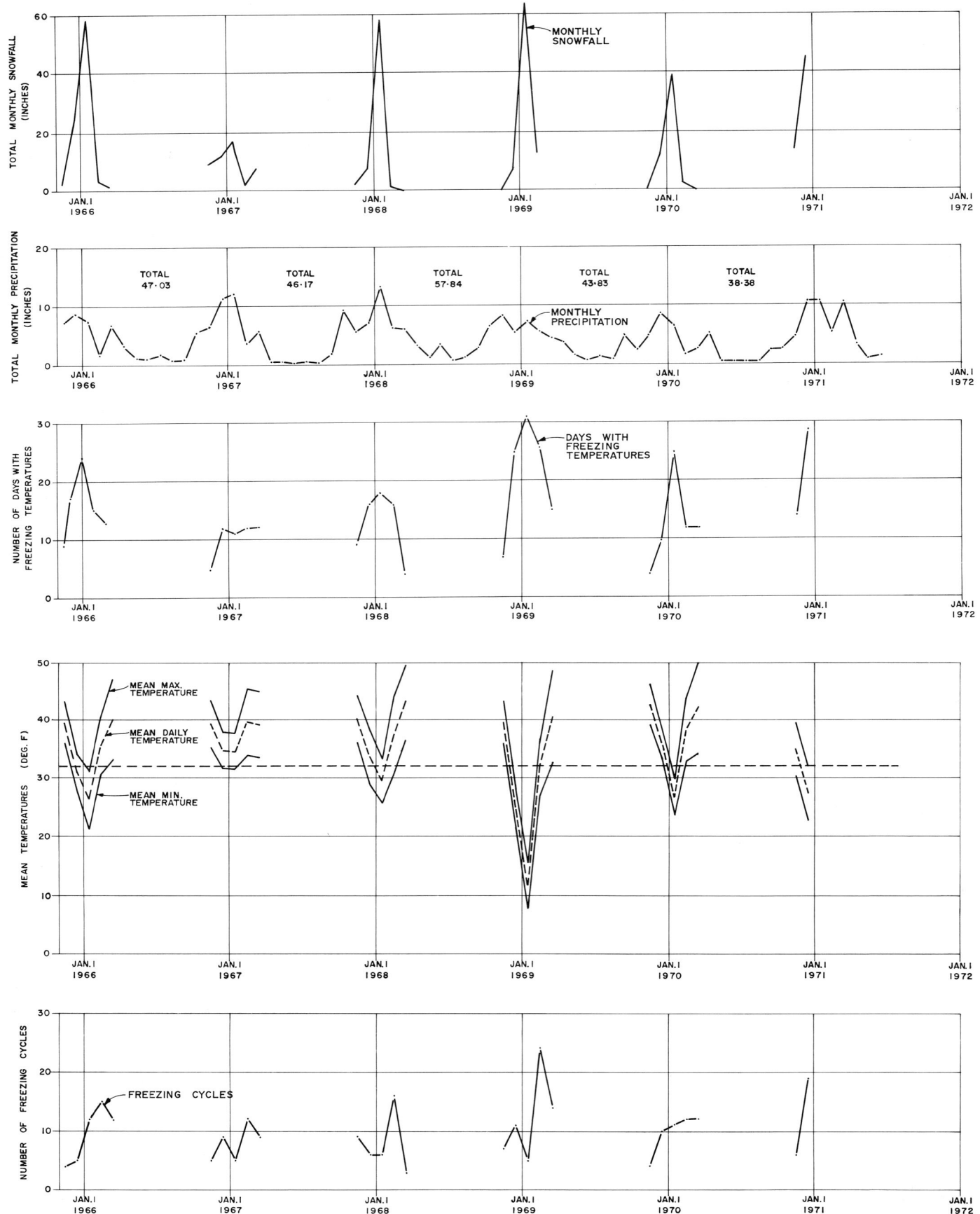

Figure 15. Climate data from 1966 through 1972 for the Hell's Gate area (after Piteau and others, 1976).

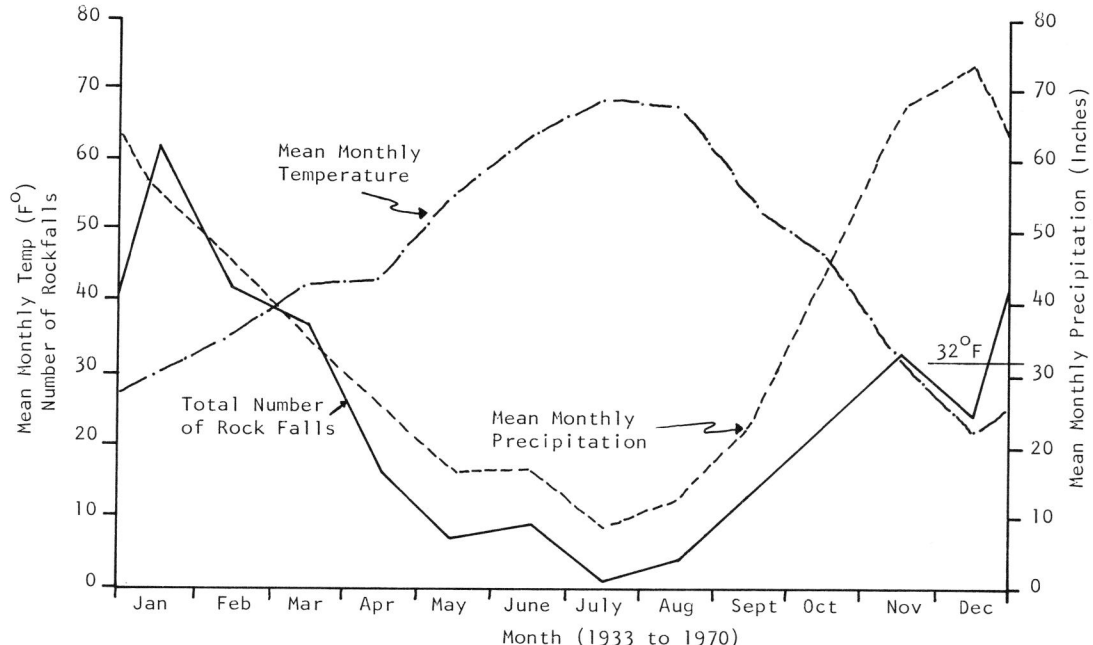

Figure 16. Number of rockfalls, mean monthly temperature, and mean monthly precipitation in the Fraser Canyon for 1933–1970 (Peckover, 1975).

Slopes in Rock. The major rock-cut slopes of the CNR for the most part are confined to localities affected by severe lateral erosion by the river, that is, the central reach of the canyon. Here, because of the steepness of most of the rock faces and the absence of natural benches or other catchments, rockfalls are the major concern. Unfortunately, the rocks have been subjected to considerable blasting damage as well as the usual problems associated with toe unloading and reduction in lateral support. A low-energy black powder, which has a high gas content, was probably used during the original CNR excavation work in 1913–14. Rock damage due to the excessive gasses during blasting is significant. Uneven contours, overhangs, excessive shattering, cracks, and dislodged blocks commonly facilitate ice jacking by freeze-thaw action. Rockfalls are the consequence.

ALLUVIAL FANS AND THEIR EFFECTS ON STABILITY

Alluvial Fan Development

The tributary valleys from which the alluvial fans in the Fraser River originate were essentially formed preglacially or interglacially by stream erosion and were then modified by glacial action during various ice advances. The fan development began during late glacial and early postglacial times, probably when a large quantity of ice was still present and the Fraser River was cutting its channel through glacial debris (Tipper, 1971).

Because even now the river is not well enough established to remove the alluvial fan material, the fans force the river into the bank opposite them. An armor plate or riprap of large boulders tends to develop around the alluvial fan's upstream periphery. This riprap resists erosion and helps to divert the river flow into the opposite bank in the restricted canyon area. Hence, slopes opposite the fans have been severely oversteepened because they have been subjected not only to the normal unloading due to postglacial vertical downcutting of the Fraser channel but also to intense lateral scouring.

The material that composes the alluvial fans in the canyon is transported there from glacial deposits (an assortment of erratics, drift, and outwash) and mechanical mass-wasting debris (talus, scree, slide debris, and other slope-forming materials that find their way from the bedrock walls into the bottoms of the tributary valleys).

At the beginning of the glacial waning period, the walls of the tributary valleys were essentially bare of vegetation, and erosion was rapid until the forest cover and other weathering- and erosion-retarding agents became effective. Once the growth of vegetation was established in these valleys,

TABLE 3. METEOROLOGICAL DATA FOR FRASER CANYON

Station	Elevation (ft)	Temperature				Precipitation				Avg monthly snowfall (in.)
		No. of yr recorded	Max (°F)	Avg (°F)	Min (°F)	No. of yr recorded	Max (in.)	Avg (in.)	Min (in.)	
Hope	152	18	103	49	−9	18	77.20	61.74	35.81	5.15
Hell's Gate	500	6	101	47	−10	6	61.17	47.21	24.00	9.25
Lytton	574	30	104	50	−6	34	26.43	17.42	10.20	4.52

mass movements such as mudflows, debris flows, rock slides, frost action, and general sloughing — although still active — were considerably reduced. Glacial debris, because of its ease of erosion and relatively limited supply, was quickly dissipated in the more susceptible areas. More protected localities and areas with extremely deep and (or) tougher, less erodable deposits eventually sustained abundant vegetation.

The amount of material transported to the fans during postglacial times probably decreased in a somewhat exponential manner. Hence, within a relatively short period, the bulk of the aggradational transportation process was completed, and the fan geometry relative to aggradation essentially became static. Then followed a degradational process — fan dissection by the tributary stream (Bull, 1968). Erosion of the up-river part of the fan by the main river may also have been significant, but depended on the velocity and capacity of the river. Many of the fans in the canyon have been only minimally eroded by the Fraser River, but several fans have been truncated, and some have been almost completely destroyed.

The severe lateral erosion that is operating now opposite alluvial fans has acted in most of these same localities during all of the previous interglacial intervals. Hence, the present steep slopes opposite the alluvial fans are the product of erosion after all four glaciations.

Age and Activity of Alluvial Fans

The time of development of most of the alluvial fans in the Fraser Canyon can be determined from a dated volcanic ash marker horizon. The Mazama ash layer, which owes its origin to a volcanic eruption 6,600 yr ago, was identified in various alluvial fans in the Thompson, Bonaparte, and Similkameen valleys by Ryder (1969). The mean depth to the Mazama ash layer below the fan surface is only approximately 7 ft (2.2 m). The alluvial fans investigated by Ryder (1969) in these three valleys appear to be of essentially the same age as those of the Fraser Canyon. Therefore, there is good reason to believe that this depth to the Mazama ash layer would apply to the fans in the Fraser Canyon. These findings indicate that the alluvial fans were formed mostly before 6,600 yr ago. Probably the supply of immediately available glacial materials in the valleys was virtually exhausted at that time, and revegetation had begun to retard other mass-wasting processes in the tributary valleys.

At the present time, therefore, the size of the alluvial fans in the canyon is probably not increasing. The amount of deflection of the river by the fans will thus remain essentially constant, and the rate of lateral erosion will probably not increase with time. However, because of the inability of the river to remove the alluvial fans, lateral erosion opposite them will continue, particularly during periods of high water level.

Because of the extremely confined conditions in the canyon, the slopes are all the more sensitive to any anomalous variations in river-channel geometry. Had the Fraser River flowed through a broad open valley with an alluvial plain, the alluvial fans at confluences would have had negligible effects on the slope stability of the opposite bank.

SUMMARY AND CONCLUSIONS

Geology

The Fraser Canyon between Lytton and Hope consists of three distinctly different areas as follows: (1) Lytton to Boston Bar — relatively broad valley, heavily timbered slopes, gravel terraces, few rock outcrops, soft sedimentary and schistose rocks, and moderate river gradient; (2) Boston Bar to Yale — constricted and gorgelike, numerous bare outcrop slopes, hard rocks, minimal gravel terraces, and steep river gradient; and (3) Yale to Hope — relatively broad flaring valley, heavily timbered slopes, sheared friable granitic rocks, and moderate river gradient.

The Fraser Canyon lies between the Cascade and Coast Mountains in a pronounced zone of faulting. Major faults have been the sites of differential weathering, glacial scouring, and intense erosion. Regional uplift has accentuated the erosion process, which has taken place faster than the canyon slopes could adjust to the changing conditions. As a result of the lag between the natural oversteepening process and the tendency of the slopes to regain stability, some slopes are, geologically speaking, in an active state of failure.

There is evidence of about 75 postglacial and (or) interglacial landslides in the canyon. Some are very large and have caused major river diversions by blocking the river and forcing it to develop new channels. Other large landslides have blocked the river temporarily.

Factors Controlling Regional Slope Stability

The most important factors controlling the regional slope stability are (1) the presence of alluvial fans and bends in the river, (2) freeze-thaw cycles, (3) excavation and fill that changes slope angles, and (4) the lithologic and structural characteristics of the canyon.

Alluvial fans and river bends have caused slope-stability problems on a regional basis. Deflection of the river into the opposite bank has caused severe lateral erosion that resulted in extensive oversteepening and undermining of the toe of the canyon slope. Debris fans at the toes of major postglacial landslides also appear to be very significant river deflectors as well.

Incidents of rockfalls and other slope failures which have been recorded along the CNR show that about 66% of all incidents occur opposite alluvial fans and river bends; such locations amount to only about 31% of the total canyon length considered. The average number of incidents per mile occurring in stretches affected by alluvial fans and river bends is 5.6 and 3.3 times greater, respectively, than is recorded in stretches not affected by either of these conditions.

A strong correlation exists between mapped postglacial landslides and stretches with alluvial fans and river bends. About 60% of the postglacial landslides were in these stretches because they are affected by severe lateral erosion.

Although only about 20% of the east side of the canyon consists of postglacial landslide areas, about 48% of all recorded incidents of instability occurred in these areas. This is about 3.3 times greater than the average number of incidents recorded in areas not having postglacial landslides.

About 86% of all recorded incidents of slope failure occurred in the late fall and winter months when precipitation and frequencies of freeze-thaw cycles are at a maximum. The Fraser Canyon lies in a zone which has one of the highest freeze-thaw frequencies in the country.

Effects of man are also significant. Dumping excavated material onto the upper part of the postglacial landslides or talus accumulations oversteepens the slope if it was previously in a state of equilibrium, whereas excavation of the lower part of the same feature unloads the slope and removes its support. Slow translation takes place, and large failures may develop. Furthermore, ground-water regimes have been changed by construction, and changes of surface drainage in some instances resulted in debris slides and mudflows in concentrated drainage channels. Also, blasting damage of steep rock faces in combination with frost action has resulted in numerous rockfalls and in development of overhangs in certain areas.

Overall effects on the geomorphology of the canyon are also dependent to a large extent on the lithologic and structural characteristics of different parts of the canyon.

Alluvial Fans

Indications are that most of the alluvial fans were formed during late glacial and early postglacial times and that the fans, at least in terms of aggradation, are essentially in a dormant state. Hence, the degree of river deflection around the nose of these fans — and, accordingly, the rate of river erosion — should not increase in time. The river will, however, continue to erode and undercut the slope opposite the alluvial fans.

ACKNOWLEDGMENTS

I thank the Canadian National Railway for their permission to publish certain information contained herein and for their cooperation during the course of this work in the Fraser Canyon. In particular, I thank E. R. Scales, G. Roesler, M. Robertson, N. L. McLeod, and R. Bailey of CNR. I am also grateful to the British Columbia Department of Highways for providing background information in certain areas. Detailed review comments by C.F.S. Sharpe and a general overview by K. C. McTaggart and W. H. Mathews are gratefully acknowledged.

REFERENCES CITED

Armstrong, J. E., Crandell, D. R., Easterbrook, D. J., and Nobel, J. B., 1965, Late Pleistocene stratigraphy and chronology in southwestern British Columbia: Geol. Soc. America Bull., v. 76, p. 321–330.

Borden, C. E., 1965, Radiocarbon and geological dating of the lower Fraser Canyon archaeological sequence: Radiocarbon and Tritium Dating, 6th Internat. Conf., Pullman, Wash., June, p. 165–178.

Bull, W. B., 1968, Alluvial fans: Jour. Geol. Education, v. 16, no. 3.

Cairnes, C. E., 1944, Geology of Hope area, British Columbia: Canada Geol. Survey Map 737A.

Camsell, C., 1920, The origin and history of the great canyon of Fraser River: Royal Soc. Canada Proc., sec. 4, p. 45–59.

Cleveland, G. B., 1971, Regional landslide prediction: California Div. Mines and Geology, Federal Insurance Administration, June, 33 p.

Duffell, S., and McTaggart, K. E., 1952, Ashcroft map-area, British Columbia: Canada Geol. Survey Mem. 262, 122 p.

Fraser, J. K., 1959, Freeze-thaw frequencies and mechanical weathering in Canada: Arctic, v. 12, p. 40–53.

Fraser River Board, 1958, Preliminary report on floor control and hydroelectric power in the Fraser River basin: Lands, Forests and Water Resources Dept., Victoria, B.C., 100 p.

Mathews, W. H., 1972, Geology of the Vancouver area of British Columbia: Internat. Geol. Cong., 24th, Montreal 1972, Field Excursion A05-C05, p. 47.

McTaggart, K. C., and Thompson, R. M., 1967, Geology of the Northern Cascades in southern British Columbia: Canadian Jour. Earth Sci., 4, p. 1199–1228.

Monger, J.W.H., 1969, Hope map-sheet, west half, British Columbia: Canada Geol. Survey Paper 69-47, Map 12, p. 75.

Peckover, F. L., 1975, Treatment of rock falls on railway lines: Am. Railway Eng. Assoc. Bull., 653, p. 471–500.

Piteau, D. R., McLeod, B. C., Parkes, D. R., and Lou, J. K., 1976, Overturning rock slope failure at Hell's Gate bluffs *in* Voight, B., ed., Geology and mechanics of rockslides and avalanches: Amsterdam, Elsevier, chap. 10.

Read, P. B., 1960, Geology of the Fraser Valley between Hope and Emory Creek, British Columbia [M.Sc. thesis]: Vancouver, Univ. British Columbia.

Reiche, P., 1950, a survey of weathering processes and products: Univ. New Mexico, Geol. Survey Pub. No. 3, 95 p.

Ryder, J. M., 1969, Alluvial fans of post-glacial environments [Ph.D. thesis]: Vancouver, Univ. British Columbia, 432 p.

Tipper, H. W., 1971, Glacial geomorphology and Pleistocene history of central British Columbia: Canada Geol. Survey Bull. 196.

MANUSCRIPT RECEIVED BY THE SOCIETY SEPTEMBER 7, 1976
MANUSCRIPT ACCEPTED SEPTEMBER 17, 1976

Printed in U.S.A.

6
Complex mass-movement terrains in the western Cascade Range, Oregon

FREDERICK J. SWANSON
School of Forestry, Oregon State University, Corvallis, Oregon 97331

DOUGLAS N. SWANSTON
Forestry Sciences Laboratory, Pacific Northwest Forest and Range Experiment Station, Corvallis, Oregon 97331

ABSTRACT

A variety of mass-movement processes interact to form complex mass-movement terrains in the western Cascade Range in Oregon. Slow, deep-seated (>5 m depth) processes of creep, slump, and earthflow operate simultaneously and sequentially, resulting in unstable conditions that may initiate rapid, shallow (<5 m) soil mass movements on hillslopes and debris torrents in stream channels. This combination of mass-movement processes supplies large volumes of sediment to streams and determines the geometry of the channel and valley floor.

Creep movement in western Oregon has been monitored at rates as high as 15 mm/yr. Relative movement between discrete blocks in the Lookout Creek earthflow ranges up to nearly 10 times as fast. Movement rate accelerates during periods of high moisture availability.

Geomorphic observations and tree-ring analysis indicate that mass-movement terrains may have histories spanning centuries and possibly millennia.

INTRODUCTION

Much of the literature on mass movement of soil and rock materials has focused on catastrophic landslide events. In many areas, the subtle, slow mass-movement processes of creep, slump, and earthflow may account for more erosion in the long term.

In the steep forest lands of the western Cascade Range in Oregon, areas of creep and earthflow activity form complex mass-movement terrains that pose important land-management problems. Road construction and timber harvest adversely influence slope stability and increase mass-movement activity, which in turn influence the stability of roads and the quality and quantity of timber produced from unstable lands. Impacts of creep and earthflow activity are most dramatic in the mountain-stream environment where earth movement shapes or displaces the stream channel and delivers large quantities of sediment and large organic debris.

These considerations have led to the examination of the processes, histories, movement patterns, and stream impacts of several complex mass-movement terrains in the Pacific Northwest.

Mass-movement terrains involve a great variety of processes that transport earth materials downslope by gravity. In the western Cascade Range, mass-movement phenomena need to be considered as an entire geomorphic continuum, as they act together in parallel and serial fashion. The mass-movement processes include a number of types of landslides, as well as related creep and certain stream-channel processes.

DEFINITION OF PROCESSES

Creep

Creep is the process of slow, downslope movement of mantle materials in response to gravitational stress. The mechanics of creep have been investigated experimentally and theoretically by a number of workers (Terzaghi, 1953; Goldstein and Ter-Stephanian, 1957; Saito and Uezawa, 1961; Culling, 1963; Haefeli, 1965, Bjerrum, 1967; Kojan, 1968; Carson and Kirby, 1972; and others). In a purely rheological sense, creep movement occurs as quasi-viscous flow under shear stresses sufficient to produce permanent deformation but too small to result in discrete failure. Mobilization of the soil mass is primarily by deformation at grain boundaries and within clay minerals. Both interstitial and absorbed water appear to contribute to creep movement by opening the structure within and between mineral grains, reducing friction within the soil mass. This permits a remolding of the clay fraction, transforming it into a slurry that lubricates the remaining soil mass.

Under field conditions, local variations in soil properties, degree of weathering, and clay and water content lead to variations in creep movement. Rapid creep may result in discrete failures and development of tension cracks, pressure ridges, radial crack patterns, and lobate features that grade into slumps, earthflows, and debris avalanches. Together, these processes and landforms produce the complex mass-movement terrain that characterizes much of the western Cascade Range.

Movement Rate and Occurrence. Movement rates monitored with borehole inclinometers at sites in complex mass-movement terrain in the western Cascades and Coast Ranges of Oregon and northern California range from 7.9 to 15.2 mm/yr at the base of the active creep zone (Swanston and Swanson, 1976). Movement is quite variable within this range, regardless of parent material or geographic setting.

The depth over which movement is active is also variable, depending largely on degree and depth of weathering, subsurface structure, and soil-water content. The depth of significant movement for all monitered sites ranges from 1 to 15 m and, on the basis of well-log data (D. N. Swanston, unpub. data), appears to be associated primarily with an abrupt increase in hardness and decrease in water content of the mantle materials. This well-defined, creep-zone boundary may correspond either to the lower limit of the saprolite zone or of the creep-prone bed-rock units.

Creep is generally the most widespread of all mass-erosion processes. It operates at varying rates in clayey soils at slope angles as low as two or three degrees. In small watersheds, developed in cohesive soils, creep may operate over more than 90% of the landscape. As a result of this activity, a continuing supply of soil material is dumped into the stream from encroaching banks and small-scale bank failures.

The quantity of soil delivered to a stream by creep-related processes may be quite large. For example, assuming that creep of soil material with a dry unit weight of 1,600 kg/m^3 advances a 2-m-high streambank at 10 mm/yr (conservative estimates for watersheds in volcaniclastic materials within the western Cascade Range in Oregon), approximately 64 t/km will be supplied to the channel annually. During high stream-flow, this material is carried into the stream by direct water erosion of streambanks and local bank slumping. In areas characterized by low stream flow with only occasional storm flows, creep may fill the channel with soil and organic debris, and the stream water may move by subsurface flow and piping within the channel filling. Only during storm periods is flow great enough to open the channel and remove the stored debris, resulting in the periodic discharge of very high sediment loads and occasional debris torrents when debris dams fail. Such a mechanism is a dominant mode of sediment transfer into intermittent and first-order streams in the creep-dominated areas of the Pacific Northwest.

Earthflow

In local areas where creep-induced shear stress exceeds the strength of soil and rock material, failure occurs and slumps and earthflow landforms are developed (Varnes, 1958). Simple slumping takes place as rotational movement of a block of earth over a broadly concave slip surface, involving very little breakup of the moving material. Slow earthflow occurs where the moving material slips downslope and is broken up and transported either by a flowage mechanism or by a complex mixture of translational and rotational displacement of a series of blocks (Varnes, 1958).

Earthflows have been described by Varnes (1958), Schlicker and others (1961), Wilson (1970), Colman (1973), Swanson and James (1975), and others. In the western Cascades, these features may range in area from less than one hectare to more than several square kilometres. The zones of failure occurs at depths from a few metres to several tens of metres below the surface. Commonly, there is a slump basin with a headwall scarp at the top of the failure area. Lower ends of earthflows are typically incised by streams draining the movement terrain, and in some cases earthflows move directly into large streams. Transfer of earthflow debris to stream channels may take place by small-scale slumping, debris avalanching, gullying, and surface erosion. Therefore, the general instability set up by an active earthflow initiates erosion activity by a variety of other processes.

Movement Rate and Occurrence. Movement rates of earthflows vary from below detectable levels to metres per day. In parts of the Pacific Northwest, many earthflow areas appear to be presently inactive (Colman, 1973; Swanson and James, 1975). Areas of active movement may be recognized by fresh ground breaks at shear and tension cracks and by tipped, split, and bowed trees.

Movement rates may be highly variable both in time and space, even within a single earthflow (Colman, 1973; Swanson and James, 1975). Presence and absence of open tension cracks and degree of disturbance of vegetation on earthflows indicate that some parts may move rather rapidly, whereas other areas appear to be temporarily stabilized. The history of individual earthflow landforms can span thousands of years. This is indicated by age estimates based on radiometric dating of included wood, comparison of modern mass-movement erosion rate with the total volume of material removed from an earthflow terrain, presence of 7,000-yr-old Mazama ash on pre-existing earthflow landforms and characteristics of drainage development over earthflow surfaces (Swanson and James, 1975; Swanson, unpub. data). Variations in earthflow activity may occur in response to changes in moisture availability on the scale of major climate change as well as shorter term seasonal and storm events. Progressive erosion by fluvial and mass-movement processes alter the configuration of landforms and the distribution of mass above the failure plan. These changes may effect the locations and rates of activity in a mass-movement terrain.

FACTORS INFLUENCING MASS-MOVEMENT PROCESSES AND LANDFORMS IN THE WESTERN CASCADES

Geologic, hydrologic, and vegetative factors control the spatial distribution and movement rates or frequency of occurrence of natural mass-erosion processes on forest lands in the western Cascade Range in Oregon.

Geology and Soils

The western Cascades are generally composed of Tertiary lava flows and volcaniclastic and intrusive rocks, which in many areas have undergone extensive alteration and weathering to form clay-rich soil and saprolite (Peck and others, 1964). Peck and others (1964) have mapped several major stratigraphic units including the Oligocene and lower Miocene Little Butte Volcanic Series and the upper Miocene Sardine Formation (Fig. 1). Little Butte volcanic rocks dominate the Western Cascades and consist of lava flows, variably altered ash flows, and laharic and epiclastic materials. In general, the overlying lava and ash flows of the Sardine formation are less altered. Pliocene to Holocene volcanic materials along the eastern margin of the range are composed predominantly of unaltered lava flows.

Bedrock materials, and the weathering products derived from them, influence both the distribution and morphology of terrains shaped by slow, deep-seated mass-movement processes. The most unstable landscapes in the western Cascades tend to be loacted in areas of altered volcaniclastic materials in the Little Butte Series and Sardine formation (Swanson and James, 1975). For example, in the H. J. Andrews Experimental Forest in the central western Cascades, more than 25% of the area underlain by volcaniclastic rock is mantled with recognized active or presently inactive earthflow landforms (Swanson and James, 1975). Less than 1% of the area of younger basalt and andesite flow rock exhibit earthflow landforms.

The soils developed on volcaniclastic materials tend to be deep, fine textured, and poorly drained on the gentler slopes. These conditions are particularly conducive to creep and earthflow activity. Where the volcaniclastic rocks have been extensively altered, Peath and others (1971) and Youngberg and others (1975) have identified highly unstable soils containing high concentrations of expandable clays. Soils derived from lava flows are generally stonier, coarser textured, better drained, and much more stable than soils derived from volcaniclastic bedrock.

Contrasting physical characteristics of bedrock units commonly play an important role in shaping complex mass-movement terrains. For example, in the Andrews Forest, Swanson and James (1975) noted that steep headwall scarps of slumps and earthflows occur at contacts where resistant flow rocks cap incompetent volcaniclastic rocks. In areas of earthflow activity where there is no capping resistant rock, headwall scarps are absent and earthflows form broad depressions of low relief (Colman, 1973). Creep and earthflow movement over irregular bedrock surfaces, commonly resulting from a mix of rock types of differing resistance, will result in an irregular landscape that is a subdued expression of the subsurface topography.

Climate

The western Cascades are characterized by a maritime climate consisting of wet, mild winters, and dry summers. Mean temperatures range from 6° to 10°C, and precipitation ranges from 100 to 200 cm/yr with 75% to 80% falling between October 1 and March 31 (Franklin and Dyrness, 1973). Precipitation and snowfall increase and mean temperature decreases with elevation. Middle elevation sites, which include large areas of potentially unstable terrain, are subjected to periods of high intensity precipitation and snowmelt during major storm events.

Rates and quantities of precipitation and snowmelt control the presence or absence of active piezometric levels in the subsurface. Moisture content and piezometric level affect the weight of the soil mass and control the development of positive pore pressures. Moisture content acts to reduce the resistance of the soil mass to sliding by mobilization of clay structures primarily through adsorption of water onto clay minerals. A rising piezometric level increases pore pressure, thereby reducing the frictional resistance of the soil mass along the failure surface. Such conditions are commonly believed to accelerate creep and earthflow movement (Swanston, 1969; Swanston and Swanson, 1976).

Vegetation

Slope stability in the western Cascades is controlled in part by the extensive cover of forests of Douglas fir, western hemlock, and other species. Forest vegetation exercises some control over the amount and timing of water reaching the soil and the amount held in storage as soil water and snow. Forests regulate hydrology through a combination of interception, evapotranspiration, and influence on snowmelt rate (Rothacher, 1963, 1971, 1973; Anderson, 1969; Harr, 1976, and others). In general these hydrologic influences of vegetation are thought to enhance slope stability (Gray, 1970; Swanston and Swanson, 1976).

Plant roots play a crucial role in the stability of slopes by contributing to the shear strength of soil. Roots add strength to the soil by vertical anchoring through the soil mass into fractures in the bedrock and by laterally binding the soil across potential zones of failure. In shallow soils both effects may be important. In deep soils the vertical rooting factor is negligible, but lateral anchoring and reinforcing may remain important.

Modification of Forest Cover

Removal of forest cover by fire or the activities of man will lead to decreased rooting strength and an altered hydro-

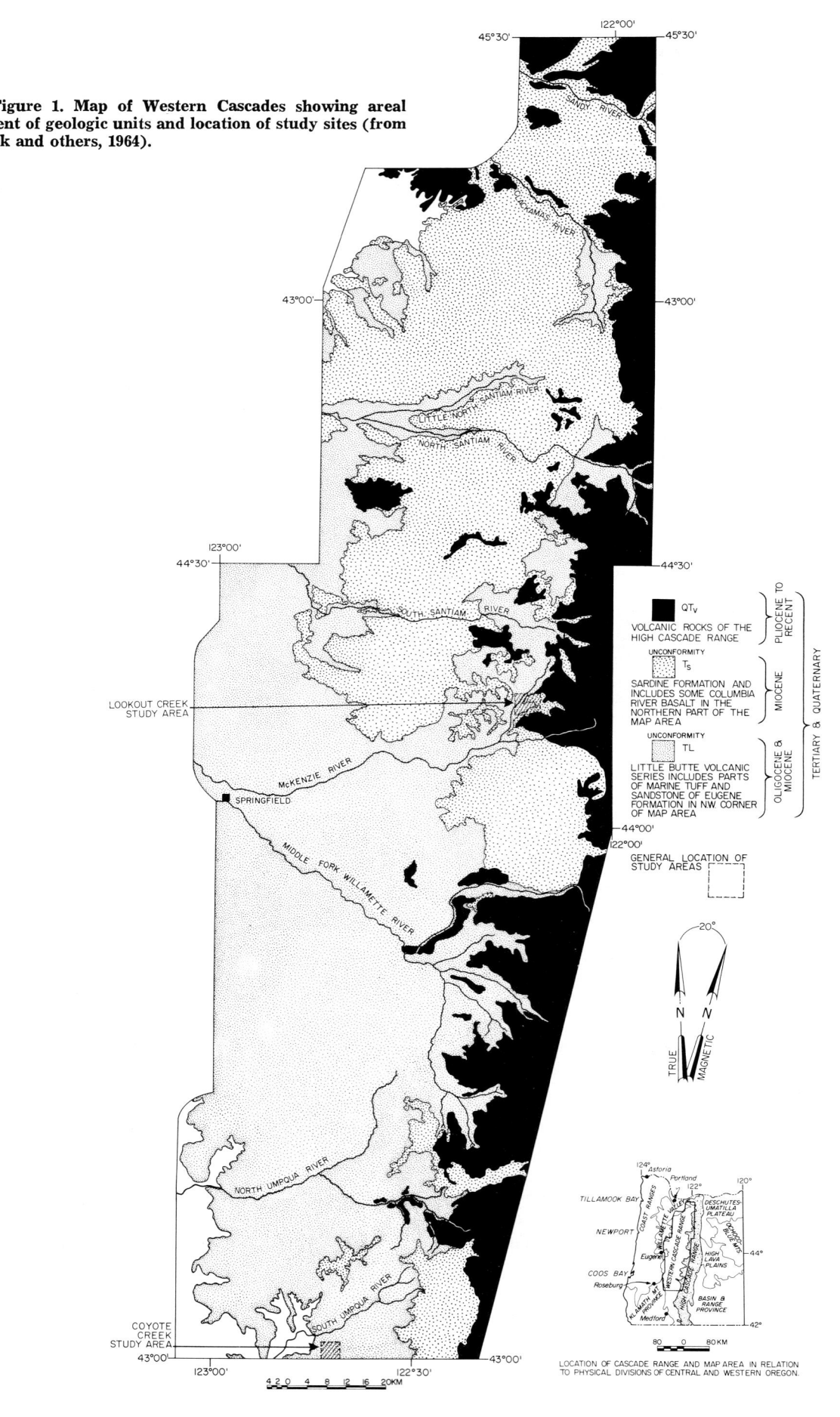

Figure 1. Map of Western Cascades showing areal extent of geologic units and location of study sites (from Peck and others, 1964).

logic regime at the site. Research in Japan (Kitamura and Namba, 1966, 1968), Alaska,[1] and British Columbia (O'Loughlin, 1974) has described a period of greatly reduced soil strength attributable to loss of rooting strength following timber cutting when there has been significant decay of root systems. Reduction in rooting strength may lead to increased creep and earthflow activity.

The hydrologic impacts of the loss of forest cover include modification of annual soil-water budget and changes in peaks of soil moisture and piezometric levels during periods of storm runoff. These modifications in the soil-water regime may result in prolonged periods of active creep and earthflow movement during a single season or reactivation of dormant terrain. Water-yield studies in experimental watersheds in Oregon (Rothacher, 1971; Harr, 1976) suggest that the hydrologic effects of deforestation may continue for more than a decade after cutting.

RELATIONSHIP AMONG PROCESSES: TWO CASE STUDIES

Creep and earthflow processes function as primary mechanisms of soil transport into streams in the western Cascades. The ultimate delivery of soil from slow, mass-movement forms of transport to the fluvial system commonly occurs as rapid, shallow-soil mass movements at stream banks. All of these mass-erosion processes interact with one another. Creep may be a precursor of earthflow activity, and the two processes may operate simultaneously on a column of soil. Both of these forms of deep-seated failure develop instabilities in the soil mantle that directly contribute to the initiation of rapid, shallow-soil failures by debris avalanches, slides, flows, and torrents (Varnes, 1958; Swanston and Swanson, 1976). Consequently, areas of active, deep-seated mass failure form complex mass-movement terrains, where a variety of erosion processes are interactive.

These relationships among processes, rates of movement, and histories of processes in complex mass-movement terrains are illustrated by analysis of two areas of ongoing research in the central and southern parts of the western Cascade Range.

Coyote Creek Mass-Movement Site

The Coyote Creek mass-movement site lies in the South Umpqua Experimental Forest, which is located in the South Umpqua Falls region, as defined by Kays (1970), about 64 km southeast of Roseburg, Oregon (Fig. 1). About a fourth of the region is underlain by deeply weathered, clay-rich rhyodacitic volcaniclastic rocks and subordinate lava flows of the Little Butte Volcanic Series (Kays, 1970). Relatively small parts of the region on the eastern and western edges are capped by andesitic flows of the overlying Sardine formation.

Creep and active and dormant earthflows are common throughout the region and constitute dominant erosion processes under natural conditions. Where present, the contact zone between Sardine flows and the underlying volcaniclastic rocks of the Little Butte Volcanic Series serves as a point of origin of extensive creep terrain and slump-earthflow features. Locally, however, creep and earthflow activity are controlled by interbedded rhyodacite flows and beds of massive welded tuff within the Little Butte Volcanic Series. Such is the case at the South Umpqua Experimental Forest.

The mass-movement study site is located on a 50-ha watershed (watershed 3) on the Coyote Creek drainage within the experimental forest. Active creep and minor earthflow activity are occurring throughout the watershed except for a prominent bluff of massive, welded-tuff breccia that bisects the drainage, separating the basin into two parts (Fig. 2). Above the bluff a history of slump-earthflow activity is indicated by sharply defined headwall scarps and associated earthflow landforms developed below a capping welded-tuff layer. These features encompass most of the upper part of the watershed. The earthflow areas are characterized by hummocky relief that terminates abruptly at the upper edge of the bluff. The earthflow zone on the west side of the upper watershed exhibits no observable evidence of active movement. The eastern earthflow zone exhibits numerous signs of active movement, including undrained sag ponds, tension cracks, and surface-water flow into pipes and underground channels resulting from inward-creeping channel banks. Soil and debris are spilling over the bluff into a perennial channel draining the eastern side of the lower basin. There are no perennial streams above the bluff. The bluff clearly serves as a lower boundary to creep and slump-earthflow activity in the upper part of the watershed and has a major impact on movement and distribution of ground water in the watershed. On the eastern side at least part of the ground-water flow into the watershed is trapped above this welded-tuff layer in the zone of active earthflow above the bluff. Above the bluff on the west, water is apparently able to pass through the tuff layer, and numerous springs at the base of the bluff feed water into deeply weathered clay-rich volcaniclastic bedrock into the lower basin.

Active creep and slumping in the lower part of the watershed is indicated by hummocky ground, tension cracks, and active slumping and debris avalanching along both sides of the stream channels. Locally, backward-tilted benches having long, linear depressions or trenches below headscarps mark the intersection of shear zones with the slope surface (Fig. 2). The area between the east and west perennial-stream branches exhibit some excellent examples of these backward-tilted benches. This area is composed of at least three massive slump blocks, beginning just below the central bluff and extending to the apex of the stream branching. These large-scale blocks are part of the complex mass-movement terrain moving downslope in a northerly direction above a welded-tuff layer that floors the watershed. Bank slumping as the result of mass-movement activity is pervasive along the channel from a point 20 m below the bluff to the confluence with the eastern branch. Throughout this section of stream, bank slumping supplies large volumes of sediment and organic debris to the

---

[1] D. N. Swanston and W. J. Walkotten, "Tree rooting and soil stability in coastal forests of southeastern Alaska" (Study No. FS-NOR-1604:26 on file at PNW Forestry Sciences Laboratory, Juneau, Alaska).

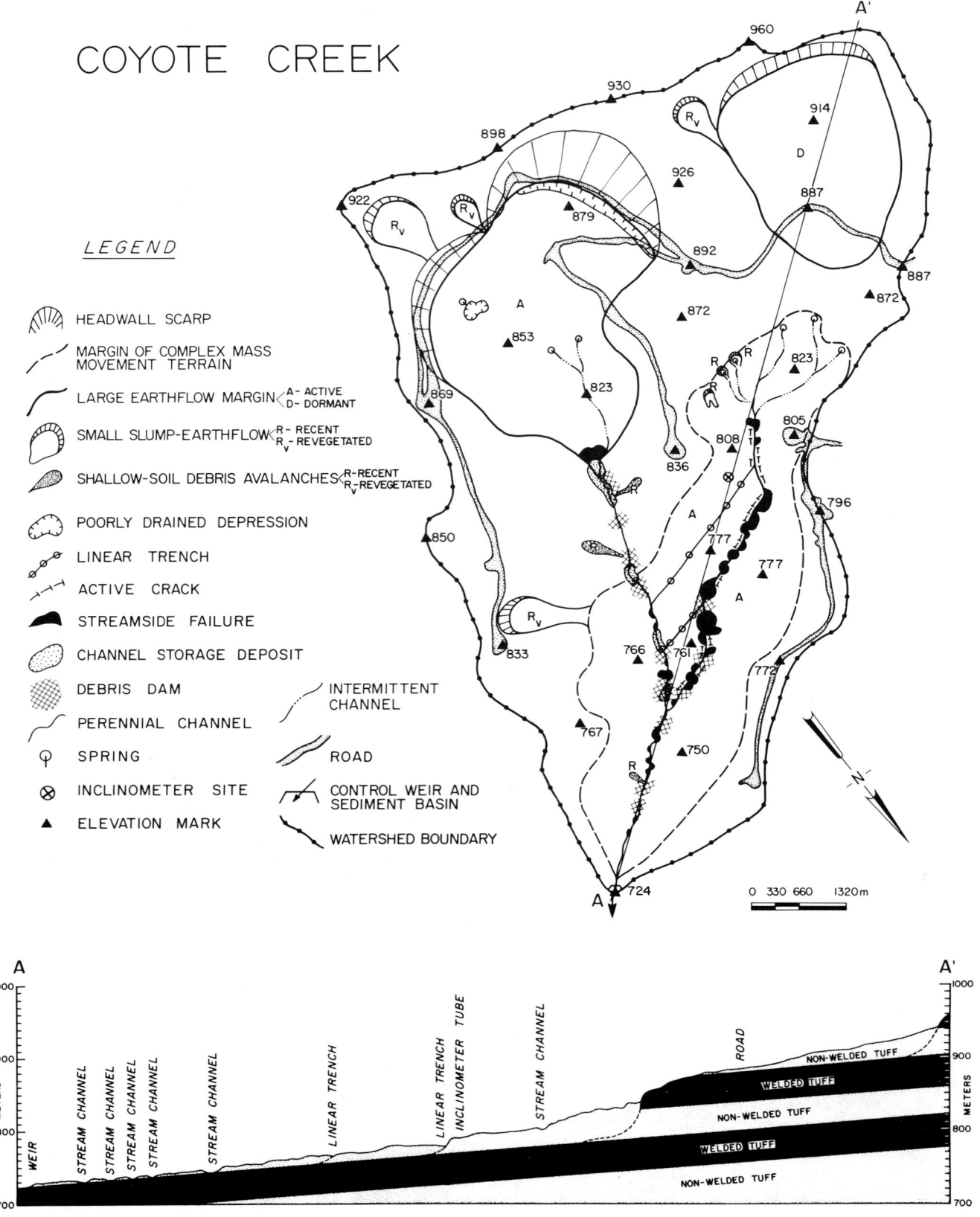

Figure 2. Map of Coyote Creek watershed 3 showing location of earthflows, complex mass-movement terrain, and associated channel features.

channel for later removal by streamflow.

Since 1966, both streamflow and sediment discharge have been monitored in the mass-movement study basin and three adjacent watersheds. The principal purpose is to determine water yield and nutrient outflow under forested conditions and following several types of management activities.[2] Two roads were constructed in watershed 3 in 1971, and the watershed was logged by clearcutting in 1972.

The annual bedload per metre of perennial stream channel was estimated from the volume of sediment removed from a weir at the mouth of the watershed. From 1966 to 1970 this was less than 0.01 m^3/m·yr (Table 1; R. L. Fredricksen, 1976, personal commun.). In 1971, sediment yield dramatically increased to approximately 0.05 m^3/m·yr as the result of unusually heavy winter precipitation and possibly increased runoff following construction of logging access roads. The bedload materials were derived primarily from debris avalanching and slumping along the banks of the stream draining the watershed. In 1972, the first year after the entire watershed was clearcut and also a year of exceptionally heavy rainfall, bedload movement tripled over prelogging and road-building levels to an estimated volume of 0.16 m^3/m·yr. During this period, two new debris avalanches reached the channel from midslope, possibly resulting from reduction of rooting strength of vegetation following logging. Part of this overall bedload increase is due to surface erosion of severely scarified soil resulting from piling of slash after logging. However, the greater part can be directly linked to mass movements along the channel.

Reconnaissance of the area and dissection of the weir deposits immediately after a major storm during the winter of 1972 exposed layering of poorly sorted sedimentary materials separated by zones of organic matter defining short periods or pulses of heavy sediment deposition and channel flushing. Such pulses probably result from repeated episodes of slumping and debris avalanching into the channel above the weir. A survey of the channel above the weir showed that nine new bank slumps and debris avalanches had occurred as a result of the storm. Two of the slumps were each large enough to provide the volumes of material necessary to fill the weir basin (approximately 5.4 m^3). Since 1972, bedload yields have been much lower but still substantially greater than prelogging levels. This reflects accelerated bank erosion and removal of sediment stored in the channel owing to increased peak flows. A detailed survey of the watershed in 1974 revealed over 50 sites of active bank slumping and debris avalanching along the active stream channel with slump volumes ranging from 5.6 to 350 m^3. Much of this material moved into the channel, diverting it or causing water to flow beneath the surface. At least 12 debris dams have blocked the channel, leading to temporary storage of from 45 to 1,340 m^3 of alluvium, soil, and organic debris behind each dam. In 1974 the total volume of material available for stream trans-

---

[2]Study 1602-10, "A study of the effects of timber harvesting on small watersheds in the Sugar pine, Douglas fir area of southeastern Oregon" (USDA Forest Service, Forestry Sciences Laboratory, Corvallis, Oregon).

TABLE 1. ESTIMATED ANNUAL BEDLOAD EXPORT FROM COYOTE CREEK WATERSHED 3

Year	Precipitation (cm)	Total bedload volume (m^3)	Volume/unit stream length (m^3/m)	Basin condition
1966	120.5	3.9	0.0091	Forested
1967	118.5	0.5	0.0012	Do.
1968	87.6	0.6	0.0014	Do.
1969	110.9	0.2	0.0005	Do.
1970	116.3	1.3	0.0030	Do.
1971	155.7	21.9	0.051	Roads
1972	153.3	69.9*	0.163	Clearcut
1973	89.5	3.2	0.0074	Do.
1974	156.5	46.4	0.108	Do.
1975	122.6	7.7	0.018	Do.

*Minimum estimate: basin overflowed.

port was estimated to be 3,100 m^3. Of this, 1,090 m^3 was stored as slump blocks and 2,010 m^3 was stored behind debris dams.

Inclinometer tube measurements in the soil and deeply weathered volcaniclastic rocks in the lower half of the watershed (Fig. 2) indicate movement along the plane of maximum deformation in a N50°E direction toward the mouth of the valley. Surface movement is approximately 10.5 mm/yr and movement at the base of the active movement zone at a depth of approximately 6.0 m is 9.1 mm/yr. This movement exhibits strong seasonal variation with most movement taking place during the fall and winter rainy period when maximum soil-water levels occur (Fig. 3). The similarity of the movement rates at the surface and the base of the movement zone suggests a fairly uniform rate of deformation of approximately 10 mm/yr above a lower boundary layer corresponding to the saprolite contact or the top of an intervening layer of relatively resistant welded tuff or flow rock.

The stream banks in the watershed average at least 1 m in height and show about the same magnitude of active creep and slumping into the stream on both sides of the channels. Thus, assuming an average mass-movement rate of 10 mm/yr, soil material is supplied to the stream channel in the lower part of the watershed at an annual rate of 0.02 m^3/m of channel. There are approximately 430 m of active perennial-stream channel in watershed 3. Therefore, approximately 8.6 m^3 of soil are made available annually for stream transport by creep-related processes. The average annual yield since 1972 has been 31.8 m^3/yr. This suggests that about 27% of the annual yield is being supplied by creep-related processes. The remainder is supplied from earthflow movement, surface erosion, debris avalanching, or reworking of channel deposits.

Lookout Creek Earthflow

The Lookout Creek earthflow site (sec. 30, T. 15 S., R. 6 E.) is a complex mass-movement terrain in the H. J. Andrews Experimental Forest approximately 90 km east of Eugene, Oregon (Fig. 1). Much of the earthflow has developed in a

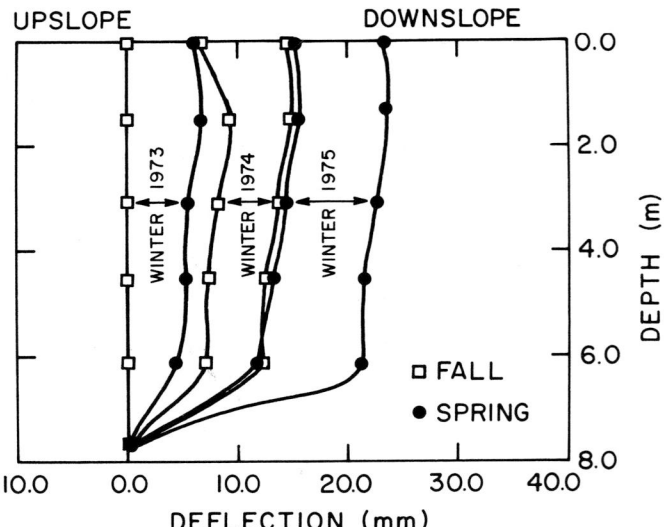

Figure 3. Apparent seasonal deformation along N70°E direction recorded in inclinometer tube installed in complex mass-movement terrain at Coyote Creek. Movement is directly plotted from field data and has not been projected into the plane of maximum deformation (N50°E). Note minor upslope movement of tube during summer of 1974 owing to settlement in hole.

variety of volcaniclastic rocks, but the headwall occurs in an area of capping basalt flows.

The area experiences average annual precipitation of more than 240 cm, falling mainly between October and May. A wet snowpack persists from December through April.

Most of the earthflow terrain is covered with a mixture of old-growth Douglas fir and western red cedar, 300 to 500 yr old, and a stand of the same species that developed following wildfire in the mid-1800s. About 1.5 ha of the forest at the earthflow was clearcut logged in 1968, and an equal area along the lower east side was clearcut in 1961.

The earthflow covers an area of approximately 20 ha on a south-facing valley wall. The flow extends from a rocky headwall at an elevation of 1,010 m downslope 900 m into Lookout Creek at a 790-m elevation (Fig. 4).

Topography on the earthflow surface is very irregular because of earthflow movement and stream erosion processes. Open cracks as much as 1.5 m wide have developed in areas of active tensional or shear deformation. Active crack systems bound major blocks or subunits of the earthflow that are presently undergoing differential movement (Fig. 4). There are also numerous inactive cracks defined by scarps or linear depressions. It appears that differential shear or tensional movement of less than about 1 cm/yr does not produce open, conspicuously active cracks, because litterfall, surface erosion, and growth of vegetation are effective in obscuring fine-scale features of ground breaks.

Drainage patterns on the earthflow are very irregular owing to frequent disruption by earth movement (Fig. 4). In several instances, streams have been channeled along tensional and shear cracks. Discontinued gullies, unusual features in the western Cascades, have developed along restricted stream reaches. Surface-water movement has been altered by the formation of poorly drained depressions on the uphill side of rotational slump blocks and where drainage has been obscured by fallen trees.

Movement History and Rates

A part of the history of earthflow movement may be learned from geomorphic and dendrochronologic observations, as well as by direct measurement. Various observation methods yield information on the age of the earthflow movement, rate of relative motion between discrete blocks within the earthflow, absolute movement of portions of the earthflow relative to stable reference points, and creep deformation within individual earthflow blocks.

Arrays of 4 to 9 stakes were established on several areas of the earthflow where shear and tensional movement appear to be very active (locations shown in Fig. 4). The stake arrays

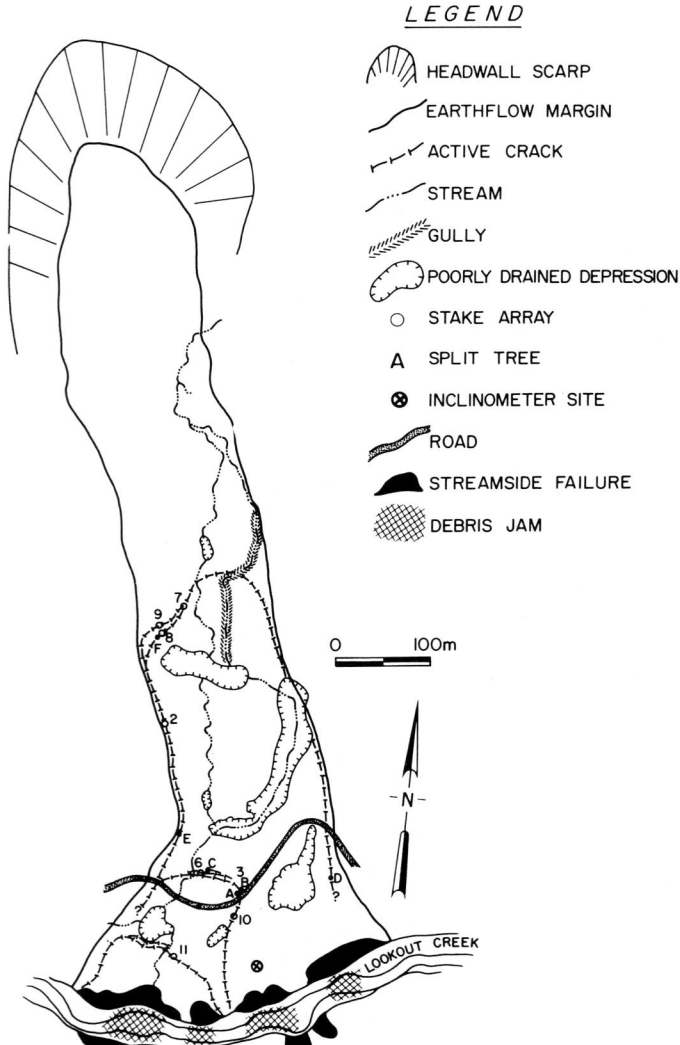

Figure 4. Map of Lookout Creek earthflow. Mapped by G. Lienkaemper using compass and range-finder. See also Tables 2 and 3.

recorded the relative motion between separate blocks within the earthflow and, in the case of set 2, between a block of earthflow and adjacent, relatively stable terrain.

Movement data for several stake arrays are shown in Table 2. Limited observations suggest that differential movement rates vary from 1.8 to 9.8 cm/yr. At arrays, which have been in for two years, movement for the 1974–1975 wet season (October through March) was only 50% to 70% of the 1975–1976 wet-season movement. Some of this difference may reflect the 8% lower precipitation of the 1974–1975 wet season and heavy precipitation early in the 1975–1976 wet season. (Cited precipitation data were collected at a meteorology station 8.5 km west of the earthflow.)

The influence of moisture on earthflow movement is seen more clearly in readings taken every 1 to 4 mo between August 1975 and June 1976. A plot of movement rate (centimetres per month) versus precipitation rate (centimetres per month) for the periods of observation clearly shows accelerated mass movement during periods of high-moisture availability (Fig. 5). Maximum rates of differential earthflow movement were recorded during the December 7, 1975, to February 9, 1976, period of highest precipitation. Early in the wet season, movement rates were low even for precipitation of 27 cm/mo because of very low antecedent moisture conditions. During spring snowmelt, late in the wet season, movement rate was high relative to precipitation because of the presence of residual water from earlier in the wet season, plus incoming water from snowmelt. In the month following May 18, differential movement at the three stake arrays was negligible.

An inclinometer tube installed near the toe of the earthflow has been used to record deformation within an individual block of the earthflow adjacent to Lookout Creek (Fig. 6). The tube (located in Fig. 4) reveals downslope movement of 8.25 mm/yr at the surface along a plane of maximum deformation oriented approximately S82°E. Near the base of the movement zone at a depth of 4.3 m, deformation is occurring at an average rate of 2.3 mm/yr to approximately S60°E. Creep deformation is occurring throughout the profile with

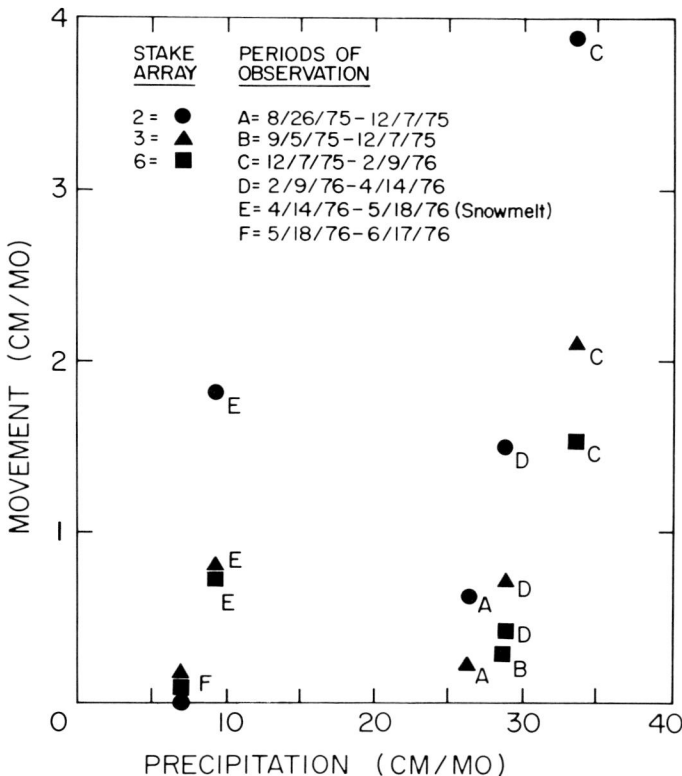

Figure 5. Rate of precipitation versus relative earthflow movement rate at several sites on the Lookout Creek earthflow for wet season of water year 1976.

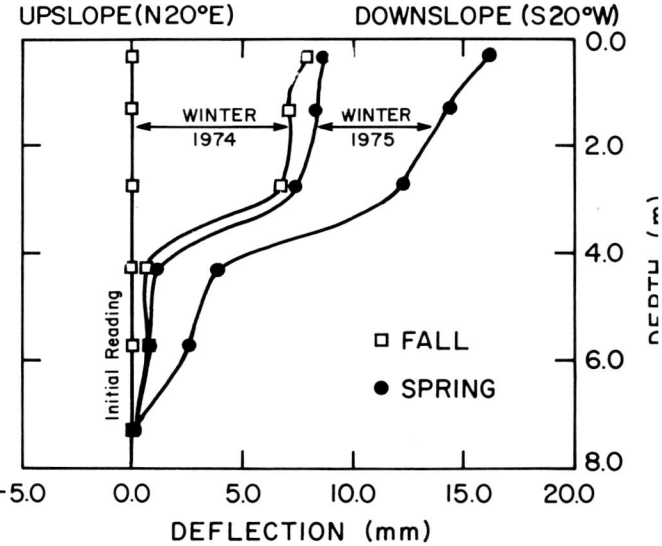

Figure 6. Apparent seasonal deformation along S20°W direction recorded in inclinometer tube installed in Lookout Creek earthflow. Movement is directly plotted from field data and has not been adjusted into the plane of maximum deformation. Note minor upslope movement of tube during summer of 1975 owing to settlement of tube.

TABLE 2. MOVEMENT DATA FROM STAKE ARRAYS ON LOOKOUT CREEK EARTHFLOW

Stake array	Wet seasons			
	1974–1975		1975–1976	
	10/11/74–8/25/75	12/12/74–8/25/75	8/25/75–5/18/76	9/5/75–5/18/76
2	..	7.4	12.4	..
3	5.0	..	7.6	..
6	..	..	..	5.8
7	..	..	..	3.0
8	..	..	..	4.8
9	..	..	..	5.5
10	..	..	..	7.6
11	..	..	..	1.8

Note: Differential movement in centimetres. Precision of measurement is approximately ±2 mm. Locations of stake arrays shown in Figure 4.

probable discrete failure at about 4 m. This may be related to incipient streamside slumping and probably accounts for the difference in orientation of the plane of maximum movement activity, with the fastest movement occurring during the months of high precipitation.

Unlike the Coyote Creek site, which was logged before mass-movement studies began, the Lookout Creek earthflow still carries many trees and the historical record they embody. Numerous broadly curved old-growth Douglas fir trees on the earthflow indicate that the area has been active for at least the past several centuries.

Several of the active tension and shear cracks on the earthflow are straddled by trees split up the center to as much as 7 m above ground level. These trees have grown scar tissue onto the wound, and counts of the annual rings in the post-split wood yield estimates of the time since the split and differential earth movement began to develop. Measurement of the distance that fragments of a tree have been pulled apart at ground level may be used to make estimates of rates of differential movement between adjacent blocks (Table 3; Fig. 4). The splitting rates are generally only minimum estimates of differential earth-movement rates, because most of the trees are not tightly coupled to the blocks on opposite sides of the crack. Estimated rates of tree splitting, ranging from 0.3 to 9.3 cm/yr, are in general agreement with short-term rates of differential earth movement on the basis of direct measurements. Dated periods of tree splitting began as much as 74 years ago. Overall earthflow activity probably began well before that time.

Geomorphic considerations also suggest that earthflow movement was initiated at least several centuries ago. Cross profiles of the floor of Lookout Creek valley above, within, and below that earthflow area reveal a constriction of the valley bottom by about 30 m through the area of direct earthflow influence (Fig. 7). At a movement rate of 10 cm/yr, which is reasonable in light of dendrochronologic and direct measurements of movement, it would take about 300 yr to achieve the existing valley-floor geometry. Extensive bank cutting on the south side of Lookout Creek indicates the earthflow pushed Lookout Creek against the south valley wall at least several decades ago and probably much earlier.

TABLE 3. TREE SPLITTING ON LOOKOUT CREEK EARTHFLOW

Tree	Time since split initiation (yr)	Splitting rate (cm/yr)
A	≥31*	≤6.9*
B	37	1.3
C	74	0.3
D	51	1.3
E	23	9.3
F	43	2.7

Note: Location of trees noted in Figure 4.
*Minimum estimate of time since split initiation. Therefore, splitting rate is a maximum estimate.

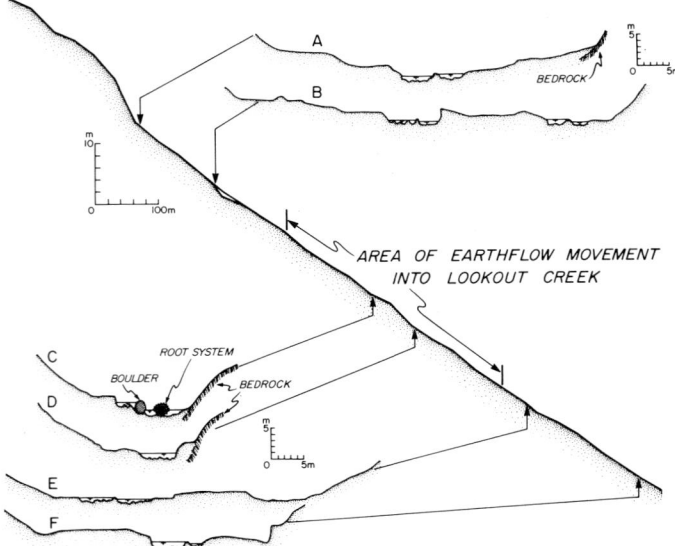

Figure 7. Long profile and cross profiles (A-F) Lookout Creek valley floor in the vicinity of the earthflow. Long profile based on field measurements of channel gradient over 30-m channel sections.

The complete history of the earthflow most likely extends over a period much longer than three centuries.

Impacts on Lookout Creek

The Lookout Creek earthflow has a variety of impacts on Lookout Creek, both in the immediate area of the earthflow and for hundreds of metres downstream. The erosional history and landforms in the area of stream-earthflow interface are determined by the relative rates of earthflow processes that tend to close the channel and by fluvial processes that tend to erode and open the channel.

Movement of the earthflow has forced Lookout Creek against the south wall of the valley. Consequently, in plan view the channel has a broadly arcuate pattern shaped by the convex front of the earthflow (Fig. 4). Above and below the earthflow, the gross geometry of the channel pattern is the result of fluvial processes and changes in flow direction owing to accumulations of large organic debris. Outside the area of earthflow influence, the valley floor includes a low terrace and floodplain as much as 55 m wide and 1 to 4 above the active channel (Fig. 7). These broad, valley-floor cross profiles (Fig. 7) contrast markedly with the narrow, steep-walled channel and adjacent slopes characteristic of the stream reach through the active earthflow (Fig. 7).

The long profile of Lookout Creek is remarkably straight through the area of earthflow impact and for more than 200 m both upstream and downstream (Fig. 7). Impact of the earthflow on this long profile is not clear. If there were no earthflow impact, the channel might have a more typical concave long profile. Earthflow movement may have caused local increases in stream-bed elevation as a result of high levels of sediment and large organic debris inputs to the stream in the area of direct impact and immediately downstream. This effect could also lead to some increased deposition and decreased gradient just upstream from the earthflow.

Large quantities of mineral sediment and organic debris enter Lookout Creek by slumping, shallow debris avalanches, and root throw. Surface erosion and sloughing from undercut banks and debris-avalanche scars appear to be of significant but secondary importance because of rapid revegetation of bare soil areas. Most of the input comes from the earthflow side of the stream, but several small debris avalanches have occurred in response to stream undercutting of the opposite bank (Fig. 4).

Sediment input to the stream by bank failure appears to be an episodic process caused by a series of factors. The earthflow gradually constricts the channel, narrowing it at a rate measured in centimetres per year. The banks of the constricted channel become progressively more susceptible to undercutting and failures triggered by infrequent floods. Earth movement also tips the massive old-growth trees in the streamside area, making them more susceptible to windthrow, which may in turn trigger stream-bank failures. During extreme precipitation, high rates of subsurface water movement at the distal end of the earthflow also increase the potential for shallow failures where the front of the earthflow encroaches on the stream. The earthflow itself may experience a brief acceleration of movement as a result of prolonged periods of heavy precipitation. This combination of storm event and progressive long-term buildup of stress results in the occurrence of episodes of streamside failures.

Seven streamside debris avalanches and slides larger than 75 m^3 have occurred along Lookout Creek in the earthflow area since about 1950. The age of brushy and herbaceous vegetation, which covers much of the slide-scar surfaces, indicates that most of the erosion from these sites appears to have occurred in the severe storms of the winter of 1964–1965. Measurements of the geometry of slide scars show that approximately 5,000 m^3 of soil and rock debris has been supplied to Lookout Creek by the inventoried slides. Some of this material was probably moved into position for potential input to the stream by earthflow movement occurring more than 25 yr ago.

The annual rate of buildup of potential sediment supply to the stream may be estimated by assuming that the 300-m-long, 9- to 16-m-high front of the earthflow is encroaching on the stream at some assumed rate. A movement rate of 10 cm/yr would move 340 m^3/yr of earthflow debris to the stream. At this rate it would take about 15 yr to recharge the streambank areas with 5,000 m^3 of material available for input into the stream area. These calculations suggest that rather continuous earthflow may make material available for pulses of sediment input to the stream every few decades.

Downstream Impacts

During extreme floods large quantities of organic and inorganic debris may be flushed out of the section of earthflow-constricted channel and moved downstream for hundreds of metres as catastrophic torrents of debris. In the course of this movement the debris torrents damage or destroy vegetation on the floodplain and low terraces and expose mineral soil in the stream banks. Where a torrent stops, it sets up an area of sediment and large organic debris deposition that may cover several hectares. Such a chain of events was last triggered from the lower end of the Lookout Creek earthflow in the winter of 1964–1965.

Several characteristics of the area of interaction between earthflows and streams make them common sites for the triggering of debris torrents. The high sediment and organic matter inputs to the stream from the earthflow result in high concentrations of potentially mobile material temporarily stored behind loosely structured debris dams in the stream channel through the earthflow. Failure of debris dams during high flow may initiate debris torrents. Bank failures may lead to development of short-lived dams that are quickly overtopped and cut away, resulting in pulses of debris movement downstream. The narrowness of the valley bottom through the earthflow area restricts the lateral movement of the stream and reduces its ability to bypass obstructions in the channel. Stream energy during high flow is focused on obstructions until they are moved downstream. Where the valley floor is broader, the stream may spread out, flowing around and through debris deposits and slowly reworking the material over the course of decades and centuries.

SUMMARY

Areas of active creep and earthflow in the western Cascade Range form complex mass-movement terrains. Creep motion initiates discrete failure, locally resulting in slump and earthflow development. Thereafter, the processes operate together. Slow movement of large masses of earth develop unstable sites where smaller, rapid failure occurs.

A single mass-movement terrain may include a number of units moving at quite different rates. Recorded creep rates range from undetectable levels to 15 mm/yr. Observed differential movement between blocks within the Lookout Creek earthflow range to 10 cm/yr. Movement rate is regulated by moisture availability; periods of high precipitation and (or) snowmelt correspond with times of accelerated movement. Bedrock geology determines the characteristics of deep-seated mass-movement features. Extensively altered, deeply weathered volcaniclastic material is most prone to creep and earthflow activity. In areas of rock types with contrasting physical characteristics, bedrock configuration determines the distribution and morphology of mass-movement landforms.

Complex mass-movement terrains supply large quantities of sediment and organic debris to streams, thereby altering channel geometry. A comparison of sediment discharge to mass-movement activity and channel storage suggests that creep may account for as much as 27% of the annual sediment discharge from Coyote Creek. Although deep-seated processes may operate at slow rates, sediment input to streams may occur as infrequent pulses during major storms.

Geomorphic observations and tree-ring analysis indicate that the history of individual mass-movement terrains in the westen Cascades may have spanned centuries and possibly millennia.

ACKNOWLEGMENTS

We thank S. H. Wood, H. Schlicker, and E. H. Muller for helpful discussions.

This work was supported in part by National Science Foundation Grant BMS-7602656 to the Coniferous Forest Biome, U.S. Analysis of Ecosystems (International Biological Program).

REFERENCES CITED

Anderson, H. W., 1969, Snowpack management, *in* Snow, seminar of Oregon State University: Oregon Water Resources Research Inst., p. 27–40.

Bjerrum, L., 1967, Progressive failure in slopes of overconsolidated plastic clay and clay shales: Am. Soc. Civil Engineers Proc., Jour. Soil Mechanics and Found. Div., v. 93, p. 1–49.

Carson, M. A., and Kirkby, M. J., 1972, Hillslope form and process: London, Cambridge Press, 475 p.

Colman, S. M., 1973, The history of mass movement processes in the Redwood Creek basin, Humboldt County, California [M.S. thesis]: University Park, Pennsylvania State Univ., 151 p.

Culling, W.E.H., 1963, Soil creep and the development of hillside slopes: Jour. Geology, v. 71, p. 127–161.

Franklin, J. F., and Dyrness, C. T., 1973, Natural vegetation of Oregon and Washington: U.S. Dept. Agriculture, Forest Serv. Gen. Tech. Rept. PNW-8, 417 p.

Goldstein, M., and Ter-Stepanian, G., 1957, The long-term strength of clays and deep creep of slopes: 4th Internat. Conf. Soil Mechanics and Found. Eng. Proc., v. 2, p. 311–314.

Gray, D. H., 1970, Effects of forest clearcutting on the stability of natural slopes: Assoc. Eng. Geologist Bull., v. 7, p. 45–67.

Haefeli, R., 1965, Creep and progressive failure in snow, soil, rock, and ice: 6th Internat. Conf. Soil Mechanics and Found. Eng. Proc., v. 3, p. 134–148.

Harr, R. D., 1976, Forest practices and streamflow in western Oregon: U.S. Dept. Agriculture, Forest Serv. Gen. Tech. Rept. PNW-49, 18 p.

Kays, M., 1970, Western Cascades volcanic series, South Umpqua Falls region, Oregon: Ore Bin, v. 32, p. 81–94.

Kitamura, Y., and Namba, S., 1966, A field experiment on the uprooting resistance of tree roots (I): 77th Mtg. Japanese Forest Soc. Proc., p. 568–570.

——1968, A field experiment of the uprooting resistance of tree roots (II): Proc. 79th Meet. Japanese Forest Soc., p. 360–361.

Kojan, E., 1968, Mechanics and rates of natural soil creep: Proc. 1st Session Internat. Assoc. Eng. Geol., Prague, p. 122–154.

O'Loughlin, C. L., 1974, The effects of timber removal in the stability of forest soils: Jour. Hydrology (NZ), v. 13, p. 121–134.

Peath, R. C., Harward, M. E., Knox, E. G., and Dyrness, C. T., 1971, Factors affecting mass movement of four soils in the western Cascades of Oregon: Soil Sci. Soc. America Proc., v. 35, p. 943–947.

Peck, D. L., Griggs, A. B., Schlicker, H. G., Wells, F. G., and Dole, H. M., 1964, Geology of the central and northern parts of the western Cascade Range in Oregon: U.S. Geol. Survey Prof. Paper 449, 56 p.

Rothacher, J., 1963, Net precipitation under a Douglas fir forest: Forest Sci., v. 9, p. 423–429.

——1971, Regimes of streamflow and their modification by logging, *in* Krygier, J. T., and Hall, J. D., eds., Forest land uses and the stream environment: Corvallis, Oregon State Univ., p. 40–54.

——1973, Does harvest in west slope Douglas-fir increase peak flow in small forest streams: U.S. Dept. Agriculture Forest Serv. Research Paper PNW-163, 13 p.

Saito, M., and Uezawa, H., 1961, Failure of soil due to creep: 5th Internat. Conf. Soil Mechanics and Found. Eng. Proc., v. 1, p. 315–318.

Schlicker, H. G., Corcoran, R. E., and Bowen, R. G., 1961, Geology of the Escola State Park landslide area, Oregon: Ore Bin, v. 23, p. 85–90.

Swanson, F. J., and James, M. E., 1975, Geology and geomorphology of the H. J. Andrews Experimental Forest, western Cascades, Oregon: U.S. Dept. Agriculture Forest Serv. Research Paper PNW-188, 14 p.

Swanston, D. N., 1969, Mass wasting in coastal Alaska: U.S. Dept. Agriculture Forest Serv. Research Paper PNW-83, 15 p.

Swanston, D. N., and Swanson, F. J., 1976, Timber harvesting, mass erosion and steepland forest geomorphology in the Pacific Northwest, *in* Coates, D. R., ed., Geomorphology and engineering: Stroudsburg, Pa., Dowden, Hutchinson and Ross, Inc., p. 199–221.

Terzaghi, K., 1953, Some miscellaneous notes on creep: 3rd Internat. Conf. Soil Mechanics and Found. Eng. Proc., v. 3, p. 205–206.

Varnes, D. J., 1958, Landslide types and processes, *in* Eckel, E. B., ed., Landslides and engineering practice: Natl. Research Council, Highway Research Board Spec. Rept. 29, p. 20–47.

Wilson, S. D., 1970, Observational data on ground movements related to slope instability: Jour. Soil Mechanics and Found. Div., Am. Soc. Civil Engineers Proc., v. 96, p. 1521–1544.

Youngberg, C. T., Harward, M. E., Simonsen, G. H., and Rai, D., 1975, Nature and causes of stream turbidity in a mountain watershed, *in* Bernier, B., and Winget, C. H., eds., Forest soils and forest land management: Quebec, Laval Univ., p. 267–283.

MANUSCRIPT RECEIVED BY THE SOCIETY SEPTEMBER 7, 1976
MANUSCRIPT ACCEPTED SEPTEMBER 17, 1976

CONIFEROUS FOREST BIOME CONTRIBUTION NO. 246

PART 3
Specific and Local Studies

7

Landslides and the weathering of granitic rocks

PHILIP B. DURGIN
*Pacific Southwest Forest and Range Experiment Station, Forest Service, U.S. Department of Agriculture, Berkeley, California 94701
(stationed at Arcata, California 95521)*

ABSTRACT

Granitic batholiths around the Pacific Ocean basin provide examples of landslide types that characterize progressive stages of weathering. The stages include (1) fresh rock, (2) corestones, (3) decomposed granitoid, and (4) saprolite. Fresh granitoid is subject to rockfalls, rockslides, and block glides. They are all controlled by factors related to jointing. Smooth surfaces of sheeted fresh granite encourage debris avalanches or debris slides in the overlying material. The corestone phase is characterized by unweathered granitic blocks or boulders within decomposed rock. Hazards at this stage are rockfall avalanches and rolling rocks. Decomposed granitoid is rock that has undergone granular disintegration. Its characteristic failures are debris flows, debris avalanches, and debris slides. Saprolite is residual granitic rock that is vulnerable to rotational slides and slumps. As a granitic rock mass progressively decomposes, the critical slope angle decreases, allowing slope failures throughout its weathering history. Failures in granitic rock are more abundant during the advanced stages of decomposition. Therefore, landslides are most common in the humid tropics, where intense chemical weathering occurs. Identification of the granitoid's weathering stage will help the engineering geologist evaluate the slope-stability hazards of an area.

INTRODUCTION

Granitic rocks are not associated with landslides in the minds of most people, yet huge and numerous slides have occurred in some granitic areas. Batholiths surrounding the Pacific Ocean basin provide good examples of landslides in granitic rocks. These batholiths are associated with active subduction zones; therefore, the granitic terranes are relatively young and topographically rugged. Granodiorite is the predominant rock type of these plutons, and true granite is rare (Roddick, 1974). Granitic masses are relatively resistant to decomposition, so they commonly occur as mountainous erosional remnants. Nevertheless, granitoids undergo progressive physical, chemical, and biological weathering that weakens the rock and prepares it for mass movement. Rainstorms and earthquakes then trigger slides at susceptible sites.

The minerals of granitic rock weather according to this sequence: plagioclase feldspar, biotite, potassium feldspar, muscovite, and quartz. Biotite is a particularly active agent in the weathering process of granite. It expands to form hydrobiotite that helps disintegrate the rock into grus (Wahrhaftig, 1965; Isherwood and Street, 1976). The feldspars break down by hyrolysis and hydration into clays and colloids, which may migrate from the rock. Muscovite and quartz grains weather slowly and usually form the skeleton of saprolite. Some granitoids undergo hydrothermal alteration, but this review of the literature does not specifically consider them.

The physical appearance and properties of granitic rock change as weathering progresses. Several investigators have divided the transition into stages (Ruxton and Berry, 1957; Deere and Patton, 1971; Clayton and Arnold, 1972). This paper uses a four-stage classification of weathering products (Fig. 1): (1) fresh rock, (2) corestones, (3) decomposed granitoid, and (4) saprolite. All four stages may exist on an individual slope, but geographically the later stages of weathering increase in occurrence where there are high temperatures and ample precipitation. Fresh rock contains a maximum of 15% weathered material that forms in the joint system. The corestone stage ranges from 15 to 85% weathered rock enclosing remnants of fresh rock. Decomposed granitoid consists of 85 to 100% weathered disintegrated rock that can be broken down into granules. Saprolite is a fine-grained residual rock that generally has an upper lateritic layer.

FRESH ROCK

The susceptibility of fresh bedrock to rockfalls, rockslides, or block glides depends upon a number of conditions, includ-

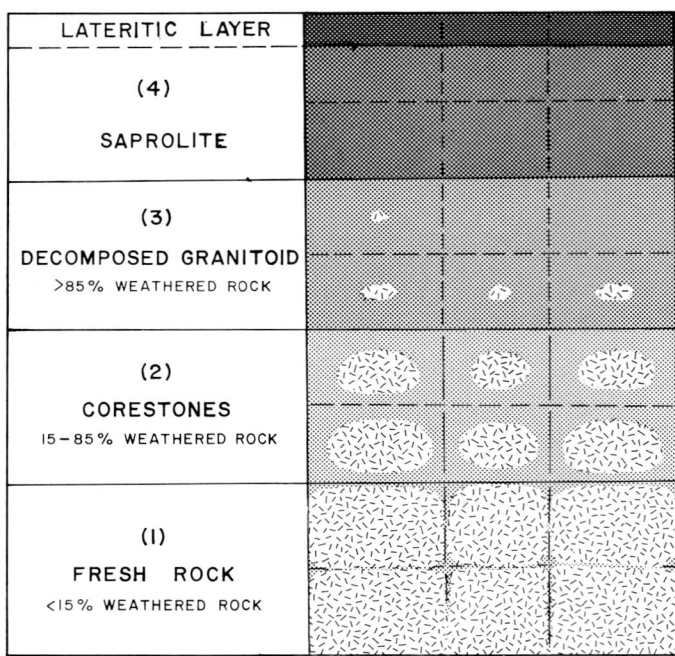

Figure 1. Stages of granitic rock weathering.

ing (1) the angle of shearing resistance of the jointed rock, (2) the effective cohesion, and (3) the seepage pressures of water in the joints (Terzaghi, 1962b). A simple but useful method of studying rockslide occurrence is to consider only the angle of inclination of rock joints. The critical slope angle for rocks with a random joint pattern is about 70° (Terzaghi, 1962b), but sheeting in granite often creates joints that are subparallel to the ground surface, with critical slope angles as low as 30° to 40°. Slabs of sheeted granite may slide down a slope as block glides (Terzaghi, 1962a). As weathering progresses, decomposed granite forms along the joint planes and decreases the effective cohesion while increasing the void ratio. Ground water seeps through these permeable zones (Le Grand, 1949) and generates hydrostatic pressures in the joints.

The Emerald Bay landslide, bordering Lake Tahoe in the Sierra Nevada of California, has undergone two rockslide avalanches and occasional rockslides. The failure originated in fresh granite having fracture planes dipping 30° to 40° downslope toward the lake (McCauley, 1975). The first rockslide avalanche is attributed to road construction across the slope; the second was triggered by a period of very high rainfall. In contrast, the rockslide avalanche of Cerro Condor-Seneca in Peru occurred on a slope of 37° during an abnormally dry season (Snow, 1964). That slide apparently was triggered by desiccation and cracking of the clay joint fillings in the granodiorite.

Debris avalanches or debris slides can occur when fresh granite is near the surface and is covered by organic debris, colluvium, or soil (Fig. 2). Smooth-surfaced granite can reduce the effective friction angle with the overlying material. In additon, water flows through the thin permeable zone above the fresh bedrock, producing pore-water pressures and seepage forces during rainstorms. This process has been observed in steeply sloping forest soils of British Columbia and Alaska, where piezometric levels approached the ground surfaces during periods of high rainfall or snowmelt (O'Loughlin, 1972; Swanston, 1967).

Debris avalanches and slides on fresh granite have also been noted in the Eastern United States, particularly in the Appalachian Blue Ridge Mountains (Williams and Guy, 1973; Scott 1972), and in the White Mountains of New Hampshire (Flaccus, 1958; Pirsson and Rice, 1911).

Evidence of rockslides and rockfalls is common in the geologic record. Talus and scree slopes are accumulations of rock debris that originated with rockfalls or rockslides.

CORESTONES

The weathered zone develops outward from the joints and isolates blocks or boulders of fresh rock to form corestones. In some areas, corestones become remnants on the ground surface and may roll down slopes during rainy periods, causing extensive damage (Barata, 1969; Da Costa Nunes, 1969). The type of slide most characteristic of this weathering stage is the rockfall avalanche. It consists of a large rock mass that falls off a steep slope, generating a stream of fast-moving debris. Earthquakes commonly trigger this type of slide by dislodging the partially decomposed rock.

Slide Mountain in Western Nevada offers an example of a rockfall avalanche in granodiorite (Thompson and White, 1964). Intermittent slides have occurred at the scar on this mountain since Pleistocene time. The bedrock has been crushed extensively, and joints dip parallel to the slope. The composite slide consists of about 96,000,000 m³ of debris — one of the largest on record.

The Mount Huascaran slide in Peru showed that if population centers are in the path of a rockfall avalanche, the results can be catastrophic. The Peruvian earthquake of 1970 — the greatest disaster in the history of the Western Hemisphere — killed about 70,000 people. The quake triggered a slide off a cliff face of partially weathered granodiorite. A stream of debris sped down the valley at 400 km/h, killing 2,000 people in one town and 19,000 in another. Eyewitnesses reported that the debris swept downslope "with a deafening noise and was everywhere accompanied by a strong turbulent airblast" (Ericksen and others, 1970). Browning (1973) concluded that because of its high velocity and movement as a single mass, the Mount Huascaran slide must have moved almost without friction over an air cushion. Arguing against this interpretation, Hsü (1975) suggested that the granitic blocks were dispersed in a dense interstitial mud and flowed down the valleys.

Some deposits of rockfall avalanches are identifiable in the rock record. For example, granitic breccia of a massive landslide that occurred during Miocene time was found in the Las Vegas Valley of southern Nevada (Longwell, 1974). The landslide debris from Slide Mountain still can be seen. The deposits indicate the number of slides, and their degree of weathering provides a clue to the time of their occurrence.

DECOMPOSED GRANITOID

At the stage of decomposed granitoid, granular disintegration takes place, and crystals become increasingly detached from each other. Erosion of the material produces grus. Rainstorms usually bring on the failures that generate debris flows and debris avalanches, although dry debris slides also take place. The most widespread slide problems have been reported at this stage.

The biotite and feldspar weather first, causing microfractures and pores to form. Ground water can then leach out the resulting colloids and clays. As the granitoid decomposes, it decreases in bulk density and shear strength (Matsuo and others, 1968). A saturated decomposed granite has essentially no cohesion. Its angle of internal fraction depends on the grain-size distribution, but for design purposes it is assumed to be about 35° (Lumb, 1962; Gonsior and Gardner, 1971).

The weathering front in granitoids is abrupt, and water can be perched above the fresh impermeable bedrock. Subsurface water drains through the decomposed granitoid above the fresher rock. The seeping water removes very fine particles by solution and mechanical eluviation (Ruxton and Berry, 1957; Ruxton, 1958). After rainfall, seepage forces can be especially strong on steep slopes, contributing significantly to instability. Although seepage forces are usually neglected in analyzing slides by the infinite-slope method, Hartsog and Martin (1974) included them in their computer program for landslide prediction. Their analysis was specifically designed for granitoids of the Idaho batholith.

The degree of saturation of the decomposed granite is also important when the factor of safety is about one. When saturation reaches 100%, the apparent cohesion is zero, and pore-water pressures may cause failure. If the material is dry, however, the apparent cohesion is also zero, and dry sliding may follow. Dry debris slides occur on the granitic batholiths of southern California and are particularly common after summer wildfires (Krammes, 1965). Masses of decomposed granitoid slide into ephemeral stream channels, and debris flows may remove them during heavy winter rainstorms. These flows surge down channels and can cause extensive damage to homes (Scott, 1971).

Shallow debris avalanches have occurred in the Klamath Mountains batholiths of northern California and southern Oregon (Fig. 3). Tree roots growing in the decomposed granitoid help stabilize the slope. At some sites, rainstorms have triggered failures a few years after roadbuilding and clearcutting of forests.

Japan has undergone recurring disasters in its granitic areas. During such events, landslides and surface erosion injected slugs of decomposed granite (*masa*, in Japanese) into rivers. One such disaster occurred in 1964 at the Kamo-Daito area on the island of Honshu. A map of the weathering stages for this area shows that the landslides were concentrated in the decomposed granitoid and the saprolite (Oyagi, 1968). The decomposed granitoid had a type of immunity to failures once slides had occurred, because the weakest material failed and stronger ones remained. The saprolite had the potential for repeated landslides since the failures exposed rock having a low shear strength. The mapping also indicated that the depths of weathering stages were related to long-standing erosional base levels. Weathering continued to penetrate the rock until it reached the elevation of nearby streams.

Rio de Janeiro, on the east coast of South America, also provides examples of failures in decomposed granitoid. The city's highest recorded rains fell in the summer of 1966 and 1967. These storms resulted in tens of thousands of landslides on bedrock that is predominantly gneiss and granite. The landslides devastated a large area, and in one village about 1,000 people were killed. Jones (1973) concluded that the storm laid waste to "a greater landmass than any ever recorded in geologic literature." He provided striking photos as evidence.

Figure 2. Debris avalanche on fresh granitoid near Sawyers Bar, California. Water is flowing on surface of rock.

Figure 3. Debris avalanche in decomposed granitoid near Sawyers Bar, California.

Although several types of failures such as rockslides, rockfall avalanches, slumps, and rapid creep occurred, debris avalanches and debris flows were particularly pronounced (Barata, 1969). The thin decomposed mantle was commonly stripped off the slope to form a herringbone pattern in some areas and large shallow scars in others.

Earthquakes can effectively trigger shallow debris avalanches in decomposed granitic material. A statistical analysis of slides in the Bewani and Torricelli Mountains of New Guinea indicated that the distribution of failures in granitic rock was directly related to the epicenters of two earthquakes with magnitudes 7.9 and 7.0 (Simonett, 1967).

Landslide debris associated with this weathering stage is rarely found in the rock record because decomposed granite is highly erodible. Surface erosion produces gullies in the failure scars and discharges grus into the river system. Therefore, it is difficult to determine if a deposit of decomposed granite is the result of mass movement or surface erosion.

SAPROLITE

The final stage of weathering results in a whitish to brownish saprolite overlain by a red lateritic layer, which develops deeper with time. Residual granitoid consists predominantly of quartz, muscovite, and kaolinite. The void ratio increases until overburden pressures force the pore spaces to decrease as the rock compresses. This advanced stage of weathering is most common in the tropics, but it even occurs in the Klamath Mountains and other forested temperate regions, where leaching of the rock is more important than surface erosion.

A slope failure characteristic of saprolite is the slump or rotational slide. The rupture surface is no longer controlled by a fresh rock boundary as in earlier stages, but forms a circular failure surface. Ground water drains through the failure surface and relict joints, precipitating iron and manganese. The shear strength of these seams is usually one-half to two-thirds that of the saprolite (St. John and others, 1969). This type of landslide is most suited to traditional slope-stability analyses that assume rotational failures. Therefore, engineering geologists can usually determine the factor of safety for specific slopes in residual rock.

The degree of saturation strongly affects the strength of residual granitoid. In unsaturated saprolite the apparent cohesion can be as high as 2,000 g/cm^2, but it will drop to zero when saturated (Lumb, 1965). At some construction sites, deep cuts have encountered the water table in residual granitoid and produced rotational slides (Deere, 1957). Ground water can seep out at the base of the slope, causing piping that undermines the slope. The saturated cohesionless material has little shear strength, and failures may occur in the saprolite as well as along joint planes.

Parts of Rio de Janeiro have rotational slides in saprolite, but man usually contributed to their occurrence (Barata, 1969). Residents of the city destabilized slopes by excavating saprolite, adding surcharges to the tops of hills, and removing man-made structures that provided support to the slopes. When the heavy rains fell in 1966 and 1967, failures occurred in areas that had been made marginally unstable.

Hong Kong is underlain by a mantle of residual granitoid that is about 30 m thick (Lumb, 1965). A rainstorm in 1966 caused the most disastrous mass movements ever recorded in that area (So, 1971). The large, deep-seated failures involved rotational sliding and slumping. Observers reported that a great deal of subsurface water emerged from the slope coincident with failure. So (1971) inventoried the mass movements and concluded (99% confidence level) that they occurred in greater numbers in woodlands than in other types of vegetation. The tree roots probably had no stabilizing effect on deep slides. In addition, woodland soils have infiltration capacities that conduct rainfall to the subsurface rather than allowing overland flow.

Topographic evidence for ancient slumps and rotational slides is commonly observable in aerial photos. If the saprolite is excavated, slickensides in seams of iron and manganese provide evidence of past movement.

DISCUSSION

The number of landslides in granitic rocks compared with other lithologies depends on factors that promote decomposition, such as climate and erosional history. The humid tropics is the site of the most intense chemical weathering, and as a result many of the problem areas in granitoid, such as Hong Kong and Rio de Janeiro, are also within that zone. Rhodes (1968) compared the landslides in granitic rock to ten other lithologies in humid tropical New Guinea. He found that silicic igneous rocks had the most landslides per unit area. On the other hand, Radbruch and Crowther (1973) indicated that in California, granitoid has one of the lowest rates of slope failure of any rock type.

The engineering properties of granitic rock change as weathering continues. Granitoids break down progressively from massive blocks to a deep layer of clay-size particles. Therefore, the disciplines of both rock mechanics and soil mechanics are useful for investigating the slope stability of such materials. The shear strength and critical slope angle decrease as a granitic rock mass weathers. Merritt (1972) described how knowledge of the characteristic critical slope angle at each weathering stage was helpful in a construction project on intrusive rock in Colombia.

Man is more influential at the later stages of granitic weathering because he makes greater use of the gentler slopes, and he can excavate the material more easily. Each stage of weathering is susceptible to specific slope-stability hazards. If the stage of weathering is identified at a site, it will provide clues to the engineering properties of the material and help the engineering geologist predict the slope-stability hazards of proposed actions.

REFERENCES CITED

Barata, F. E., 1969, Landslides in the tropical region of Rio de Janeiro: Internat. Conf. Soil Mechanics and Foundation Engineering, 7th, Mexico City 1969, Proc., v. 2, p. 507–516.

Browning, J. M., 1973, Catastrophic rock slide, Mount Huascaran, north-central Peru, May 31, 1970: Am. Assoc. Petroleum

Geologists Bull., v. 57, p. 1335–1341.

Clayton, J. L., and Arnold, J. F., 1972, Practical grain size, fracturing density, and weathering classification of intrusive rocks of the Idaho batholith: U.S. Dept. Agriculture Forest Service Gen. Tech. Rept. INT-2, 17 p.

De Costa Nunes, A. J., 1969, Landslides in soils of decomposed rocks due to intense rainstorms: Internat. Conf. Soil Mechanics and Foundation Engineering, 7th, Mexico City 1969, Proc., v. 2, p. 547–554.

Deere, D. U., 1957, Seepage and stability problems in deep cuts in residual soils, Charlotte, N. C.: Am. Railway Engineering Assoc., Proc., v. 58, 738–745.

Deere, D. U., and Patton, F. D., 1971, Slope stability in residual soils: Panamerican Conf. Soil Mechanics and Foundation Engineering, 4th, Caracas 1971, Proc., p. 87–170.

Ericksen, G. E., Plafker, G., and Concha, J. F., 1970, Preliminary report on the geologic events associated with the May 31, 1970, Peru earthquake: U.S. Geol. Survey Circ. 639, 25 p.

Flaccus, E., 1958, White Mountains landslides: Appalachia, v. 32, p. 175–191.

Gonsior, M. J., and Gardner, R. B., 1971, Investigation of slope failures in the Idaho batholith: U.S. Dept. Agriculture Forest Service Research Paper INT-97, 34 p.

Hartsog, W. S., and Martin, G. H., 1974, Failure conditions in infinite slopes and the resulting soil pressures: U.S. Dept. Agriculture Forest Service Research Paper INT-149, 32 p.

Hsü, K. J., 1975, Catastrophic debris streams (Sturzstroms) generated by rockfalls: Geol. Soc. America Bull., v. 86, p. 129–140.

Isherwood, D., and Street, A., 1976, Biotite-induced grussification of the Boulder Creek Granodiorite, Boulder County, Colorado: Geol. Soc. America Bull., v. 87, p. 366–370.

Jones, F. O., 1973, Landslides of Rio de Janeiro and the Serra das Araras escarpment, Brazil: U.S. Geol. Survey Prof. Paper 697, 42 p.

Krammes, J. S., 1965, Seasonal debris movement from steep mountainside slopes in southern California: U.S. Dept. Agriculture Misc. Pub. 970, p. 85–88.

Le Grand, H. E., 1949, Sheet structure, a major factor in the occurrence of ground water in the granites of Georgia: Economic Geology, v. 44, p. 110–118.

Longwell, C. R., 1974, Measure and date of movement on Las Vegas Valley shear zone, Clark County, Nevada: Geol. Soc. America Bull., v. 85, p. 985–990.

Lumb, P., 1962, The properties of decomposed granite: Geotechnique, v. 12, p. 226–243.

——1965, The residual soils of Hong Kong: Geotechnique, v. 15, p. 180–194.

Matsuo, S. Nishida, K., and Yamashita, S., 1968, Weathering of the granite soils and its influence on the stability of slope: Kyoto Univ. Fac. Eng. Mem., v. 30, pt. 2, p. 85–93.

McCauley, M. L., 1975, The Emerald Bay landslide [abs.]: Assoc. Engineering Geologists, 1975 Ann. Mtg., Program, p. 36.

Merritt, A. H., 1972, Slope stability in tropically weathered diorite, in Cording, E. J., ed., Stability of rock slopes: Symp. Rock Mechanics, 13th, Urbana, Illinois 1971, Proc., p. 625–641.

O'Loughlin, C. L., 1972, An investigation of the stability of the steep land forest soils in the Coast Mountains, southwest British Columbia [Ph.D. dissert.]: Vancouver, Univ. British Columbia, 147 p.

Oyagi, N., 1968, Weathering-zone structure and landslides of the area of granitic rocks in Kamo-Daito, Shimane Prefecture, in Studies of the mechanism and foreknowledge of landslides in weathered areas of granite rocks (Rept. I): Reports of Cooperative Research for Disaster Prevention, no. 14, p. 113–127 (English translation by J. Arata and P. Durgin, 1976, available from National Agricultural Library, Washington, D.C.).

Pirsson, L. V., and Rice, W. N., 1911, Geology of Tripyramid Mountain: Am. Jour. Sci., 4th ser., v. 31, p. 269–274.

Radbruch, D. H., and Crowther, K. C., 1973, Map showing areas of estimated relative amounts of landslides in California: U.S. Geol. Survey Misc. Geol. Inv. Map I-747, scale 1:1,000,000.

Rhodes, D. C., 1968, Landsliding in the mountainous humid tropics: A statistical analysis of landmass denudation in New Guinea: Office of Naval Research, Geography Branch, Tech. Rept. 4, 77 p.

Roddick, J. A., 1974, Circum-Pacific plutonism–Foreword: Pacific Geology, v. 8, p. i–ii.

Ruxton, P. B., 1958, Weathering and subsurface erosion in granite at the piedmont angle, Balos, Sudan: Geol. Mag., v. 95, p. 353–377.

Ruxton, B. P., and Berry, L., 1957, Weathering of granite and associated erosional features in Hong Kong: Geol. Soc. America Bull., v. 68, p. 1263–1291.

Scott, K. M., 1971, Origin and sedimentology of 1969 debris flows near Glendora, California: U.S. Geol. Survey Prof. Paper 750-C, p. C242–C247.

Scott, R. C., 1972, The geomorphic significance of debris avalanching in the Appalachian Blue Ridge Mountains [Ph.D. dissert.]: Athens, Univ. Georgia, 185 p.

Simonett, D. S., 1967, Landslide distribution and earthquakes in the Bewani and Torricelli Mountains, New Guinea, in Jennings, J. A., and Mabbutt, J. M., eds., Landform studies from Australia and New Guinea: Canberra, Australian Natl. Univ. Press, p. 64–84.

Snow, D. T., 1964, Landslide of Cerro Condor–Seneca, Department of Ayacucho, Peru, in Trask, P. D., and Kiersch, G. A., eds., Engineering geology case histories, No. 1–5: Boulder, Colo., Geol. Soc. America, p. 243–248.

So, C. L., 1971, Mass movements associated with the rainstorm of June 1966 in Hong Kong: Inst. British Geographers Trans., v. 53, p. 55–65.

St. John, B. J., Sowers, G. F., and Weaver, Ch. E., 1969, Slickensides in residual soils and their engineering significance: Internat. Conf. Soil Mechanics and Foundation Engineering, 7th, Mexico City 1969, Proc., v. 2, p. 591–597.

Swanston, D. N., 1967, Soil-water piezometry in a southeast Alaska landslide area: U.S. Dept. Agriculture Forest Service Research Note PNW-68, 17 p.

Terzaghi, K., 1962a, Dam foundation on sheeted granite: Geotechnique v. 12, p. 199–208.

——1962b, Stability of steep slopes on hard unweathered rock: Geotechnique, v. 12, p. 251–263.

Thompson, G. A., and White, D. E., 1964, Regional geology of the Steamboat Springs area, Washoe County, Nevada: U.S. Geol. Survey Prof. Paper 458-A, 52 p.

Wahrhaftig, C., 1965, Stepped topography of the southern Sierra Nevada, California: Geol. Soc. America Bull., v. 76, p. 1165–1190.

Williams, G. P., and Guy, H. P., 1973, Erosional and depositional aspects of Hurricane Camille in Virginia, 1969: U.S. Geol. Survey Prof. Paper 804, 80 p.

MANUSCRIPT RECEIVED BY THE SOCIETY SEPTEMBER 7, 1976
MANUSCRIPT ACCEPTED SEPTEMBER 17, 1976

Printed in U.S.A.

8

Problems with Lake Albany "clays"

JAMES R. DUNN
GEORGE M. BANINO
Dunn Geoscience Corporation, 5 Northway Lane North, Latham, New York 12110

ABSTRACT

The deposits of varved clay in glacial Lake Albany have a long history of instability. Recent failures have been both natural and man-made. Slumping, creeping, and rotational sliding of the clay strata are ubiquitous and typical of these deposits. The affected area lies along New York's Hudson River from the Albany-Troy area at the north almost to Poughkeepsie at the south, a distance of 209 km. Despite a long history of problems, some construction activity is still done improperly. Analyses associated with environmental impact statements may reduce potential problems, but a properly promulgated general study of the problem might be a better approach.

INTRODUCTION

The varved clays or rhythmites of glacial Lake Albany have been a source of environmental difficulties for hundreds of years. The problems have ranged from poor foundation drainage, to surface creep, to major rotational slides which have resulted in loss of life and severe property damage. Recent problems with the clays have caused new apartment houses to be abandoned. The problems that are associated with the varved clays or rhythmites of glacial Lake Albany are similar to those found in many areas.

The purposes of this chapter are to review some of the typical problems that have occurred, to describe some of the remedial activities that have been successful, and to suggest some ways to detect potentially hazardous conditions.

PLEISTOCENE AND HOLOCENE GEOLOGIC HISTORY

Glacial Lake Albany occupied an area from a little south of Kingston, New York, northward to a little north of Whitehall, New York, a length of over 209 km (130 mi). Deltas consisting of sands with some gravel were formed at various places in the lake, and well-defined beaches are still visible in many areas. Islands consisting of rock, of lodgment till, and of occasional eskers and kames consisting of sand and gravel rose above the water surface. By far the predominant sediment associated with the lake from Albany south was varved clay. The clay once covered most topographic features up to a present elevation of 76 to 91 m (250 to 300 ft) at the north and about 46 m (150 ft) at the south. The history of Lake Albany and the sediments deposited within it is complex (see Woodworth, 1905; Newland, 1916; Stoller, 1920; Ruedemann, 1930; LaFleur, 1965).

The debris dam which blocked the south end of the lake gave way apparently in stages as the lake drained. The Mohawk and Hudson Rivers started new channels in the lake sediments. Meanwhile, the fine-grained sediments dried out, as they became exposed to air, and wind picked up and transported available loose sediments. Local areas, particularly in the vicinity of Albany and northward, became covered with sand dunes. Loess was also locally deposited as a thin sheet.

Vegetation finally covered the unconsolidated sediments and anchored them. Except for areas burned by Indians for agricultural and hunting purposes, that was the condition of the area when the Europeans first settled.

NATURE OF THE LAKE ALBANY "VARVED CLAYS" OR ARGILLACEOUS SILTS

Much of the clay is in the broad category of "rock flour" and is largely feldspar and quartz. Mather (1843) gave the following weight percentages for the components of the clays: water of absorption, 4.25; organic matter, 1.17; sulfate of lime, 1; silicates, 69.02; peroxide of iron and alumina, 17.24; potash, 0.14; carbonate of lime, 4; and magnesia, 3 (total = 99.82).

The ratio of Al_2O_3 to SiO_2 suggests that true clay minerals

Figure 1. Lake Albany clays, exposed by brick manufacturing operation about 15 km south of Albany, New York. Note the typical "bread crust" surface.

Figure 2. Rotational sediment slide in varved clays that occurred in 1968 along Normans Kill, Slingerlands, southeast of Albany, New York. (Photo by R. Burns.)

may predominate. Schock (1963) described the clay-mineral component of the sediments as being primarily a septachlorite.

TOPOGRAPHIC ANALYSIS

The presence of lake silts and clays is readily determined in most areas by the extensive plains which are at almost constant elevations. In addition, exposed lake sediments dry with "bread-crust" surface and erode to the typical pattern shown in small scale in Figure 1. The dissected surface is clearly visible on topographic maps. Topographic analysis is, thus, a quick way of determining the probable presence of varved deposits in advance of field work. Lodgment till, sand and gravel, and outcrop areas are also easily detected in the Hudson River valley by topographic map analysis.

Figure 3. Rotational sediment slide in varved clays, North Greenbush, just south of Troy, New York, in 1960. (Photo by R. Burns.)

EXAMPLES OF PROBLEMS WITH LAKE ALBANY SEDIMENTS

The instability of Lake Albany sediments has long been known. Mather (1843, p. 32), speaking of landslides in the lake sediments, said "such facts are often made public when they have been attended by the destruction of life, or property; but thousands of such occurrences have taken place, as geologic observations attest, which were either unknown, or of which no record has been preserved."

Probably the single major catastrophic slide occurred in Troy, New York, on January 1, 1837. About 200,000 short tons of debris traveled about 250 m (800 ft), killing 5 people, 16 horses, and destroying several buildings (Mather, 1843). It is interesting to note that the potential danger was signaled in the summer of 1836 when a smaller slide occurred in the same area.

Figures 2 and 3 are of recent rotational sediment slides which occurred along local streams in the Albany area. Such slides may be the result of the purely natural process of

Figure 4. Rotational sediment slide that occurred in 1971 below an apartment house under construction, Thompson Street, Troy, New York.

Figure 5. Part of toe of Thompson Street rotational sediment slide in 1971. The disrupted street to the left of the large trees is at the toe; the trees to the left of the sidewalk are holding the slide.

Figure 6. Thompson Street in 1976. The metal mesh fence at the right edge of Figure 5 is barely visible in front of the shrubbery above and to the left of the car. Note absence of the large trees in Figure 5 and the undulating nature of the street.

oversteepening a slope by undercutting the toe by stream erosion. However, experience has shown that loading the top of a steep slope with as little as 1.5 m of fill can trigger a slide (R. Burns, 1976, personal commun.).

Figures 4, 5, and 6 show the results of recent rotational sediment slides in South Troy, New York. The apartment buildings involved in both cases were abandoned as unsafe about 1971 (before occupancy) because of the development of rotational slide fractures alongside and under the buildings. In both cases, the top of a steep slope was loaded, and this probably triggered the incipient sliding. In addition, along Thompson Street at the toe of the slide, sediments frequently moved onto the street and had to be removed from the base of the slope. Loading the top and unloading the base of such a slope are, of course, classic mechanisms leading to slope instability. Figure 7 shows Thompson Street as it was in 1976. The undulating nature of the street suggests continued instability. Several large trees have been removed, a factor which could make the base of the slide less stable.

Figure 8 is of a somewhat hummocky, flat surface adjacent to the Thompson Street apartments; this surface resulted from a previous rotational sediment slide. Figure 9 is a similar surface near the slide shown in Figure 3.

TOWARD REDUCING FUTURE PROBLEMS

The causes of slides of the Lake Albany sediments are classical, and much of the problem has long been understood (Mather, 1843, p. 32–33). Identifying potentially dangerous areas is easily done through topographic analysis in the field, by photo interpretation, by analysis of contour maps, by an adequate test-boring program, and by engineering analysis of the probable position of the surfaces of potential slides. Such analysis is relatively straightforward and could have been done almost as well 133 years ago by Mather as it could be done now by a competent soils engineer or geologist.

The major problem appears to be to make certain that such

Figure 7. Apartment house at Hudson Street, Troy, New York; Never occupied because of danger of sliding of clays.

Figure 8. Flat area represents an old slide block across Thompson Street from an abandoned apartment building.

Figure 9. Flat area represents an old slide block near location of North Greenbush, New York, slide that occurred in 1960.

analyses are made at all. The lack of success in taking the action required to reduce slide problems is mutely attested to by the vacant apartments in Troy. Furthermore, potential dangers exist not only in the vicinity of Troy and Albany, but at other areas in the Hudson River valley. Heavy precipitation or ill-planned construction activities can still cause many additional slides.

One way of assuring a reduction of hazards related to the varved clays is the use of the environmental impact statement. For instance, potential problems with rotational sediment slides were reduced by such an analysis of a second-home development along a lake created by damming Murderers Creek 3.2 km north of Hudson, New York, on the west side of the Hudson River. The placement of homes on lots was laid out to reduce dangerous conditions by not allowing construction on areas within potential slide blocks.

Perhaps a better alternative to the environmental impact statement approach would be for the problem to be analyzed for the entire Hudson River valley. Areas of potential slide hazard could be outlined and ways of reducing slide hazards summarized. Such a program would assure the identification of problem areas (whereas environmental impact statements are restricted to specific sites), and the entire program might not cost much more than one or two environmental impact statements. Publishing the results of the study and making certain that it was placed in the hands of local authorities would be a vital second step. A cooperative effort involving various governmental personnel in engineering and planning could minimize problems in the future.

ACKNOWLEDGMENTS

We are indebted to Richard H. Burns, who read and criticized this paper, and to Dr. Robert G. LaFleur, who acted as an outside reviewer.

REFERENCES CITED

LaFleur, R. G., 1965, Glacial geology of the Troy, N.Y., Quadrangle: New York State Mus. and Science Ser., Map and Chart Ser. no. 7.

Mather, W. W., 1843, Geology of New York. Pt. I, comprising the geology of the first geological district, Albany: Albany, Nat. History New York, 653 p.

Newland, D. H., 1916, Albany molding sand: New York State Mus. Bull. 187, p. 107–115.

Ruedemann, R., 1930, Geology of the capital district (Albany, Cohoes, Troy, and Schenectady quadrangles), with a chapter on glacial geology by John H. Cook: New York State Mus. Bull. 285, 218 p.

Schock, R. N., 1963, Geology of the Pleistocene sediments, Troy North quadrangle, New York [M.S. thesis]: Troy, N.Y., Rensselaer Polytechnic Inst.

Stoller, J. H., 1920, Glacial geology of the Cohoes quadrangle: New York State Mus. Bulls. 215 and 216, 47 p.

Woodworth, J. B., 1905, Ancient water levels of the Champlain and Hudson Valleys: New York State Mus. Bull. 84, 265 p.

MANUSCRIPT RECEIVED BY THE SOCIETY SEPTEMBER 7, 1976
MANUSCRIPT ACCEPTED SEPTEMBER 17, 1976

9
Large submarine slide in Kayak Trough, Gulf of Alaska

BRUCE F. MOLNIA
PAUL R. CARLSON
TERRY R. BRUNS
U.S. Geological Survey, 345 Middlefield Road, Menlo Park, California 94025

ABSTRACT

Analyses of high-resolution seismic profiles have revealed the presence of a well-defined, massive submarine slide located at the north end of the Kayak Trough in the northern Gulf of Alaska. This slide is about 18 km long and 15 km wide, has a volume of about 5.9 km^3, and has moved down a 1° slope. Sediment from the upper 2 m of the slide consists of low-strength, greenish-gray clayey silt. Morphologically this slide is a classic example, with a well-preserved pull-apart scarp in the headward regions, a well-developed toe, disrupted internal bedding, and hummocky surface topography.

The age of the slide is unknown, but its clearly defined morphology in an area of very high sedimentation (7.5 to 15 m/1,000 yr) suggests that it is extremely young. The slide may have been generated by intense storm activity or earthquake-triggered, prolonged ground shaking. Laboratory and shipboard observations indicate that the slide sediments are very weak (peak shear strength of 0.02 kg/cm^2) with high water content. Other areas of the Gulf of Alaska are known to have thick accumulations of similar sediment, although a comparable slide has not been observed. The characteristics of this slide are indicative of problems that will have to be surmounted if this or similar areas in the Gulf of Alaska are to be used for pipeline emplacement or platform siting for petroleum production. High seismic risk makes the problems even more severe.

INTRODUCTION

The U.S. Geological Survey is investigating environmental hazards in the Gulf of Alaska (Fig. 1) as a prelude to possible offshore petroleum exploration and development. One of the most significant types of hazard investigated is submarine slumping and sliding. Historically, failure of marine and deltaic sediments has occurred in numerous places around the Gulf of Alaska (Coulter and Migliaccio, 1966; Lemke, 1967; Reimnitz, 1966, 1972; National Academy of Sciences, 1971). New high-resolution seismic profiles, bottom photographs, underwater television videotapes, side-scan sonar records, and sediment samples from the vicinity of the Copper River delta have revealed evidence of previously unknown submarine slumping and sliding.

Across the entire prodelta facies of the Copper River delta, a large amount of sediment appears to have slumped seaward. The affected area, about 15 km wide, extends from Controller Bay on the east to Hinchinbrook Island on the west, a distance of approximately 110 km, and encompasses an area of about 1,600 km^2. Seismic profiles southwest of Controller Bay in the Kayak Trough show that part of this area is a massive submarine slide. Clearly seen on these profiles are (1) a scarp at the head of the slide, (2) a well-defined toe, and (3) disrupted structures within the slumped mass of sediment. The scarp has a maximum relief of nearly 10 m, and the toe of the slide is 20 m thick about 2 km from the distal end. The slide is about 18 km long and as much as 15 km wide. The slide is elliptical in plan view and oriented north-south. The total volume of material in this massive slide is about 5.9 km^3. It is difficult to assess accurately the maximum displacement of sediment downslope, but analyses of profiles suggest that the magnitude of displacement could be as much as 6 km in the headward regions of the slide, where the scarp is present. Below the scarp area, it is impossible to tell whether the displacement is a few metres or many kilometres.

GEOLOGIC SETTING

Kayak Trough, the site of the submarine slide, is a glacially scoured bedrock depression with its long axis paralleling Kayak Island and trending northeast. The trough is more than 42 km long and 23 km wide, with depth to bedrock as much as 350 m below sea level. On the floor of the trough the sediment fill reaches a maximum thickness of about 150 m.

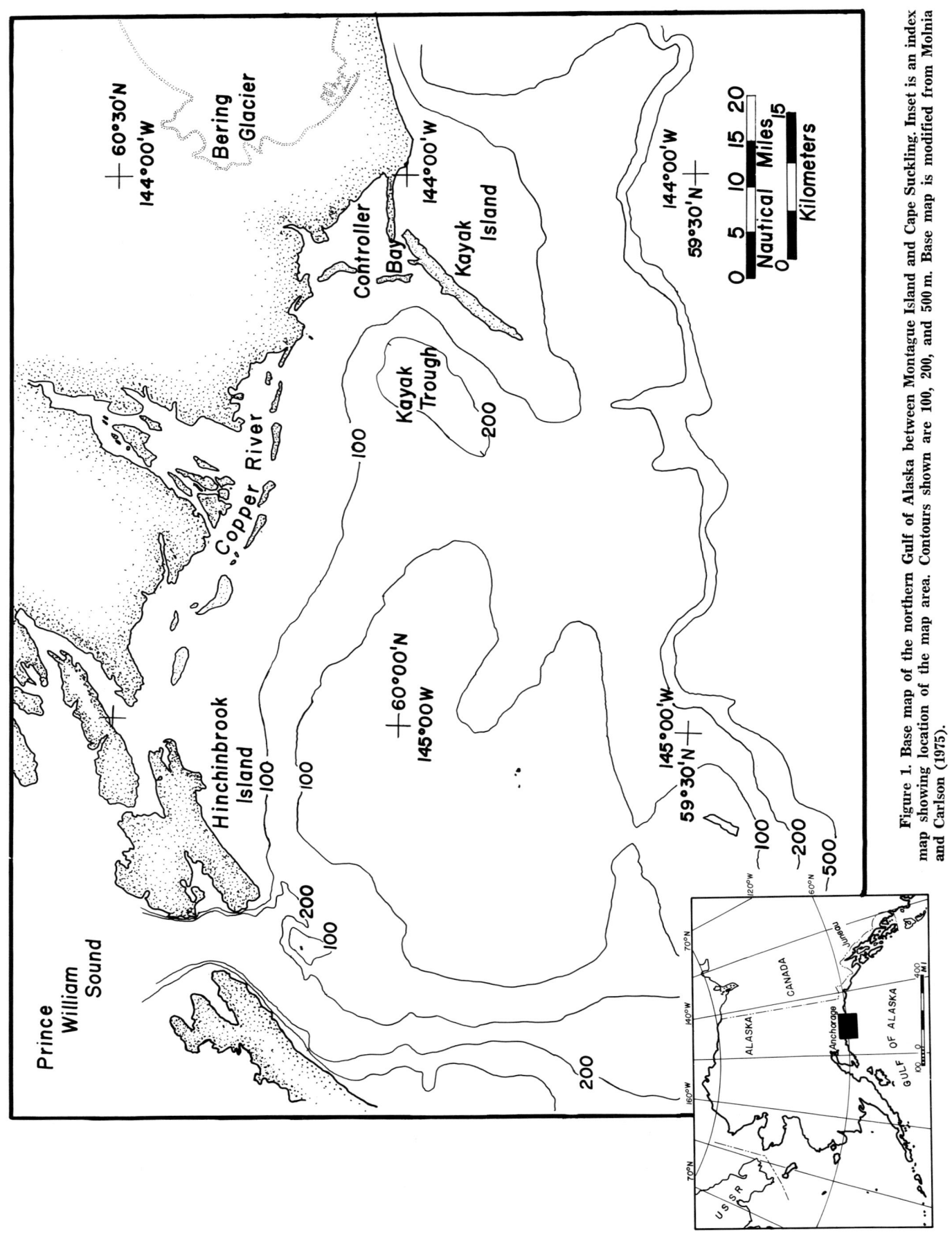

Figure 1. Base map of the northern Gulf of Alaska between Montague Island and Cape Suckling. Inset is an index map showing location of the map area. Contours shown are 100, 200, and 500 m. Base map is modified from Molnia and Carlson (1975).

It is in this thick deposit that the massive slide is located.

The Kayak Trough was probably cut by a large glacier generated in the Chugach Mountains during Pleistocene time. The Bering Glacier, with its terminus 46 km from the head of the scarp, is a present-day remnant of this much larger glacier. The major glacier probably occupied the trough during Wisconsinan time, but it is unlikely that the trough was occupied by ice during the Neoglacial event (Austin Post, 1976, oral commun.). Hence, the Kayak Trough has been ice-free for at least the past 10,000 to 20,000 yr. The presence of 150 m of sediment suggests a sedimentation rate in the Kayak Trough of 7.5 to 15 m/1,000 yr.

MORPHOLOGY OF THE SUBMARINE SLIDE

More than 1,000 km of high-resolution seismic profiles were collected in the vicinity of the Kayak Trough (Fig. 2) on cruises of the R/V *Thomas G. Thompson* (1974), NOAA *Surveyor* (1975), R/V *Acona* (1976), and R/V *Sea Sounder* (1976). Underwater television videotapes as well as side-scan sonar profiles of the slide area were made on the *Sea Sounder*. Underwater photographs were obtained during the *Acona* cruise.

Bottom samples were collected from the NOAA *Townsend Cromwell* cruise (1975) as well as from those of the *Acona* and *Sea Sounder* (Fig. 3). Five seismic lines have been selected which illustrate the morphology of the submarine slide. These lines (*Thompson* 33, 61, and 63 and *Acona* 559 and 563) are shown in Figures 4 through 8.

Thompson line 33 (Fig. 4) is a north-south line about 18 km long that approximately bisects the slide. The northernmost part of the scarp is located in 42 m of water, whereas the toe is in 227 m, a gradient of 1.1%. The northern third of the slump has as many as five major scarps, each more than 5 m high. The lower third of the line has a gradient of less than 0.1%. Beyond the toe, the basin shoals, indicating that the slide had enough momentum to carry it past the deepest part of the trough. The thickness of the disturbed material is about 20 m just north of the toe and about 120 m south of the scarp.

Acona line 559 (Fig. 5) is located about 4 km to the east of and parallel to line 33. From scarp to toe, the line is about 13.5 km long. As with line 33, the toe is located beyond the deepest part of the basin. The morphology of the slide along line 559 is very similar to that on line 33. The major difference is the presence of a broad (0.6 km) pull-apart graben about 3 km below the northernmost scarp. In the graben area the water depth changes from 69 to 93 m, and there is evidence of sediment block rotation. The corresponding area on line 33 has a relief of 8 m and is about 100 m wide.

Acona line 563 (Fig. 6) is an east-west line that approximately bisects the slide. There is a marked asymmetry in the profile, because there is a well-developed scarp on the eastern side (in 54 m of water), whereas the western limit (in 195 m of water) tapers to a thin toe. The slide along line 563 is about 14 km wide. The surface of the lower third is more broken and hummocky than in the two north-south lines, although the relief is similar.

Thompson line 61 (Fig. 7) runs diagonally across the slide from southeast to northwest. The northwest end of the line shows a well defined scarp in about 65 m of water, very similar to that seen in the two north-south lines. The toe, at the southeast end of the line, terminates abruptly on an uphill slope in about 210 m of water. There is a much broader area of disturbed sediment here than on any of the other lines, but this may be a function of the angle at which the line crosses the slump. The thickness of disturbed sediment is similar to the other profiles. The hummocky character of the surface is well developed, as is the case with the previous lines. The slump is about 14.5 km long in this segment.

Thompson line 63 (Fig. 8) is 11 km long and trends diagonally from northeast to southwest. This line shows a major section of the scarp to the northeast in water about 75 m deep. A well-defined graben occurs in about 135 m of water, 3.5 km below the scarp. The graben is about 150 m wide. The toe at the southwestern part of the line ends in about 210 m of water on a gentle uphill slope. The thickness of disturbed sediment approaches a maximum of 120 m. As with the previous lines, rotated and discontinuous subbottom reflectors are present, indicating rotation of blocks of sediment.

Figure 9 is a section of a side-scan sonar profile on a north-south crossing of a part of the main scarp. Individual steplike scarplets can be seen, each marking a separate failure plane. Underwater television pictures from the scarp area show similar failures truncating what appear to be parallel bedded layers. The television pictures showed angular broken blocks lying at the base of individual scarplets.

The elliptical nature of the slide, as well as its relation to the Kayak Trough and surrounding topography, can be seen on the composite physiographic diagram of the slide area (Fig. 10). The diagram indicates that the slide has a slightly irregular southern margin. Examination of the numerous profiles has not indicated a well-formed lobate toe. The volume of material moved during the sliding was measured by placing a grid over the slide mass and determining the thickness for each known area. The total calculated volume of material that moved is 5.9 km^3.

SEDIMENT OF THE SLIDE AREA

The character of the sediment of the upper 2 m of the Kayak Trough slide was determined from samples obtained with grabs (Van Veen and Shipek) and corers (box and gravity) and from underwater photography. In general, it is a greenish-gray clayey silt with a high water content.

Sediment samples collected with a box corer from the upper 50 cm of the surface of the slide consisted of greenish-gray clayey silt that was seemingly structureless. However, x-radiographs of the sediment revealed contorted bedding, some cross-bedding, and chaotic mixtures of irregular fragments of various shapes. The sediment was extremely weak, as shown by laboratory tests with a vane-shear apparatus that yielded a peak shear strength of 0.02 kg/cm^2. These tests were run about two months after collection, and the samples could have been weakened by possible remolding from jostl-

Figure 2. Map showing location of high-resolution seismic lines in the vicinity of Kayak Trough. The five broad, numbered lines are described in the text. Dashed line is the 100-m isobath.

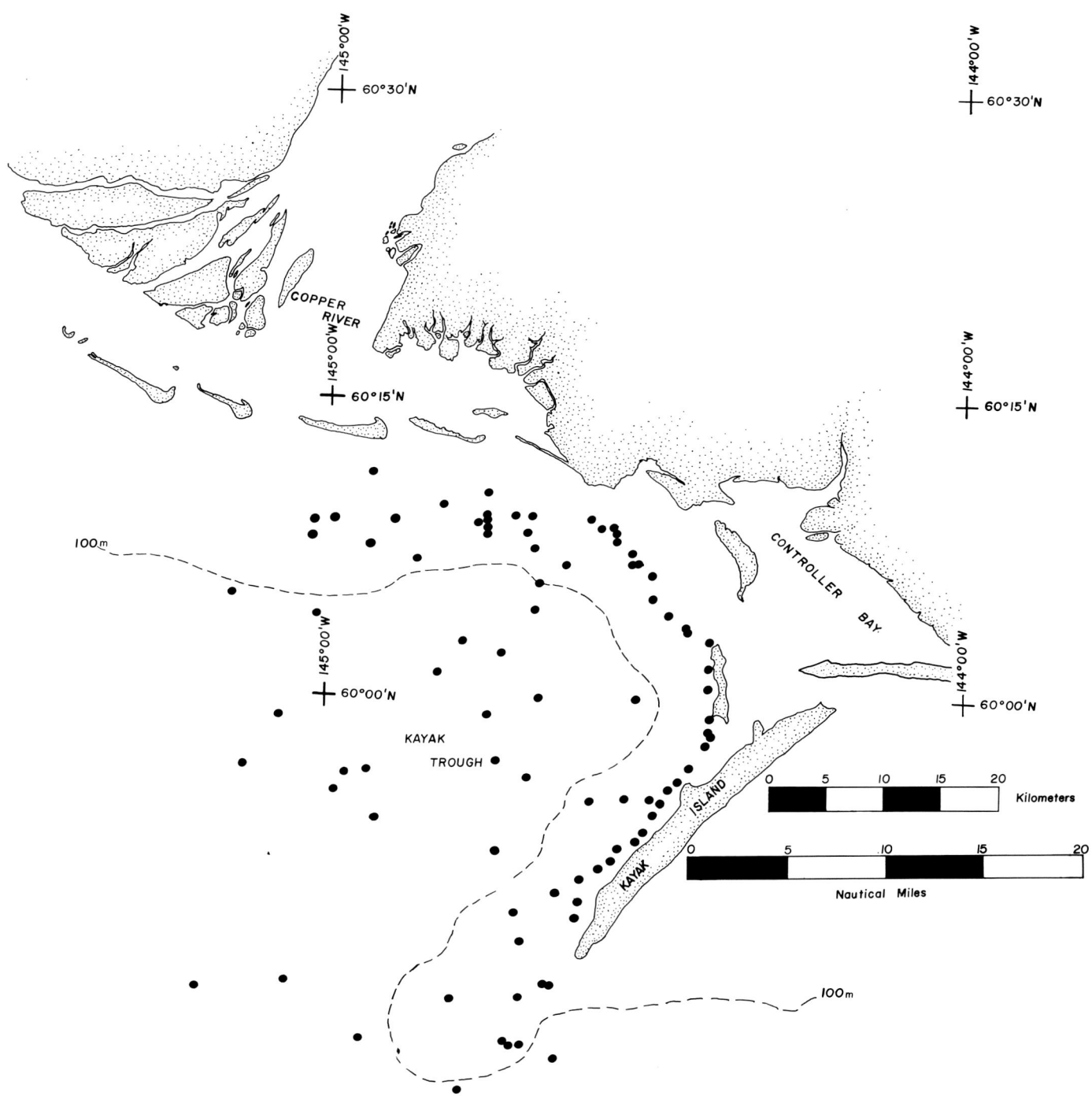

Figure 3. Map showing location of surface-sediment sampling sites in the vicinity of Kayak Trough. Dashed line is the 100-m isobath.

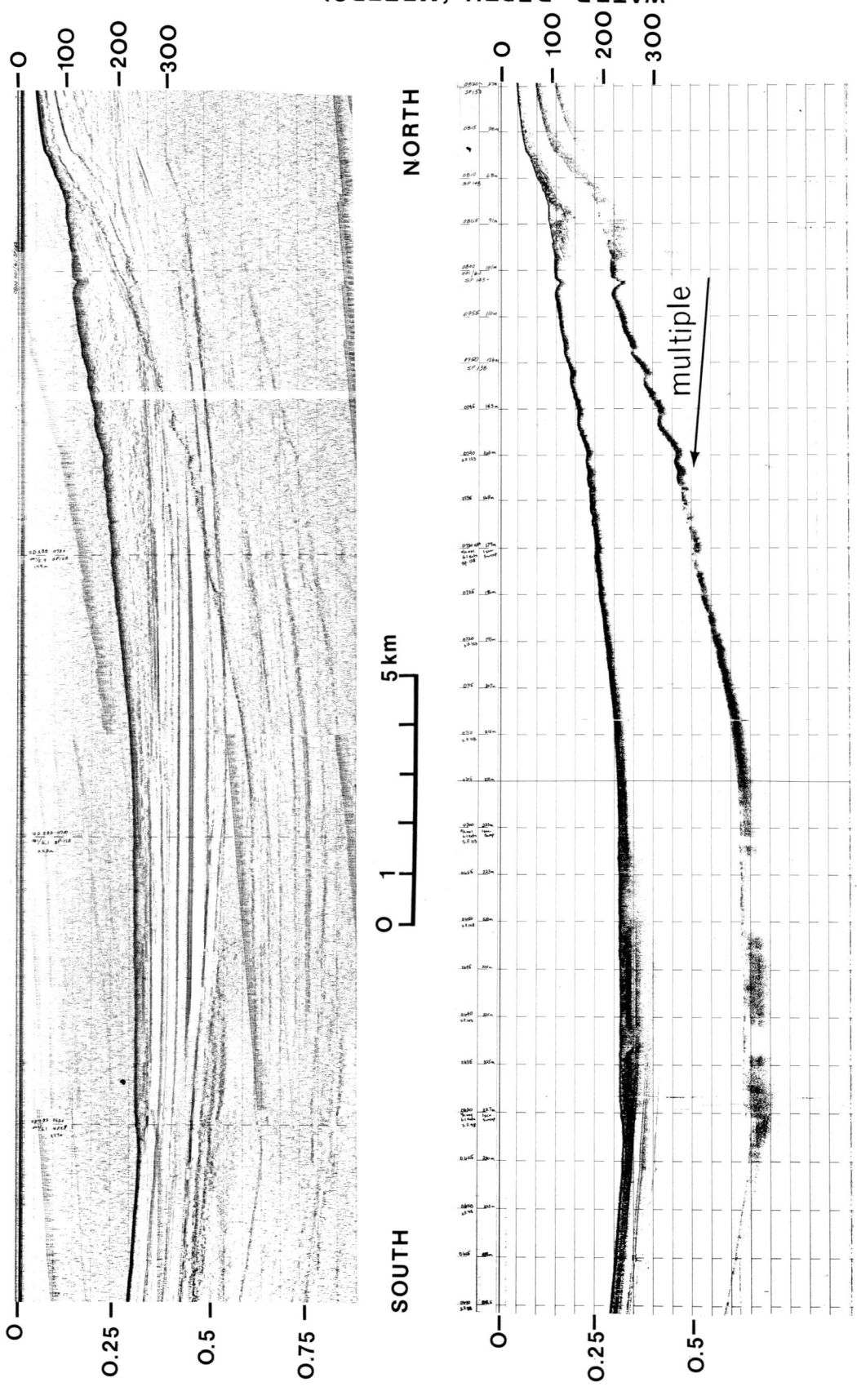

Figure 4. High-resolution seismic profile (upper) and 3.5-kHz echo-sounding profile (lower) of the Kayak Trough slide along *Thompson* line 33. Vertical scale is shown in both depth and time. Horizontal scale is in kilometres.

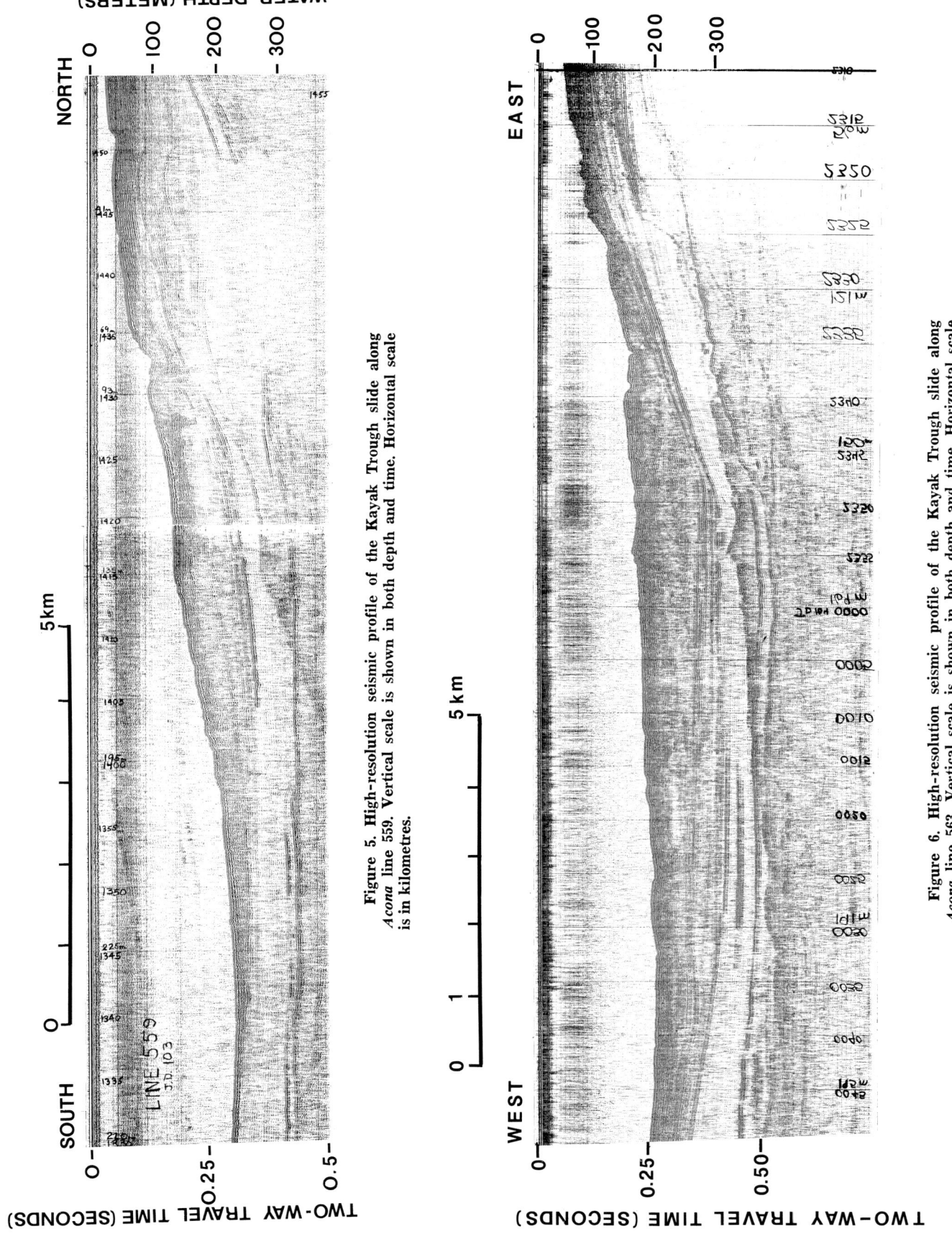

Figure 5. High-resolution seismic profile of the Kayak Trough slide along *Acona* line 559. Vertical scale is shown in both depth and time. Horizontal scale is in kilometres.

Figure 6. High-resolution seismic profile of the Kayak Trough slide along *Acona* line 563. Vertical scale is shown in both depth and time. Horizontal scale is in kilometres.

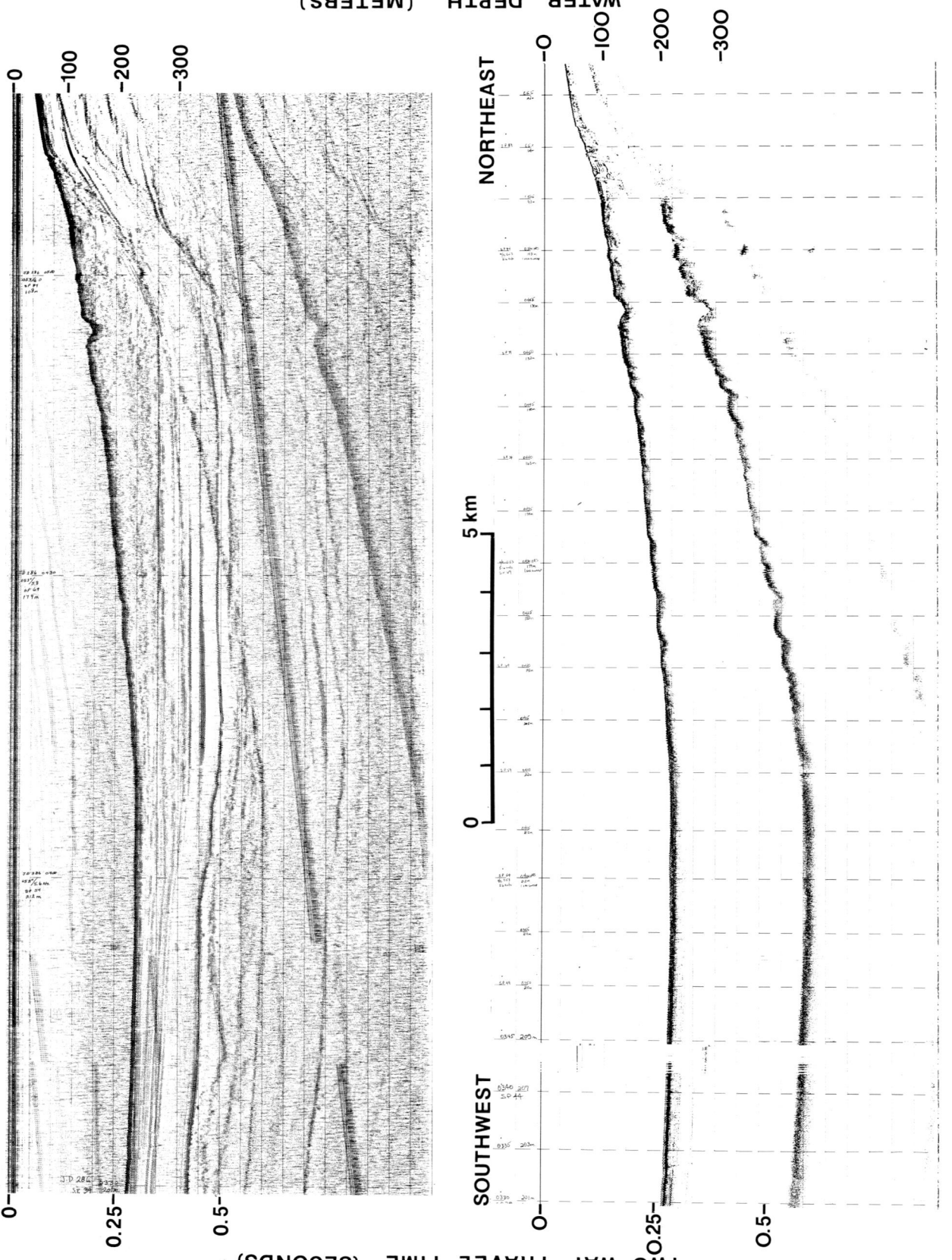

Figure 7. High-resolution seismic profile (upper) and 3.5-kHz echosounding profile (lower) of the Kayak Trough slide along *Thompson* line 61. Vertical scale is shown in both depth and time. Horizontal scale is in kilometres.

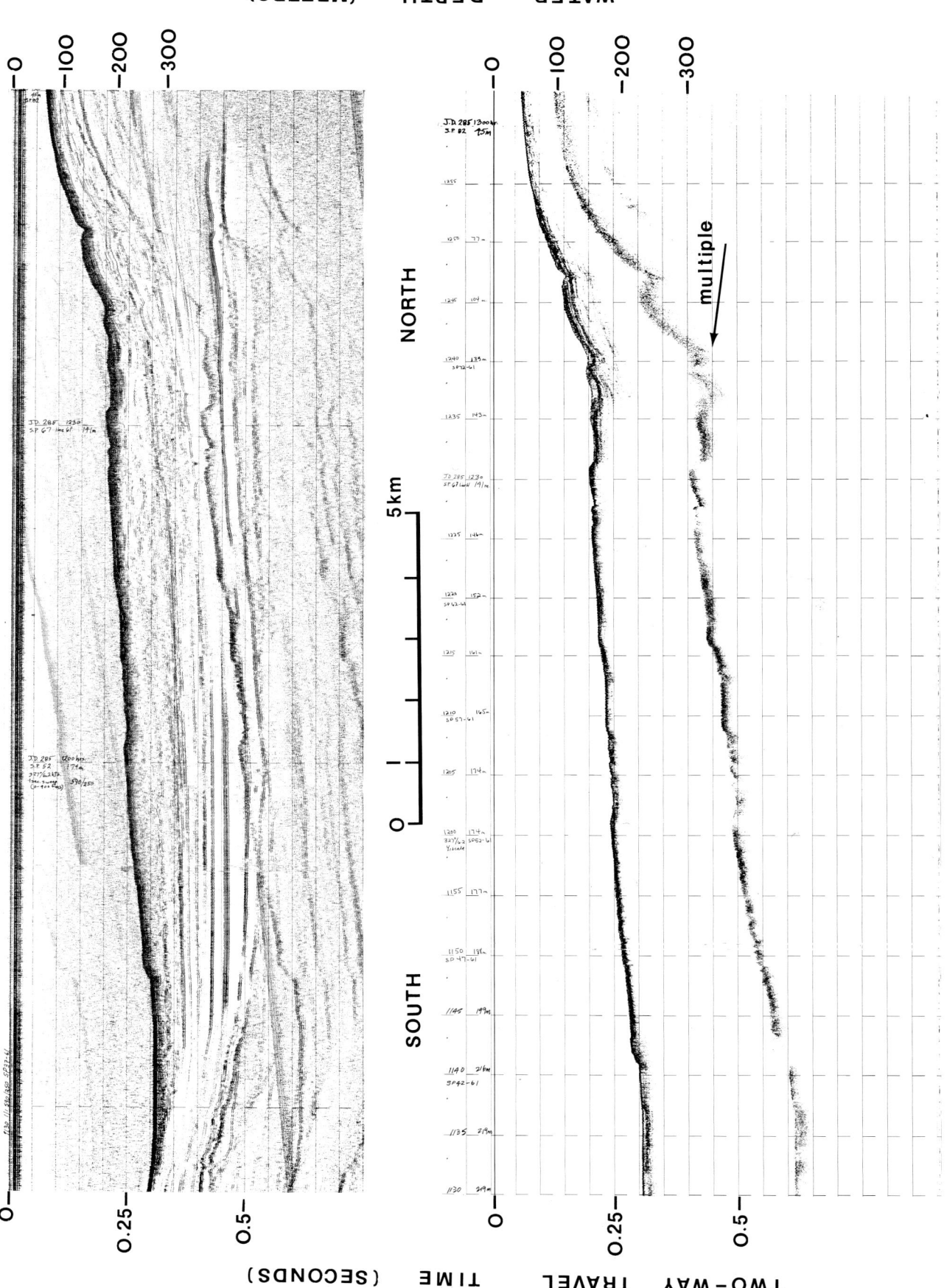

Figure 8. High-resolution seismic profile (upper) and 3.5-kHz echosounding profile (lower) on the Kayak Trough slide along *Thompson* line 63. Vertical scale is shown in both depth and time. Horizontal scale is in kilometres.

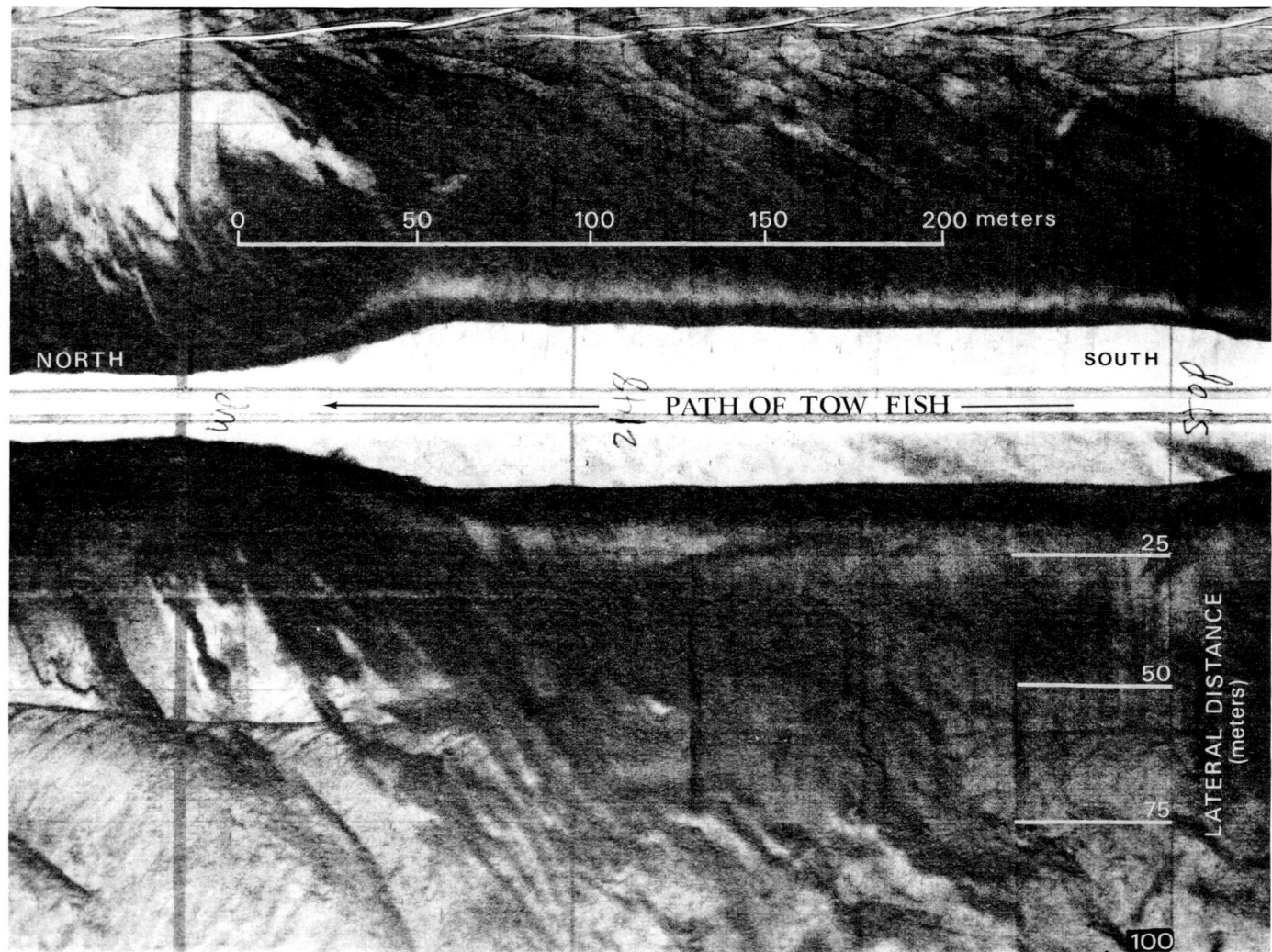

Figure 9. Part of a side-scan sonar record across the main scarp of the slide. The path of the ship and the sonar tow fish is from right to left (south to north). The side-scan sonar fish "sees" to either side but not directly below its path. The distance from the parallel lines (labeled "path of tow fish") to the heavy black profile lines on either side represents the height of the fish above the bottom. The distance from the center lines to the bottom edge of the lower profile is 100 m. The distance along the path of the tow fish is about 350 m. The side-scan sonar profile does not give a quantitative value for the height of the individual scarplets, but the varied tones and shadows confirm their presence.

ing in transit. However, the extremely weak nature of some of the samples was underscored by sediment flowage when the box cores were opened onboard ship immediately after collection.

DISCUSSION

Age and Triggering Mechanism of the Slide

The age of the slide is unknown, but its fresh surface with apparently no overlying undeformed sediment, especially in an area of such high sedimentation, suggests that it must be quite young. Samples that were radiographed did not show undisturbed layering overlying the contorted bedding. None of the seismic profiles shows any undisturbed bedding at the surface, but the high-resolution seismic method used has a resolution of 1 to 2 m at the water-sediment interface, which would obscure any bedding in that interval. To generate a possible maximum age for the slide, we assumed that the modern sedimentation rate is the minimum calculated for postglacial time (7.5 m/1,000 yr), and that there is a 2-m-thick undisturbed sedimentary section covering the slide mass that is not seen in seismic profile. The computed age for the slide in this case would be about 270 yr. The thinner the undisturbed layer, or the greater the sedimentation rate, the younger the slide. If in fact there is no undisturbed sediment covering the slide mass (as the radiographed samples suggest), then the age of the slide could be much younger.

Possible triggering mechanisms for the slide include seismic disturbances and agitation by storm waves. Numerous earthquakes of magnitude 8 and greater have affected the slide area in the past and can be expected in the future (Plafker and others, 1975). If earthquakes are the triggering mechanism, then the most likely historic occasions for the generation of the slide are the 1964 Alaska earthquake or the 1899-1900 Yakutat series of earthquakes (Tarr and Martin, 1912). An attempt was made to locate pre-1964 bathymetry in the Kayak Trough, but only a few soundings were found, all by lead line, and it was concluded that no meaningful pre- and post-1964 earthquake comparison could be made. The 1964 earthquake did generate many landslides in the Gulf of Alaska. Among these were slides at Cordova, Whittier, Valdez, Seward, Homer, and Anchorage (Hansen and others, 1966; Reimnitz, 1966, 1972).

If agitation by storm waves is the triggering mechanism, it would be very difficult to pinpoint a particular storm event as the cause. Storms with waves whose heights approach 10 m are common (Searby, 1969), and larger storms with wave heights of 15 m and greater occur less frequently (Pierson and others, 1955).

Implications for Petroleum Development

The Kayak Trough area is unique in the Gulf of Alaska only in that a major slide has already occurred. Many other areas, both on the shelf and in glacially scoured submarine valleys, are known to have thick sediment accumulations of modern, fine-grained sediment with high water content on slopes similar to those in the Kayak Trough (Carlson and Molnia, 1975; Carlson, 1976). In particular, some of these potentially unstable sediment masses are within the area leased in April 1976 for petroleum exploration. If subsequent exploratory drilling results in the discovery of economical quantities of hydrocarbons, then platform sites and pipeline corridors will have to be established in a manner most suited to exploit the offshore resources. The engineering associated with development on such sediment masses presents some rather complicated problems. How does one lay a pipeline over material with virtually no bearing strength?

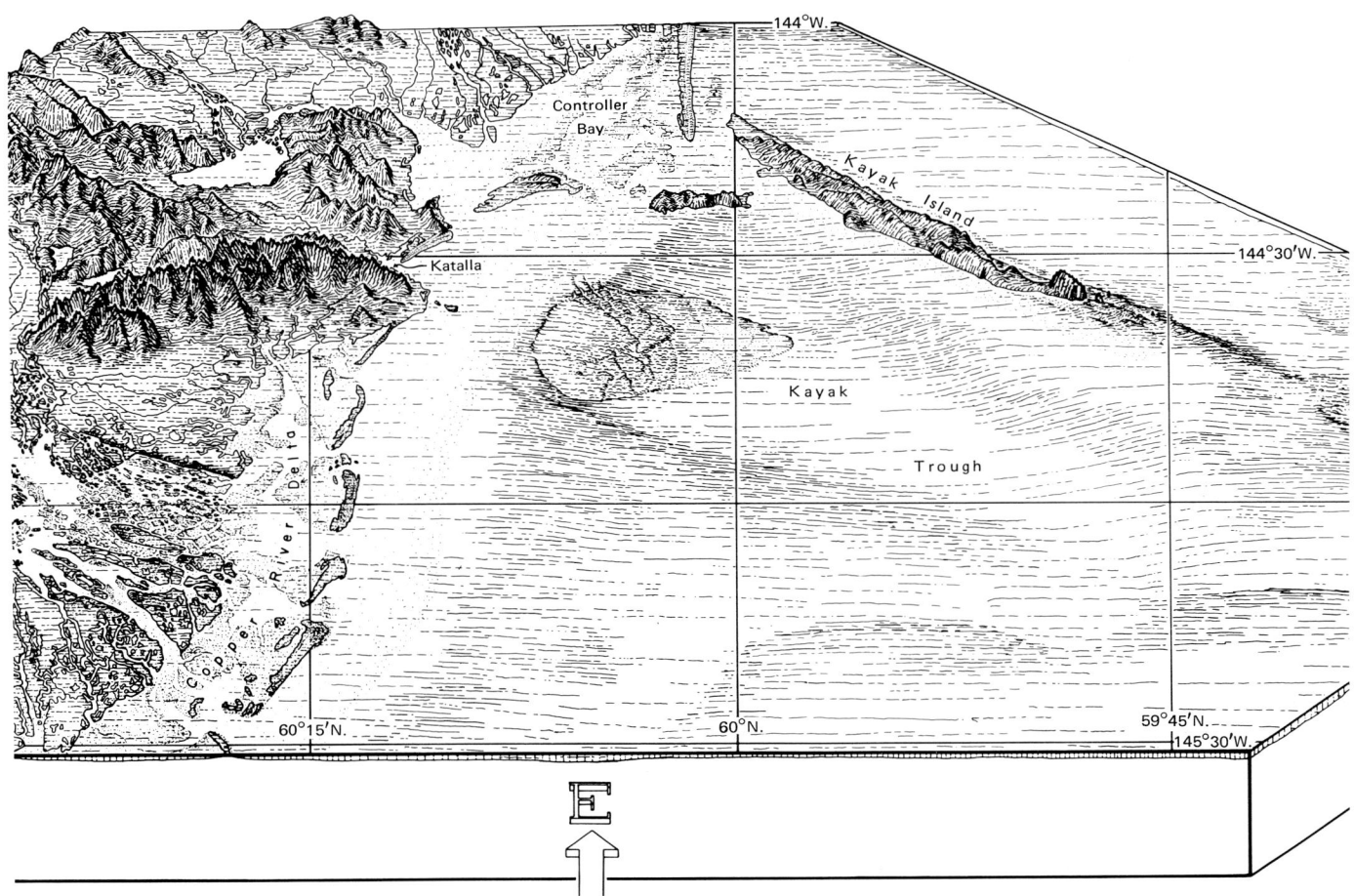

Figure 10. Physiographic diagram of Kayak Trough showing massive submarine slide. Orthographic drawing by Tau Rho Alpha; vertical exaggeration, 3:1. Bathymetry based on U.S. Geological Survey soundings (Molnia and Carlson, 1975; von Huene and others, 1975). Topography from U.S. Geological Survey maps (Cordova, 1959, and Middleton Island, 1955; original scale 1:250,000).

How can the hazards of slow creep or (even worse) massive sliding after emplacement of structures be overcome? Although these problems have been dealt with elsewhere, as in the Gulf of Mexico, the associated high seismic risk in the Gulf of Alaska makes the problem even more severe.

SUMMARY

A large submarine slide 18 km long and 15 km wide with a volume of 5.9 km^3 is located in the Kayak Trough. The slide is very young, perhaps triggered as recently as 1964. The sediment in the slide mass has very low shear strength (0.02 kg/cm^2 peak shear strength) and high water content. The environment of Kayak Trough is not unique in the Gulf of Alaska. Other areas in the gulf have similar sediment bodies and slopes. Kayak Trough is unique only in that a slide has already occurred.

ACKNOWLEDGMENTS

We thank D. A. Condra, J. H. Cudnohufsky, J. C. Hampson, S. C. Kittelson, and L. H. Wright for their assistance with laboratory analyses, data work-up, and drafting. We also appreciate the critical reviews and suggestions of E. C. Buffington, P. L. Hilpman, and T. L. Vallier.

REFERENCES CITED

Carlson, P. R., 1976, Submarine faults and slides that disrupt surficial sedimentary units, northern Gulf of Alaska: U.S. Geol. Survey Open-File Rept. 76-294, 27 p.

Carlson, P. R., and Molnia, B. F., 1975, Preliminary isopach map of Holocene sediments, northern Gulf of Alaska: U.S. Geol. Survey Open-File Rept. 75-507.

Coulter, H. W., and Migliaccio, R. R., 1966, Effects of the March 27, 1964, earthquake at Valdez, Alaska: U.S. Geol. Survey Prof. Paper 542-S, 32 p.

Hansen, W. R., Eckel, E. B., Schaem, W. E., Lyle, R. E., George, W., and Chance, G., 1966, The Alaskan earthquake, March 27, 1964: Field investigations and reconstruction effort: U.S. Geol. Survey Prof. Paper 541, 111 p.

Lemke, R. W., 1967, Effects of the earthquake of March 27, 1964, at Seward, Alaska: U.S. Geol. Survey Prof. Paper 542-E, 43 p.

Molnia, B. F., and Carlson, P. R., 1975, Base map of the northern Gulf of Alaska: U.S. Geol. Survey Open-File Map 75-506, scale 1:500,000.

National Academy of Sciences, 1971, The Great Alaska Earthquake of 1964: Washington, D.C., Natl. Research Council Comm. on Alaska Earthquake, pt. A, 834 p.

Pierson, W. J., Neumann, G., Jr., and James, R. W., 1955, Practical methods for observing and forecasting ocean waves by means of wave spectra and statistics: U.S. Naval Oceanog. Office H.O. Pub. 603, 84 p.

Plafker, G., Bruns, T. R., and Page R. A., 1975, Interim report on petroleum resource potential and geologic hazards in the outer continental shelf of the Gulf of Alaska, Tertiary province: U.S. Geol. Survey Open-File Rept. 75-592, 74 p.

Reimnitz, Erk, 1966, Late Quaternary history and sedimentation of the Copper River delta and vicinity, Alaska [Ph.D. thesis]: San Diego, Univ. California, San Diego, 160 p.

——1972, Effects in the Copper River delta, in The great Alaska earthquake of 1964: Oceanography and coastal engineering: Washington, D.C., Natl. Acad. Sci., p. 290–302.

Searby, H. W., 1969, Coastal weather and marine data summary for the Gulf of Alaska–Cape Spencer westward to Kodiak Island: U.S. Dept. Commerce Environmental Sci. Services Admin. Tech. Memo 8, 30 p.

Tarr, R. S., and Martin, L., 1912, The earthquake of Yakutat Bay, Alaska: U.S. Geol. Survey Prof. Paper 69, 135 p.

von Huene, Roland, Molnia, B. F., Bruns, T. R., and Carlson, P. R., 1975, Seismic profiles of the offshore Gulf of Alaska Tertiary province, R/V *Thompson* Sept.–Oct., 1974: U.S. Geol. Survey Open-File Rept. 75-664, 16 p.

Manuscript Received by the Society September 7, 1976
Manuscript Accepted September 17, 1976

10

Large, Holocene low-angle landslide, Samar Island, Philippines

JOHN A. WOLFE
Schoenike, Wolfe & Associates, 5133 Richmond Avenue, Suite 1, Houston, Texas 77027

ABSTRACT

Samar Island, Philippines, is located on the eastern side of the central or Visayan Islands. Here the Talavera limestone, locally as much as 400 m thick, of Miocene age unconformably overlies an argillized lower Tertiary tuff. It grades into and is fringed by upper Miocene to Pliocene clastic rocks called the Barili Formation.

By the end of Pleistocene time, the Talavera limestone had developed karst with maximum relief of 300 m. Internal drainage saturated the argillized tuff, and earthquake mobilization caused a limestone block 18 × 25 km to glide northward on a slope of 0.6° for about 5 km. This beheaded the delta in Maqueda Bay, moved the drainage divide from the center to within 1 km of the west coast, and dammed the Ulot and Tubig Rivers. Behind the dams, estuary-like lakes were formed and filled with sediment. Subsequent headward erosion of these rivers cut gorges and drained the lakes.

There would be no way of controlling a slide of this size, and some engineering projects could actually cause conditions allowing such a slide to be triggered by an earthquake. This illustrates that in a region of composite hazards it is necessary to consider features an order of magnitude larger than would otherwise be of concern.

INTRODUCTION

During a reconnaissance of central Samar Island (Fig. 1), Philippines, many drainage anomalies were found to correlate with a displaced block of limestone 18 × 25 km, possibly as large as 135 km³, which is interpreted as being the result of a Holocene low-angle landslide.

Samar Island, the third largest in the Philippine archipelago, is located in the eastern Visayas, the name given to the group lying between the two largest islands, Luzon and Mindanao. The area, which has an average annual rainfall of about 500 cm, is covered with tropical rain forest and lies in the path frequently followed by typhoons.

GEOLOGY

The Philippine Islands are composed of two geologic provinces (Gervasio, 1964; Santos-Yñigo, 1966): the stable region, consisting of Palawan, Mindoro, and adjacent islands, and the mobile region, consisting of the remainder of the archipelago, one of the most active earthquake zones in the world (Alcaraz, 1968). The stable region contains rocks dating back to Paleozoic time. The mobile region is largely composed of younger Quaternary, Pliocene, and Miocene sedimentary rocks, intrusions, volcanic rocks, and ophiolites intermixed with and overlying "basement" that consists of metamorphosed and igneous rocks of pre-Miocene to Cretaceous age.

Stratigraphy

Basement Rocks (BC and 1b in Figs. 2, 3). This sequence consists of pyroclastic rocks, volcanic breccia, andesite, and basalt, locally intruded by diorite and quartz diorite. The breccia and pyroclastic rocks are generally altered, in some places chloritized and in others argillized. This breccia is the most common rock in the basement. There have been no age determinations, but these rocks antedate Miocene time extending back to Eocene and Cretaceous times (Gervasio, 1964).

A major unconformity occurs on top of the basement rocks. Deep weathering resulted in development of clay.

Carbonaceous and Clastic Beds (2b in Figs. 2, 3). This unit unconformably overlies the basement and consists largely of brown to gray to carbonaceous clay and includes development of a lignitic coal bed. There are thin units of graywacke, some of which are cross-bedded. Because the clastic deposits grade into the overlying marl, this unit is probably of middle

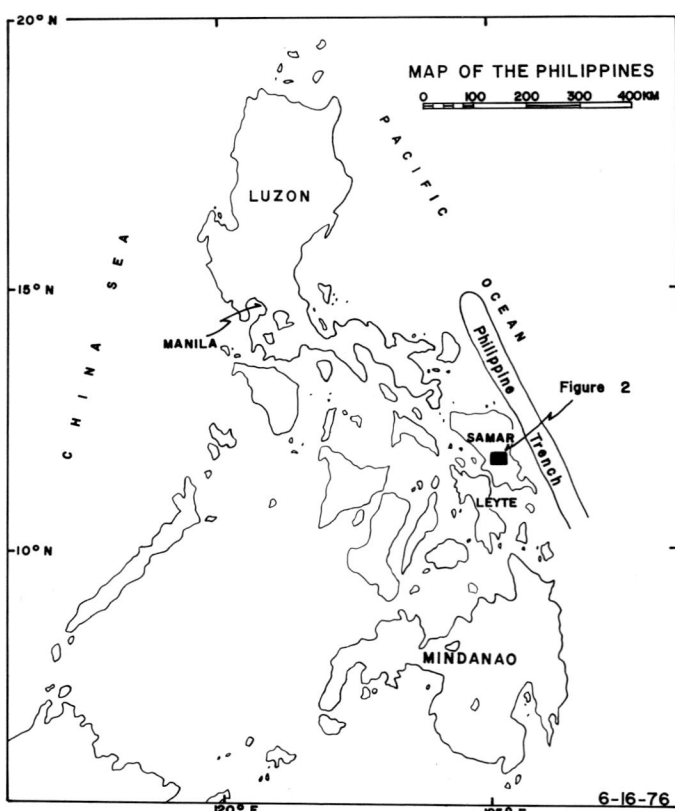

Figure 1. Map showing location of Bagacay landslide area, Samar Island, Philippines.

Miocene age. The clay of this unit is thought to have been the lubricant of the gravity slide.

Talavera Limestone (2c in Figs. 2, 3). The lower member of this formation is a marl about 5 m thick. Fossils from this bed were dated as upper to middle Miocene by Filipinas Gonzales (1965, personal commun.). The upper, massive reef limestone reaches a maximum of 800 m in thickness. In the area studied, it is about 400 m thick. Extensive karst has developed on it.

Barili Formation (3 in Figs. 2, 3). This formation conformably overlies the Talavera and in part appears to grade into it by facies change. It is considered to be of late Miocene to Pliocene age. It consists of sandstones, mudstones, and turbidites reaching 1,500 m in thickness and forms the outer parts of the present island. The underlying Talavera limestone has been superposed on it on the eastern side by the landslide.

Quaternary Deposits (Q in Figs. 2, 3). In addition to small areas of alluvium where the rivers flow into the sea, there are clastic deposits containing organic material thought to have formed on lake bottoms after the Holocene landslide dammed the rivers near them.

Structure

Just east of Samar lies the Philippine Trench, one of the deepest spots in the Pacific. Leyte Island, which adjoins Samar on the southwest, is bisected by the northwest-striking Philippine rift (Gervasio, 1971), a left-lateral fault that can be recognized throughout the country. There are major faults on Samar that strike northeast and appear to be conjugate features of the rift. One of these, lying southeast of the block, may have triggered the landslide. The rift is still active.

Geomorphology

If Samar and Leyte were 10 m higher, they would form one island with a low-lying central valley. If 100 m lower, Samar would be two islands. This topographic low extends from Wright on the west coast to Taft on the east coast and is a structural low between two large, deeply eroded anticlines of basement rock that lie to the north and south.

Prior to the movement of the gravity block, the drainage divide is thought to have been near the center of the island (Fig. 2). A normal drainage pattern had developed north of the low zone. The limestone block, however, had only internal drainage. This is believed to have kept the underlying clay saturated. The limestone at this point is 400 m thick, and the block has rough dimensions of 18 km east to west by 25 km north to south. A gargantuan karst has developed on this limestone, with some of the pinnacles rising 300 m above the adjoining collapsed sink holes.

Drainage anomalies were noted and provided the basis for this study. The present drainage divide is shown in Figure 2. It is 1 km from the west coast and 45 km from the east coast. At its lowest point it is less than 100 m in elevation. The Ulot River flows southerly in a normal dendritic manner until it encounters the limestone, where it turns 110° to the east, enters a gorge, drops rapidly from 70 to 10 m, and then flows as a meandering stream into the Pacific Ocean. Above the sharp bend, there are broad flat valleys filled with alluvium that are similar to filled and raised estuaries.

Similarly, on the eastern side of the island, the Tubig River flows northerly, draining a part of the southern area of basement rocks. Several flat valleys resembling estuaries are noted. This river flows north until it encounters the limestone barrier, turns sharply to the east where it has cut a gorge, and drops rapidly to sea level.

In Maqueda Bay on the west coast, there is what appears to be a substantial delta and low swampy region that is too large for the small stream now flowing to the bay at this point. Farther north along this bay, very steep drainages are cutting into the divide, which is less than 100 m in height. The steepest of these streams has a 5° gradient. In this area that is composed of shale and soft sandstone bedrock and that has an annual rainfall of 500 cm, these streams should rapidly cut down and capture some of the drainage on the western side of the northern basement zone, the headwaters of the Ulot River.

To the south of the limestone lies an area that is virtually inaccessible. When studied on aerial photographs before the concept of a landslide was suggested, an area adjoining the block was mapped as a "strip plain" (Corby, 1951). The overlying sedimentary rocks appear to have been uniformly removed from a zone about 5 km wide, and into this zone streams are just beginning to erode. The stream patterns sug-

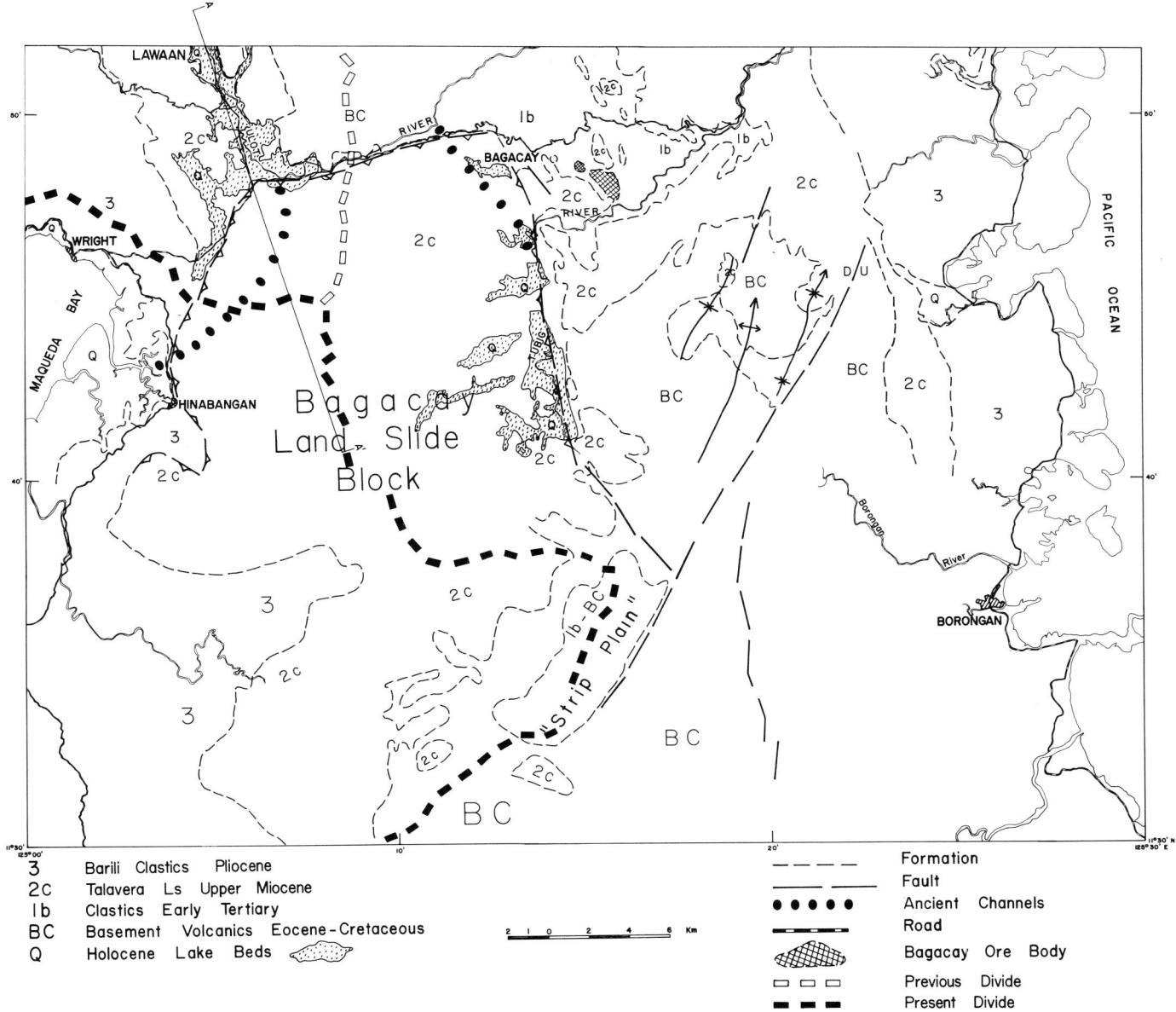

Figure 2. Geology of Bagacay landslide, Samar Island, Philippines.

gest that the plain itself is softer than the limestone removed from it and that it could not have been formed by normal erosion. This plain is cut along its southeastern edge by a northeast-striking fault that could be related to the movement of the landslide. One block of limestone 3 km long by 1.5 km wide can be seen astride this fault at its southern end. The slide appears to end at the block. This may be a remnant that did not glide northward or a piece that moved only slightly. The elevation of the upper end of the strip is approximately 300 m higher than the northern end (the toe) of the limestone block. Thus a topographic difference of 300 m in 25 km gives a northerly dip of 0.6°. Because the limestone has overridden the former stream, the true angle may be slightly steeper than this.

The shape of the "strip plain" still conforms quite closely to the shape of the southern end of the limestone.

On the west side of the limestone block, the middle Miocene Talavera limestone rests *above* the younger upper Miocene to Pliocene Barili clastic deposits. This was previously mapped as a thrust fault. On the north side of the limestone, the movement plane was observed at several points. Striations are developed on the toe of the slide on a large scale, dipping southward from 10° to 35°. It is thought that the slide crossed the low divide and rode upward over the north side of the

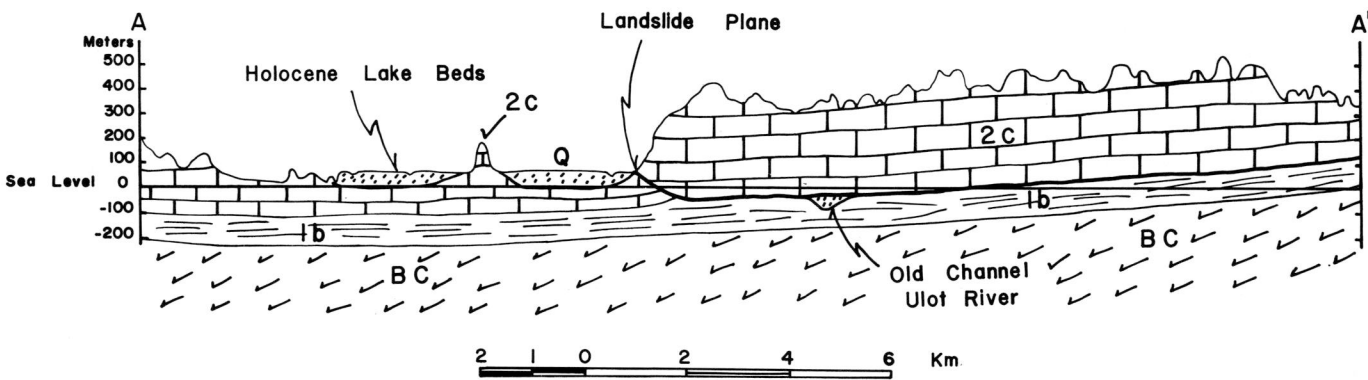

Figure 3. Inferred cross section (A-A') through toe of slide shows sediments in lake dammed by Bagacay landslide.

valley to an elevation of 30 to 40 m above the old divide at the highest points, accounting for the southerly dip of the toe of the slide plane.

INTERPRETATION OF FEATURES

Deposition of a large block of limestone upon a formation containing much clay was followed by vertical uplift. The clay was thoroughly saturated through internal drainage within the limestone. There existed a potential smooth surface to the north with a slope of 0.6°. Erosion in the low valley to the north of the limestone had probably reduced the highest point on the ancient divide to 50 m or less. During movement on one of the faults, possibly the Philippine rift or a conjugate feature south of the present location of the limestone, an earthquake fluidized the saturated clay and permitted the limestone to slide to the north, blocking the drainage of the ancient Ulot River and leaving its delta with no source. The resulting lake developed an outlet to the east, where it had to cut a gorge to reach sea level. Similarly, the Tubig River was blocked from its ancient channel into the present Ulot River. As the result of differential movement between the eastern edge of the block, the limestone split into segments on the eastern margin and tension valleys were formed.

Movement on the western margin may have been 3 to 4 km but on the eastern edge was probably 6 km. Water rising behind the limestone dam resulted in lakes and formed lacustrine sedimentary deposits that now resemble raised estuaries. A small stream became the outlet of the Tubig Lakes; it cut down to form a gorge, dropped 200 m to sea level, and drained the lakes.

Whereas it has not been possible to date accurately the slide that formed these features, the presence of the "strip plain" with clay on it and only a slight amount of headward erosion into it indicates that it cannot be very old. On the other hand, the Ulot and Tubig Rivers have cut gorges headward for several kilometres where large volumes of water are running off. The best estimate is 2,000 ± 1,000 yr.

The question arises of the nature of the movement of the block and with it the question of whether this is truly a short-lived event that was suddenly triggered into motion (thus a landslide) or a creeping phenomenon and thus more closely approximating a thrust fault. Elliot (1976) has shown that it requires about 760 m of rock to reduce the shear stress sufficiently to overcome gravity; the limestone here probably was not more than half this thickness. He further implied that a slope angle of the surface should be approximately 3° or steeper to initiate thrusting. Here the surface slope probably did not reach one-fourth of this minimum. He further implied that the term "thrust fault" should be restricted to blocks about 10 km thick and 100 km long that have moved 80 to 250 km. Thus conditions were not favorable to initiate thrusting, and the size of the block, although large for a landslide, was small for thrust faulting.

From the positive point of view, many of the features of the area are characteristically very short-lived, particularly in an area of very heavy rainfall. These are (1) very short streams with 5° gradients that eat into a divide composed of poorly consolidated clastics but which have as yet made virtually no headway toward recapture of the Ulot drainage; (2) the "strip plain" at the south end of the block that is apparently composed of argillized volcanic clastic deposits into which streams are just beginning to erode; (3) the fact that the Ulot River, which is of substantial size, was not able to maintain its course into Maqueda Bay around the toe of the slide (a strong indication); and (4) the presence of gorges with steep slopes between flat-lying lake beds and broad meandering channels near sea level, showing that streams could not maintain their grades around the limestone as it advanced and had to cut new channels when waters backed up above the limestone. The Ulot is still in the stage of steep, fast rapids in the gorge, and the Tubig still has falls along its course. Thus it is believed that the conditions found could only be formed by sudden damming of the rivers by a landslide.

CONCLUSIONS

A landslide involving over 100 km³ was triggered by an earthquake about 2,000 yr ago and moved northward on an angle of 0.6° for a distance of about 5 km. Given similar conditions, there is no way that a slide this size could be prevented or its effects minimized. A large reservoir constructed above such a block might contribute to initiating such a movement by saturating an argillaceous substratum.

The only protection that could be had from such a slide is to avoid placing major structures in its path, although there would be no way that people could be prevented from building homes in an area such as this.

The engineering implication here is that one should look beyond obvious hazards and consider features possibly an order of magnitude larger than those normally dangerous if a combination of hazards might exist.

ACKNOWLEDGMENTS

I thank E. Bayley and D. Coates for their very helpful suggestions that materially improved the paper.

REFERENCES CITED

Alcaraz, A., 1968, Crustal unrest in the Philippines: Philippine Geologist, v. 22, p. 163–170.

Corby, G. W., 1951, Geology and oil possibilities of the Philippines: Rep. Philippines Dept. Agriculture and Nat. Resources Tech. Bull. 21, 363 p.

Elliot, D., 1976, The motion of thrust sheets: Jour. Geophys. Research, v. 81, p. 949–970.

Gervasio, F. C., 1964, A study of the tectonics of the Philippine archipelago: Internatl. Geol. Cong., 22nd, New Delhi 1964, Proc., v. 4, p. 582–607.

——1971, Geotectonic development of the Philippines: Geol. Soc. Philippines Jour., v. 25, p. 18–38.

Santos-Yñigo, L., 1966, Island arc features of the Philippine archipelago: Philippine Geologist, v. 20, p. 79–92.

MANUSCRIPT RECEIVED BY THE SOCIETY SEPTEMBER 7, 1976
MANUSCRIPT ACCEPTED SEPTEMBER 17, 1976

Printed in U.S.A.

11
Martinez Mountain rock avalanche

CARL G. BOCK
Bechtel Associates Professional Corporation, 600 5th Street N.W., Washington, D.C. 20001

ABSTRACT

The Martinez Mountain rock avalanche, tentatively dated as early Holocene, comprises approximately 3.8×10^8 m^3 of granitic gneiss that broke away from the mountainside at elevation 1,926 m and flowed to elevation 49 m. The length of the slide debris is 7.6 km, and the width ranges from 1,070 to 1,370 m.

Rock in the area surrounding the slide is closely jointed, with northwest-trending faults cutting under the slide mass. Two dominant sets of high-angle joints intersect at the crown of the slide and are believed to have controlled formation of the broad V-shaped scarp. A third joint set dips downslope at 15° to 30°, and it locally controlled the basal surface of the slide.

Because of the crystalline nature of the rock involved in sliding, water was probably not a significant factor in the triggering mechanism; rather, it is thought that a major seismic event caused the slide. Such seismic triggering is consistent with historic occurrences of massive rockslides and in the seismically active Coachella Valley such an event would provide the energy needed to disaggregate the rock and start it moving.

An understanding of the structural geology and probable triggering mechanism provides a basis for comparison with similar geologic terranes and assessment of the potential for catastrophic landsliding that might affect urban developments.

INTRODUCTION

The Martinez Mountain rock avalanche is located on the east flank of the Santa Rosa Mountains of southeastern California about 8 km west of Valerie (Figs. 1, 2, 3). The Santa Rosa Mountains comprise the south end of an elevated block of crystalline rocks that flank the southwest side of the Coachella and Imperial Valleys. The major valley is a structually depressed trough, once occupied by the Gulf of California, and now contains the Salton Sea.

The first published map showing the slide was by Dibblee (1954). The map compiled by Rodgers (1965) showed the slide at a large scale but without details of its structure. An oblique aerial photograph, in which the slide is recognized and described, was published by Shelton (1966). The slide was visited several times by R. P. Sharp (1967, oral commun.), who made notes concerning the geology. However, the slide has not received the attention given the Blackhawk landslide (Woodford and Harriss, 1928; Shreve, 1968a) 105 km to the northwest, on the north side of the San Bernardino Mountains.

From the standpoint of engineering geology, as well as of public safety in urban planning, the slide has current significance. If the geologic factors and structural relationships that caused the slide can be determined, such information can then be used to identify other hazard-prone areas in this rapidly developing region along the mountain front. In mountainous terrain, catastrophic slope failures can occur downslope of large faults and (or) from oversteepened slopes on upthrown fault blocks.

GEOLOGIC SETTING

The Coachella and Imperial Valleys occupy a southeast-trending structural depression, 24 to 48 km wide and more than 193 km long, between the San Andreas fault zone on the northeast and the San Jacinto and Santa Rosa Mountains on the southwest.

The Santa Rosa Mountains and the San Jacinto Mountains to the northwest form the southwest side of the Coachella Valley, a structural depression more than 25 km wide and more than 80 km long (Fig. 2). The valley is bordered on the northeast by the Little San Bernardino Mountains, which rise to elevations of more than 1,525 m. The most prominent structural features of the Coachella Valley are the active San Andreas fault and associated faults, which roughly follow the trend of the northeast side of the depression. The

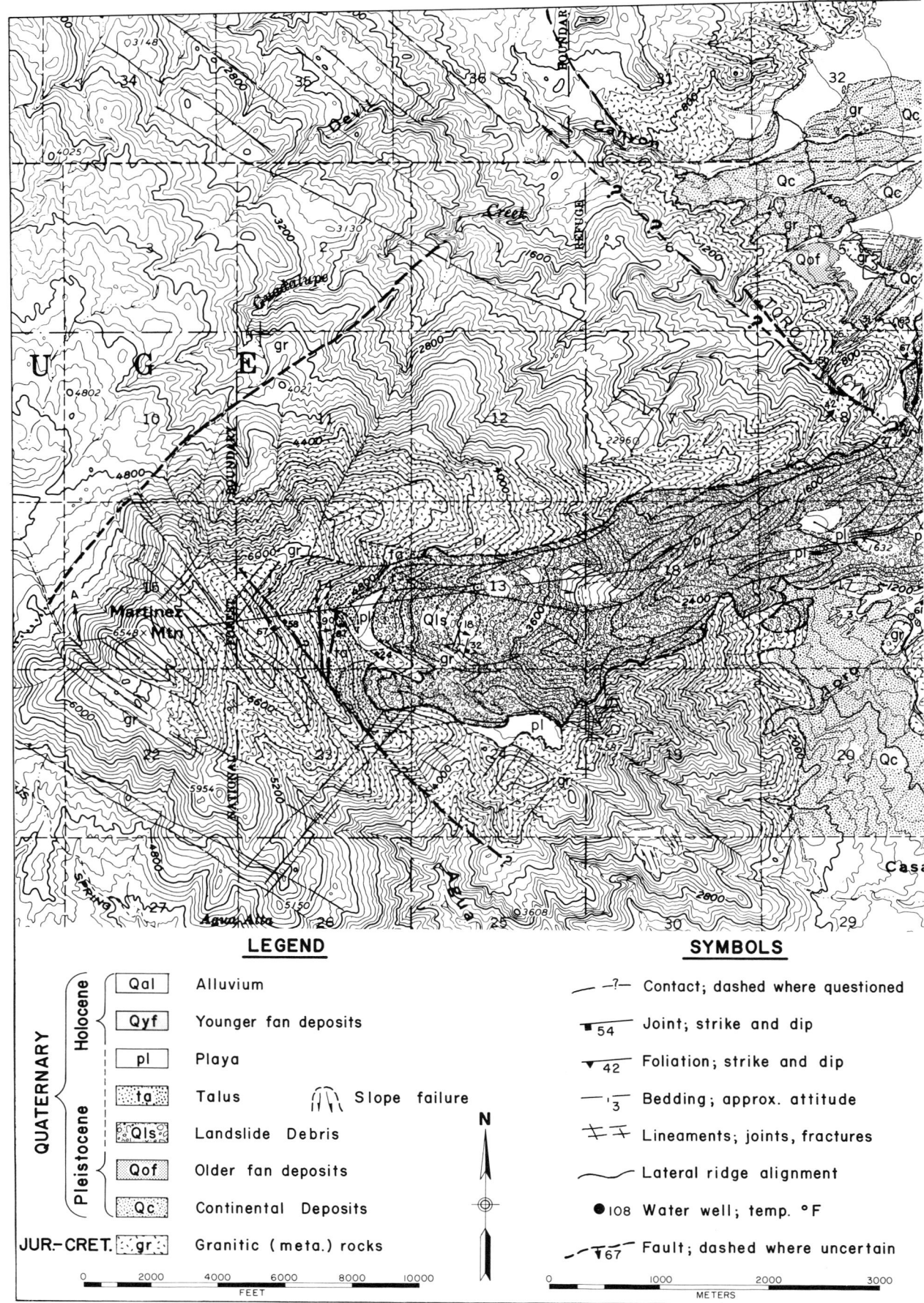

Figure 1. Geologic map of Martinez Mountain

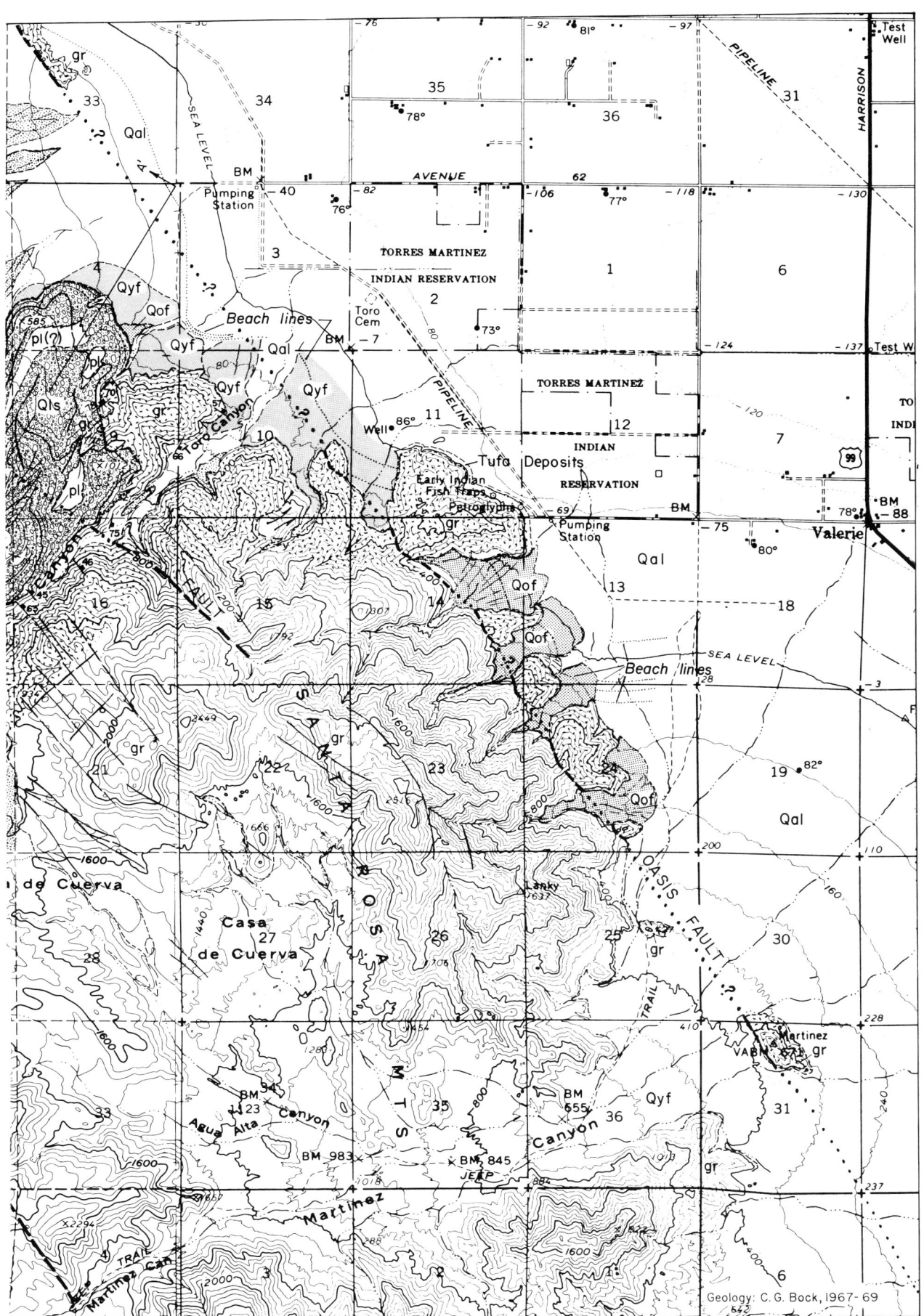

rock avalanche west of Valerie, California.

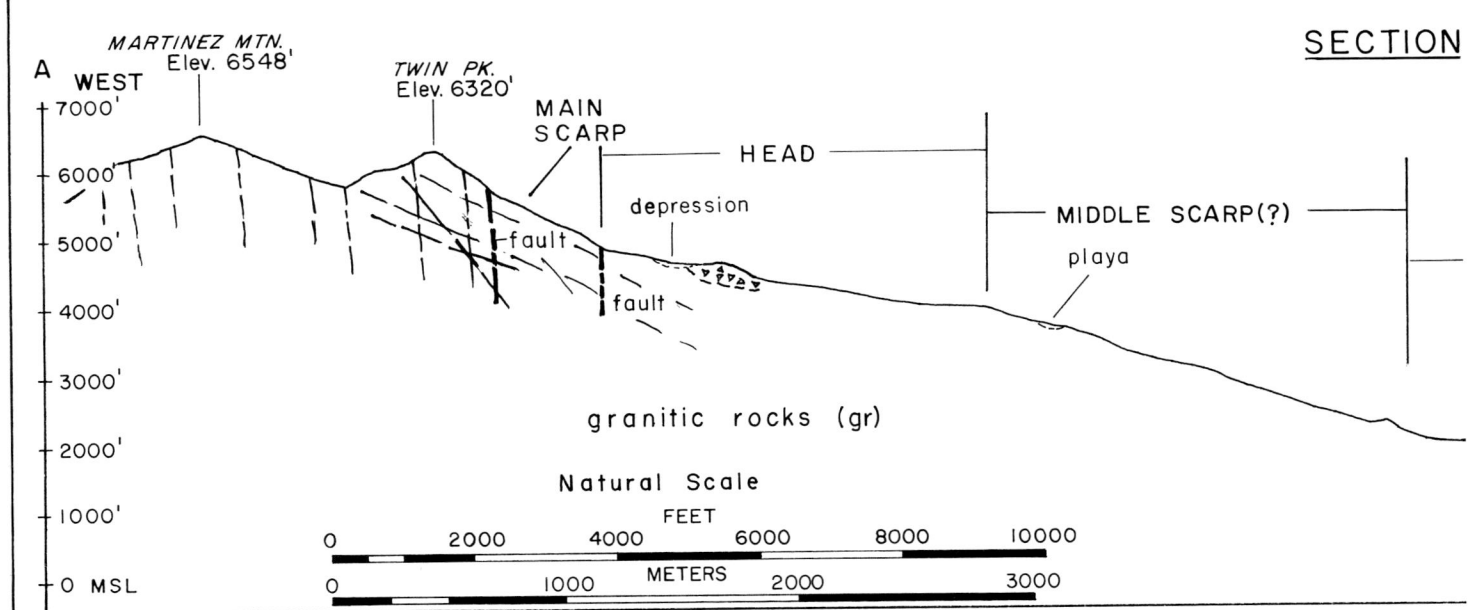

Figure 1 (continued). Cross section A-A'

San Jacinto and associated faults, southwest of the Santa Rosa Mountains, compose a seismically active belt having the same southeasterly trend as the structural grain of southeastern California.

Carbon dioxide wells, bubbling mud pots, hot springs, and thermal wells are found along the southeast end of the sea in the vicinity of the San Andreas fault as evidence of continuing tectonic activity in the area.

The region is thoroughly faulted and seismically active; historic surface displacement has occurred at Imperial (El Centro) (1940), Superstition Hills (1951), and again at Imperial (1966). These surface ruptures (Bonilla, 1967), as well as those caused by the Borrego Mountain earthquake (Allen and others, 1968), have occurred along the southwest side of the Imperial Valley, which is south of the Salton Sea and in the southern part of the great structural depression. The faults have a northwest trend and are generally subparallel with the San Jacinto fault system, which is located along the west flank of the Santa Rosa Mountains.

The Coachella and Imperial Valleys are underlain by Quaternary and Tertiary sedimentary rock as thick as 6,100 m (Dibblee, 1954), which rests on Jurassic and (or) Cretaceous granitic intrusive rocks.

Geologic Mapping

Geologic mapping was done at a scale of 1:24,000 (enlarged from 1:62,500) for the main slide mass, and many structural interpretations of surrounding areas were made by using aerial photographs. All calculations of slope angles, distances, heights, and volumes are only within the accuracy of the 1:62,500-scale map. Short field trips were made during the winters of 1967, 1968, and 1969, and the completeness of the geologic structural information shown is in proportion to the seven days total spent in the field.

Special features of the slide shown in Figure 1 are located according to the U.S. Geological Survey system for locating water wells, with an additional division of each alphabetic 40-acre plot into four equal quadrants numbered as shown in Figure 4.

ARCHAEOLOGY

The toe of the slide is about 4.4 km northwest of early Indian fish traps and petroglyphs (11-Q and R, in Fig. 4). The fish traps are stone rings along the shore of ancient Lake Cahuilla (1600 B.P. to 300 B.P.; Hubbs and others, 1960), into which the Indians drove fish as wind-generated waves washed into the rings. Fragments of brick-colored pottery are found along the surface of the slide from the desert floor almost to the head. The pottery shards are sometimes found in clusters, with enough pieces fitting together to form part of the curved surface of an olla. This evidence of Indian habitation suggests that Indians used the slide as a route to higher elevations for pinyon nuts, game, and, perhaps, fresh water. The playas on the slide surface may reveal additional artifacts if carefully investigated by archaeologists.

STRUCTURAL RELATIONSHIPS

The mountainside that contains the slide is composed of coarse-grained to very coarse-grained, light-gray to gray gra-

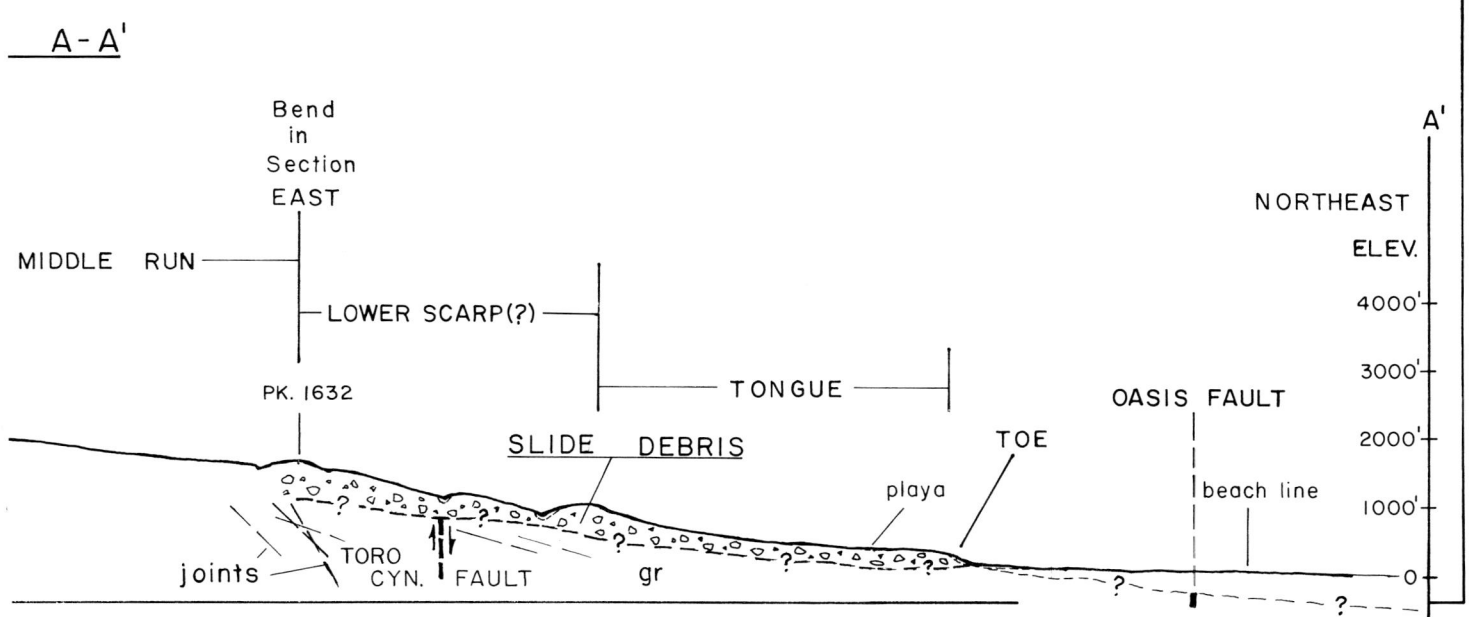

of Martinez Mountain rock avalanche.

nitic gneiss that is closely jointed, fractured, and locally faulted. Lineations indicating joints, fractures, and possible faults are clearly discernible on aerial photographs and topographic maps of the area (Figs. 1, 8). The granitic gneiss has been intruded by brown rhyolite dikes and coarse-grained pegmatite dikes; these structures are seen in Toro Canyon beneath the slide debris. South of the slide, structures (dikes?) seen on aerial photographs (Fig. 8) form prominent northwest-trending lineations across the ridges.

A prominent lineation along changes in drainages suggests the continuation of the Toro Canyon fault (Fig. 1), to both the northwest and southeast. Structure resembling foliation has been mapped in section 8. The in situ rock is so closely jointed and fractured that from a distance, it gives the appearance of having been diced. Low-angle, downslope-dipping joints are exposed in canyon walls.

Two sets of intersecting joints controlled the formation of the scarp near the summit of Martinez Mountain. These joints are apparent on aerial photographs; one set has an average strike of N35°W, the other about N35°E. A generally north-striking set of joints cuts the others and dips steeply (80° to 90°) eastward (Fig. 5). At least one of these north-south joints in the scarp area showed crushed rock in a zone at least 0.3 m wide; this may be the result of down-dropping during sliding or of a pre-existing fault. Just below the scarp crown (14-M4, Fig. 1) joints trend N52°E, 67°NW and N43°W, 58°NE.

In a large gully that cuts through the slide (13-P2, Fig. 1), 9 to 12 m of debris consisting of gray granitic gneiss cobbles and boulders in a compact, earthy matrix rests on closely fractured, jointed, and foliated metaigneous rock (Fig. 6). One joint set strikes N12°E and dips 32°E (downslope);

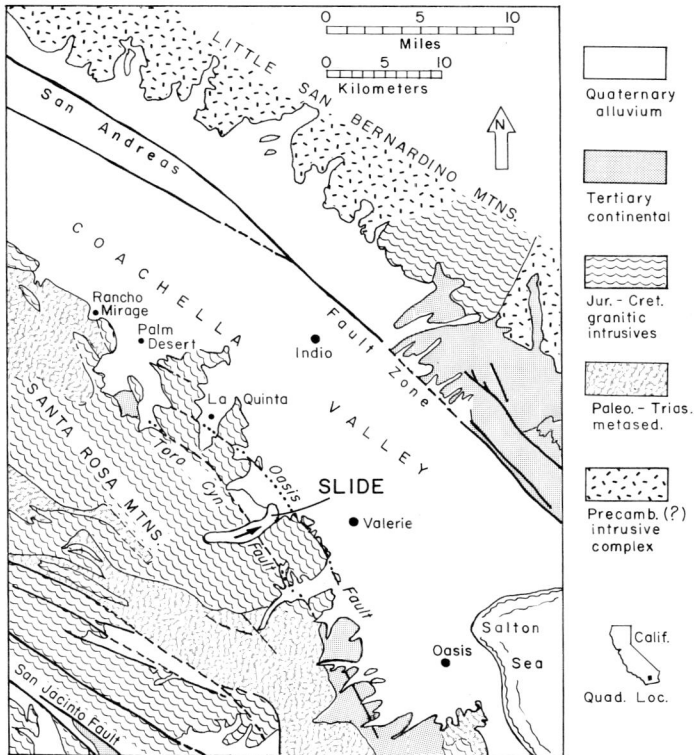

Figure 2. Generalized geologic map of Coachella Valley, showing location of Martinez Mountain rock avalanche (after Dibblee, 1953).

Figure 3. Martinez Mountain rock avalanche, showing alluvial fan from Toro Canyon (right center) and tufa deposits and old beach lines (lower left) of ancient Lake Cahuilla. View is west to Martinez Mountain.

the slide surface attitude at this point is N16°E, 18°E or a slope of 1:3.

Slide Fault

Most of the floor of Toro Canyon is covered with fresh, gray rounded and subrounded boulders and finer grained alluvial debris, but in places along its northwest wall, the basement rocks have been exposed by erosion. A fault cutting the basement rock is marked by dense, tough, silty grit, that forms a thin, high-angle gouge zone that is generally more than 2.5 cm thick. The fault can be traced intermittently for several hundred metres and has a strike paralleling the bearing of the canyon. The sense of fault displacement was not determined from the limited exposures (16-B2, Fig. 1).

Toro Canyon Fault

A northwest-trending fault zone is well exposed in Toro Canyon beneath the slide debris (16-B2, Fig. 1; Fig. 7). The fault is believed to be throughgoing and may be the continuation of the fault shown several kilometres to the north by Rodgers (1965). Where the fault zone cuts across Toro Canyon, it is 4.4 m wide and displaces the granitic basement complex. The gouge zone is 1 m wide and consists of rust-stained, crushed granitic gneiss material in which the feldspars have been sheared and altered to a clayey matrix. The gouge is dense and tough, with the crushed rock forming a grit;

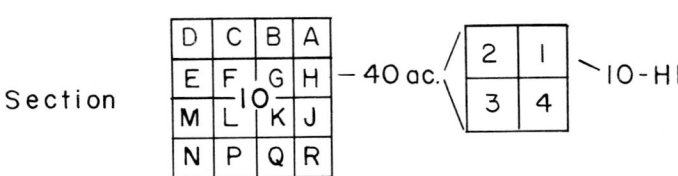

Figure 4. System used to locate special features on geologic map (Fig. 1).

Figure 5. Closely jointed zone 2.5 m wide in scarp area (14-L4, Figs. 1, 4), striking north.

Figure 6. Angular to subangular slide debris overlying shattered and closely fractured basement complex; exposure in gully near head of slide (14-P3, Figs. 1, 4).

Figure 7. Slide debris overlying basement complex and Toro Canyon fault (contact between light and dark rock at right).

original rock structure is faintly discernible. The gouge zone has been eroded, producing a slot 1 to 1.5 m deep between the two fault blocks. Slickensides along this contact indicate that the last movement was essentially vertical; the amount of displacement was not determined. Locally, the fault zone has a strike of N20°W and dips 63° to 68° northeastward.

Rock on the downstream side of the fault zone is shattered, and many jagged, unweathered spirelets stand 1 or 2 m in relief. High-angle pegmatite dikes, 10 to 15 cm wide, cut the rocks upstream of the fault zone and strike subparallel to the system of northwest-trending faults and joints.

ROCK AVALANCHE

The Martinez Mountain rock avalanche is in a seldom-visited area that is accessible by automobile from Avenue 62 (Fig. 1) across a major flood-control dike that borders the base of the mountains. Because of the immensity of the slide, it is best seen from the air, on maps, or from a distance. It can be easily delineated on the 1:62,500 Palm Desert (California; lat 33°30′, long 116°15′) U.S. Geological Survey quadrangle map, and it is a striking feature on aerial photographs (Fig. 8).

The rock avalanche broke away from the east face of a twin peak (elevation 1,926 m), just 914 m east of Martinez Mountain (elevation 1,996 m) summit.

Morphometry

The slide has an elevation differential from toe to head of about 1,340 m. If the vertical distance of the scarp above the head of the slide is included, another 520 m should be added to give a total elevation difference of 1,860 m. The axial length of the slide from toe to head is 7.6 km, and if the horizontal distance of the scarp is included, the total length is about 8.4 km. The width of the debris ranges from 1,070 to 1,370 m, and preliminary observations show that the thickness ranges from 12 to 15 m near the head to 50 to 55 m at the toe. The slide mass covers about 790 ha or 7.87 km^2.

The slide has several important changes in general slope angle, as given in Table 1 (see Fig. 1, cross section).

Course of Movement

From the scarp head, the direction of sliding was due east for the first 2.9 km, coursed N62°E for the next 3.4 km, and took a terminal direction of N15°E for the last 2.1 km. The pre-existing topography controlled the direction of the slide, and it is assumed that the debris followed drainage courses for at least part of the way. The slide mass had enough momentum to make a 60° change of course, and it

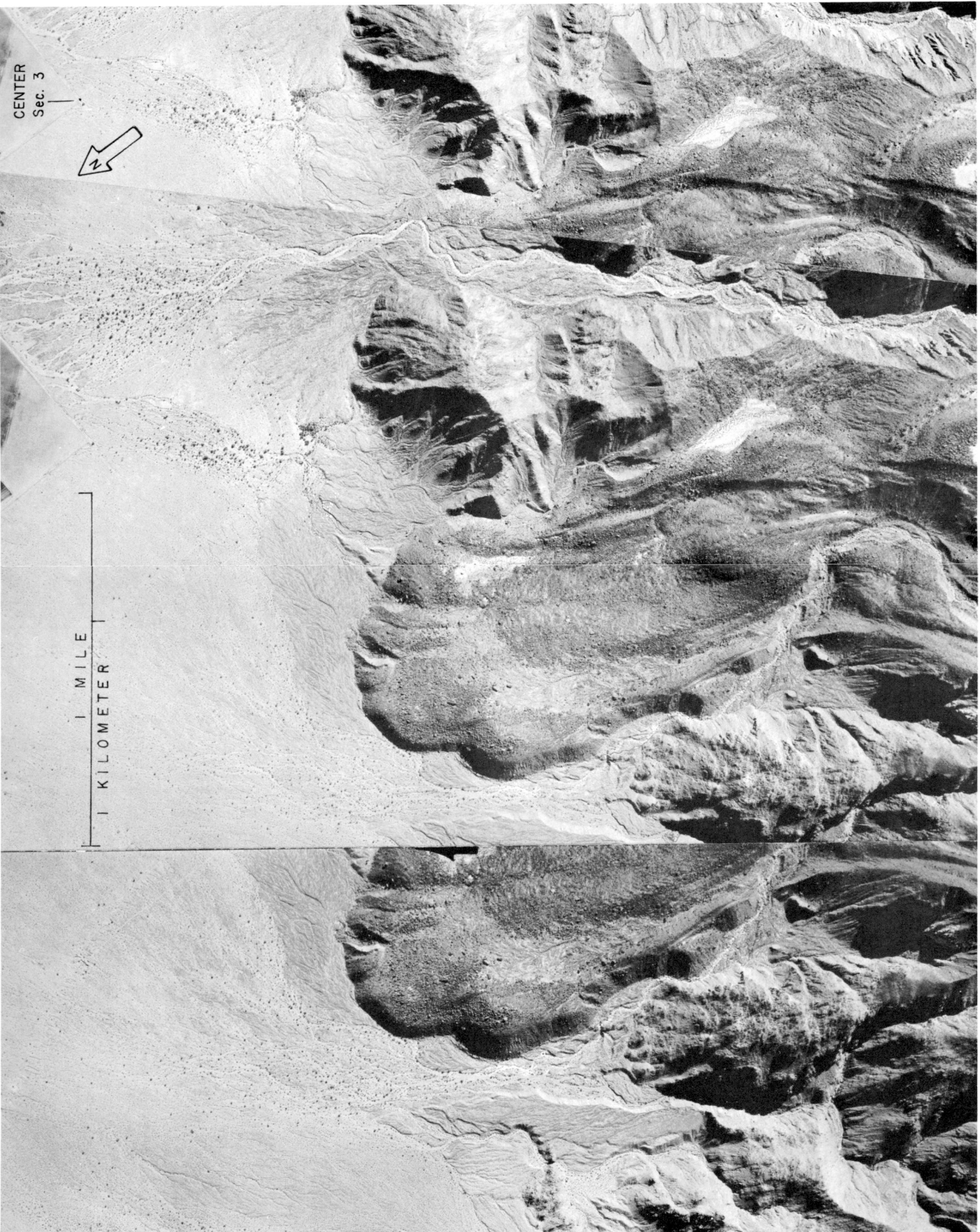

Figure 8. Stereo pairs showing tongue of Martinez Mountain rock avalanche; note downcutting by stream through slide, long lateral ridges, and alluvial fan deposits. Beach lines are visible as wave-cut notch on alluvial fans.

TABLE 1. SLOPE ANGLES, LENGTHS, AND HEIGHTS OF MARTINEZ MOUNTAIN ROCK AVALANCHE

Feature	Slope (V:H)	Remarks
Scarp	1:1.6	305 m elevation differential
Head	1:6	For 1,676 m length
Middle scarp(?)	1:3	For 1,829 m length
Middle run	1:9	For 1,372 m length
Lower scarp(?)	1:5	For 1,219 m length
Tongue	1:12	For 1,524 m length
Toe	1:1.9	49 to 55 m high
Overall (crown to toe)	1:4.5	1,878 m high, 8.4 km long

overshot into Toro Canyon. The material came to rest, forming a symmetrical lobe that projects 915 m from the mountain front out onto the gently sloping desert floor.

Volume

Calculation of the slide volume is based on measurement of the thickness in one location and contours throughout the rest of the area of accretion. The volume of depleted material should be about the same as that accreted, but the unknown preslide topography prevents even a good estimate of depleted material. Undoubtedly, erosion has removed some material but not enough to significantly change the order of magnitude of the volume.

The double-end area method was used to compute the volume of a trough-shaped wedge whose width was that of the slide and whose depth was approximated from cross sections drawn at every 120-m increment above sea level. A conservative estimate of the total volume of slide debris, based upon these assumptions, is 3.8×10^8 m^3.

Slide Topography

The slide mass can be distinguished easily by linear and lobate topographic features, subrounded rocks in the debris which contrast with the angular nonslide rock, and the general surface texture of a mass that has undergone movement by a complex interaction of sliding and flowage. Several episodes of sliding occurred, as evidenced by successively smaller debris lobes superposed on each other. Preliminary study shows three episodes, but the time interval between the three has not been determined—it could have been hours or years. Long, sharp-crested lateral ridges were formed along the periphery of the slide. Where debris was pushed up against existing slopes, closed depressions were created that have since become playas (Fig. 9). These playas are generally elongate, some are crescent shaped, and several are irregularly shaped because of topographic undulations. They are floored with fine to coarse silty sand and commonly have large boulders on the periphery which have rolled down the surrounding lateral ridge slopes.

A suggestion of the original topography is given by the steep slope (scarp?) in the east half of section 13 (Fig. 1) and the northeast quarter of section 17. Although northeast-trending ridges and canyons were topographic controls that apparently guided the moving mass, they have been greatly subdued by the enormous shearing and grinding action of the avalanche. Probably most of the major topographic changes in slope are a reflection of the buried topography.

Details of Slide

The lower part of the slide is littered with thousands of subangular boulders, 3 m or greater in long dimension, distributed almost equidistant from each other. They are unbroken, fresh, hard granitic gneiss rocks and sit on the surface of the slide; some boulders are perched precariously on slopes as steep as 1:1.3 (37°). Several megaliths are as long as 17 m and stand more than 6 m high; a smaller one measured 8.5 m × 5.5 m × 11 m. Some small, isolated sheets of rock, several tens of metres in area, appear to have moved during the slide with relatively little disturbance; blocks still fit together loosely with some infilling of openings by brown silty sand and grit.

Toro Canyon, along the southeastern edge of the slide, trends N50°E, which is the same orientation as one set of joints interpreted from aerial photos. The head of the canyon may have been partially filled with slide debris at one time, as this was the point where the slide changed course. If the slide debris filled the canyon, it has been removed by stream erosion, as there is no trace of the slide mass on the south wall. The northeast-trending ridge that borders Toro Canyon on the north had enough prominence and stability to change the course of the slide about 60°.

Erosion in Toro Canyon (16-C3, Fig. 1) has exposed the base of the slide, and the nature of the contact can be seen (Figs. 10, 11). A measured section (Fig. 12) shows the slide mass overlying alluvial rock deposits. The contact is the slide surface, and it is a prominent undulating surface that has well-preserved striae and slickensides on the underside. The general dip of the surface at this point is 15° to 17° southwest, but the slide was moving northeastward and up over a bedrock high.

Figure 9. View northeastward from elevation 1,158 m, showing playas and lateral ridges. Coachella Valley is in distance.

Figure 10. Eastward view down Toro Canyon, showing base of Martinez Mountain rock avalanche; highest part of vertical exposure above slide surface is about 26 m. Toro Canyon fault is located immediately to right of white spot in photograph.

The slide debris is tan to light brown in weathered exposures and dull gray in fresh exposures. It is composed of granitic gneiss particles, ranging in size from fine gravel to cobbles, in a silty, gritty matrix. The material is dense to moderately hard and slightly cemented. Many near-vertical joints, several tens of metres high, are present in this comparatively fine-textured material and result in block falls into Toro Canyon.

The underlying alluvial rock deposit is composed of gray (unweathered?) subround to subangular granitic gneiss particles ranging in size from fine gravel to boulders 1 m in diameter in a silty granitic grit matrix. The texture of the alluvial rock deposit is in striking contrast to the overlying slide debris and suggests a difference in origin and mode of deposition. Further study of the alluvial rock deposit may show that it is old fan-deposit material and may correlate with the continental deposits (Qc) south of the slide in sections 17 and 20.

The alluvial rock deposit rests on gray medium-grained to coarse-grained granitic gneiss along what appears to be an erosional surface. The bedrock is slightly weathered and closely fractured but appears to be in place and not part of the slide.

At the head of the slide, boulders ranging in size from 0.4 to 0.8 m³ form deposits of talus and postslide debris fillings. The relatively smooth appearance of the talus is in sharp contrast to the open-jointed, blocky, craggy in situ rock of the scarp. Rock in the scarp and crown area is a speckled black and white medium- to coarse-grained biotite hornblende granitic gneiss, which appears fresh and unvarnished.

The slide occupies a drainage area that is little larger than the slide itself. A continuous drainage course extends from the scarp down to the Toro Canyon fault essentially along the axis of the slide. At this point, the canyon is 24 to 30 m deep and trends northward along the west margin of the slide.

The downcutting must have been extremely rapid, because the canyon cuts across a well-defined surface feature (8-J1, Fig. 1) that was an older drainage course that never became well established. This drainage course leads to a plume-shaped playa(?) that drains to two separate eroded notches in the toe of the slide.

Another canyon has been eroded down along the contact between bedrock and the east side of the slide. Stream erosion has cut into the toes of slopes that are covered with desert-varnished boulders, leaving a light-colored fresh erosion scar along the canyon sides. It is evident that erosion is still a rapid and significant process even in this arid environment.

Desert Varnish

All boulders on the slide surface from elevation 49 m to at least 976 m are coated with brown to dark-brown desert varnish. The texture of most boulders is smooth; the development of the varnish leaves a shiny surface. R. P. Sharp (1967, oral commun.) has mapped "old" and "new" slide breccia based on the degree of varnishing; the "old" breccia has less varnish because it is more disintegrated by weathering and does not have the surface that will maintain the desert varnish. It is probable that surface weathering of the coarse-grained rock proceeds more rapidly at higher elevations because of temperature extremes, more moisture, and the mechanical action of freezing and thawing. With detailed mapping of the "old" and "new" breccias, the age relationship of the different episodes of sliding might be determined; however, that was not within the scope of this effort.

Calcareous Deposits

Throughout the distal part of the slide, calcareous (caliche) deposits near the top of the slide debris bind the angular sand, gravel, and boulders together, forming irregular lenticular layers. The caliche forms a light-tan to buff rim on both sides of the canyon (9-G3 and 9-F4, Fig. 1). Near-vertical joints (5-Q4) in the granitic rock contain white, calcareous rodlike structures 2 to 3 cm in diameter. The rods are surrounded by coarse granitic grit that is well cemented.

MODE OF FAILURE

The Martinez Mountain slope failure is classified as a rock avalanche after the definition of Varnes (1958) and Mudge (1965). It was a downward, rapid movement of granitic gneiss that initially failed along surfaces of weakness and joint sets at the head and created its own slide surface en route. The failure probably began as a combination rockfall-rockslide and rapidly became a rock avalanche. The rock failed along near-vertical joint surfaces that intersected other joints that dipped out of the natural slope. The failing-sliding rock created lateral ridges along its margins, slid up the slopes of ridges in its path, changed course, and terminated as a lobate tongue on the desert floor. There is no strong evidence that water played an important role in movement; the morphology at the toe indicates no excess of water escaping as the moving mass became immobilized.

While many factors probably contributed to slope failure, the mechanism required to initiate such a catastrophic slide needs explanation. Unlike soil and bedded sedimentary

Figure 11. Closeup of slide debris overlying coarse alluvial deposits that overlie granitic gneiss basement complex.

rocks, which are highly susceptible to pore-water pressure, the blocky, granitic rock was probably free draining. The absence in this crystalline rock of clay seams, continuous bedding planes, and slide-prone contacts (that is, water-bearing sandstone overlying clay shale) eliminates three factors usually contributing to slides in soft rocks.

Some special set of structural conditions had to exist in the particular area in order for the slide to occur there rather than elsewhere. The contributing factors probably present before failure (see Figs. 13, 14) were (1) closely fractured and jointed rock, (2) steep natural slopes, (3) planar surfaces oriented out of slope at an angle flatter than the natural slope angle, and (4) existing faults in the preslide area and possible movement along them. Of course, gravity is an ever-present force that affects slope stability. The uplift of the Santa Rosa–San Jacinto Mountain block has led to slope steepening that has created an impressive east-facing escarpment. As discussed by Terzaghi (1950), tectonic stresses that develop from tectonic movements lead to large-scale deformation of the Earth's crust which affects all materials and causes an increase in slope angles and shearing stresses.

The only factor still to be considered is friction (shearing strength), which is a force resisting gravity. In the absence of weathering, the friction between rock blocks probably

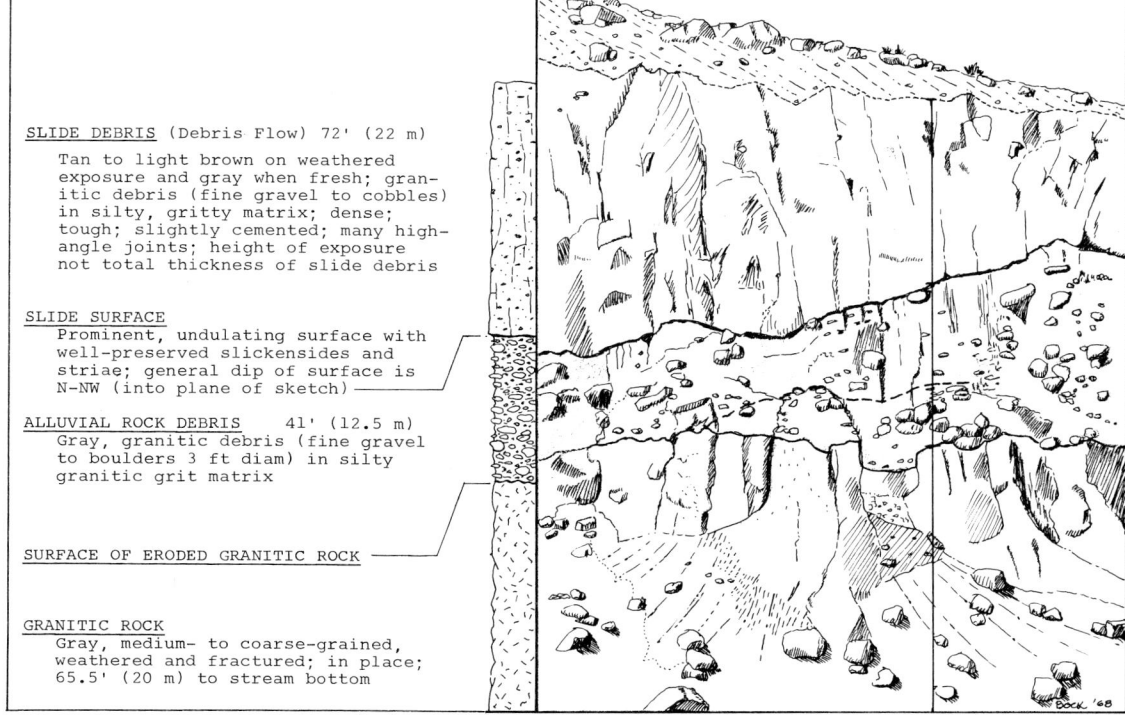

Figure 12. Photosketch of measured section of slide exposed in Toro Canyon (not maximum thickness); vertical line in sketch shows approximate location of section.

SLIDE DEBRIS (Debris Flow) 72' (22 m)
Tan to light brown on weathered exposure and gray when fresh; granitic debris (fine gravel to cobbles) in silty, gritty matrix; dense; tough; slightly cemented; many high-angle joints; height of exposure not total thickness of slide debris

SLIDE SURFACE
Prominent, undulating surface with well-preserved slickensides and striae; general dip of surface is N-NW (into plane of sketch)

ALLUVIAL ROCK DEBRIS 41' (12.5 m)
Gray, granitic debris (fine gravel to boulders 3 ft diam) in silty granitic grit matrix

SURFACE OF ERODED GRANITIC ROCK

GRANITIC ROCK
Gray, medium- to coarse-grained, weathered and fractured; in place; 65.5' (20 m) to stream bottom

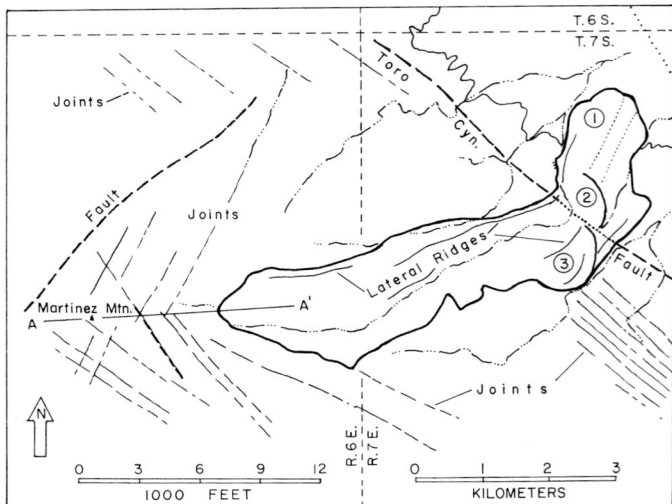

Figure 13. Joint patterns controlling mode of failure of rock avalanche; numbers indicate time sequence of successive slide lobes.

does not change perceptibly, and the effect of gravity does not change except during a major earthquake. A force must be imparted to the static rock mass to start movement. It is hypothesized that a seismic event, with its epicenter in the immediate Martinez Mountain area, was the initiating force.

In a tectonically active area such as that considered here, transitory earth stresses occur frequently in the geologic time sense. Transitory earth stresses as discussed by Varnes (1958) include earthquakes and blasting. In regard to earthquakes, he says, "Earthquakes have triggered a great many landslides, both small and very large and disastrous. Their action is complex, involving both increase in shear stress and, in some examples, decrease in shear strength. They produce horizontal accelerations that may greatly modify the state of stress within slope forming materials" (p. 43). Because of the open-jointed nature of the rock composing the mountain slopes, a strong seismic shock would literally throw the closely jointed rock from place. The reaction to violent shaking would be of the same type experienced by masonry walls and chimneys during earthquakes.

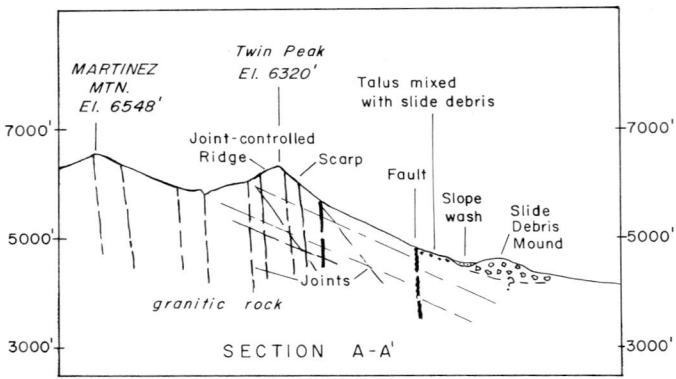

Figure 14. Joint patterns controlling breakaway of rock from mountainside.

MECHANISM OF MOVEMENT

In a discussion of the mechanism of movement, it is prudent to review the work of Shreve (1968a, 1968b), Kent (1966), and Hsü (1975). The question not satisfactorily answered is whether rockslides and rock avalanches slide or flow; I believe they move through a complex interaction of both sliding and flowing as these terms are understood in the strict sense. Shreve hypothesized the air cushion lubrication-sliding mechanism, Kent proposed flowage because of fluidization by air trapped in the mass, and Hsü also favored flowage after review of the Elm slide (Heim, 1882) and laboratory experiments.

Before the argument continues further, consider that of 35 slides evaluated by Hsü (including one on the Moon) little attention is given to rock type, rock quality, geologic structure, factors contributing to (historic) slides, and seismicity of the area. Rather than try to make all terrestrial catastrophic landslides "flow" a certain distance depending on the fahrböschung (ratio of height of fall to length traveled), these slides should be classified according to (1) rock type—igneous, metamorphic, sedimentary; (2) rock quality—fresh and unweathered, slightly weathered, deeply weathered; (3) geologic structure—dip slope, jointed-fractured rock, faulted zones, bedded or foliated rock; (4) factors contributing to (historic) slides—ground water, snow-ice load, mining, blasting and excavation, erosion of soft beds, prolonged heavy rainfall; and (5) seismicity of the area—frequency and intensity of seismic activity.

Discounting lunar slides from this study because of gravity and "atmospheric" differences, it would seem that a bedded sequence on a smooth, continuous dip slope could travel farther than a blocky, hard-rock mass, falling the same height, but moving along a much more irregular and difficult surface.

Factors contributing to length of travel include (1) slope angle of fall, (2) texture and relief of surface along slide path on mountain slope, (3) ability of material to break down upon particle impact to produce material approaching soil-flow consistency (dependent on degree of weathering and compressive strength of rock), (4) straightness of slide path, (5) length of unobstructed slope (runout) at base of mountain slope, (6) increase in inertia from seismic acceleration, and (7) water content (viscosity). But none of this decisively determines whether the mass slides or flows. The topographic features of the Martinez slide definitely show flow characteristics (lateral ridges miles long), and yet the basal surface shows undulating narrow- and broad-troughed grooves and striations parallel to the direction of movement. Do flows have the ability to cause slickensides, or can this only happen during sliding? Does the upper part flow and the bottom part slide, and why didn't the Martinez slide continue farther across the unobstructed desert floor? Was it sliding on an air cushion, was it flowing because of entrained air, or was it a viscous mass that sometimes flowed and sometimes slid?

Consider the case of coal, gravel, or sand moving down a steeply declined chute: is it sliding or flowing? When it leaves the end of the chute, it travels freely through the air but maintains its cross-sectional form and consistency until it impacts upon another surface that causes it to deflect and change from sliding to flowing. However, within the mass sliding down the chute, surges can be seen that must be the result of flowage.

The rock involved in the Martinez Mountain rock avalanche is so closely jointed that it does not have the structural integrity to slide as a coherent sheet. The particles would have to violently rearrange themselves during movement in response to topographic obstructions and frictional resistance along the margins of the mass; this would be flowage.

Much grinding and abrasion, from impacting of boulders, took place and produced the grit that is part of the matrix. The pre-existing topography was rugged, with small rock pinnacles and sharp-crested ridges. The force of the slide, combined with the seismic shock, carried these rocks from joint surfaces, and they were floated downslope on the surface of the moving debris. The slide material was thoroughly disaggregated, although composed of every size particle from rock dust to megaliths. The presence of continuous lateral ridges suggests that the material flowed in long runs, and, of course, there had to be some air, at least for an initial few seconds, that aided the movement and promoted flowage. The air between rock particles became filled with grit and dust as the mass slid downslope. Such a mixture of air, dust, and fine particles would act as a fluid medium that would cause the coarse debris to flow rather than slide. This idea was expressed by Varnes (1958, p. 35). The impacting rock blocks caused some size reduction of the smaller rock pieces by crushing, grinding, and abrasion. As the entire mass moved downslope, there was also internal movement that segregated some of the megaliths to the top of the slide debris. This phenomenon can be roughly demonstrated by shaking a silt-sand-gravel mix back and forth in a pan and watching the coarser material "float" to the surface. The finer fraction of the slide debris still contained some air but was compacted by the overlying weight as it came to rest.

The finer fraction material then became the debris material seen at the base of the slide in Toro Canyon (9-C1, Fig. 1). When compacted, this material essentially becomes coherent and retains slickensides and grooves at the base as evidence of its movement and direction. Although almost any type of slope failure can produce slickensides, striations, and undulating grooves at its base, these features are evidence that the moving mass was in contact with the underlying surface, at least during the last few seconds of movement. I interpret these slickensides to mean that, if only along the bottom-most part, the material slid and was coherent enough to permit these features to be formed. There is some evidence for both mechanisms—sliding and flowing—in considering the mechanism of movement of rock avalanches. The movement must have been in surges or multi-episodal to produce the multiple subparallel lateral ridges. As each surge or slide moved along, friction along the lateral margins slowed movement there, while the axial part continued to move. When the axial mass moved on, support was lost at the lateral margins and the material failed downward in shear, producing steep-sided, symmetrical ridges (Fig. 15). The distal part of the slide flared out only slightly as it left the confining rock walls and swept out onto the desert floor. However, by this time the debris was very viscous, as shown

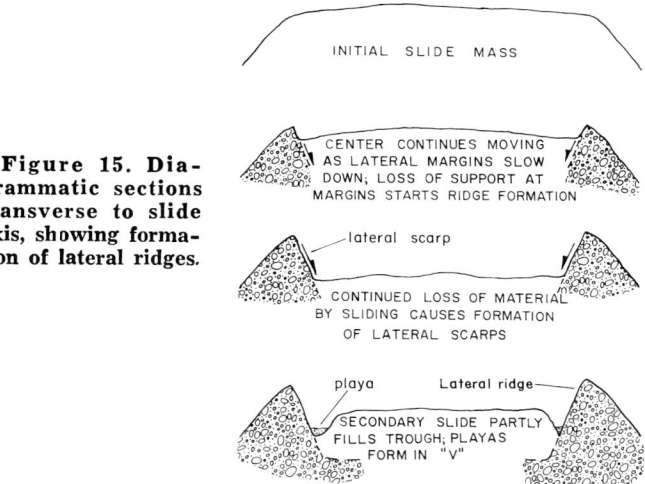

Figure 15. Diagrammatic sections transverse to slide axis, showing formation of lateral ridges.

by the steep (37°) toe slope and the unusually symmetrical shape of the tongue.

AGE OF SLIDE

No organic materials have been found in the debris which could provide radiocarbon dates for the age of the rock avalanche. The well-developed desert varnish on the boulders has formed since the slide occurred, as the underside of the boulders are unvarnished. Attempts to use varnish, or the thickness of varnish, as an indicator of age are unreliable. As stated by Engel and Sharp (1958, p. 516), "The existence of a heavy varnish coating is not necessarily an indication of great antiquity, and varnish is not a good measure of time unless the various local factors pertinent to its formation and preservation can be evaluated."

The toe of the slide is approximately 37 m higher than the highest prominent beach line of ancient Lake Cahuilla, which is believed to have existed from 1600 B.P. to as recently as 300 yr ago (Hubbs and others, 1960). This beach line, at elevation +44 ft (+13 m), is probably the 300-yr shoreline and is clearly discernible as a wave-cut notch around the rock promontory (sec. 11, T. 75 N., R. 7 E.) 1.6 km southeast of Toro Canyon.

The coarse alluvial material discharged from Toro Canyon forms a crude alluvial fan that has been incised by the intermittent stream flow from the canyon. The drainage pattern is almost braided, with many crisscrossing flow lines. Two ages of fans are delineated on aerial photographs at the mouth of Toro Canyon. The older fan is higher, darker colored (desert varnish), and coarser textured than the younger, which is topographically lower, lighter colored, and set into the older fan material. Both of these fans show the continuation of the wave-cut notch from farther southeast, following the contour of elevation +44 ft (+13 m), and are therefore older than the last high stand of Lake Cahuilla, which occurred 300 yr ago. The slide appears to overlie older fan material; further study may show that it even overlies younger fan material.

The slide is tentatively dated as early Holocene because

of the remnant of older alluvium that projects from beneath the slide toe (4Q, Fig. 1). The material existing beneath the slide in the Toro Canyon exposure (Fig. 7) may be the same age.

SLIDE POTENTIAL IN OTHER AREAS

The towns of Rancho Mirage, Palm Desert, and La Quinta, northwest of the Martinez Mountain slide (Fig. 2), are situated in re-entrants in the mountain front to afford seclusion and privacy. The main hazard used to be damage from floods, but storm-water channels and berms have alleviated this problem.

The potential for catastrophic megaslides in these areas cannot be properly evaluated without study of geologic structure in the surrounding mountain slopes. However, maximum relief above these towns is on the order of only 610 m, differing considerably from the 1,829-m relief of Martinez Mountain. Northwest-trending structures, probably joints, are easily interpreted from inspection of the 1:62,500-scale Palm Desert quadrangle, and northeast-draining streams probably follow another joint set. This structure is especially prominent southwest of La Quinta and continues to the summit of Sheep Mountain, elevation 1,567 m, about 7.2 km away. Other faults traverse the area and may still be potentially active. The effect of seismic activity on the jointed rock in this area can only be surmised.

CONCLUSIONS

Large catastrophic landslides can occur in tectonically active areas where elevated fault blocks produce oversteepened slopes in rock that has been weakened by fracturing and faulting during crustal movement. These slides result from a combination of geologic structure favorable to sliding and seismic shock, which causes a part of the mountainside to literally become disaggregated.

In the downward movement of the rock mass, some of the rock particles become smaller through impact and abrasion. The dispersion of air, dust, and grit between moving blocks of rock causes the mass to flow, although well-preserved striations along the basal surface may be evidence of the complex interaction of flowing and sliding.

From the standpoints of engineering geology and urban geologic hazards, the Martinez Mountain rock avalanche affords a case history that can be used to assess the safety of other intermontane valleys where urban development is encroaching up to the mountain front.

ACKNOWLEDGMENTS

I identifed the slide on aerial photographs (Fig. 8) in 1967 while engaged in a ground-water study being conducted by Bechtel Corporation for the Coachella Valley County Water District (CVCWD). At that time, no description of the slide had been published, although it had been photographed and visited and its configuration mapped by others.

I thank R. P. Sharp for his discussion with me of his personal observations of the slide and for his encouragement to pursue this work. Aerial photographs were loaned both by Sharp and by Lowell O. Weeks, general manager and chief engineer for the CVCWD. I also appreciate the encouragement of John S. Shelton and Charles W. Jennings. I am indebted to R. P. Sharp and E. H. Muller for their critical review of this manuscript and the many helpful suggestions they made. Assistance and backing have also been provided by many people from the several Bechtel companies, particularly C. R. McClure, P. Karpa, and R. H. Weight.

REFERENCES CITED

Allen, C. R., Grantz, Arthur, Brune, J. N., Clark, M. M., Sharp, R. V., Theodore, T. G., Wolfe, E. W., and Wyss, Max, 1968, The Borrego Mountain, California, earthquake of April 9, 1968: A preliminary report: Seismol Soc. America Bull., v. 58, p. 1183–1186.

Bonilla, M. G., 1967, Historic surface faulting in continental United States and adjacent parts of Mexico: U.S. Geol. Survey Interagency Rept., Reactor Siting Research 1, Table 1.

Dibblee, Thomas W., Jr., 1954, Geology of the Imperial Valley region, California: California Div. Mines Bull. 170, p. 21–28.

Engel, C. G., and Sharp, R. P., 1958, Chemical data on desert varnish: Geol. Soc. America Bull., v. 69, p. 487–518.

Heim, Albert, 1882, Der Bergsturz von Elm: Deutsch. Geol. Gesell. Zeitschr., v. 34, p. 74–115.

Hsü, K. J., 1975, Catastrophic debris streams (Sturzstroms) generated by rockfalls: Geol. Soc. America Bull., v. 86, p. 129–140.

Hubbs, Carl L., Bien, George S., and Suess, Hans E., 1960, La Jolla natural radiocarbon measurements: Am. Jour. Sci., Radiocarbon Suppl., v. 4, p. 197–223.

Kent, P. E., 1966, The transport mechanism in catastrophic rockfalls: Jour. Geology, v. 74, p. 79–83.

Mudge, M. R., 1965, Rockfall-avalanche and rockslide-avalanche deposits at Sawtooth Ridge, Montana: Geol. Soc. America Bull., v. 76, p. 1003–1014.

Rodgers, T. H., 1965, Geologic map of California, Santa Ana sheet: California Div. Mines and Geology, scale 1:250,000.

Shelton, John S., 1966, Geology illustrated: San Francisco, W. H. Freeman and Co., p. 424.

Shreve, Ronald L., 1968a, The Blackhawk landslide: Geol. Soc. America Spec. Paper 108, p. 47.

——1968b, Leakage and fluidization in air-layer lubricated avalanches: Geol. Soc. America Bull., v. 79, p. 653–658.

Terzaghi, Karl, 1950, Mechanics of landslides; in Paige, S., chairman, Application of geology to engineering practice (Berkey volume): Boulder, Co., Geol. Soc. America, p. 83–123.

Varnes, David J., 1958, Landslide types and processes, in Eckel, E. B., ed., Landslides and engineering practice: Natl. Research Council, Highway Research Board Spec. Rept. 29, p. 20–47.

Woodford, A. O., and Harriss, T. F., 1928, Geology of Blackhawk Canyon, San Bernardino Mountains, Calif.: California Univ. Pubs. Geol. Sci., v. 17, p. 265–304.

Manuscript Received by the Society September 7, 1976
Manuscript Accepted September 17, 1976

PART 4
Engineering Geology and Highway Engineering

12
Utiku landslide, North Island, New Zealand

MARTIN L. STOUT
Department of Geology, California State University, Los Angeles, California 90032

ABSTRACT

The Utiku landslide in the central part of North Island, New Zealand, is a composite failure in Pliocene siltstones, part of a regional east-northeast-trending sedimentary belt across the entire island. The Utiku slide, a small portion of a larger, ancient landslide complex, was reactivated in 1964 and has caused considerable damage to the North Island Main Trunk railway line and State Highway 1, both major transportation routes between Wellington and Auckland.

The Utiku landslide is unique in that movement is almost parallel to the strike of the underlying siltstone. The lower, planar slip surface is a thin montmorillonite layer, and the principal driving force in the upper part of the slide is believed to be primarily from seepage forces developed in wide tension cracks.

Several different stabilization measures are possible to reduce or eliminate movement. Although some remedial measures have been attempted and others have been recommended, the landslide continues to move, and eventually a larger portion of the ancient slide may be reactivated.

INTRODUCTION

The Utiku landslide, covering about 18 ha (45 acres), is one of several large landslides in a broad belt of Pliocene siltstones in the central part of North Island, New Zealand (Fig. 1). The Utiku slide is of particular concern because its continuing movement affects the North Island Main Trunk railway line and State Highway 1, both major routes connecting Wellington and Auckland.

An engineering geological review and study of the Utiku landslide was undertaken by me while on a Senior Research Fellowship tendered by the National Research Advisory Council of New Zealand. The purpose of the fellowship was to conduct and review slope stability studies and programs in New Zealand. Most of the information in this paper is taken from a report prepared for the New Zealand Geological Survey in 1971.

Scope

Although no subsurface exploration was done for this study, the boring logs and cores were made available from previous small-diameter borings, drilled either by the New Zealand Railways Department or the Ministry of Works in conjunction with roading. The logs from two large-diameter borings were also studied, although limited data were available from the in-hole inspections, as casing with viewing slots was used. A number of unpublished "in-house" reports for railways and roading purposes were also reviewed.

The field investigation was limited to reconnaissance studies of Utiku and other nearby landslide areas and to sampling of clay on the exposed slip surface at Utiku. A number of aerial and other photographs in the photo library of the New Zealand Geological Survey were also studied.

HISTORY OF MOVEMENT AND EXTENT OF DAMAGE

The Utiku slide was reactivated in September 1964, and the first reported damage was to a county road crossing the outer part of the slide (Fig. 2). In one month, damage to the railway line was reported, and within a year, State Highway 1 above the railway line had been damaged. The history of movement between 1964 and early 1969 is well documented by Ker (1970).

To replace the county road, which was beyond repair, a new road was constructed to connect with the Rangitikei Bridge (Fig. 3), and subsequent movement on the Utiku slide opened several tensional cracks in the roadcuts.

Although not done entirely as a result of slide movement, a new alignment for State Highway 1 was completed in 1967 (Fig. 4). As part of this construction project, a number of

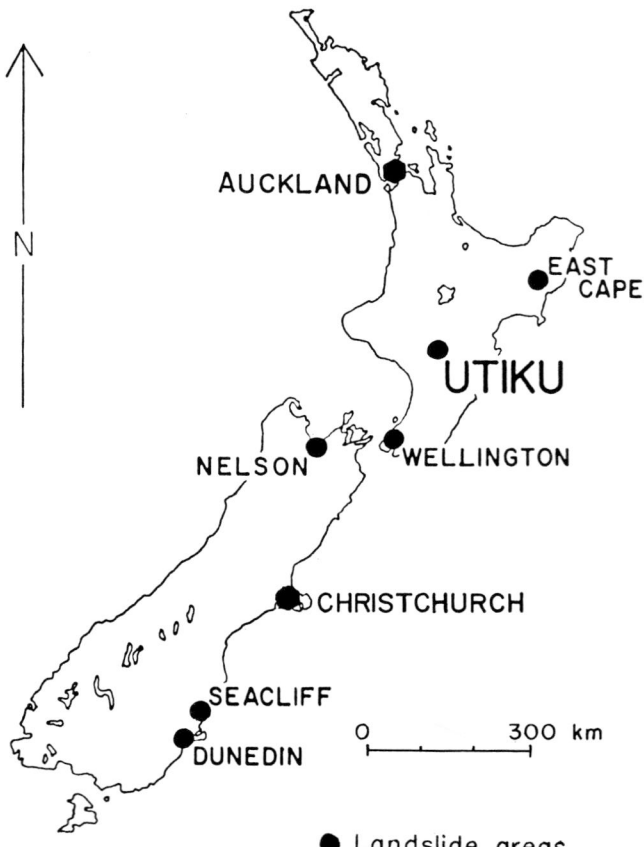

Figure 1. Locality map of Utiku landslide, North Island, New Zealand.

borings and drilled horizontal drains were placed to determine the slip surface geometry and reduce pore pressures within the upper part of the slide mass. In addition, the New Zealand Forest Service planted several hundred Monterey pine *(Pinus radiata)* trees in an effort to stabilize the moving mass. By 1971, however, extensive movement was still continuing across the railway line, and subsidence cracks were affecting about 15 m of the northbound lane of State Highway 1.

In general, movement has occurred predominantly between July and October (winter months) of each succeeding year, and movement generally ceases from November through June. In this sense, it is a seasonal landslide, and the cause of present movement is related largely to precipitation. There are no data to suggest that the landslide will stop without further stabilization efforts. It is critical to note that the presently active slide mass does not include the entire ancient slide mass, and with further movement of the lower portion, the upper ancient landslide mass may be reactivated.

GEOLOGIC SUMMARY

The Utiku slide is about 8 km south of Taihape, a rail and trucking center in central North Island (Fig. 1). The region is in a belt of southerly and southwesterly dipping siltstones and sandstones (locally called papa) of late Tertiary age, mostly Pliocene. Except for a northeasterly trending fault block of older graywackes to the west, the Pliocene siltstones extend across the entire width of North Island, and similar engineering properties and problems are common in most of these Pliocene sedimentary rocks (Hawley and Riddolls, 1975). One suspected landslide in this belt (on the east coast of New Zealand) covers about 168 km² north of Napier, making it one of the largest landslides known in the world.

Most of the Pliocene sedimentary rocks are exceptionally stable in the Taihape area, supporting vertical or nearly vertical natural slopes about 60 m high along the nearby rivers. In artificial slopes (batters), the rocks are usually stable at 0.5:1 (horizontal:vertical) or steeper slopes (Fig. 5). Thus, in a regional stratigraphic sense, the presence of large, massive landslides may be surprising, but an analysis of the Utiku slide provides insight into the broad belt of landsliding in these rocks across all of North Island. Many of the probable landslide areas I saw are so large that some local geologists question that interpretation, particularly where movement along a bedding surface does not change the overall stratigraphy but certainly changes the local geomorphology.

Ultrathin montmorillonite layers, usually with illite interbedded with the normally stable siltstone, is the dominant parameter defining the slip surface on the Utiku landslide and others in that area. The clay layers, at most only a few millimetres thick, have not been recognized in the upper Tertiary of North Island. However, they are commonly marked by vegetation (Fig. 6). Another helpful field criterion suggesting clay layers underlying thin soil cover is grass that stays green for a few days longer in the summer (Fig. 7).

At least three clay layers are present in the Utiku area: two in the underlying Taihape Siltstone and one in the overlying Mangaweka Siltstone (Fig. 8). Several other clay layers are in the Taihape Siltstone between Utiku and Taihape and in the Mangaweka Siltstone to the south; again the layers are usually marked by vegetation and (or) block glides where local erosion has removed bedding-plane support. The Utiku clay layer defines a stratigraphic boundary between the underlying Taihape Siltstone and the overlying Utiku Sandstone. The Utiku clay is well exposed in the slip surface of the Utiku slide in the cliff above the Hautapu River (Fig. 2), in a railroad cut just above the slide, and in a highway cut above the railroad cut (Fig. 9).

ENGINEERING GEOLOGY

Slip Surface and Structural Controls

The slip surface is believed to be a composite surface, essentially planar on the lower, southeastern two-thirds of the slide and rotational on the higher, northwestern third (Fig. 3). The planar portion of the surface is only locally modified and can be well seen in the upper cliff face above the Hautapu River (Figs. 2, 4). This slip surface was also probably encountered in large-diameter shaft 2 and apparently in a number of other borings where the disturbed slide

Figure 2. Utiku slide from southeast, showing damage to county roads. Note the arcuate tension cracks in upper right, with "undisturbed" slide debris above. The planar Utiku clay layer is well exposed in cliff above the Hautapu River, which flows to the lower left. The same clay underlies the slide in foreground. Photo taken September 6, 1965, by S. N. Beatus shows earlier location of State Highway 1.

debris is highly sheared, easy to penetrate, and the drilling resistance of undisturbed bedrock is much greater. Trigonometric solution of several three-point problems on the planar slip surface gives an average strike of N35°W and a dip just over 8° (probably varies between 7° and 10°) to the southwest. This parallels regional bedding in the area, and it is believed that the planar part of the slip surface is basically a bedding plane, defined originally by the thin clay layer.

The present dominant direction of movement is nearly parallel to the strike of the clay layer. Because of this, the slip surface as seen in a center of mass section is a very low angle surface, between 3° and 4° to the southeast (Fig. 3). There is probably some minor variation in this surface, particularly near the Hautapu River where slopes on the slip surface seem to increase to around 7° to 10°.

I believe that the Utiku slip originally developed largely as a dominantly rotational, but composite failure in siltstone with the lower portion of the failure arc following the very planar clay layer. The upper part of the slip surface could have been nearly vertical when original movement occurred. At this time, the slide could have been described as an "across-dip" or "along strike" slide—it was certainly not a typical bedding-plane failure in the sense that movement was downdip. This original movement probably occurred several thousand years ago before the Hautapu River had cut down to its present level, and the slide has most likely had periodic movement since then. The evidence for this ancient movement comes from at least a 15-cm (6-in.) layer of oxidized Mangaweka Siltstone along joints probably associated with the original movement. Similar oxidation zones have de-

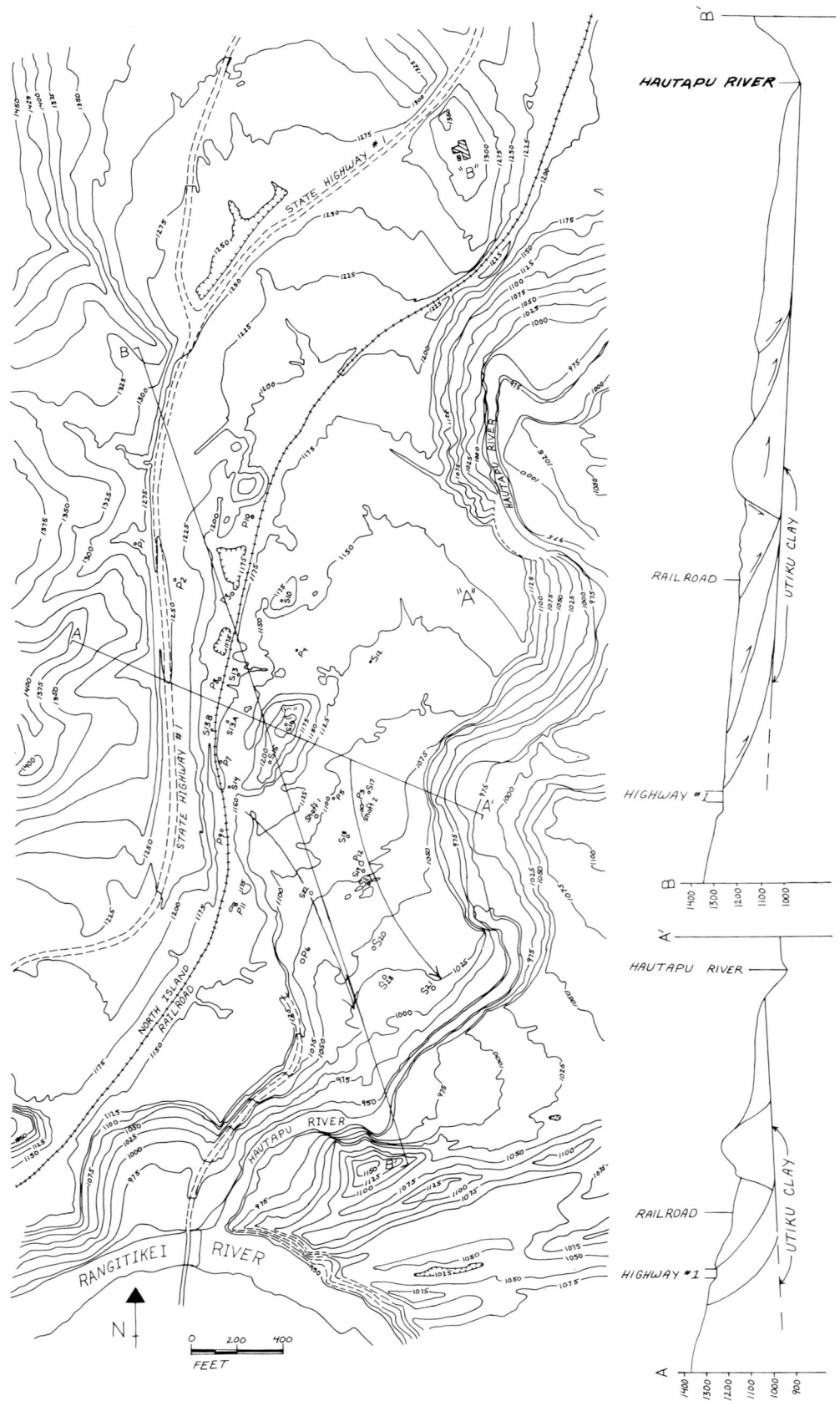

Figure 3. Map and section of Utiku landslide. Map reduced from New Zealand Department of Lands and Survey Map 3117. Contours and elevations in feet. S = triangulation standard; P = piezometric boring.

Figure 4. Later photo oriented similar to Figure 2 shows relocated highway and newly planted Monterey pine trees. The high knob just below the railway is believed to be underlain by relatively undisturbed siltstone. The arcuate topography in the center of the slide, emphasized also by a dirt road, suggests that the lower part of the slip is moving as a separate mass. Total horizontal movement in the lower area since 1965 is about 29 m (95 ft); total horizontal movement of the massive knoll above is 7 m (23 ft). Planar clay layer is visible at lower right. Photo taken on December 11, 1968, by L. Homer.

veloped in siltstones of the Capistrano Formation in southern California where they were involved in landsliding about 17,000 yr ago (Stout, 1969). The flat terrace surface in the region around and including the upper portion of the Utiku slide is not covered by the Aokautere Ash, dated at 20,000 yr B.P. (Te Punga, 1971, personal commun.). Thus stream planation of the terrace and the incision of the Hautapu River channel have been eroded since that time, establishing a maximum age for original movement.

The active tensional cracks in roadcuts along the Rangitikei River Road probably extend down to and are terminated by the same clay layer. The shear strength of the undisturbed siltstones above the Utiku clay in this area appears to be relatively high, but frictional drag caused by renewed movement in the Utiku slide has further opened tear joints in the nearby cliff. The Utiku clay also extends to the northeast, where the uniform southwesterly sloping surface ("A" in Fig. 3) northeast of the slide is probably a stripped structural surface formed by removal of material above the clay layer and is now covered by soil and some terrace gravels.

To the north, structural projections and mapping show the same clay layer in a new railroad cut and in highway cuts (Fig. 9). Some movement of the ridge at "B" (Fig. 3) is suggested by what are interpreted as compressional gravity folds (Stout, 1965) in the pasture to the southwest (Fig. 10). These gravity folds are marked by ridges and troughs in the

Figure 5. Typical artificial slope in roadcut (batter) in Mangaweka Siltstone, south of Utiku slide. Lack of continuous clay beds in most of this formation makes these cuts very stable. View to south.

otherwise uniformly sloping pasture. One private residence with a septic system has been constructed on this potentially unstable ridge, and special care should be taken in this area to avoid more rapid movement. Careless trenching in the pasture could upset the present near-equilibrium of the slope.

To the southeast across the Hautapu River, a broad erosional surface now covered with slide debris and terrace gravels aligns favorably with the clay horizon under the Utiku slide and is believed to be another stripped structural surface on the Utiku clay (Fig. 11). To the south, in the apex area bounded by the confluence of the Hautapu and Rangitikei Rivers (Fig. 3), large massive blocks of siltstone have apparently moved southwesterly on the same clay, opening large tension cracks (now valleys and closed depressions). Structural projections indicate that the Utiku clay could be as much as 38 m below river level at the confluence of the two rivers, although the depth seems more reasonable between 15 and 23 m.

It has not been possible in this limited study to evaluate in a quantitative way the effects of stress release in the Taihape Siltstone. Joint patterns at the Utiku slide appear to be parallel to the marginal scarps of ancient slide movement where tension produced by sliding is greatest. Spalling, or exfoliation in blocks of siltstone, does not account for well-developed linear joints near and above the highway. No joints similar to these have been found along the Hautapu River between the Utiku slide and the town of Taihape, suggesting that regional stress release by erosional removal is not a factor in their formation.

X-ray analyses of a clay sample taken from the slip surface of the Utiku slide in the cliff above the Hautapu River showed relatively high concentrations of montmorillonite. X-ray fluorescence and infrared studies by the New Zealand Geological Survey indicated that the clay was most likely a sodium-rich montmorillonite.

Although surface-movement data are not always reliable because of local tensioning and collapse, it is interesting to determine the slope vectors for triangulation points, established on the slide by the New Zealand Lands and Survey Department. Movements shown in Table 1 are cumulative since 1965–66 through January 1971, although some stations were destroyed or eliminated during or before this time.

Most of these data appear to agree well with the idea of

Figure 6. Natural slopes just north of Utiku in the Taihape Siltstone show clay layer by line of vegetation. Test pits, dug by hand, showed existence of clay layer, which is not usually exposed at surface. View to west.

Figure 7. Nearly flat-lying clay layers in Mangaweka Siltstone just south of Utiku are emphasized by darker, greener grass in center of photo. View to south.

a composite slip surface (rotational and planar) as described above. The most recent data from standard 20 indicate that the movement direction of the standard has changed from a southeasterly direction to almost a true south direction. In so doing, the vector angle of movement has increased as expected as the direction of movement gets closer to the direction of dip. This is also true for the two newly established stations 27 and 28.

It should be noted from Table 1 that standards 15 and 16 on the top of the highest knoll ("C" in Fig. 3) within the slide mass consistently show dominantly horizontal movements. This knoll is believed to be at or near the point where the planar slip surface connects with the rotational slip surface. The coherence of materials in this block and the lack of any large-scale vertical movement within it suggest that the knoll and underlying material represent relatively undisturbed and unbroken siltstone moving on what is probably a thin zone of sheared material derived from the montmorillonite layer. Although recent slides have occurred on all sides of the knoll, bedding can still be seen in the southeasterly face, and in April 1971, a vertical face about 6.1 m (20 ft) high was exposed on the southeastern side of the knoll. This interpretation of basically undisturbed siltstone underlying the knoll is important for proposed stabilization and needs to be tested by drilling, if appropriate.

On the basis of rates and direction of movement and surface configuration, the slide appears to be dividing into two distinct parts (Figs. 2, 4). The lower part of the slide now appears to be active most of the year.

Nature of Material in Utiku Slip

As described above, at least part of the moving mass consists of relatively undisturbed siltstone. Most topographic highs in the slide are probably underlain in large part by this type of material. In addition, many of the borings placed by New Zealand Railways almost certainly encountered similar blocks below the flatter topography of the slide. These also suggest a simple, planar slip surface. Possibly, and more importantly, the coherent blocks in the slide mass are probably not saturated, except in joints or cracks, reflecting the very tight, relatively impermeable, in situ silty sandstone. Materials between these blocks are broken, crushed, and saturated, and in some cases contain very loose soil and organic debris from near-surface cracks.

Ground-water Conditions

In 1969, 14 borings for piezometric purposes were installed in and around the landslide. In addition, daily precipitation data are obtained at a nearby station. Ground-water has been of major concern in the past, and despite various attempts to dewater the upper portions of the slide, movement has continued. A study of the piezometer measurements relative

TABLE 1. TRIANGULATION DATA FROM STATIONS

Standard	Horizontal (cm)	Vertical (cm)	Angle (°)
10	193.0	8.6	2.5
12	904.2	7.6	<1
13	439.4	29.5	4
13A	467.4	193.0	22
13B	251.5	152.4	31
13C	147.3	71.1	26
14	403.9	398.8	44.5
15	683.3	11.7	<1
16	716.3	9.1	<1
17	1,511.3	91.4	3.5
18	1,391.9	231.1	3
19	1,699.3	129.5	4
20	2,905.8	434.3	8.5
21	1,346.2	127.0	5.5
22	1,140.5	198.1	10
27*	449.6	61.0	7.5
28*	721.4	88.9	7

*1970 only

to movement suggests that present movement could be due entirely to seepage forces, developed in the sheared and jointed siltstone. The slip surface remains saturated for the entire year (except for small marginal areas that are dried in summer), and movement is initiated when certain water levels are reached.

Periods of slide movement versus piezometer readings showing critical ground-water elevations for 1969 and 1970 are shown on Table 2.

Several anomalies are found in a study of the movement and piezometric readings. In 1969, movement apparently ceased while water levels were very high in all borings; they remained high for 1.5 to 2 mo after movement ceased. Conversely, movement did not seem to affect water levels in any piezometer, with the possible exception of piezometric boring (P.B.) 2. In 1970, most piezometric values were rising in a more normal manner when movement was initiated and were dropping when movement stopped. Again, movement did not seem to affect water levels in any of the piezometers.

Despite the horizontal drains in the area near it, P.B. 1 has consistently shown high seasonal fluctuations. It is not known whether the nearly constant ground-water level at P.B. 4, which is just below the surface, is due primarily to poor permeability or intense subsurface drainage from the above area. Inspection of the nearby channel in April 1971 showed active flowage in parts of the channel, but the flow varied from place to place. If rates of flowage can be documented in this area, it would be of value to ascertain the degree of success for possible surface dewatering in this area.

Although detailed flowage studies have not been made, the principal sources of subsurface water in the Utiku slide are believed to be (in decreasing order of discharge) (1) rainfall on and through porous materials overlying the Utiku clay on the plain to the north and northeast of the slide, (2) rainfall directly on the slide, and (3) subsurface percolation largely through joints and sheared materials above highway. The relative contribution of the last two are probably very close and ranking could be reversed; however, even taken collectively, they may not equal the first one. Draining an area of approximately 93 ha (230 acres), water percolates through terrace gravels, oxidized silty sandstone, and landslide debris to the impermeable clay layer, from which the path of movement invariably takes it to the Utiku slide. Several springs are on the plain above the slide to the north, and all 9 borings placed on the highway alignment by the Ministry of Works encoun-

Figure 8. Montmorillonite layer and zeolites(?) form planar ledge in roadcut along Rangitikei River Road about 3 km southeast of Utiku slide. The slip surfaces of several nearby slides are formed on this layer in the Mangaweka Siltstone. View to south-southwest.

Figure 9. New roadcut on State Highway 1 shows Utiku clay marked by vegetation line (dashed) and small failure just above it. View to northeast.

tered free water at shallow depths with highly saturated materials below.

Ground-water percolation through the slide mass is believed to be largely through interconnecting cracks and broken zones, rather than intergranular movement through siltstone. This is indicated by relatively dry holes in undisturbed material or in firm blocks within the slide mass. Water losses (as noted on the driller's logs) and water entering borings appear to be concentrated in certain zones. Logs of the inspection shafts noted water entering only at cracks or joints.

The two large-diameter shafts are used as well points for dewatering. In the case of shaft 2, a piezometer (P.B. 13) about 3.05 m (10 ft) away has shown a progressive lowering of water level with pumping since November 1967 — 6.7 m (22 ft) lower by the end of 1970. When pumping ceased temporarily, ground-water levels at P.B. 13 rose, indicating relatively good permeability in this area. This is probably due to the large number of tensional joints in this area shown on 1965 air photographs. However, P.B. 5, located about 36.6 m (120 ft) from dewatering shaft 1, shows no similar history, despite continued pumping. Thus positive dewatering by vertical shafts is very difficult because successful dewatering is largely a function of locating the active, interconnecting cracks. With continued movement, active cracks can easily become inactive, except possibly in the areas of increased tension.

Inspection of shafts 1 and 2 in April 1971 showed another maintenance problem with vertical shafts in moving materials. The casing in shaft 1 had been deformed about 40.6 cm (16 in.) from plumb in 20.4 m (67 ft)—"dragging" of the lower portion suggests that an arcuate slip surface (not the basal slip surface) has developed around 12.2 to 18.3 m (40 to 60 ft) below the surface. This distortion broke all casing joints. At least one seam was opened far enough to allow siltstone "blocks" to fall into the shaft and make the pump inoperative. The pump line was also highly distorted, and the shaft was filled with silty mud to 21.3 m (70 ft) in mid-April 1971. Shaft 2 is tilted about 20.3 cm (8 in.) from 21.3 m (70 ft) to water level at 23.2 m (76 ft). This is believed to reflect movement on the basal slip surface at 23.3 m (76.5 ft).

No sand or silt "boils" were noted in any of the surface ponds in October or November 1970. The fine silt particles may not be conducive to this type of buoyancy, so it is not possible to know whether ground water is under any pressure.

Equilibrium of Slide and Shear Strength

Movement of the slide appears to be largely a function of piezometric levels, which are directly related to precipitation values. Seasonal fluctuations in ground-water levels for the past two years strongly support this conclusion. Rates and times of movement indicate that the slip is very close to equilibrium or, in a stability sense, has a factor of safety close to 1.00. This suggests that even minor remedial measures could allow partial or even possibly complete stabilization under normal climatic conditions.

The low-angle, basically planar surface over which most of the slide debris is moving suggests that movement will probably continue on a sporadic basis indefinitely, and sliding may ultimately involve the presently inactive portion of the ancient landslide. On the basis of the present subsurface data, there

Figure 10. Northerly view of pasture shows gravity folding due to rock creep above the Utiku clay. Highway cut at upper left is shown in Figure 9. Ridge with house has moved toward camera, forming ridges in pasture. Subsurface drainage of this entire area must pass through Utiku landslide.

is no indication that the slide will in time buttress itself, such as might occur in a purely rotational failure. The fact that the railway embankment is close to and at the head of the presently active slide, and that the embankment is continually built up, will not allow a significant reduction in driving forces on the rotational portion of the slip surface for movement to cease.

A center of mass section was constructed for the moving slide for a preliminary stability analysis. Assuming a factor of safety equal to 0.85 (condition of movement) and a cohesion of 97.7 g/cm² (200 psf) for material on the slip surface, the angle of internal friction (ϕ) was calculated to be 3.4°. With a factory safety equal to 1.0 (condition of temporary stability) and the same cohesion, ϕ is equal to 4.3°.

Figure 11. View to southeast across Hautapu River (flowing to right) shows Utiku clay underlying another large slide area. Note 5-yr-old pine trees as Utiku slide in foreground moves them into river.

TABLE 2. PIEZOMETER READINGS

Piezometer	G.L.*	1969			1970		
		Low†	High§	Change	Low†	High§	Change
1	1,274	1,253	1,260	7	1,249	1,260	11
2	1,247	1,206	1,215	9	1,208	1,221	13
3	1,177	1,162	1,166	4	1,162	1,166	4
4#	1,136	1,134	1,135	1	1,133	1,136	3
5	1,099	1,099	1,093	3	1,087	1,092	5
6	1,062	1,053	1,057	4	1,050	1,059	9
7A	1,172	unknown	1,161	8**	1,153	1,162	9
7B	1,171	unknown	1,157	5**	1,153	1,160	7
8	1,185	unknown	1,179	6**	1,175	1,181	6
9	1,172	unknown	1,142	7**	1,137	1,142	5
10	1.192	unknown	1,153	6**	1,168	1,172	4
11	1,174	unknown	1,122	5**	1,116	1,122	6
12	1,075	unknown	1,070	6**	1,065	1,072	7
13	Continuous lowering of ground water by pumping						

Note: All values have been rounded to the nearest foot to correspond with contours on map (Fig. 3).
*G.L. values are not corrected to show vertical movements.
†Low values just prior to movement.
§High values during or just after period of movement.
#Piezometer 4 shows an unusually small amount of fluctuation in both 1969 and 1970.
**Change based on low-level readings in October–November 1969.

This calculated value agrees well with residual shear values on some montmorillonitic clays in southern California.

Stabilization Recommendations

Stabilization measures depend to some degree on future plans for the area around the slide. The assumption is made that realignment of rail and highway to avoid the area is not feasible at this time. Certain measures could bring about immediate stability with minor future movement; others could achieve long-term stability, but with significant movement before final stability is achieved, and, of course, all have separate economic considerations. In general, the measures which are considered include (1) dewatering — surface and subsurface measures; (2) buttressing — compacted fill buttress at toe, shear key (stabilization trench) of compacted fill within slide mass, use of rock anchors in selected localities, and use of "shear pins" (cast-in-drilled-hole caissons); and (3) chemical treatment.

Most stabilization measures considered in this report deal primarily with stabilization of the area containing the railway and the highway generally above shaft 1. All recommendations are based on available data and are intended to be used as a guide for further planning and economic and cost studies. More specific recommendations could be given, after appropriate field work, when a decision has been reached on one particular approach.

Dewatering

Although serious attempts have been made to control surface and subsurface drainage in the higher areas of the ancient slide, piezometric levels remain high. My experience in southern California with horizontal borings in silty and clayey rocks suggests that these are rarely successful on a long-term basis because of plugging during drilling, clogging of perforations (either by debris or crystallization), and inability to predict routes of water flowage in such rocks and thus to tap these areas during drilling. As discussed earlier, vertical shafts have continued maintenance problems, particularly where active movement is experienced, and additional problems at Utiku include the location of intersecting permeable zones. Vertical wells also have a definite limit of influence. Conditions at Utiku between P.B. 5 and shaft 1 have already been cited as an example of the inefficiency of this approach on a long-term basis.

There is little doubt that if piezometric levels could be significantly lowered and maintained at the lower levels (as shown in Table 2) within the actively moving part of the slide, movement would essentially cease. However, the permanence of this cessation depends entirely on the effectiveness of the dewatering system, rates of precipitation, and ground-water movement.

Methods to achieve adequate dewatering are probably among the most economical to construct but also are the least secure from a stability viewpoint.

Probably the most effective dewatering method for the moving slide is interception of subsurface water by positive control and maintenance of existing critical drainage courses and trenching to intercept and drain the higher ground-water levels. Both require continued maintenance. On the lower and flatter areas of the slide, open channels, with slide slopes no steeper than 1.5:1 (horizontal:vertical) could easily be cut to intercept and drain the adjacent slide debris. If constructed during periods of low water levels, ease of access for equipment should be optimum, although still difficult; otherwise, drag lines might have to be used. Spoil from any operation of this type should be placed on the southern side of the channel. For cost-estimating purposes, a minimum of approximately 731.7 m (2,400 ft) of open trench is needed. Depth of trenches should be approximately 4.6 to 6.1 m (15 to 20 ft), with at least a 3.05-m-wide (10-ft-wide) channel. Locations depend in part on future planning for stability purposes. If buttressing is considered in the future, dewatering troughs should be placed to aid future construction. Probably the most critical area for trenching of this type is a southeasterly trending channel from near P.B. 4 to the Hautapu River cliff, which would be about 304.9 m (1,000 ft) in length.

In some of the higher and steeper parts of the slide, particularly around the railway, open trenches are not feasible, as these would be narrow and subject to blockage by side slope failures. In these areas, existing canyons could be cleaned out by earthmoving equipment and a moderate flow gradient established. Trenches 61 cm (24 in.) wide could then be cut to appropriate depths with a backhoe and subdrains placed in them. Stability of the trench walls during construction would be marginal at best.

Subdrains should consist of a filter of at least 3,721 cm^2 (4 ft^2) of 2-cm (0.75-in.) gravel or similar coarse granular material (possibly imported from nearby terrace deposits) around accordian-style, 10-cm (4-in.) perforated plastic pipe (to prevent immediate destruction with movement). It is estimated that at least 222.6 m (730 ft) of this type of subdrain is necessary in the area just below the railroad tracks, and these subdrains should outflow into one of the open channels discussed above. In addition, another 518.3 m (1,700 ft) of this type of subdrain could be used above the railroad if a satisfactory method of passing the water beneath the railroad embankment can be developed. Otherwise it might be necessary to have a separate dewatering system for this area. In this regard, all existing culverts beneath the railway embankment should be carefully maintained, as water collecting just upslope from the railway in the active slide area is also in the least desirable area from a stability viewpoint.

Although dewatering of the central and upper reaches of the slide, as discussed above, could conceivably bring about stability, the factor of safety would probably not be significantly increased. It is believed that in order to gain maximum benefits, dewatering measures across the entire slide mass and adjoining areas should be considered. These measures might include the following:

1. A westerly trending trench extending from near the existing outfall from the surface-drainage control system to and under the railroad tracks. This trench would be approximately 213.4 m (700 ft) long and at the railroad tracks about 19.8 m (65 ft) deep. The purpose of the trench would be to intercept a significant quantity of ground water from the plains to the north of the slide and divert it into the Hautapu River. Because of the southwesterly dip of strata in the area, much of the trench would be in firm siltstone.

2. In order to drain more effectively the upper reaches of the ancient landslide and thus prevent further water penetration into the active slide mass, one or two vertical borings along the western side of State Highway 1 near the area of present subsidence could be considered. Although these have all the problems of maintenance mentioned earlier, they would not be in the presently active slide mass, although there is no guarantee that they would remain so. Their locations should be carefully measured in the field to intersect both the steeply inclined joints in that area, as well as the lower angle ancient slip surfaces, in order to obtain maximum regional drawdown. It does not seem practical to consider further horizontal drains in this area, as they would have to be lower than the existing drains and thus be drilled through actively moving slide debris. The vertical borings would need to be 24.4 to 30.5 m (80 to 100 ft) deep, and water should be conveyed into the existing surface-drainage system. Sustained pumping in this area would probably cause considerable settlement in the adjacent roadway embankment; once drawdown was achieved, however, subsidence should cease unless slide movement is still active.

An adit has been considered beneath the slide debris to dewater the actively moving mass. Although sound in principle, it appears to have a number of construction problems and questions, not the least of which is the permanence of the method by which water will be transported from the slide to the adit across the active slip surface. Even with this problem solved, this method contains the same reservations as those with horizontal drains, because an adit is not considered practical through the slide debris, and some type of flexible drain would have to be used.

Buttressing

All buttressing estimates depend largely on the desired factor of safety and strength values assumed for the stability analysis. Although no soil testing has been performed, it is possible to make some general yardage estimates from this study. Soil testing, however, is necessary before final specifications can be outlined in detail.

Compacted-Fill Buttress at Toe

This scheme would place a compacted-fill buttress across and along the toe of the slide. Unless the buttress could be used also as an earth-fill dam, it would require designed culverts to allow passage of the Hautapu River, which would probably have to be of exceptional size in order to take peak flowage. The effects of proposed power schemes downstream along the Rangitikei River would also have to be considered.

Several problems in addition to the culvert are raised with this scheme. It is felt that the slide mass has been undergoing tension in the central part since 1965 and is now almost two separate slides; thus a buttress at the toe of the more active slide would not necessarily have immediate effects at the railway. Several years of minimum subsidence could be expected at the railway before final stability was achieved. On the other hand, construction is probably the easiest and least hazardous of all measures designed to provide total stabilization. This is the only measure that would provide permanent adequate stabilization for the entire slide mass, including the area with tension joints along the Rangitikei River Road. In addition, the buttress would also stabilize a part of the slope on the eastern side of the Hautapu River.

No stability calculations were made for this scheme. For

general cost purposes, it is estimated that placement of approximately 200,000 m³ (260,000 yds³) would increase the stability by approximately 30%. This would assume fill in the canyon area up to an approximate elevation of 1,000 ft. Culvert construction would be an additional expense. The quantity of fill can be reduced, depending on the desired factor of safety and strength of fill material. The buttress would have to be designed to provide stability by itself, because the opposite steep bank of the Hautapu River is part of another large slide mass, and even though it appears stable at present, no additional shear stress should be placed on it.

The advantages of this scheme are that long-term stability of the entire moving mass is assured, including the cliffs and roadcuts of the Rangitikei River Road. Fill to be placed in this type of buttress needs to be placed but once (not stockpiled and moved a second time as in some shear keys), thus reducing somewhat the earthmoving costs. Fill in the buttress could consist of any nearby materials. The need for subdrains behind the buttress would depend largely on materials used in buttress construction.

Shear Key in Central Area

A shear-key stabilization trench in the central part of the slide mass, designed to remove all slide debris and be keyed into firm bedrock, would achieve almost immediate stability at the railway line with only minor future subsidence. However, the amount of earthworks for such a scheme is large, and hazards would be present during construction.

Some preliminary stability calculations have been made on the active slide mass on the basis of the following assumptions: slip surface clay cohesion = 97.7 g/cm³ (200 psf), $\phi = 3.4°$; fill material in shear-key cohesion = 488.3 g/cm³ (1,000 psf), $\phi = 28°$; shear-key material will be drained by a system of subdrains, and piezometric values during time of construction will not significantly rise above the levels of 1969 or 1970. On the basis of these data, a northwesterly trending shear key near shaft 1 to the cliff above the Hautapu River would have to have a base width of 30.5 m (100 ft) normal to the direction of movement to increase the stability by 30%. Total excavation yardage depends largely on stability of side slopes, but assuming 1:1 side slopes, yardage would probably be around 122,000 m³ (160,000 yd³). This assumes an actual shear-key width at the slip surface of 24.4 m (80 ft) with a shear-key orientation 37° oblique to the principal direction of movement. Dewatering devices such as electro-osmosis, freezing, or dewatering pumps in lines would almost certainly have to be employed to complete construction successfully. In addition, stability of side slopes during construction of shear keys is always a hazard, as well as large-scale movement of the entire mass above the shear key.

Again, a shorter or narrower shear key would reduce the overall yardage as well as decrease the factor of safety, unless soil cement or other high strength materials were used as fill materials.

Some thought might be given to a shorter shear key constructed near shaft 2 and extending outward to the cliff above the Hautapu River. This could be constructed after the dewatering troughs had been in existence (and only if the troughs are not adequate) and would probably provide enough local resistance to stop movement of a portion of the slide. Without any known strength values, it would be difficult to assess what effect, if any, this would have on the remaining slide mass, and it would also present the same hazards during construction.

Rock Anchors and Shear Pins

On the basis of the tentative conclusion that the bedrock material below the knoll containing triangulation standards 15 and 16 consists of basically undisturbed siltstone, consideration could be given to the possibility of "pinning" or stabilizing the underlying block of siltstone. The topographic high allows the thicker section of siltstone to exert a high normal force on what is apparently a slip surface of very low angle. The use of rock anchors would tend to increase this normal force on the slip surface, resulting in stabilization of the material upslope from the block. The principal disadvantage of this scheme, besides the cost, is that stabilization of this knoll does not provide stabilization of the slide material just to the north.

Rock-anchor specifications provided by BBR New Zealand Ltd.[1] suggest that anchors are feasible and may be easier to install than shear pins (cast-in-drilled-hole caissons), which have the disadvantage of requiring a curing time to become effective. Rock anchors would need to be 30.5 to 36.6 m (100 to 120 ft) long (depending on amount of surface preparation and in situ shear strength of siltstone below the slide). Spacing would depend largely on the desired factor of safety and strength of in situ rock. Use of cable in rock anchors would allow an adequate time for concrete curing, even if the slide were active. Movement of the slide mass could, in effect, "pre-stress" the rock anchors.

Shear pins would probably need to be 6.1 to 9.1 m (20 to 30 ft) long, depending on the actual thickness of the zone of movement, although the drilled holes would have to be 30.5 to 36.6 m (100 to 120 ft) deep and seem less practical than the rock anchors.

Both of these remedial methods require good subsurface data on the character and strength of materials below the knoll. These are feasible only if the material beneath the knoll is similar to the material exposed in the nearby cliff above the Hautapu River.

Other knolls to the north might also be considered for similar treatment, and a larger part of the slide might then be controlled by this method.

Further subsurface exploration would show whether any of the knolls to the north would be suitable for similar remedial measures in an attempt to bring the entire slide mass under control.

---

[1] The BBRV system presents a unique method of post-tensioning concrete and is ideally suited to bridges and other structures requiring large and concentrated cable forces. The system was developed between the years 1941 and 1945 by four Swiss engineers, Birkenmaier, Brandestini, Ros, and Vogt, whose initials form the name BBRV. *BBR New Zealand Ltd. is the sole licensee* for the system in New Zealand, and a fully comprehensive service is available at all times.

Chemical Treatment

Preliminary analysis of the clay minerals involved in the slip surface has indicated that some of it is a sodic montmorillonite. Thus it is theoretically possible to induce a base exchange reaction which in turn will increase the stability. A commonly used method in southern California is to inject quick lime or cement by grouting or, if conditions allow, mix by machine and spread as a compacted fill. Other chemicals may also be used (Arora and Scott, 1974). The resulting increase in shear strength usually shows up in shear test values in 2 to 3 days and reaches a reliable value within 7 days. Of course, the character of ground water in the area also needs to be studied.

In order to ascertain whether such treatment is possible at Utiku, it is suggested that the clay materials on the slip surface be subjected to direct shear tests to determine "residual" shear strength and then saturated in varying concentrations of Ca-rich solutions, such as could be developed in ground water. Periods of saturation, up to and including several weeks, should be used.

Use of Vegetation to Stabilize Deep-Seated Landslide

The pine trees that were planted on the slide seem to have had no effect on the rate of movements. Because the slide is so deep seated, the only apparent effect could be partial removal of surface waters by transpiration, and this appears to be a small volume in comparison to the overall quantities involved. Stability of this type of slide should not be expected with any type of vegetative cover.

CONCLUDING STATEMENT

The Utiku slide, affecting major transportation routes on North Island, New Zealand, should be partially or totally stabilized because continued movement may reactivate a much larger slide mass, which would be much more difficult to control.

REFERENCES CITED

Arora, H. S., and Scott, J. B., 1974, Chemical stabilization of landslides by ion exchange: California Geology, v. 27, p. 99–107.

Hawley, J. G., and Riddolls, B. W., 1975, Soft rock engineering in the central North Island of New Zealand: Australia–New Zealand Conf. on Geomechanics, 2nd, p. 201–206.

Ker, D. S., 1970, Renewed movement on a slump at Utiku: New Zealand Jour. Geology and Geophysics, v. 13, p. 996–1017.

Stout, M. L., 1965, Gravity folds in the Modelo Formation, western Los Angeles County, California: Geol. Soc. America Bull, v. 76, p. 967–970.

——1969, Radiocarbon dating of landslides in southern California and engineering geology implications, *in* Schumm, S. A., and Bradley, W. C., eds., United States contributions to Quaternary research: Geol. Soc. America Spec. Paper 123, p. 167–179.

Manuscript Received by the Society September 9, 1976
Manuscript Accepted September 17, 1976

13
Engineering geology of the Woodstock rockslide, New Hampshire

BRIAN K. FOWLER
Geological Engineer, Pike Industries, Inc., Tilton, New Hampshire, 03276

ABSTRACT

On November 7, 1972, a rockslide of approximately 13,000 m³ fell into the uncompleted northbound lane of a section of Interstate 93 in the town of Woodstock, New Hampshire. The rockslide forced almost total curtailment of construction operations at the site of the failure and necessitated a total redesign of the highway section through the rock cut.

The redesign of the 44 × 213 m rock cut began with a detailed geologic mapping of the structure of the rock in the vicinity of the failure, because such a study had not been undertaken during the preliminary design phases for the project. Several weak structural trends were established for the cut area, with the most disadvantageous system striking parallel to the roadway and dipping into the open cut. This system was also troublesome because many of the joint surfaces associated with it were filled with a mylonitic material of little structural competence.

A 2:1 (vertical:horizontal) slope was specified for the rock cut following this study, rather than the 8:1 slope as had been originally designed. The neatline for the presplit along this new slope was carefully placed on the mountainside to take best advantage of the existing structure of the rock mass.

As excavations proceeded along this new slope, a second set of mylonitic joints was encountered on the new presplit face. Construction was again halted while this set was structurally analyzed and a second redesign for the section formulated.

The second redesign was based upon the gravity-mass concept and involved the following engineering and construction: (1) excavation of a rock bench at the top of the anticipated sliding failure parallel to the newly exposed mylonite; (2) placement of a system of high-strength steel tendons into the toe of the bench to further resist the anticipated failure; (3) installation of spot rock bolts along the crest of the bench to actively secure the intact mass of the bench itself; (4) drilling of an extensive system of rock drains along the toe of the cut; and (5) installation of an instrumentation system to monitor the success of the redesign.

INTRODUCTION

On November 7, 1972, at 2:30 a.m., approximately 13,000 m³ of rock slid out of a recently excavated rock cut on Interstate 93 in the town of Woodstock, New Hampshire, about 97 km north of the state's capital of Concord along the Pemigewasset River (Fig. 1). The northbound lane, which was under construction and which had been used as a haul road from the southern end of the project, was completely buried by the rockslide. Although hauling operations were underway at the time of the slide, none of the contractor's personnel was injured, and damage was confined to the loss of two track drills and a compressor which were being stored in the area.

The rockslide caused a considerable construction delay as well as reappraisal of the design for the roadway through the rock cut. The original section called for two lanes of interstate highway to be built into the mountainside, with the northbound lane being founded on ledge and with the southbound lane being carried on a viaduct structure over an existing state route along the river's edge (Fig. 2). Because of the large quantity of rock excavation proposed under this design, an 8:1 (vertical:horizontal) slope was selected.

The occurrence of the rockslide cost the state and the contractor many months of construction delay while the project was redesigned and the mountainside in the vicinity of the slide was thoroughly studied and reinforced. Immediately following the slide, further excavation on the southbound lane was halted, and the contractor proceeded, under force account, to clean up the debris. The Federal Highway Administration and the New Hampshire Highway Department turned to the redesign work, retaining the firm of Haley and Aldrich, Inc., for technical advice regarding the various aspects of the redesign and reinforcement.

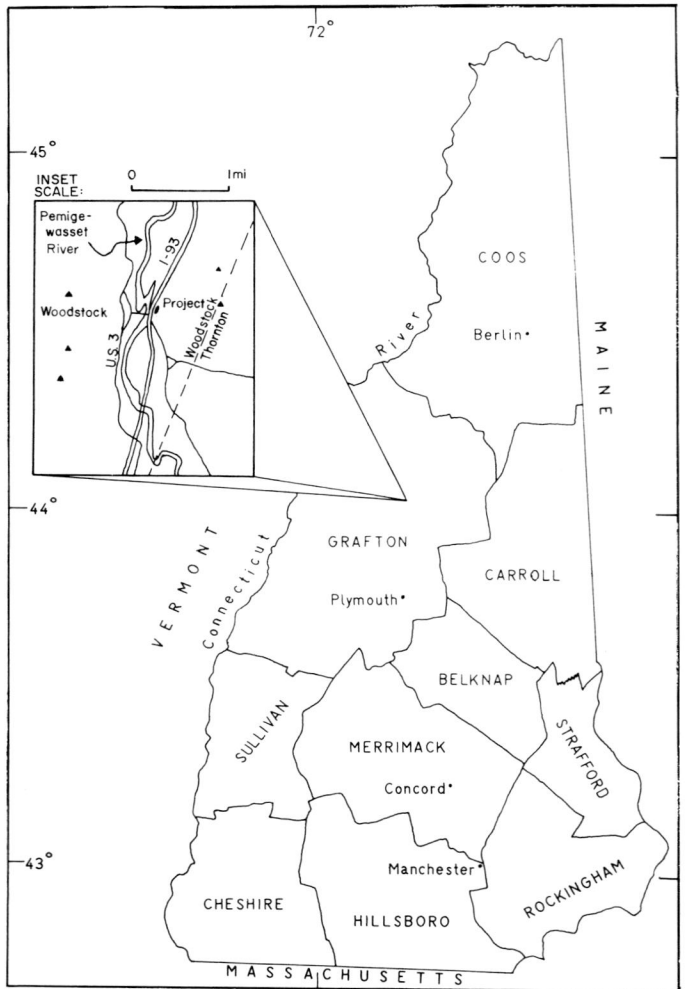

Figure 1. Project location map, Woodstock, New Hampshire.

Ultimately, the southbound viaduct was discarded in favor of a lane built on an embankment held on the mountainside by a 6-m retaining wall at the southbound lane's closest point to the river's edge. The local state route was relocated to the opposite shore of the river, thus eliminating the need for the viaduct structure (Fig. 3). The removal of the viaduct also eliminated the need for further excavation in the vicinity of the potentially unstable northbound cut, thus preventing further changes in the stress configurations within the mountainside. An extensive rock reinforcement system was also designed to secure the northbound rock cut from further massive or local failures.

Geology

The rocks in the immediate vicinity of the rock cut are gneissic and schistose varieties of the Devonian(?) Littleton Formation, which crops out widely throughout central and northern New Hampshire. The gneissic rocks occur primarily at the southern end of the cut, grading into the schistose rocks to the north. The schistose foliation becomes quite prominent in the rocks in the area where the rockslide occurred. These country rocks are discordantly intruded by dikes of pegmatite, basalt, and andesite, which are all members of the White Mountain Magma Series of Jurassic(?) age. In some places these intrusions are nearly concordant with the foliation, but in most cases this condition is not continuous for long distances. Figures 4 and 5 illustrate the areal geology at the cut site immediately following the excavation of the northbound cut to its subgrade.

Within the schistose area of the rock cut, several layers of mylonitized material exist. These zones range in thickness from several centimetres to as much as 3.5 m at the bottom of the rock mass making up the slide (Figs. 6A and 6B). Exploratory drilling undertaken following the slide and during the redesign work indicated that these zones loosely follow the structure of the rocks surrounding them.

An extensive and detailed structural mapping program was undertaken at the site following the slide. This study was complicated by the high density of fracturing in the rock mass produced by the many and varied tectonic events that have affected this region throughout geologic time. Major structural trends were difficult to establish in the field, but subsequent office analysis revealed several weak trends as shown in Figure 7. The southeast-dipping set is approximately parallel to the schistose foliation and is extensively slickensided, while the northwest-dipping set is roughly transverse to the foliation. Most of the mylonitic zones strike approximately parallel to the northwest-dipping set, and together they strike roughly parallel to the center line of the northbound lane and dip into the open cut (Fig. 7).

A careful study of the hydrology of the rock-cut area was made to assess the existing and potential effects of ground water on the stability of the cut. Although little flow could be attributed directly to artesian sources, considerable flow was present on the face of the cut. Because of the fact that the amount of flow would respond to local rainfall very shortly after the precipitation had occurred, it was determined that this flow was created by water seeping into surface joints and openings along the crest of the cut. The persistent "lubricating" effect of this water within the mylonitic zones was probably a major contributor to the failure, in view of the fact that flow of this type through a rock mass can build up tremendous cleft water pressures along joints and behind potentially unstable blocks as well as create high pore pressures in more permeable materials within the rock mass such as the mylonites.

The cataclastic metamorphism responsible for the production of the mylonitic zones seems to have been accompanied by very little recrystallization of the mica schists making up the mylonites. Thin-section study was not undertaken on the mylonites, so firm statements cannot be made concerning the mineralogy of the mylonites or the level of metamorphism involved. However, it seems likely that if the mylonitization occurred at considerable depth, hydrothermal solutions may have contributed to the cataclastic process causing alterations to the micas and possibly the feldspars, thus facilitating the production of a sheared mylonite with a fine- to medium-grained texture similar to that of the parent mica schist. Individual fragments within the mylonites range from several

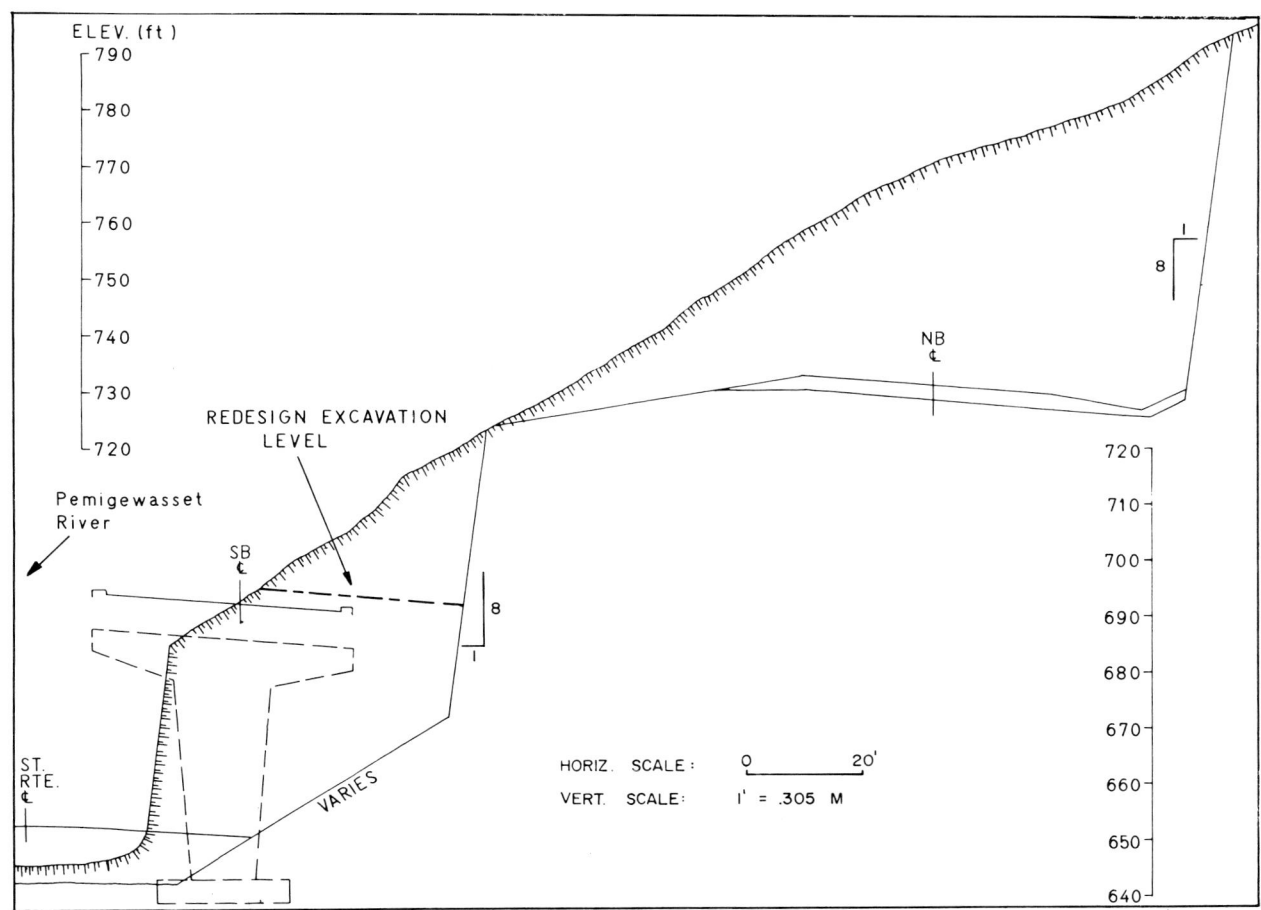

Figure 2. Original typical section of rock-cut excavation, Woodstock, New Hampshire (after Fowler, 1976). SB, southbound; NB, northbound.

millimetres in diameter to particles of dust size, which when combined with water, produce an extremely slippery substance of little structural competence. In addition to this, structural incompetence within the mylonites is further accented by the fact that the mylonites appear to weather more rapidly than the neighboring mica schist. This is probably due to the fact that the mylonitization process effectively increased the porosity within the mylonite and allowed for greater permeability, thus increasing the rate of weathering of the micas and feldspars within the mylonite itself.

Since no equipment was readily available for testing and quantifying the mechanical properties of the mylonite, it was qualitatively assumed that the material was very incompetent and could not be relied upon for structural support in any reinforcement system. In all computations and studies that took place relative to the redesign and reinforcement of the rock cut, the mylonite was assumed to have little intrinsic shear strength.

Factors Responsible for Failure

Prior to the actual failure and following the initial northbound excavation, several field examinations of the potential failure area had been made by the personnel of the New Hampshire Highway Department and the Federal Highway Administration. It was determined that the rock cut was potentially unstable and that a failure could occur at any time. Joints bounding the blocks that eventually made up the rockslide were found to have been opened, presumably by the pressure of excess blasting gases produced by overloading during the blasting operations. The slickensided surfaces of these joints were, in some cases, opened as much as 5 cm, thus severely reducing the intact strength of the rock mass. In view of this, elastic deformations that normally follow this type of massive excavation are presumed to have had very little resistance from the rock mass, thus allowing the newly exposed structure to separate and weaken easily. The overall strength of the rock mass was reduced also by the moderately convex plan of the rock cut as it lay along a slight curve at this point on the highway (see Long and others, 1966). All of this, coupled with the unfavorably dipping mylonites, created what was thought to be a very unstable situation.

It had been previously estimated, during preslide discussions, that the structures exhibiting potential instability would be most prone to failure when "lubricated" by the surface water seeping into the rock mass. It was proposed that the most dangerous time for the rock cut would be in the spring

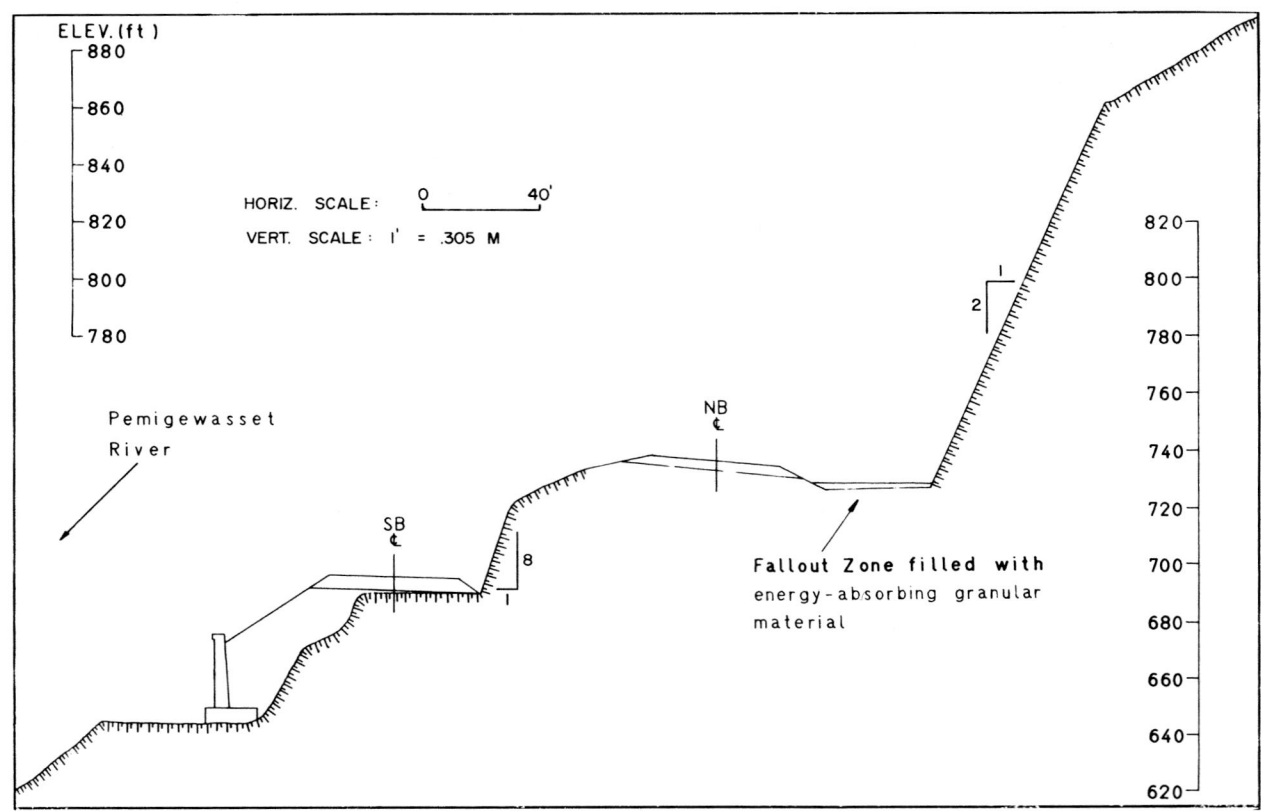

Figure 3. Redesign typical section of rock-cut excavation (after Fowler, 1976). SB, Southbound, NB, northbound.

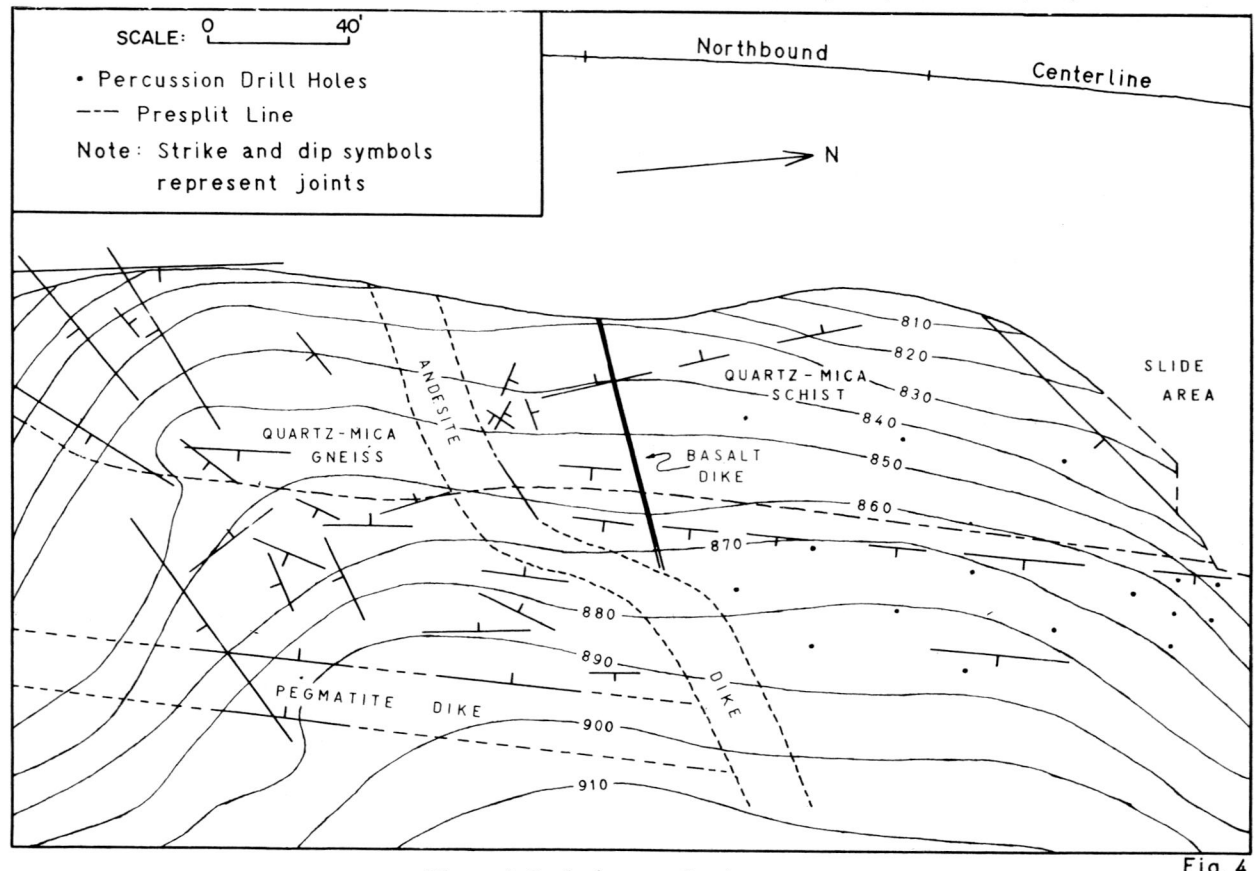

Figure 4. Geologic map of cut area.

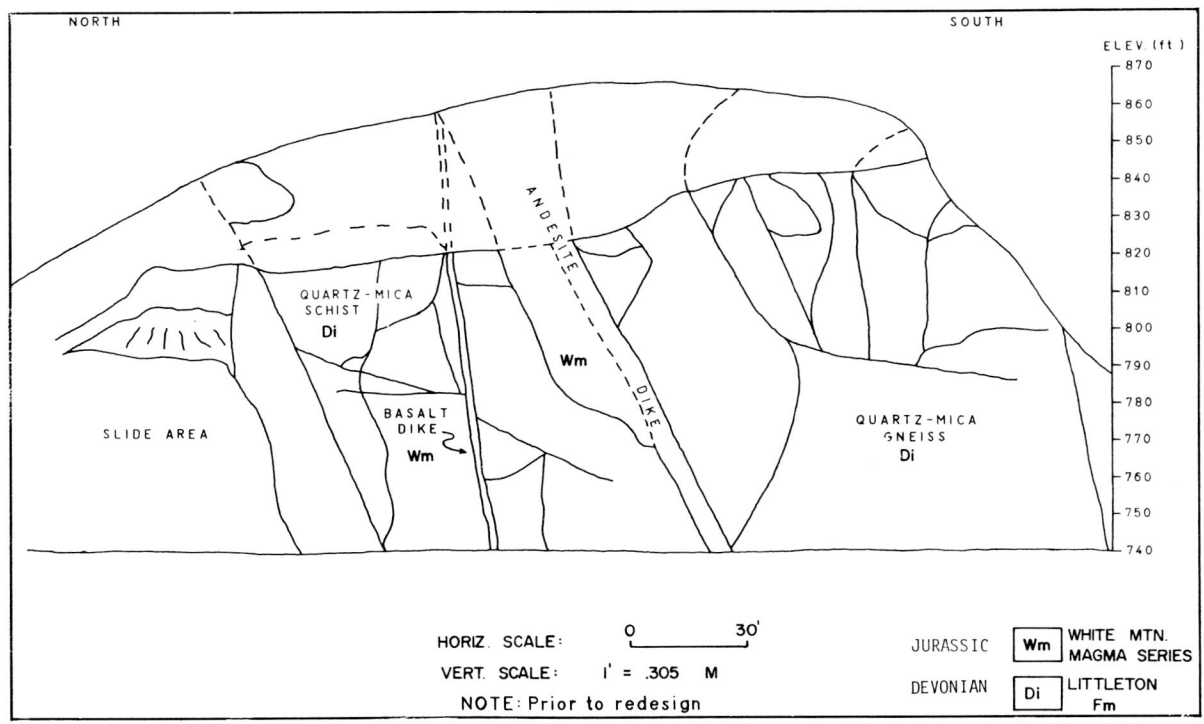

Figure 5. Schematic view of rock-cut excavation.

of the next year (1973) when the spring meltwaters would push surface flow in the rock mass to a maximum. During the two days prior to the slide, however, extremely heavy rains fell within the northern New England area, supplying the large surface runoff and the resulting "lubrication" of the already weakened rock mass that precipitated the failure.

Prior to the rockslide, as has been commonly the case in northern New England, the New Hampshire Highway Department had no established procedure for the preliminary study of rock-cut sites. This was primarily due to the fact that until this time, most major construction within the state had been completed in areas outside the mountainous northern section or in areas sufficiently large to permit location of roadways around geologic barriers requiring extensive ledge excavations. Since the rockslide, however, the value of detailed, preliminary engineering and design for rockslopes has become apparent, and these studies are now carried out as a matter of routine practice.

REMEDIAL ACTIVITIES

Immediately following the rockslide, an order was given to cease all further rock excavation on the project until an assessment of the overall stability of the rock mass could be made. It was concluded that further excavation along the construction line of the southbound lane would seriously endanger the already unstable northbound rock cut. Figure 2 shows the originally proposed southbound excavation and indicates the level to which excavation had proceeded when the order to cease was given.

It was reasoned, based upon photoelastic research (Long and others, 1966), that stresses previously distributed evenly throughout the mountainside had become concentrated at the northbound toe of the cut and, coincidentally, very close to the position of the mylonitic zones. In view of this, it seemed likely that continued excavations in either the northbound or southbound would further concentrate these undesirable stresses and would complicate the stability problem further. In addition, the structural competence of the northbound foundation was questioned, especially if structures similar to those responsible for the failure were discovered under the northbound lane and along the steep median wall above the southbound lane. It appeared that continued excavation along the southbound lane might force a more severe redesign to include the support of the northbound lane as well. Consequently, it was concluded that the northbound rockslope would be carefully re-excavated to remove large portions of the existing 8:1 slope whose stability was suspect, and that further major excavations along the southbound lane should be curtailed to prevent further deep-seated changes in the northbound cut stress configurations. The overall redesign of the cut then proceeded under the proviso that no additional southbound excavation would be allowed and that the finished southbound grade would be at or fairly near the 210-m (690-ft) elevation to which the excavations had already proceeded.

Using the detailed surface structure maps that had been prepared, a new presplit neatline was laid out along the crest of the northbound cut slope. This line closely followed the structure of the rock to take advantage of what favorably dipping joints were available (Fig. 4). The new presplit

Figure 6A. View illustrates the preslide conditions. Mylonite outcrop is visible as a crescent-shaped interruption on the presplit face. Trees along the crest of the cut average 20 m in height (after Fowler, 1976).

Figure 6B. Woodstock slide area. Slide rests on the northbound bench at elevation 225 m; southbound and older state route benches are also visible. Loader and truck on state route grade provide visual scale.

slope was designed at 2:1 (vertical:horizontal) in view of the historical success which had been obtained with this slope in this type of rock elsewhere along the interstate highway to the south and because of its favorable intersection with the joints to its rear. Provisions were made for a "fallout zone" at the toe of the cut as per design recommendations taken from Ritchie (1963). At the time this design was formulated, it was believed that the mylonitic zones occurred principally along the joint system that generally was striking to the northwest and dipping at 35° to 40° to the southwest (Fig. 7).

The southbound viaduct as originally designed (Fig. 2) was replaced by an embankment at elevation 210 m. In the area of the rock bench, a 6-m retaining wall was designed to support the embankment at the narrowest point along the river's edge. A minor realignment of the southbound curve was necessary to accommodate this retaining wall, but the blasting involved was limited to the presplitting of an additional 3 m from the median wall.

Changes were made in the presplitting specifications to ensure that a neat presplit face would be produced with as little overbreakage as possible. Since the work involved in implementing the redesign was being performed on a force-account basis, the department was able to alter its usual mode of payment for rock excavation from a per-cubic-yard basis to a per-lineal-foot-of-hole-drilled basis, which allowed the spacing of the presplit holes along the presplit neatline to

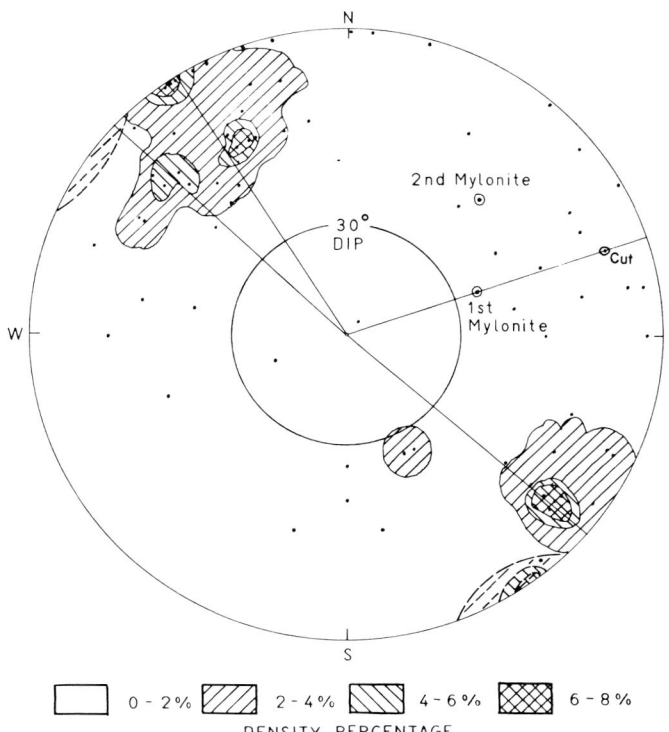

Figure 7. Equal area net of data from structural survey (lower hemisphere). Total data points represented equal 206 (after Fowler, 1976).

be regulated by the resident engineer and geologist. Original specifications in the contract called for presplit holes to be drilled along the neatline no further apart than approximately 1 m on centers and loaded with cartridge explosives designed for presplitting of no less than 40% strength at a loading ratio of not less than 0.05 kg/ft and not more than 0.23 kg/ft (0.305 m) of hole. During the force-account work, this neatline spacing was reduced to as little as 30 cm, with a spacing of 46 cm with alternately loaded presplit holes proving to be the most satisfactory. Figures 8A and 8B illustrate the results of these changes. The adjustments produced faces that were smoother, more regular, and free of "gas-opened" joints and partings.

After three "shots" had been excavated from the first lift along the new presplit slope, another unpredicted mylonitic zone appeared on the new presplit face. Its structural orientation, which differed slightly but significantly from the presumed orientation of these zones, was carefully measured (Fig. 7). The results of these observations indicated that the zone would continue to descend behind the 2:1 slope to a point approximately 3 m above the toe of the cut face, and would isolate a huge block above the northbound lane. In view of the uncertain mechanical properties of the mylonites exposed previously and the unfavorable structural orientation of this zone to the northbound lane, a second redesign of this section was undertaken to make appropriate adjustments in the slope to prevent another major rockslide, the potential size of which was estimated at 30,600 m³.

An exploratory drilling program was initiated to ascertain the precise orientation of the mylonite in this newly exposed zone. Percussion drilling equipment was employed (drilling dry), since it was already on the site and in position for the work. Intersection of the vertically drilled holes with the mylonitic material was indicated by an increase in the descent rate of the drill header and by a pronounced change in the color of the dust issuing from the hole, that is, from the dark gray of the mica schist to a tan of the weathered mylonite. Sixteen holes were drilled on a grid (Fig. 4), and the resulting depth data was structurally analyzed, yielding the exact position of the second mylonite within the rock mass as closely as possible.

Based on this data, the northbound rock cut was again redesigned to prevent the isolation of the large block and to reinforce the mountainside against any further failures. Figure 9 illustrates the redesign employed. The basis for the design is the gravity effect of the bench upon the anticipated sliding failure parallel to the second mylonite. To preserve the gravity mass, two reinforcement systems were designed. The consultant formulated a design for high-strength steel tendons grouted into holes approximately parallel to the expected failure to act as a passive measure to counteract such motion. Approximately 70 of these tendons were installed along the toe of the cut as shown in Figure 9. Spot rock bolting was designed for the upper section of the bench to actively keep the rock mass intact, thus preserving the gravity mass, and to prevent minor rockfalls onto the northbound lane. An extensive system of rock drains was installed at the toe of the cut in both the northbound and southbound lanes to reduce cleft and pore-water pressures in the vicinity of the mylonites intersected by the drains. In view of the low cost of this drainage procedure and the significant benefit obtained in dewatering the rock mass, this installation was considered to be one of the greatest contributors to the ultimate stability gained in this rock mass following completion of the project.

Concurrent with the construction implementation of the foregoing measures, an extensive instrumentation system was designed and installed by the consultant (Fig. 10). The system consisted of rectilinear extensometers (4 and 6 anchor units capable of measuring movements to ± 0.075 mm) installed at critical points in the rock mass where lateral movements seemed possible. Several mechanical extensometers were also installed at points that were considered to be not so critical but that merited observations. Electrical strain gauges were mounted on the steel tendons to measure changes in the load being placed on the tendons through their grout bonds with the rock, and load cells were installed on the exterior plates of several of the rock bolts to monitor changes in the loads being placed on the bolts following their tensioning at installation. A microseismic program was set up using holes drilled along the crest and toes of the cuts to monitor changes in the frequency of subaudible rock-noise events.

Rock-Bolt Design

The rock-bolt system used was the newly popular resin anchorage system. This system consists of a high-strength steel bar inserted into the drilled hole and set in a polyester resin.

Figure 8A. View illustrates poor presplit results prior to the revision in specifications. Note ragged face and numerous "gas-opened" joints. View looks north along the toe of the original 8:1 slope (after Fowler, 1976).

Figure 8B. View illustrates results of the presplitting following revisions in technique. Note smoother plane and reduction in overbreakage. View looks north over bench shown in Figure 9 (after Fowler, 1976).

The resin is supplied in plastic cartridges that are inserted into the hole and broken when the bolt is inserted and spun by means of an air-track drill. Once the resin has set (about 30 min under good conditions), the bolt may be tensioned, depending upon the product being used, to a maximum load of 445 kN (100,000 lb). Various rock-bolting systems were investigated in the course of the rock-bolt design, but this high-strength steel and resin anchorage system was found to be the best for this project because of its ease of installation, its flexibility in the field as to actual bolt lengths used, its high load capabilities, and its lack of future maintenance requirements.

In an effort to estimate the number and length of rock bolts required for the reinforcement of the bench under the tight cost-plus-materials basis prevailing at the time on the project, a somewhat novel approach to designing the bolting system was used in this rockslope situation, where individual block stabilization and facial compression and interlocking were desired. Figure 11 illustrates the principles and definitions used in the design procedure (Fowler, 1976). While this method is not applicable in every conceivable rockslope bolting situation, it does allow for the maximum reinforcement of the rock-cut slope with the minimum number of rock-bolt installations.

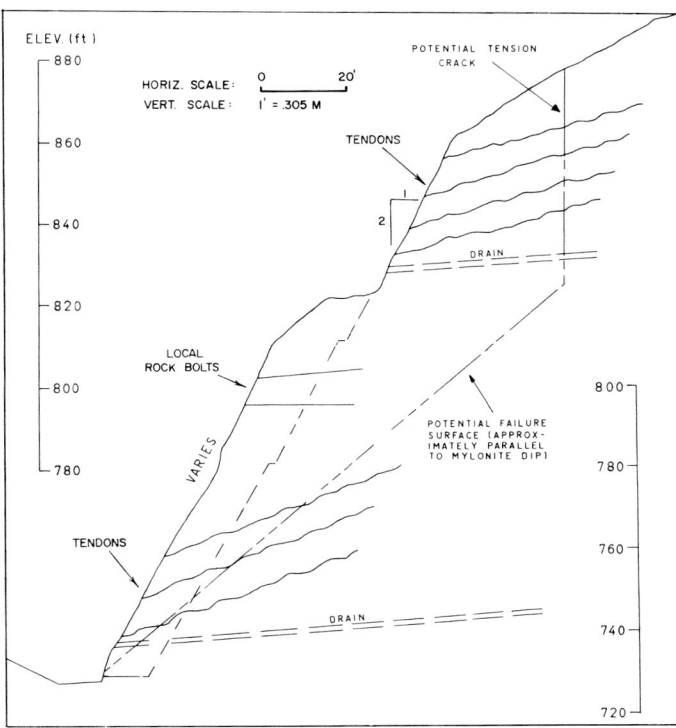

Figure 9. Typical section — second redesign (after Fowler, 1976).

It is apparent from a mechanics point of view (Fig. 11) that this method does not take best advantage of the reinforcement available from a rock bolt inserted into a single block. However, in this case, it was reasoned that because both individual block and block group stabilizations were required, and because the southwest-dipping potential failure surfaces were roughly parallel to the facial surface, the sacrifice of the available upslope tangential force generated by an installation angling slightly uphill relative to the normal to the slope would be offset in the final reinforcement by the interlocking across the slope created by the more even distribution of compression that would result. Based upon research reported by Goodman (1976), it was known that block stabilization could be adequately provided so long as the bolt was installed at an angle normal to or slightly above the normal to the cut surface, such that the value of that installation angle would not exceed the value of ϕ being used for the design. Hence, in this case, optimum individual block stabilization was sacrificed so that interlocking across the face could be used to secure the slope as a unit.

The rock-bolt design scheme is based upon the typical factor of safety relationship applied to the design of earth slopes, that is, the resisting forces divided by the driving forces in each case. Here, the driving force or sliding component equals W_S, where W_S is the component of the total block weight (W_T) acting parallel to the potential failure surface. The resisting force is equal to $W_N \tan \phi + F_B \tan \phi$,

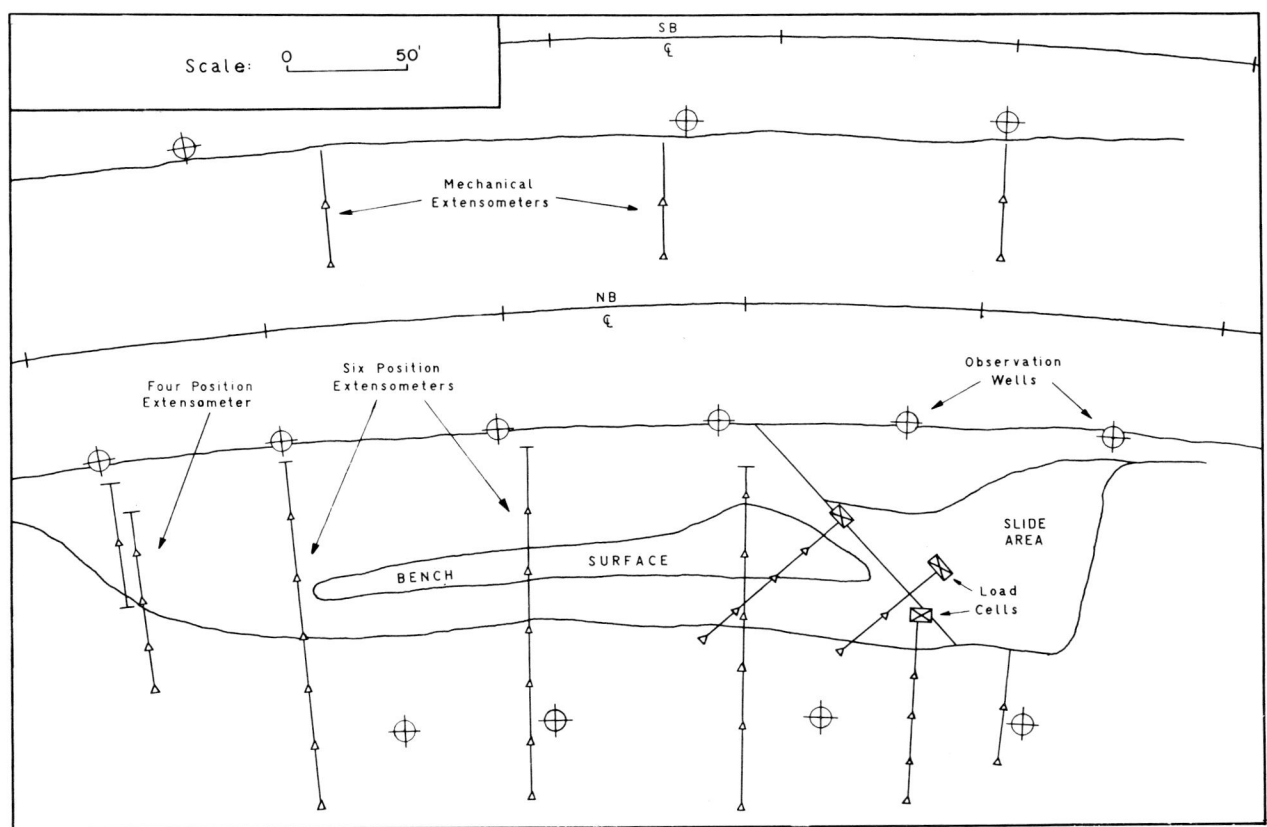

Figure 10. Instrumentation plan (from Haley and Aldrich Report No. 2). SB, southbound; NB, northbound.

where W_N is the component of W_T acting normal to the potential failure surface and F_B is the total rock-bolt force. Tan ϕ represents the coefficient of friction between the block and its underlying failure surface or the frictional resistance present within a failure zone, such as the mylonite layers. Using the factor of safety (FS) relationship,

$$FS = \frac{(W_N + F_B) \tan \phi}{W_S}. \quad (1)$$

By solving for F_B, then substituting for W_S and W_N and simplifying,

$$F_B = W_T \left[\frac{(FS)(\sin \alpha)}{\tan \phi} \right] - \cos \alpha \quad (2)$$

where α is the slope angle of the potential failure plane (Fowler, 1976).

This method makes it possible to prepare a graph showing the number of bolts required to secure a block of a given weight on a failure plane of a given slope. Figure 11 also shows an example of such a graphic presentation for a bolting system using a design tension load of 180 kN per bolt (40,000 lb). The K value shown above the graph is obtained by substituting actual design numbers in equation 2 and K_B is obtained by dividing the desired bolt load as shown. The actual number of bolts for the block in question is then calculated by multiplying

$$K_B \times W_T \quad (3)$$

where W_T is the total weight of the block in question.

A qualitative average value for ϕ was obtained from experimental results (triaxial tests) reported by Hendron (1972). The tests were run on intact specimens of schistose gneiss, the petrographic description of which was very similar to that found on the project. The test data ran from a low ϕ value of 28.0° to a high 43.0°, yielding an average of 35.5°. This value was supported by data taken from back-calculations made just after the slide occurred, which indicated that a ϕ value of approximately 35° was in force at the time of the slide. The factor of safety was set for design purposes at 1.5.

The limitations of this rationale are: (1) as the angle of the potential failure surface becomes significantly greater than the value being used for ϕ in the design, the results of equation 2 are no longer practical, and the design of an adequate bolting system for a given block depends more upon the generation of the uphill tangential forces mentioned earlier (Fowler, 1976); (2) this method assumes that the coefficient of internal friction represents all of the available shearing resistance along the potential failure plane, which tends to make the calculations conservative, because surficial cohesion and regosite interlocking are not taken into consideration (Fowler, 1976).

The foregoing design procedure established the number of bolts necessary to secure the individual blocks. The next step was to estimate the optimum locations and arrangements for the bolts to produce the desired interlocking between adjacent blocks. The bolts had been designed to be installed approximately perpendicular to the cut surface and the potential failure surface so as to spread the compressional forces generated by the bolts equally across the face in all directions. Lang (1966) suggested that the ratio of the bolt length to its spacing relative to its nearest neighboring bolt should be not less than 2. It was reasoned that the compressional influence around a tensioned rock bolt tends to spread laterally in a cone from the top of the bolt. If the bolts are spread too far apart, extensive areas of tension develop between these compressive cones, thus facilitating local rockfalls. If, however, the cones are forced to overlap by virtue of the closer spacing of the bolts, the face then assumes a maximum surficial compression and is thus secure from these local failures. This perpendicular scheme will operate effectively only if the bolts are installed perpendicular to the surface of the cut, because if they are installed in what would normally be assumed to be optimum bolting direction (uphill relative to the normal to the slope surface), wide areas of tension will develop below the bolts and facial compression will be confined to the area immediately above the bolt. This ragged distribution of tension and compression between the bolts and across the face facilitates local rockfalls especially when areas of tension coincide with discontinuities on the slope's surface (Fowler, 1976).

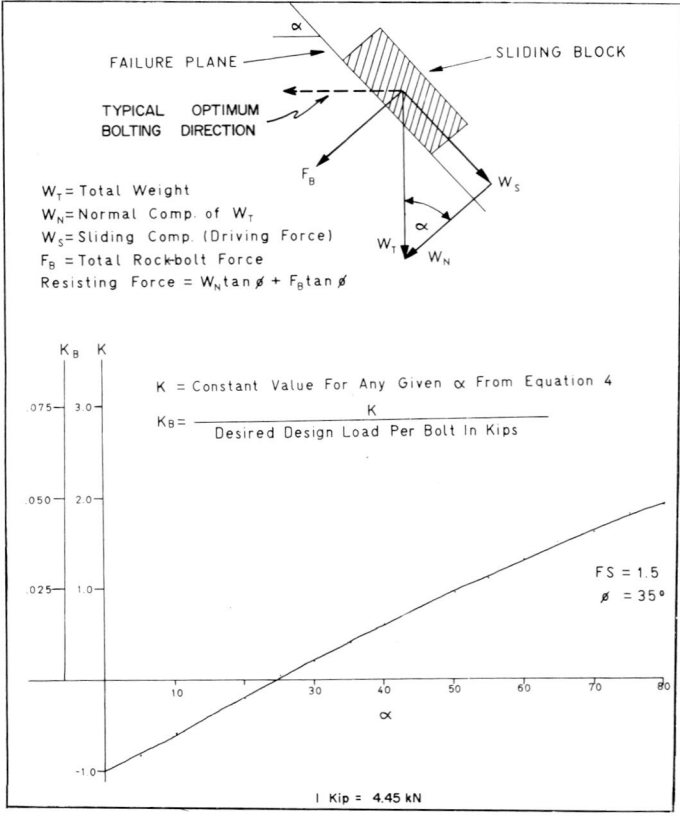

Figure 11. Rock-bolt design principle (top). Graph of 40-K bolt design data (bottom) (from Fowler, 1976).

Instrumentation Program Results

To date, data obtained from the various instruments installed in the rock cut have indicated no further significant movements in the rock mass since the installation of the reinforcement systems. Data from the microseismic program has shown ever-decreasing rates of the subaudible rock-noise events, and at this time no events are being observed. Despite some problems with the electrical systems of the various instruments caused by the extremes of the northern New England climate, all of the instruments have performed well, and they appear to indicate that the remediation has provided basic stability in the rock mass.

ACKNOWLEDGMENTS

I thank Joseph F. Guertin and Donald E. Reed of Haley and Aldrich, Inc., without whose advice and assistance much of this work would never have been completed. Thanks are also due to Robert M. Leary, Howard J. Pincus, and Donald R. Coates for their advice and criticisms of the manuscript. A special note of thanks is also due to Roger Moody, resident engineer on the project, for his patience and advice during the long months when the solutions to the problem cut were formulated.

REFERENCES CITED

Fowler, B. K., 1976, Construction redesign—Woodstock rockslide, N.H., *in* Proceedings of the conference on rock engineering for foundations and slopes, Aug. 16–18, 1976, University of Colorado: New York, Soc. Civil Engineers.

Goodman, R. E., 1976, Analysis of a sliding block on a plane — the friction circle concept, *in* Methods of geological engineering: Los Angeles, Calif., West Publishing Co., p. 237–244.

Hendron, A. J., 1972, Mechanical properties of rock, *in* Zienkiewicz, O. C., and Stagg, K. G., eds., Rock mechanics in engineering practice: New York, John Wiley & Sons, p. 124–126.

Lang, T. A., 1966, Theory of rock reinforcement: Washington, D.C., Rept. of 45th Ann. Meeting, Highway Research Board, p. 8–13.

Long, A. E., and others, 1966, Stability of high bank slopes in rock: Washington, D.C., Highway Research Record No. 135, Highway Research Board, p. 1–12.

Ritchie, A. M., 1963, Evaluation of rockfall and its control: Washington, D.C., Highway Research Record No. 17, Highway Research Board, p. 13–17.

Manuscript Received by the Society September 7, 1976
Manuscript Accepted September 17, 1976

Printed in U.S.A.

14
Relationship between morphology, hydrology, geotechnics, and vegetation on an old northern Ohio landslide

HAZEL F. KRIST
MURRAY R. McCOMAS
Department of Geology, Kent State University, Kent, Ohio 44242

ABSTRACT

The geotechnics, hydrology, morphology, and vegetation of an ancient but active landslide in the lower Cuyahoga River valley, Ohio, were studied to define mechanisms contributing to failure of the slope. Engineering laboratory tests of soils showed them to be heavily overconsolidated and prone to erosion. Hydrology of the slope was monitored by periodically measuring water levels in piezometers and wells. Fluctuations in water elevations and hydraulic gradients during the observation period were correlated with precipitation data and with effective stresses. The morphology of the slope was defined in terms of recent and relict activity and was related to the vegetation and hydrology. The results of this study indicate that an interrelationship exists between the morphology, hydrology, and vegetation of the slope. Continued slope failure is caused by the unstable nature of soils and the abundance of water entering the flow system of the slope.

INTRODUCTION

Slope-stability problems in the lower Cuyahoga River valley of northeastern Ohio have been encountered in the construction of numerous highways and have resulted in added expense and delay to highway departments and developers (Gardner, 1972; Gardner and others, 1974). The site chosen for the Ohio Turnpike (Interstate 80) bridge across the valley included an unstable slope adjacent to the planned eastern abutment. In this paper we describe this unstable slope (Fig. 1) in terms of its geotechnics, hydrology, and morphology in order to define the mechanisms producing instability.

This particular slope, referred to as the Stumpy Basin slope, is on the eastern valley wall of the Cuyahoga River in northern Summit County (Fig. 2). Located 22 km northeast of Akron, the site occupies approximately 4 ha and is surrounded by heavily wooded slopes.

The Stumpy Basin slope has been active for at least 150 yr, and it contains several aspects uncommon to the area. A large segment of the topography is mobilized and supports prairie vegetation considered unique in northeastern Ohio. Another aspect is the presence of both recent and relict scarps, seeplines, and toes. In addition, several types of mass movement are evident on the slope, including rotational slumping, slab-sliding, creep, and mudruns.

PREVIOUS INVESTIGATIONS

Among the many workers who have studied the glacial geology in and around the site are White (1953), Rau (1969), and Wittine (1970). Gardner (1972) included the Stumpy Basin slide in his investigation of slope-stability problems in

Figure 1. Stumpy Basin slope as seen from Ohio Turnpike on the east side of Cuyahoga River valley, Ohio.

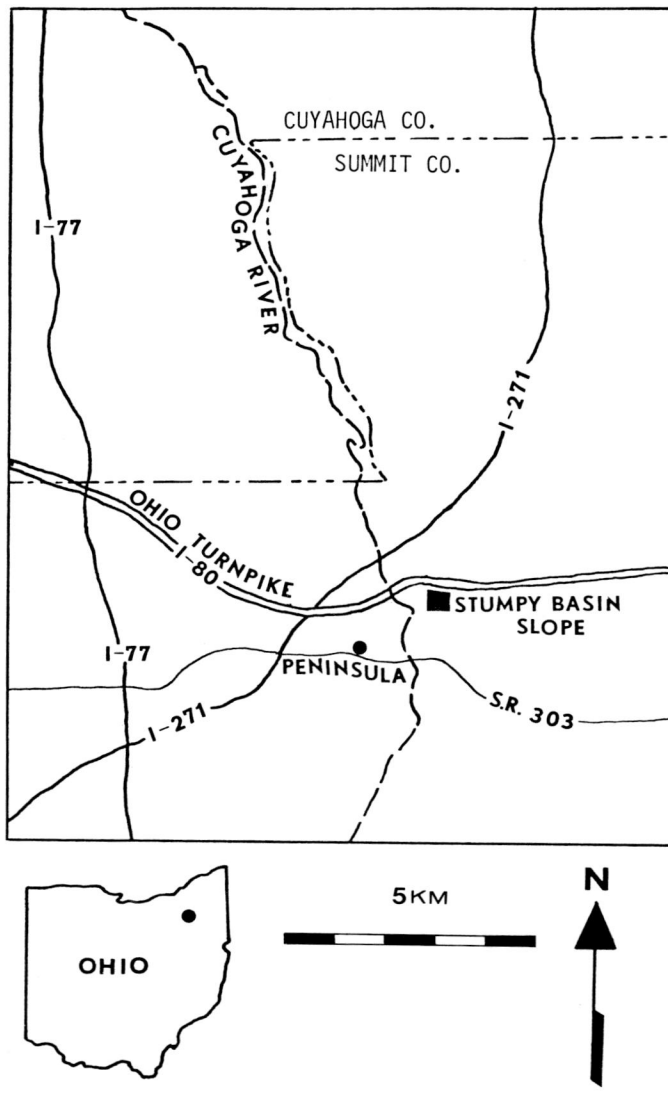

Figure 2. Location of Stumpy Basin slope.

the lower Cuyahoga River valley, but cited only a brief and general overview of its engineering geology. The slope was also discussed in Gardner and others (1974).

GEOLOGY

The bedrock underlying the study area consists mainly of fine-grained clastic material of the Mississippian and Pennsylvanian Systems, which formed a flat upland surface incised by several valleys (Rau, 1969). The Stumpy Basin slope is situated in one such preglacial bedrock valley, as is much of the present course of the lower Cuyahoga River.

The overall result of Pleistocene glaciation on the study area was the thick accumulation of sediment in the bedrock valley. The sediments consist of glaciolacustrine clays and silts, interstratified with coarse-grained outwash and river deposits. These sediments are overlain by till in many upland areas (Gardner, 1972).

The Cuyahoga River now flows concordantly and northward within the buried valley from Akron to Cleveland (Rau, 1969). The present rugged topography of the valley with its steep to moderately steep slopes is a function of the unstable nature of the valley fill, which has produced slope failure (Gardner, 1972).

The Stumpy Basin slope is composed of the valley fill. The uppermost unit present is a very hard, brown calcareous till, approximately 6 m thick. The till (Lavery?; Gardner and others, 1974) is exposed in the upper part of the slope for a lateral distance of 63 m. This unit is pebbly and contains joints with oxidized surfaces. Stiff, yellow to brown silts underlie the till and extend to the lowest part of the slope studied, accounting for a thickness of at least 30 m. The silt contains sparse quartz pebbles and is interbedded with numerous smaller units of coarse sand and gravel and distinctive layers of soft, gray clay and silt. These smaller units are varied in their lateral and vertical extent. Sand and gravel bodies range from a few metres to > 10 m in lateral dimension and from a few centimetres to roughly 3 m in thickness. The gray clay and silt layers are abundant, with thicknesses of less than 2 m. The lateral extent is unknown.

METHOD OF INVESTIGATION

Near-surface soil samples were collected with a hand auger and analyzed for their engineering properties. X-ray diffraction techniques were employed to reveal the mineralogy of the clay-sized fractions of six samples, which were centrifuged onto porous plates and air-dried. Consolidation tests (one-dimensional confined loading) were performed on two undisturbed samples, using a fixed-ring consolidometer and mechanical loading device.

Water table and hydraulic gradient fluctuations were monitored with the use of five piezometer nests and four wells for nearly one year. The location of the nests is shown in Figure 3. Wells were installed near each nest with the exception of nest 3. The maximum depth of the deep piezometers is 2.7 m, and all standpipes are polyvinyl chloride tubing. Each well is 2 m deep. Water levels were read two to three times per month with a portable electronic water-level sounder.

Alidade and plane-table mapping techniques were used in the construction of a morphologic map of the slope. The method of depicting morphology was derived from Cooke and Doornkamp (1974).

RESULTS OF LABORATORY TESTS

The predominant clay minerals present in the till, gray clay, and yellow-brown clay are illite, chlorite, and kaolinite, with illite the most abundant. Traces of calcite and dolomite and an abundance of quartz were present in all samples. On the basis of the diffraction patterns obtained, there is little mineralogic difference between the clay-sized fractions of the soils studied.

A sample from the till and a sample from the yellow-brown silt were subjected to consolidation tests. The data (given in

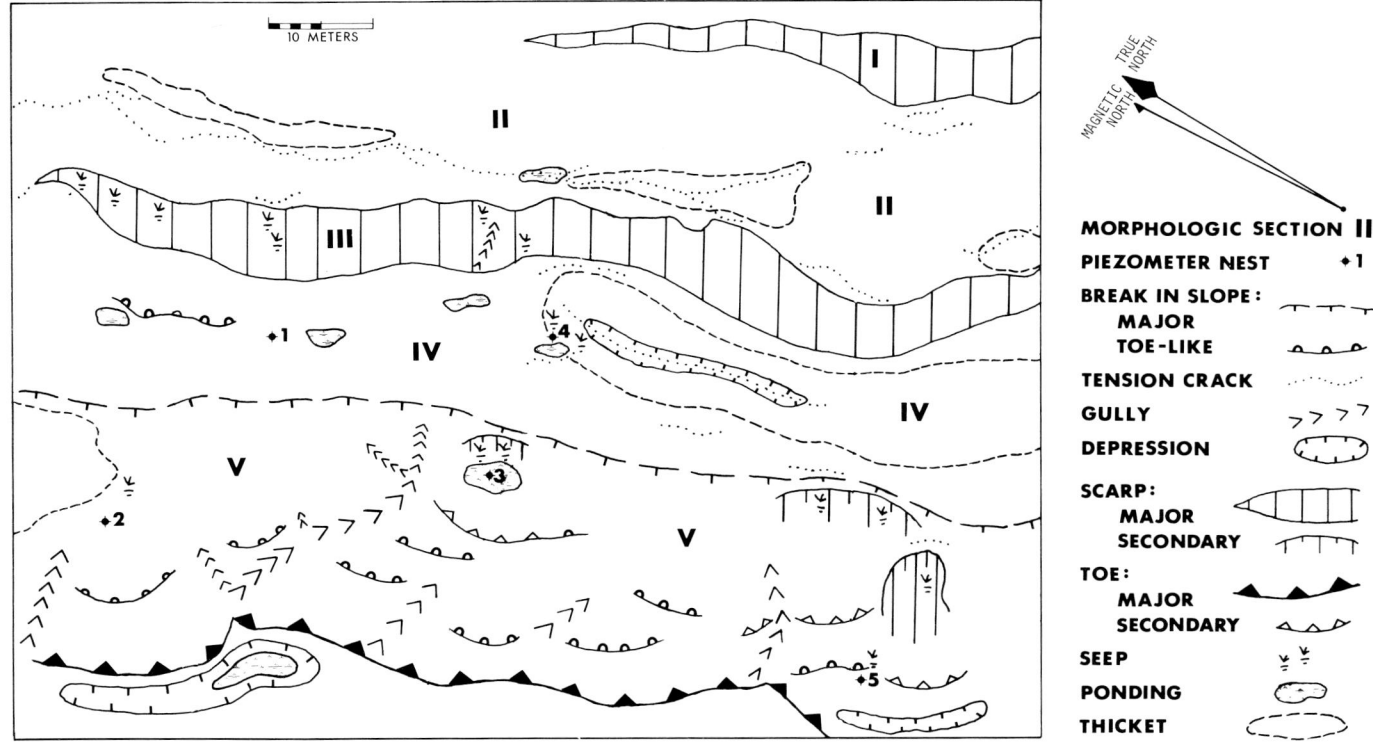

Figure 3. Morphologic map of Stumpy Basin slope.

Figs. 4, 5) indicate that both soils are heavily overconsolidated. It is not known whether the preloading pressure was caused by overburden, glacial ice, or a combination of the two. Of the two samples, the yellow-brown silt is considerably softer and less elastic than the till. Soil elasticity was estimated from the void ratio versus log pressure curves by computing the amount of strain recovered on removal of the final 10.0 kg/cm² load. The silt also has a higher initial void ratio than the till, causing its softer consistency.

HYDROLOGY

The plot of depth to water table versus date (Fig. 6) shows a striking parallelism of responses in the three wells, and most of the fluctuations correspond nicely with precipitation data. The pronounced rises are associated with > 2 cm of rainfall within the preceding few days or with the melting of accumulated snow. Drops in water-table elevation are attributed to little or no rainfall for several days prior to readings, with the exception of the large decline on August 6, 1975. This drop followed 2.54 cm of rain on August 4, 1975. The shortest distances to water table were encountered in piezometers 4 and 5 — the former located in a seepline with upward ground water flow, and the latter being the lowermost on the slope.

Effective stresses were computed using pore-pressure and hydraulic gradient data obtained from water levels. Effective stress was of concern here, because changes in this parameter elicit changes in shearing resistance of the soil. Effective stress values versus date are plotted in Figure 7. The trend produced by nest 5 appears to be a mirror image of the trend in the other nests. This reversal might be the result of a lag time required for precipitation to produce a ground-water response at lower slope elevations, where nest 5 is located. After precipitation, the nests on the upper slope show decreases in effective stress as water enters ground-water flow. Because of the low permeability of clay and silt, more time is necessary for this increased flow to reach the lower slope. Thus, while the upper nests showed decreased effective stresses, nest 5 had not yet received the effects of the rainfall. By the time nest 5 showed decreases in effective stress, the upper slope had increased effective stresses as the result of ground-water flow subsidence.

Seasonal variations in hydrology were qualitatively defined by field observations. Winter and early spring are marked by numerous and frequent mudruns on exposed scarps. Areas of active seepage often remain thawed where continuous outflux of warmer ground water retards freezing. During thaws of considerable snow accumulation, large volumes of water are released to the slope, resulting in large mudruns and springs on the scarps, followed by extensive head ponding. The slope becomes dry toward the end of spring as precipitation diminishes and vegetation requires more water. Dry conditions prevail throughout summer and early fall, as can be seen in the trends of water levels in Figure 6. During the dry period, desiccation cracks develop on the scarps and flat areas of the slope. These cracks provide avenues for the entrance of water into the flow system of the slope, but much of the precipitation is lost to evaporation and plant intake

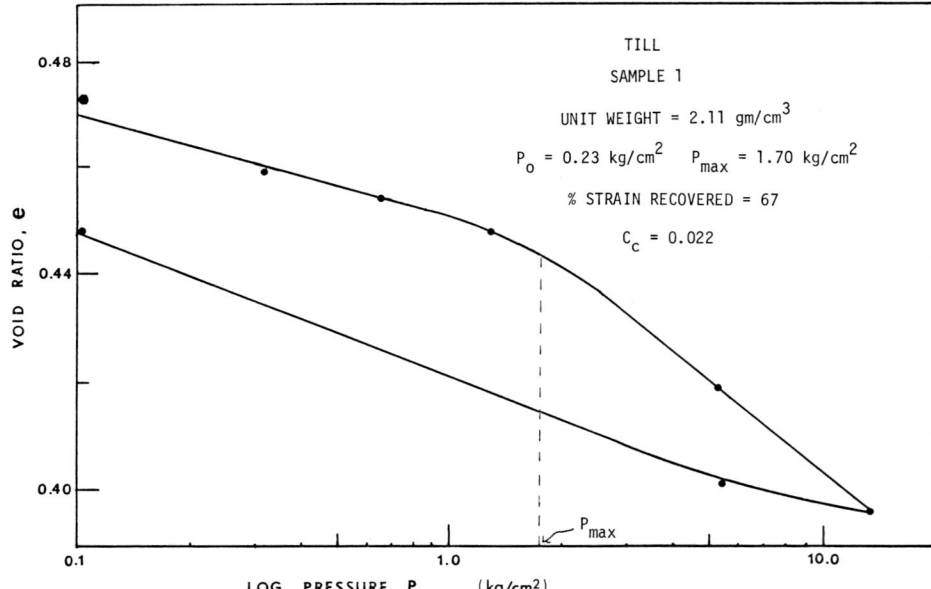

Figure 4. Consolidation test of till unit.

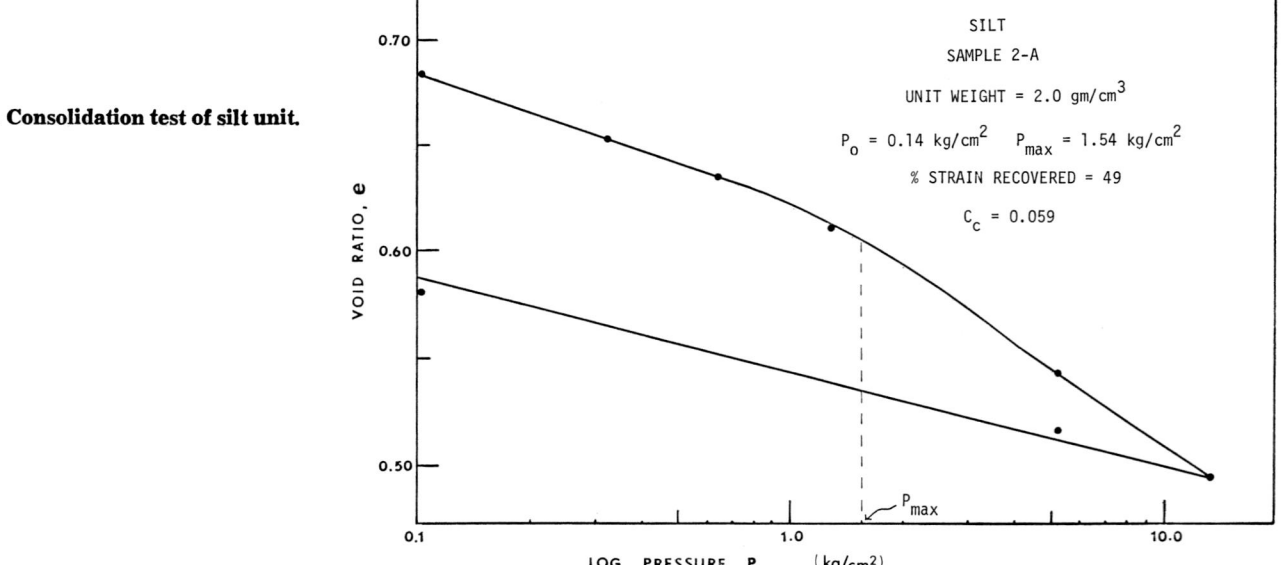

Figure 5. Consolidation test of silt unit.

during this period. Mudruns are rare during the summer owing to the loss of water to evapotranspiration. During the later part of fall, the slope becomes wetter with increased rainfall (indicated in Fig. 6) and once again reaches the mudrun stage.

VEGETATION AND MORPHOLOGY

The vegetation of the Stumpy Basin slope is unique in that several species of prairie vegetation uncommon in northeastern Ohio flourish there. Large areas of prairie grasses are present, surrounded by typical deciduous forestation. Some of the plants found on the slope are buffalo berry, Indian paintbrush, fringed gentian, Indian grass, downey arrowwood, Seneca snake root, common juniper, yellow oak, burr oak, and rough-leaved dogwood (W. A. Cusick, 1976, oral commun.). These plants and trees are typically associated with alkaline conditions. These prairie species have persisted as a result of ground displacements which have prevented the establishment of the deep-rooted deciduous trees (J. A. Herrick, 1974, oral commun.). White cedars also occur in close association with the prairie grasses (Fig. 8).

The vegetation is closely related to the morphology of the slope and consequently was used as an aid in defining this factor. For instance, white cedars growing on the slope indicate the presence of near-surface ground water, and a concentration of these trees delineates an active seep area. One piezometer nest, placed in a cedar grove, revealed upward hydraulic gradients for the most part and demonstrated that

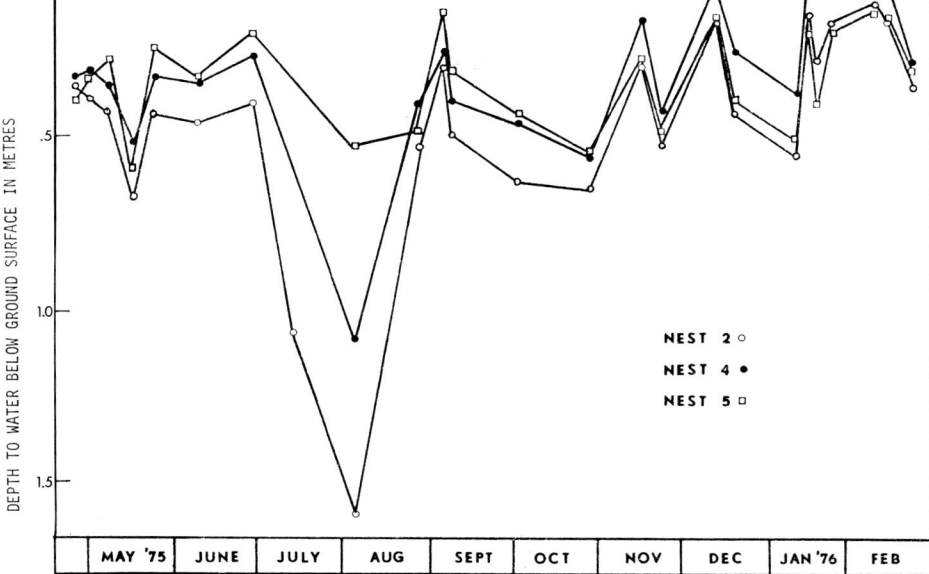

Figure 6. Graph showing water-level fluctuations with time in three shallow wells.

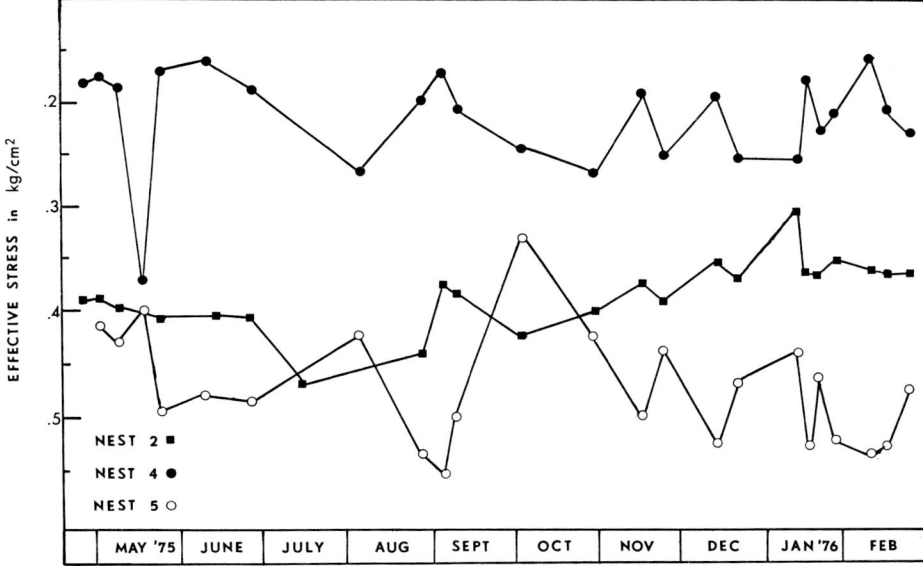

Figure 7. Graph showing change in effective stress with time.

the cedars are associated with water seepage. The abundance of prairie grass implies a fairly long period of instability, since more stable areas on and surrounding the slope permit the growth of deciduous trees. The density of vegetation or lack of it on scarps was useful in determining their relative ages.

The morphologic map of the Stumpy Basin slope is shown in Figure 3. On the basis of morphology, the slope is divisible into five narrow sections whose long dimensions traverse the slope laterally. Sections I and III are major scarps, with the till exposed in section I (Fig. 8). Section II is a hummocky sloping area with numerous tension cracks. It supports a mixture of grasses and trees. Section IV is flat and grassy and is the major head of the slump system. The final section contains the major toe of the failed slope, with three secondary or toe failures. The shallow slump (slab-slide) adjacent to nest 5 occurred in the spring of 1974, while the remaining slumps are relict. From this section to the river, topography is rolling and dissected, and it supports thick tree growth. No major movement was observed here, recent or otherwise.

MECHANISMS

The initial mechanism of slope failure was the rapid downcutting of the Cuyahoga River into the glaciolacustrine and outwash material and its consequent meandering to the east (Gardner, 1972). The meandering and erosion oversteepened the then-existing slope and removed its basal support. Stumpy Basin, constructed around 1826 for a canal boat turnaround, was built in the toe of the old landslide. For more than

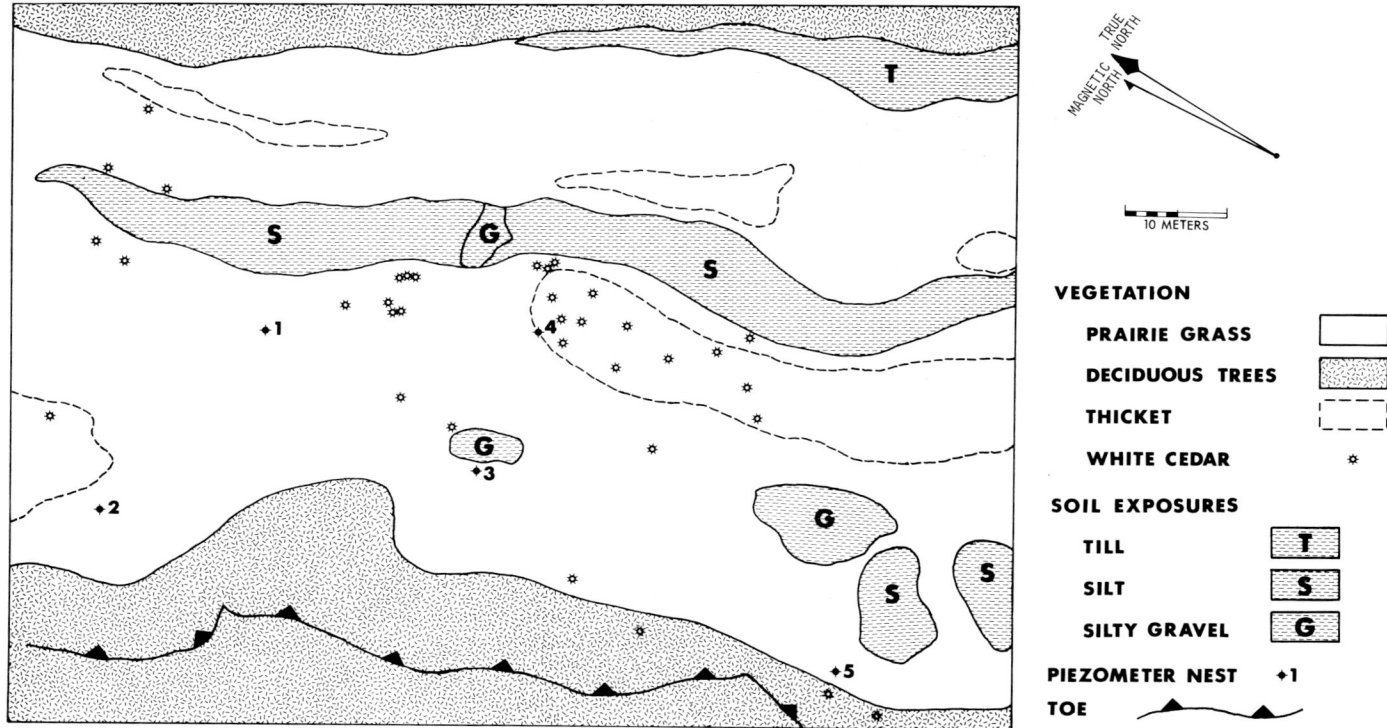

Figure 8. Map of vegetation and soils on Stumpy Basin slope.

150 yr, the Stumpy Basin slope has been adjusting to the initial oversteepening by means of progressive slumping. The slumping is accommodated by the unstable nature of the soils and the hydrology of the slope.

According to Gardner (1972), soils having a 2:1 silt-to-clay ratio are most prone to slope instability and erosion. The yellow-brown clayey silt and the gray silt have this critical ratio. The mineralogy of the soils in this case has little effect on the instability of the slope. The stress history of the soils, however, is significant and may be contributing to the slope failure. Because of their preloaded nature, the soils have excess lateral stresses in them. With the application of a large vertical load, such as 1 to 2 m of overburden or the weight of glacial ice, large lateral stresses are induced in the soil. Owing to the plastic nature of soil, these large stresses are not dissipated upon removal of the load. The horizontal stresses existing in the field on the two samples studied slightly exceed the vertical stresses, as dictated by their maximum overburden pressures and plastic indices (Lambe and Whitman, 1969). Such excess lateral stresses may be relieved only by progressive slope failure (Carson and Kirkby, 1972).

The hydrology of the slope has a profound influence on its stability. Varying levels of precipitation produced fluctuations in water-table elevations, hydraulic gradients, and, consequently, pore pressures and effective stresses. Although no major movement occurred during the observation period, we feel that the large drop in water table on August 6, 1975 (Fig. 6), was produced by a subsurface adjustment in the slide system. On that date the water levels theoretically should have been much higher, as the result of a 2- to 3-cm rain the preceding two days, rather than being the lowest levels encountered throughout the investigation. A subsurface adjustment of slippage prior to that date would most likely provide an egress for the ground water, with the corresponding drop in water table. Lambe and Whitman (1969) stated that for overconsolidated soils with an overconsolidation ratio (maximum overburden pressure to present overburden pressure) of 4 or greater, the soils will expand in response to shearing stress. The results of the consolidation tests showed that the Stumpy Basin soils have overconsolidation ratios of 7 and 11 and would therefore expand under shearing stress. Since expansion of a soil is accompanied by an increase in void ratio and water content, a subsurface slippage caused by shear stress would expand the soils and permit them to absorb water, thereby accounting for the drop in water-table elevation. The effective stresses did not show a marked change on the date in question, which can be accounted for by the slippage being deeper than the 2.5-m depth monitored. The subsurface disturbance probably occurred in the vicinity of nests 1, 2, and 4 (Fig. 2), since these piezometers showed the largest changes.

The nature of the yellow-brown silt, with its interstratification of sand and gravel, affects ground-water flow due to varying permeabilities. The flow in turn affects the hydraulic gradient, pore pressure, and effective stress in the soil near zones of variation. The soil is quite varied, as was shown in encountering lateral and vertical changes over distances of 1 to 2 m or less while installing piezometers and collecting samples. The possibility exists that excess pore pressure developed in the higher permeability zones, and quick conditions occurred. This was most likely the case in the relict slump above nest 5, because coarse granular material is abundant there.

The abundance of water available to the slope is evident in the presence of numerous seeps and ponds and the occurrence of springs and mudruns with higher levels of precipitation. Some of the water originates from an excavation depression behind the top of the slope. During construction of the Ohio Turnpike, the large area was defoliated, and the relatively impermeable till was removed to be used as fill elsewhere (Gardner, 1972). Consequently, precipitation in this area can percolate into the silts and join the flow system of the slope.

The presence of prairie grasses and tension cracks on the upper slope allows more rainfall to enter the soils than would be the case with deciduous tree cover and the absence of cracks. Ground water reaching the slope from the uplands emerges near the tops of major scarps during wet periods, is ponded, and then reabsorbed into the silts. Below the major toe, water flows underground for the most part, as is evidenced by the fact that gullies there rarely contain running water.

DISCUSSION

The Stumpy Basin slope is progressively stabilizing itself from the base to its top. All recent activity is confined to the upper segments of the slope, which are steeper than the section near the river (below the mapped area). The relatively stable lower section has a gentler slope and is covered with a thick growth of large deciduous trees. Dark topsoil and thick accumulations in the area imply that sliding activity, or the deposition of materials from landsliding, has been dormant for some time. The river is not eroding this base actively, as is indicated by the low gradient and marshy growth along its poorly defined banks. The upper slope has continued to adjust by local failure, such as the recent slump near piezometer nest 5.

As indicated in this study, the mechanisms contributing to failure of the Stumpy Basin slope are the unstable nature of the soils, their preloaded aspect, and their interaction with the abundant water available. The interrelationship of morphology, hydrology, and vegetation was evident in this study and was essential to understanding this ancient and complex landslide.

ACKNOWLEDGMENTS

We gratefully acknowledge Sigma Xi, the Scientific Research Society of North America, for awarding this study a Grant-in-Aid of Research. We also thank H. R. Collins and C.F.S. Sharpe, who critically reviewed this manuscript and contributed valuable suggestions.

REFERENCES CITED

Carson, M. A., and Kirkby, M. J., 1972, Hillslope form and process: London, Cambridge Univ. Press, 475 p.

Cooke, R. U., and Doornkamp, J. C., 1974, Geomorphology in environmental management: London, Oxford Univ. Press, 413 p.

Gardner, G. D., 1972, A regional study of landsliding in the lower Cuyahoga River valley, Ohio [M.S. thesis]: Kent, Ohio, Kent State Univ., 85 p.

Gardner, G. D., Wittine, A. H., McComas, M. R., Miller, B. B., and Manus, R. W., 1974, Engineering and Pleistocene geology of the lower Cuyahoga River valley, *in* Heimlich, R. A., and Feldman, R. M., eds., Selected field trips in northeastern Ohio: Ohio Geol. Survey Guidebook 2, 59 p.

Lambe, T. W., and Whitman, R. V., 1969, Soil mechanics: New York, John Wiley & Sons, Inc., 553 p.

Rau, J. L., 1969, The evolution of the Cuyahoga River; its geomorphology and environmental geology, *in* Cooke, G. D., ed., The Cuyahoga River watershed: Kent, Ohio, Kent State Univ. Inst. Limnology and Dept. Biology, 143 p.

White, G. W., 1953, The character and distribution of the glacial and alluvial deposits, *in* Smith, R. C., The ground-water resources of Summit County, Ohio: Ohio Div. Water Bull. 27, p. 18–27.

Wittine, A. H., 1970, A study of a late Pleistocene lake in the Cuyahoga River valley, Summit County, Ohio [M.S. thesis]: Kent, Ohio, Kent State Univ., 64 p.

MANUSCRIPT RECEIVED BY THE SOCIETY SEPTEMBER 7, 1976
MANUSCRIPT ACCEPTED SEPTEMBER 17, 1976

Printed in U.S.A.

15
Engineering geology of the slope instability of two overconsolidated north-central Texas shales

ROBERT G. FONT
Department of Geology, Baylor University, Waco, Texas 76703

ABSTRACT

The marine shales of the Eagle Ford and Washita Groups of the Cretaceous System of north-central Texas offer an excellent opportunity for the study of factors influencing development of different landslide types. These shales are exposed within metropolitan areas such as Dallas and Waco and have earned a reputation of being "problematic." Although the shales of both groups have similar histories of overconsolidation and postdepositional changes, as indicated by their engineering properties, they can fail along slopes in a noticeably different manner. Types of failures may range from the classic rotational slump to the slow earthflow. Landslide genesis and morphology are found to be largely dictated by locally dominant geologic features, such as stratigraphic position, topographic expression, and specific lithologic and morphologic characteristics. The fundamental causes that dictate the instability of the shales are directly related to these geologic features, as well as to their engineering properties and to the uncontrolled modification of natural slopes.

INTRODUCTION

In the Waco urban area, the shales of the Eagle Ford and Washita Groups have caused problems related to engineering construction. The majority of these problems arise from the extremely critical potential volume-change characteristics of the shales and from their slope instability. In this study, we are specifically interested in predicting and preventing the unstable behavior of the South Bosque and Del Rio Formations. In Waco, the South Bosque shale is recognized as the uppermost formation of the Eagle Ford Group, whereas the Del Rio formation forms part of the Washita Group (Tables 1, 2).

Four different types of slope failure have been recognized in the field: (1) rockfalls involving the chalk caprock that overlies the top of the South Bosque shale, without involving the slumping failure of the shale itself; (2) slow earthflow of the South Bosque shale, with no well-defined, single slip surface, and the associated failure of the overlying chalk caprock; (3) rotational slumping of the South Bosque shale along an arcuate slip surface and the associated failure of the locally thin overlying chalk caprock wherever present; and (4) rotational slumping of the Del Rio Clay along a well-defined, nearly cylindrical slip surface.

In order to determine the causes responsible for the slope instability, which conditions favor each failure style, and how to predict and prevent future problems, an engineering geology study has been conducted with the following objectives: (1) to determine the engineering properties of the shales for use as a guideline in future engineering endeavors, and (2) to investigate the field conditions that influence the styles of failure observed.

The basic conclusions that have emerged from the study are summarized as follows: (1) The engineering properties of the shales are indicative of their field instability. Both shales are characterized by liquid limits ranging from 60 to 80, by plasticity indices ranging from 30 to 50, and by very critical potential volume-change characteristics. They also exhibit low values of residual strength. Calculated values of sensitivity (Skempton and Northey, 1952) and liquidity index are indicative of their overconsolidated nature (Tables 3, 4, 5). (2) Types of slope failures associated with the outcrops of the shales include rockfall, slow earthflow, and rotational slump (Eckel, 1958). Whatever type occurs depends on local stratigraphic position, local topographic expression, and specific lithologic and morphologic characteristics. (3) Field evidence indicates that natural slopes ranging from 4:1 to 6:1 (horizontal/vertical) are stable. Uncontrolled modifications of natural slopes by oversteepening and overloading almost invariably result in failures.

TABLE 1. GEOLOGIC CHARACTER OF THE STRATA OF THE CRETACEOUS SYSTEM IN WACO, TEXAS

Series	Group	Formation and symbol	Depositional history	Maximum local thickness (m)	Maximum probable thickness at one time (m)
Gulf	Taylor	Taylor Marl K_{ta}	Deposited during marine transgressions and regressions	76	355
Gulf	Austin	Austin Chalk K_{au}	Deposited during marine transgression with possible fluctuations of the strandline. Unconformably underlies K_{ta}	76	89
Gulf	Eagle Ford	South Bosque shale K_{sb}	Deposited in a neritic marine environment with poor circulation. Unconformably underlies K_{au}	48	48
Gulf	Eagle Ford	Lake Waco Formation K_{lw}	Deposited in a lagoonal environment. Conformably underlies K_{sb}	24	44
Gulf	Woodbine	Pepper Shale K_{pe}	Deposited in brackish environment. Unconformably underlies K_{lw}	21	30
Comanche	Washita	Buda Limestone K_{bu}	Nearshore marine deposit. Almost completely eroded locally prior to deposition of K_{pe}	1	11
Comanche	Washita	Del Rio Clay K_{dr}	Marine regression and transgression. Restricted environment. Conformably underlies K_{bu}	26	26
Comanche	Washita	Georgetown Limestone K_{ge}	Shallow marine deposit	64	64

GEOLOGIC SETTING

Location and Previous Work

The study area lies in and around the Waco urban region in McLennan County, Texas (Fig. 1). It is bounded by lat 31°37'30" and 31°30'00" N and long 97°07'30" and 97°15'00" W.

Previous work in the area was confined to geologic studies of some of the bedrock exposed within it (O. T. Hayward, unpub. data; K. O. Seewald, unpub. data; R. S. Chamness, unpub. data; Beall, 1964; J. Ray, unpub. data; Burket, 1965; Brown, 1971) and to a few general engineering geology studies (R. G. Font, unpub. data; Font and Williamson, 1970).

Regional Structure

McLennan County is underlain by east-dipping beds of the Gulf Coastal Plain. The typical homoclinal structure of the plain is segmented in the Waco area by the Balcones fault zone, a zone of normal faults that extends from west of Uvalde to the Dallas area (Fig. 1). Major faults associated with the Balcones system in the vicinity of Waco trend north-northeast (Fig. 2) and exhibit vertical displacements of as much as 80 m. They are commonly downthrown to the east, although some are downthrown to the west (Fig. 3). Precise dating of the faulting in the vicinity of Waco is impossible, but there is no evidence of displacement along any of the faults since middle Pleistocene time (Burket, 1965).

Local Structure and Stratigraphy

The Balcones fault zone cuts across Waco in a northeast trend. Indeed, most of the engineering geology problems in the area are a direct result of this faulted zone and the lithologic variety of sedimentary rocks. Most of the rocks exposed are limestones and shales of marine origin and Cretaceous age, although some Pleistocene and Holocene alluvial deposits also occur. The most prominent geomorphic feature is the Bosque Escarpment, an obsequent fault-line scarp that cuts across the city (Fig. 2). The total Cretaceous section is about 610 m thick, but only the upper 150 m crop out in the region and are of immediate interest in engineering projects (Fig. 3). Strata in Waco strike approximately N5°E and dip eastward at less than 1°. The stratigraphic relationship, depositional history (Burket, 1965), and maximum thickness of the Cretaceous strata in the local area are summarized in Table 1. The geologic characteristics of the South Bosque shale (K_{sb}) and Del Rio Clay (K_{dr}) are in Table 2.

ENGINEERING PROPERTIES OF THE SOUTH BOSQUE SHALE AND DEL RIO CLAY IN WACO

Stress History

The South Bosque shale and the Del Rio Clay were in the past buried under a greater thickness of sediments than

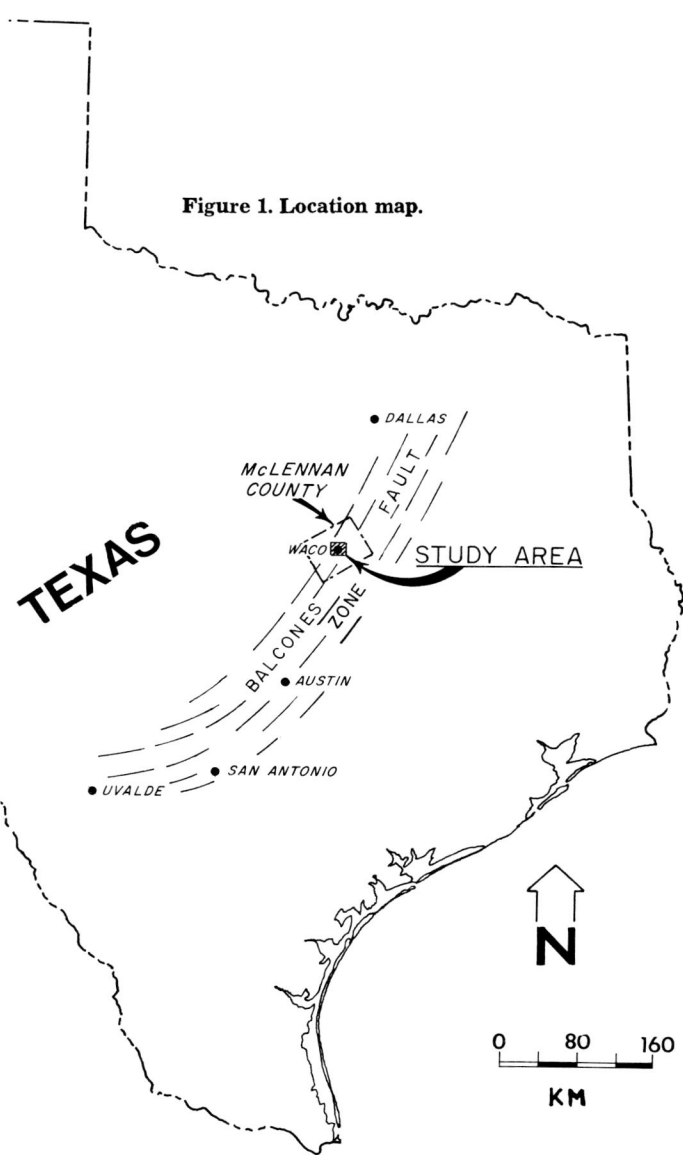

Figure 1. Location map.

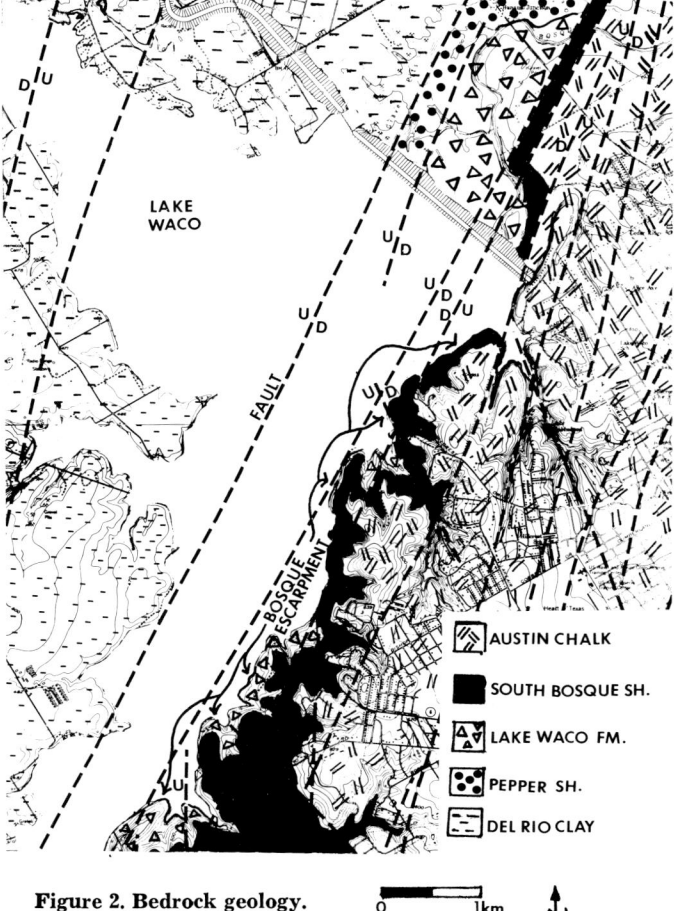

Figure 2. Bedrock geology.

they are now. Beall (1964) established that the lower Taylor Marl member of the Taylor Marl once covered the entire Waco area and extended some distance to the west; he also indicated the possibility that the entire Taylor Group could have once covered it. Estimates based upon the relatively well established geologic history of the area indicate that at least 150 m and perhaps as much as 450 m of overburden had overlain the top of the South Bosque shale. Therefore, the South Bosque shale has been subjected to a minimum pressure of 40 to 120 kg/cm². Since overburden on top of these shales is now either absent or at the most only a few metres thick, both the South Bosque shale and the Del Rio Clay are heavily overconsolidated.

Index Properties

Standard laboratory tests have been conducted to determine moisture content, Atterberg limits and indices, unit weight, clay content, calcium carbonate content, free swell, and soil classification according to the Unified System. The procedures for these tests were described by Lambe (1951), Brown (1971), and Krynine and Judd (1957).

The measured and calculated index properties of the South Bosque shale and Del Rio Clay are summarized in Tables

TABLE 2. GEOLOGIC CHARACTERISTICS OF THE SOUTH BOSQUE SHALE

Rock Unit	Description	Approximate composition*	Topographic expression	Related urban engineering problems
South Bosque shale	Dark gray to black, blocky to fissile, fissured shale that weathers blue-gray to tan. Upper 9 to 15 m are noncalcareous and give rise to most problems	Montmorillonite, 34%; illite, 7%; kaolinite, 5%; calcite, 2–8%; quartz, 11%; others, 35%	Exposed along face of slopes of Bosque Escarpment, but capped by 1 to 6 m of Austin Chalk	Slope instability High shrink-swell Inadequate for septic sewage disposal Poor foundation support strength
Del Rio Clay	Blue-gray homogeneous fractured and fissured shale that weathers light-gray to buff. Middle 10 m are less calcareous than rest. Shale is pervaded by 6 sets of high-angle fractures that strike N3°E, N30°E, N60°E, N80°E, N75°W, and N50°W	Montmorillonite, 10%; illite, 20%; kaolinite, 20%; calcite, 20–30%; quartz and others 20%	Exposed on higher divides between east-flowing tributaries of Bosque River	Slope instability High shrink-swell Inadequate for septic sewage disposal Poor foundation support strength

*Based on X-ray diffraction data.

3 and 4. In general, the tests corroborate the field behavior of the shales, indicating that they are both highly ductile, with high liquid limits, plasticity indices, and clay content.

Strength Tests

In order to prevent future stability problems in the Waco area related to the South Bosque and Del Rio shales, we must be prepared to cope with all of the possible types of failures that may occur. Since both shales are heavily overconsolidated, fractured, and fissured, it is likely that future failures could be either of the short-term type, involving the undrained shear strength, or of the long-term type, involving the drained and residual strengths (Skempton, 1964). For this reason, field and laboratory tests have been conducted to investigate the undrained, drained, and residual strengths of both shales. Field measurements of undrained shear strength have been based on test results obtained with hand-operated penetrometers and vane-shear devices. Standard laboratory tests have also been performed on saturated samples. These have included unconfined compression tests as well as consolidated-undrained, consolidated-drained, and residual direct shear tests. Shear strength parameters have been obtained for both total and effective stresses and for the residual strength condition (Table 5). (For more specific information on the test procedures, see Lambe, 1951.)

TABLE 3. ATTERBERG LIMITS AND INDICES

	Liquid limit ω_L (%)	Plastic limit ω_P (%)	Shrinkage limit ω_S (%)	Plasticity index I_P (%)	Shrinkage index I_S (%)	Activity ratio A_r	Liquidity index I_L	Potential volume change
South Bosque shale	60 to 75	26	13	34 to 49	13	0.48 to 0.60	0 to —0.5 avg —0.2 to —0.3	Very critical
Del Rio Clay	60 to 80	25 to 30	19	52 to 30	6 to 11	0.33 to 0.60	0 to —0.5 avg —0.2 to —0.3	Marginal to very critical

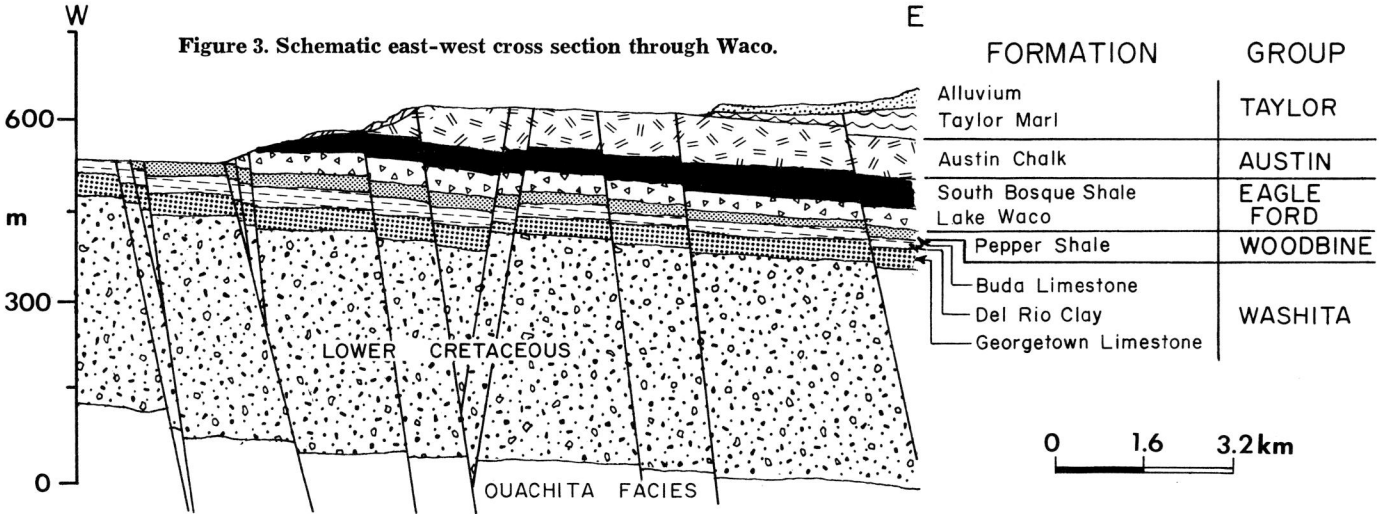

Figure 3. Schematic east-west cross section through Waco.

SLOPE FAILURES

Field evidence indicates that within the study area natural stable slopes have a horizontal to vertical ratio of between 4:1 and 6:1. Slope failures typically occur when these natural slopes are disturbed by man. Commonly, failures have resulted when construction practices have led to the removal of the toe of a slope or to oversteepening and overloading. Thus, whenever possible, natural slopes should be maintained. Steeper slopes must be properly designed and modified to ensure stability.

Del Rio Clay

The characteristic type of slope failure in the Del Rio Clay is influenced by stratigraphic position, topographic expression, and lithologic characteristics of the shale. The shale usually fails by slumping along a well-defined, cylindrical slip surface (Fig. 4). The slip surface is initiated at a tension crack that develops on top of the uncovered, unprotected slope or along the face of the slope near the top (Fig. 5). Tension cracks commonly parallel the fracture trends in the unweathered shale and suggest that the position and orientation of the fractures can exert an important influence on initiation of the slip surface. In other words, the tension crack that defines the upper end of the slip surface may develop directly from the widening of pre-existing fractures (Table 2). This widening process is enhanced by the more rapid weathering along the fracture planes.

Although pre-existing fractures can facilitate the development of tension cracks, they do not control the shape of the slip surfaces. The upper part of the slip surface may follow the steeply dipping plane of the widened crack, but its lower, flatter part is apparently not influenced by the presence of predetermined lower strength surfaces. On the other hand, cylindrical slip surfaces commonly occur in homogeneous or nearly homogeneous clay-shales (Krynine and Judd, 1957). Thus, the homogeneity of the Del Rio Clay with respect to

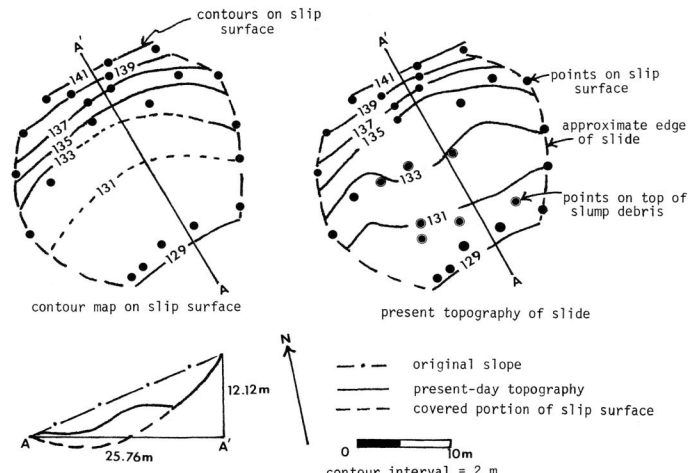

Figure 4. Topographic map and cross section of a slope failure in the Del Rio Clay.

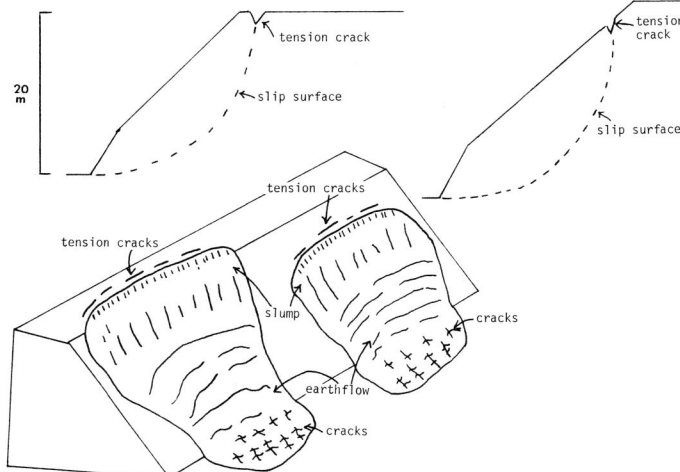

Figure 5. Typical slope failure in the Del Rio Clay.

TABLE 4. ENGINEERING SOIL CLASSIFICATION AND OTHER INDEX PROPERTIES

Rock unit	Clay content (%)	CaCO$_3$ content (%)	Dry unit weight (g/cm^3)	Classification under Unified System
South Bosque shale	70	2 to 8 in upper 9 to 15 m	1.31	CH
Del Rio Clay	90	15 to 30 in middle 10 m and upper and lower portions, respectively	1.55	CH

grain size and composition is a factor that favors development of the mapped cylindrical surfaces.

Slumping of the Del Rio Clay may also occur when two major gullies, essentially parallel to each other, become interconnected by a transverse gully or when several small, closely spaced gullies are joined (Fig. 6).

To summarize, certain geologic characteristics of the Del Rio Clay in the local area influence the failure mode of the shale. The general absence of a caprock allows the opening of tension cracks. These tension cracks may develop from the widening of already existing fractures. The homogeneous composition and grain size are favorable to the development of cylindrical slip surfaces. The highly erodable character of the shale increases the instability of the slopes.

South Bosque Shale

Stratigraphic position, topographic expression, and certain lithologic and morphologic characteristics also exert influence on the mode of failure of the South Bosque shale in the Waco area (Tables 1, 2). The weight of the overlying Austin Chalk caprock and that of man-made structures placed upon it tend to force the shale outward from the face of the slope, resulting in failure. The highly expansive characteristics of the loaded wet shale, for which expansion pressures in excess of 15 kg/cm^2 have been measured, contribute to the failure process. The wet shale is able to expand freely outwardly and subsequently contract with desiccation. Failures may assume the form of a slow earthflow (Eckel, 1958) with no single, well-defined cylindrical slip surface (Fig. 7). This type of downslope motion is a result of the slow deformation in the shale from the overlying load stress acting over long periods of time. The creeplike motion may become noticeable only after considerable strain has occurred (Zaruba and Mencl, 1969). During the initial phase of the failure, when the downslope motion is slow, the shale flows along a large number of partial slide surfaces rather than along a single, well-defined slip surface. This flowage is enhanced by the presence of the chalk caprock, which prevents the opening and widening of tension cracks on the top of the slope and inhibits the initiation of a single slip surface at this end. Under these conditions, it is easier for the shale to flow along a system of partial slide surfaces than to flow along a single surface. The blocky character of the shale provides it with an already existing system of partial slide surfaces along which sliding can take place, thus facilitating flow.

In some instances, after considerable strain has occurred, a rapid downslope motion may result. When this occurs, the shale may slump along a definable cylindrical slip surface (Fig. 8), or it may continue to flow along a system of partial slide surfaces. In any case, the overlying chalk caprock and any foundation placed upon it will also fail.

Outside Waco and in the Dallas urban area, the South Bosque shale has been observed to slump in the classic style, with rotation along a well-defined cylindrical slip surface. It should be indicated that in these cases the slopes are usually without caprock, or, when present, it is locally very thin (generally less than 2 to 3 m). Thus, local stratigraphic position and topographic expression can, indeed, play a significant role in the mode of failure observed.

Austin Chalk

Failure of the Austin Chalk caprock, and anything on top of it, commonly occurs throughout the Bosque Escarpment. As might be expected, the behavior of the chalk caprock is closely related to the instability of the underlying South Bosque shale; however, the shale does not necessarily have to fail by slumping to cause the chalk to fail. The highly fractured Austin Chalk caprock may fail as follows: rapid erosion of the South Bosque shale along the shale-chalk contact undermines the chalk caprock. Large chalk blocks break off from the main mass owing to the creeplike motion

TABLE 5. STRENGTH PARAMETERS

Rock unit	Cohesion intercepts			Angles of internal friction			Sensitivity ratio
	Consolidated, undrained (kg/cm^2)	Consolidated, drained (kg/cm^2)	Residual strength (kg/cm^2)	Consolidated, undrained (°)	Consolidated, undrained (°)	Residual strength (°)	
South Bosque shale	0.34	0.28	0	18	13	7 to 10	1.0
Del Rio Clay	0.35	0.32	0	11.5	10.5	6 to 9.5	1.2

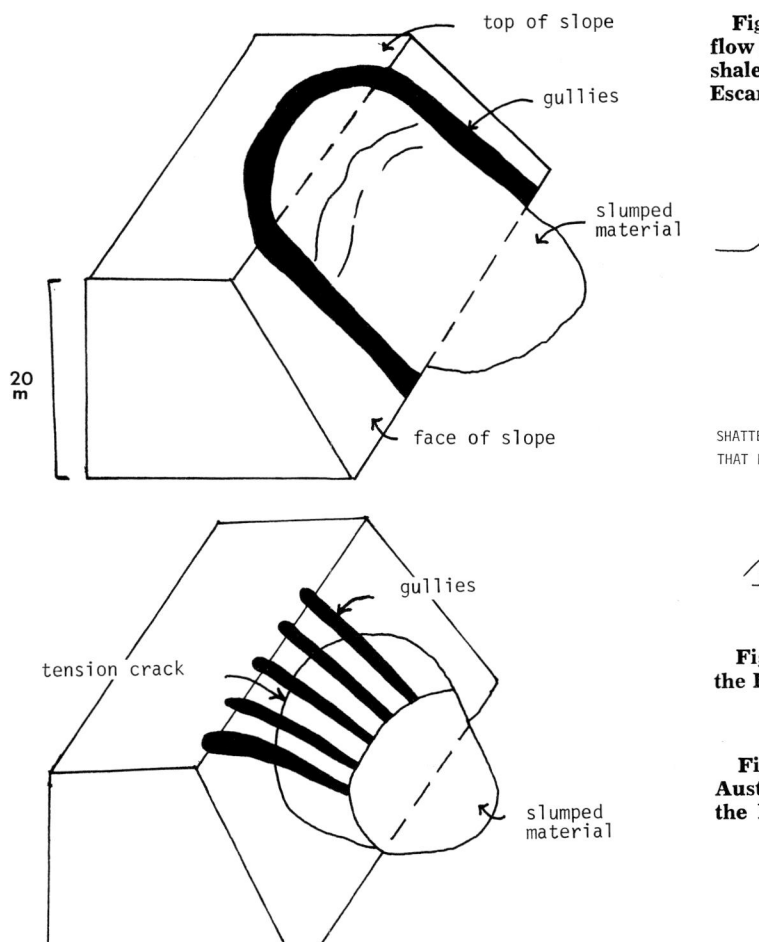

Figure 6. Relationship between gullying and slumping in the Del Rio Clay.

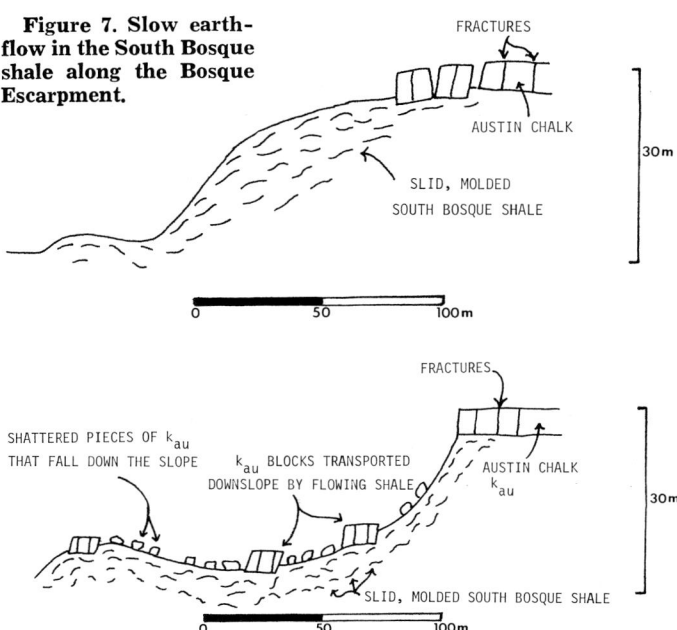

Figure 7. Slow earthflow in the South Bosque shale along the Bosque Escarpment.

Figure 8. Schematic diagram of a typical slope failure along the Bosque Escarpment in the Waco area.

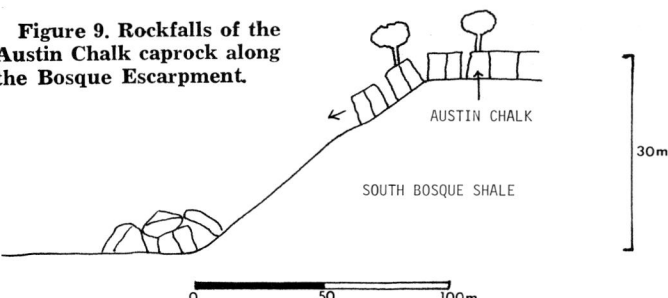

Figure 9. Rockfalls of the Austin Chalk caprock along the Bosque Escarpment.

of the underlying shale and the pressures exerted by tree roots. The overall cohesion of the rock mass is further decreased as water seeps into the fractures. The water eventually moves through the limestone and saturates the top of the shale. As more water moves through the caprock it begins to seep along the shale-chalk contact, creating a surface along which the chalk blocks can slide. Thus, rockfalls may result which involve the failure of the chalk caprock without involving the slumping failure of the underlying shale (Fig. 9).

CONCLUSIONS

Landslide genesis and morphology in the South Bosque shale and Del Rio Clay are directly related to locally dominant geologic features, such as stratigraphic position, topographic expression, and specific lithologic and morphologic characteristics. Rotational slumps tend to occur when the shale slopes lack a protective caprock or when the caprock is thin (generally less than 3 m thick). The opening and widening of tension cracks usually precede the slumping.

In some instances, the development of these tension cracks is enhanced by the already pre-existing fracture planes that pervade the overconsolidated shales. Gully development is also found to increase instability.

Slow earthflows tend to occur when the shales slopes are covered by a thick caprock. The weight of the caprock exerts a downward pressure on the shales and, with time, forces the plastic material outward from the face of the slope. The flows may be imperceptible until considerable strain has occurred. Eventually, a rapid downslope motion may result. Typically slow earthflows are characterized by flowage along a large number of partial slide surfaces rather than by movement along a well-defined slip surface. The flowage is enhanced by the presence of the thick caprock which prevents the opening and widening of tension cracks on top of the slope and inhibits the initiation of a single slip surface at this end. Under these conditions it is easier for the blocky shales to flow along a system of partial slide surfaces than to flow along a single surface.

Rockfalls of the chalk caprock can occur without necessarily involving the failure of the underlying shales. Large blocks of the highly fractured caprock material may become detached from the main rock mass as a result of the creep-like motion of the underlying shales, the pressure exerted by tree roots, and erosion and undermining of the shales below the chalk-shale contact. Water-filled fractures and water seepage along the caprock-shale contact contribute to the instability of the caprock blocks.

The instability of the South Bosque shale and Del Rio Clay is also reflected by their engineering or geotechnical properties. Both shales are characterized by liquid limits ranging from 60 to 80 and plasticity indices ranging from 30 to 50 and by very critical potential volume-change characteristics. Low residual angles of internal friction, ranging from 6° to 10° are also typical of these formations.

Natural stable slopes generally have a horizontal to vertical ratio of from 4:1 to 6:1. Unjudicious modification of these slopes will almost certainly result in failures.

ACKNOWLEDGMENTS

My special thanks go to R. R. Berg, D. W. Stearns, C. C. Mathewson, D. R. Coates, and H. J. Pincus for their support and help in writing and reviewing this manuscript. I am also indebted to J. M. Logan, W. R. Bryant, and O. T. Hayward for their encouragement.

REFERENCES CITED

Beall, A. O., 1964, Stratigraphy of the Taylor Formation (Upper) Cretaceous), east-central Texas: Baylor Geol. Studies Bull. 6, 35 p.

Brown, T. E., 1971, Stratigraphy of the Washita Group in Central Texas: Baylor Geol. Studies Bull. 21, 43 p.

Burket, J. M., 1965, Geology of Waco, in Urban geology of greater Waco—Pt. I: Baylor Geol. Studies Bull. 8, 45 p.

Eckel, E. B., ed., 1958, Landslides and engineering practice: Washington, D.C., Highway Research Board Spec. Rept. 29, NAS-NRC Pub. 544, 232 p.

Font, R. G., and Williamson, E. F., 1970, Geologic factors affecting construction in Waco, in Urban geology of greater Waco—Pt. IV: Baylor Geol. Studies Bull. 12, 33 p.

Krynine, D. P., and Judd, W. R., 1957, Principles of engineering geology and geotechnics: New York, McGraw-Hill Book Co., 730 p.

Lambe, T. W., 1951, Soil testing for engineers: New York, John Wiley & Sons, 165 p.

Skempton, A. W., 1964, Long term stability of clay slopes: Géotechnique, v. 14, p. 77–102.

Skempton, A. W., and Northey, R. D., 1952, The sensitivity of clays: Géotechnique, v. 3, p. 30–53.

Zaruba, Q., and Mencl, V., 1969, Landslides and their control: New York, Elsevier Pub. Co., 205 p.

MANUSCRIPT RECEIVED BY THE SOCIETY SEPTEMBER 7, 1976
MANUSCRIPT ACCEPTED SEPTEMBER 17, 1976

16
Engineering geology of multiple landsliding along I-45 road cut near Centerville, Texas

CHRISTOPHER C. MATHEWSON
JAMES H. CLARY*
Department of Geology, Texas A&M University, College Station, Texas 77843

ABSTRACT

Multiple landsliding in both backslopes of the Interstate 45 road cut near Centerville, Texas, has been a recurrent problem since the cut was made. Landslide events can be correlated with periods of heavy rainfall. The west backslope slide moves in a downdip (southeasterly) direction, whereas the east backslope slide moves in an updip (northwesterly) direction.

The road cut is primarily within the Eocene Queen City Formation, which at this locality is a fluvial deposit consisting of natural levee clay, crevasse-splay sands, and floodplain-marsh organic clay. The downdip impermeable floodplain clay retards the movement of ground water, thereby creating perched water tables and artesian conditions within the natural levee clay and crevasse-splay sand units updip. The positions of the two landslides opposite each other in the north end of the cut and within the levee clay and sand units are a direct result of stratigraphic and hydrologic conditions.

In engineering terms, the highway backslopes are composed of overconsolidated, low- to high-plasticity clay and silt and poorly graded sand. Results of stability analyses, using shear-strength parameters determined from laboratory testing, indicate that initial slope failure occurred as a result of the combination of high pore-water pressure and oversteepening of the slopes. Subsequent failures have occurred because residual shear strengths are lower than initial (prefailure) shear strengths and because pore-water pressure increases during periods of heavy rainfall.

INTRODUCTION

Centerville is located in the southwest corner of Leon County in east Texas halfway between Houston and Dallas (Fig. 1). The landslides occur in the backslopes on the east and west sides of Interstate 45, 1 mi (1.6 km) west of

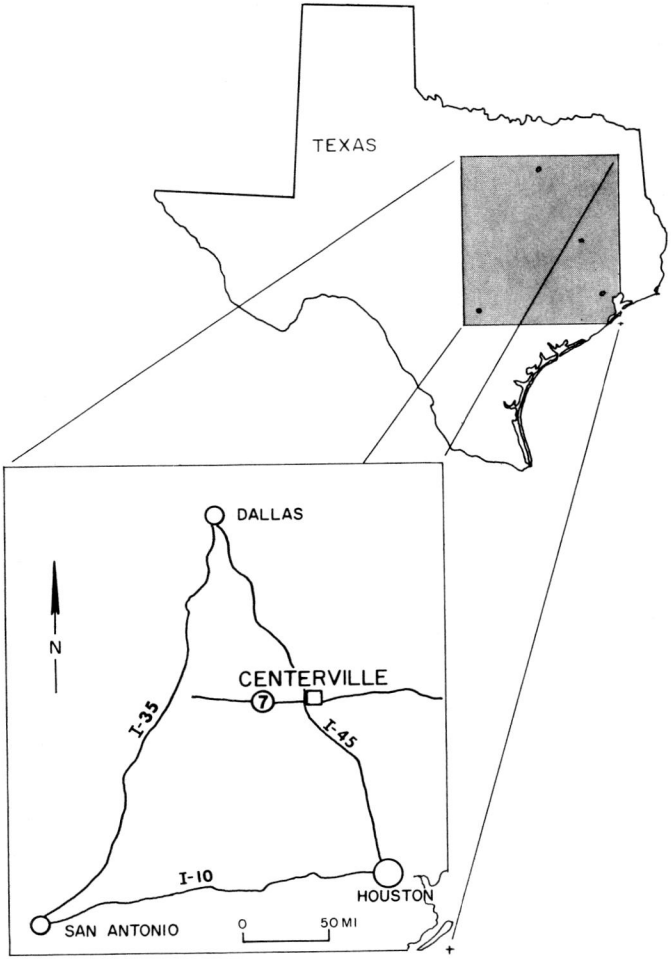

Figure 1. Location of Centerville, Texas, and the intersection of I-45 and State Highway 7.

---
*Present address: Albuquerque Testing Laboratory, 532 Jefferson, N.E., Albuquerque, New Mexico 87104.

Centerville and north of the intersection of Interstate 45 and State Highway 7 (Fig. 2).

Upon completion of the Interstate 45 road cut in 1965 and prior to highway opening, the west backslope failed. The affected area extended from just below the frontage road to about one-third of the way down the slope. To prevent continued failure, the State Department of Highways and Public Transportation (SDHPT) decreased the gradient of the failed part of the slope from 1:4 to 1:6 or 25% slope to 16.6% slope (note that in highway engineering terms the slopes are 4:1 to 6:1) and constructed a berm across the toe of the slide at a cost of about $15,000. These remedial measures were effective until late 1968 when slope failure occurred on both the east and west backslopes. The west backslope failure included the berm, and the toe extended farther downslope. At this time, SDHPT engineers decided to install slope underdrains parallel to the roadway with cutoff drains connecting and emptying into the drainage ditch below. The underdrains were completed in 1970, but in late 1973 and early 1974, further sliding occurred in both slopes after heavy rains (Fig. 3). This time the head of the west slide included part of the frontage road with the toe extending onto the southbound exit ramp. The east slide also affected a larger area than had previous sliding, but no roadways were blocked or damaged.

After the 1973 failures, the SDHPT implemented a program of investigation which consisted of the drilling and logging of 37 20-ft-deep auger holes in both backslopes. They decided to install new underdrains and retaining walls, to reshape the slopes, and to destroy the berm on the west backslope. Beginning in August 1974, a line of 20-ft-deep (6-m) auger holes spaced on 14-ft (4.2-m) centers was drilled in both slopes at the heads of the slides, and 18-in. (46-cm) steel I-beam piles were driven to as much as an additional 20 ft (6 m) in each hole. Another retaining wall consisting of 12-in. (30-cm) I-beam piles was added to the west landslide about midway up the slope in September 1975.

Installation of the underdrains began in the spring of 1975. Movement of the old slides into the underdrain trench plagued the operation, and a steel bulkhead was used to keep

Figure 3. View of the east backslope (a) and west backslope (b) landslides in February 1974.

the excavation open long enough to install the underdrains. The new underdrains, consisting of 6-in. (15-cm) slotted, galvanized steel pipe surrounded by graded filter material, empty near the toes of the slopes to the north and extend into and upslope diagonally beneath the slides. To date, more than $200,000 has been spent or appropriated by the State Department of Highways and Public Transportation for maintenance and remedial measures on the two landslides.

The 1965, 1968, and 1973 slides all occurred shortly after periods of heavy rainfall (Fig. 4). In May 1965 a total of 10.66 in. (27.08 cm) of rain was recorded. The 9.80-in. (24.89-cm) rainfall in April 1966 occurred during recon-

Figure 2. Oblique aerial photo (looking westward) of the landslide sites in the I-45 backslopes north of the State Highway 7 overpass.

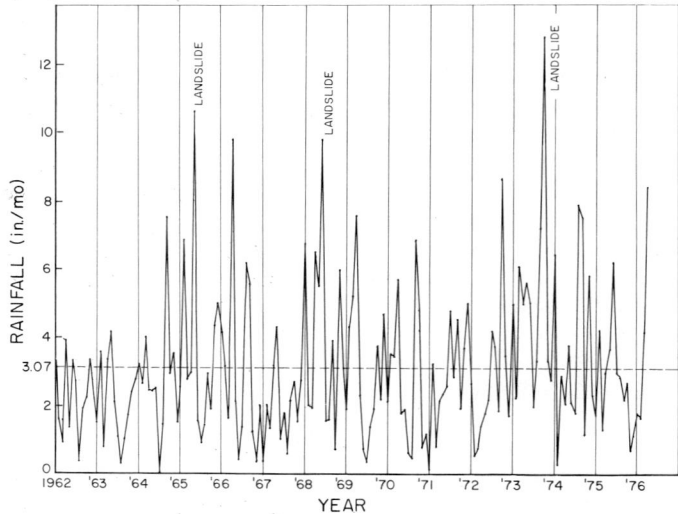

Figure 4. Monthly rainfall and landslide events at Centerville, Texas. Average is 3.07 in./mo (7.80 cm) (National Weather Service, 1960–1976).

struction of the backslope; therefore, any slide movement was probably not noticed. The 1968 landslide events followed 9.81 in. (24.92 cm) of rain. The 1968 reconstruction of the slope plus the underdrains apparently provided sufficient drainage to resist 8.68 in. (22.05 cm) of rain in October 1972. However, in October 1973 a total of 12.82 in. (35.56 cm) of rain fell, and both backslopes failed again. The 1974–75 retaining walls and underdrain system have not yet been tested by the weather.

GEOLOGY OF THE I-45 ROAD-CUT AREA

Before 1929, many contradictions concerning the lower Claiborne Group existed in the geological literature. Wendlandt and Knebel (1929) redefined the old formation names and introduced new ones to clarify these contradictions. On the basis of their definition, the Queen City Formation consists of "local clay zones, zones of sandy clay, beds of almost pure, cross-bedded, light-colored quartz sand, thin beds of lignite, and possibly two local beds of bentonitic clay." The thickness of the Queen City Formation in the east Texas basin ranges from 0 to 480 ft. According to Wendlandt and Knebel (1929), "following the shallow conditions of the sea during Queen City time, the basin subsided sufficiently again to cause conditions favorable for glauconite deposition. The Weches, which overlies the Queen City, is a remarkable deposit of rather pure clayey glauconite whose average thickness is approximately 50 ft throughout the basin proper...." Unweathered beds of the Weches Formation are greenish-blue and contain abundant fossils, but when weathered, the Weches Formation is a deep reddish-brown, and its fossils are usually destroyed. The top few inches to 4 ft (1.2 m) of the formation is a zone of laminated iron ore.

Plummer (1933) interpreted the Queen City Formation to represent a "continental fluviatile deposit laid down by meandering streams and shifting rivers on a flat coastal plain. The strata merge gulfward with shallow water beds that were in part laid down in marshes and bays, in which detritus was abundant, and in part were delta deposits in shallow water." The glauconitic fossiliferous marine beds of the Weches Formation unconformably overlie the Queen City Formation.

In 1938, H. B. Stenzel reported on the geology of Leon County, and among other observations he described numerous outcrops of the Queen City and Weches Formations. Stenzel's lithologic descriptions are consistent with the previous descriptions, and he interpreted the Queen City Formation to be basically nonmarine and the Weches Formation to represent nearshore deposition. Stenzel's thicknesses of the Queen City and Weches Formations are also consistent with those reported by Wendlandt and Knebel (1929) and by Plummer (1933).

The detailed geology of the landslide site is shown on the map (Fig. 5). Hill crests on both sides of the highway cut are capped by ledges of highly fossiliferous, glauconitic sandy clay of the Weches Formation. The part of the Weches underlying these ledges and above the Queen City Formation contact is reflected only as soil profiles.

The geologic section of the east backslope includes about 15 feet (4.6 m) of exposed Weches Formation overlying the Queen City Formation. Thirty-eight feet (11.5 m) of upper Queen City Formation was encountered in core borings in the east backslope (Fig. 6). This part of the formation consists of laminated clay and sand interbedded with thin layers and lenses of fine- to coarse-grained sand. Laterally, sand layers and lenses become interbedded with light gray clay and pinch out. Many of the sand layers are iron stained, and some exhibit high-angle cross-bedding. For the most part, however, the sand bodies show no internal sedimentary structures. The light gray clay, which occurs primarily in the north end of the road cut, shows traces of ripple laminations, slump structures, scour and fill, dessication cracks, occasional calcareous nodules, and animal or plant borings filled with sand. Bioturbation, though, is not common. Downdip, the light gray clay becomes darker gray, and the organic content increases. The dark gray clay is thinly laminated, contains finely disseminated pyrite and pyrite nodules, and exhibits an abundance of leaf imprints on bedding planes.

On the basis of core and auger holes, the sand bodies occur primarily in the north end of the road cut, and they are found higher in the section toward the southeast (downdip). However, individual sand layers as much as 1 ft (30.4 cm) thick are continuous downdip into the dark gray clay. These relationships were also observable in cuts made for the underdrains. The sand bodies are generally clean, uncemented quartz sand of relatively high porosity and permeability, as evidenced by backsapping (slope failure by flowage of the sand) of the sand after trenching. Some feldspar, volcanic glass shards, and muscovite are the most common constituents of the sand other than quartz. Results of grain-size analyses (Fig. 7) reveal the interbedded sand-clay relationships of the Queen City Formation and the change to a coarser fraction near the erosional contact of the overlying Weches Formation. Individual sand bodies within the Queen City Formation grade upward from relatively coarse- to fine-grained sand.

The Queen City Formation in the study area represents nonmarine sediments deposited on a lower coastal plain. Lithologies and sedimentary structures suggest rapidly alternating flow conditions which resulted in the deposition of thin layers of laminated clay interbedded with thin layers and lenses of fine to coarse, loose, well-sorted sand. The light gray clay layers probably represent natural levee deposits lateral to a river channel, with the interbedded sand layers being crevasse-splay deposits. Crevasse-splay deposits are formed when flood waters breach the natural levees and deposit coarse materials on the flood basin. The dark gray, highly organic clay layers probably represent poorly drained marsh or floodplain deposits that were later covered by natural levee clay and sand. A diagrammatic sketch of the geologic environment of the upper Queen City Formation is shown in Figure 8; the proposed depositional environment of the sand and clay units in the landslide area is indicated.

The Queen City Formation is truncated by the overlying Weches Formation, which was encountered in core holes CE-1 and CE-2. This subtle unconformity was also exposed during grading and trenching in the east backslope. The Weches Formation is here distinguished as a highly weathered, fossiliferous, glauconitic sandy clay with occasional limestone beds of marine origin.

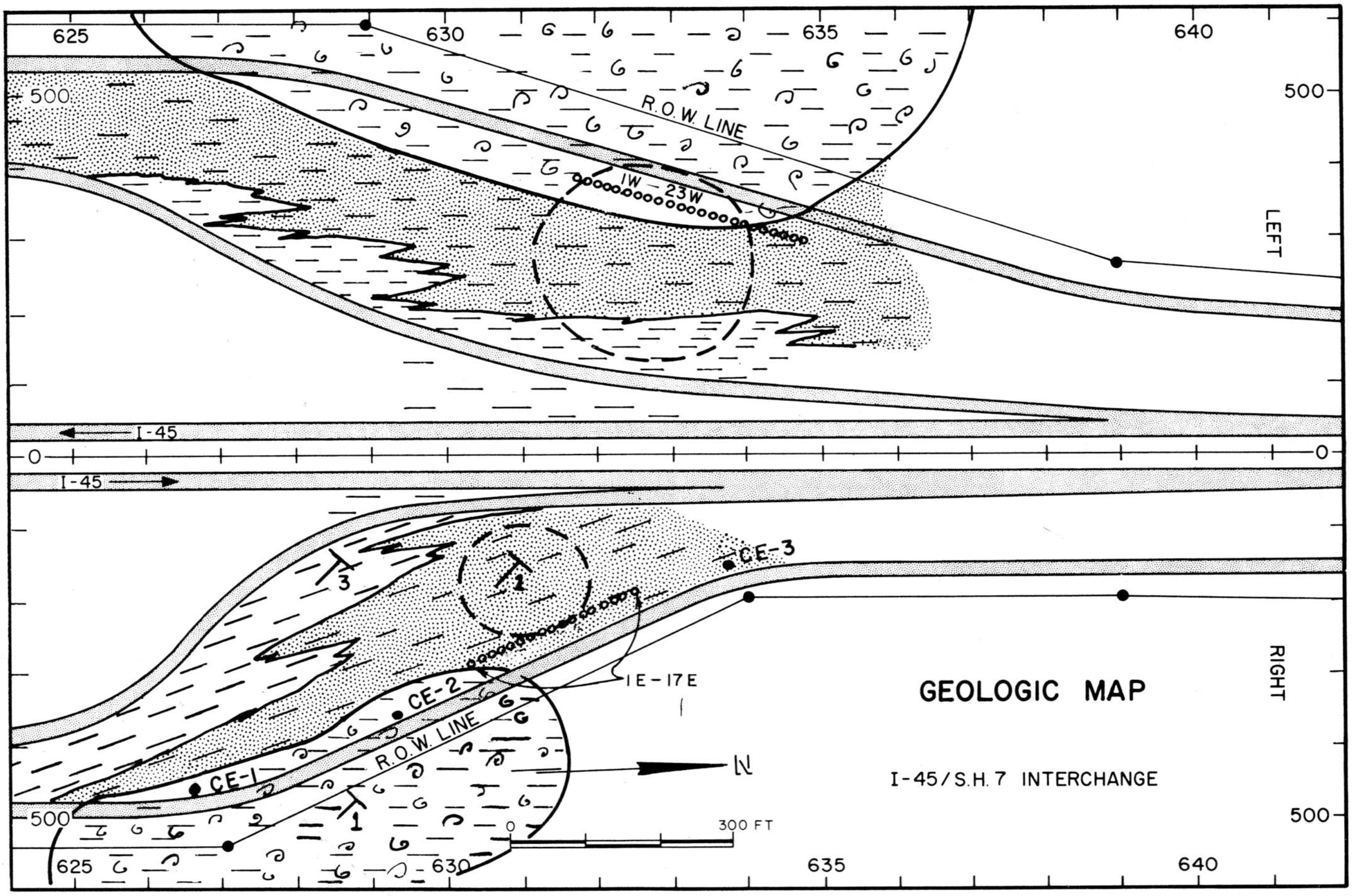

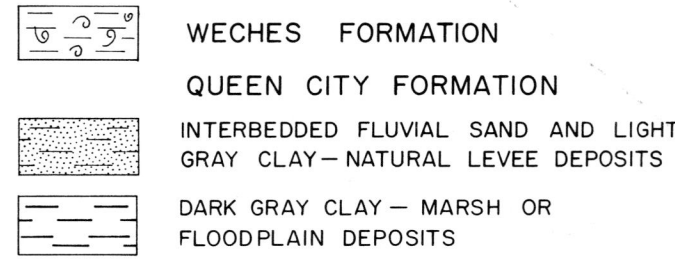

Figure 5. Geologic map of the I-45 and SH-7 interchange. R.O.W. line is right-of-way line. "Left" and "right" are highway department designations used to locate points relative to the highway centerline. Sites of the retaining walls are designated by 1W through 24W and 1E through 17E.

In geologic terms, the landslides occurred within the interbedded levee clay and crevasse-splay sand deposits primarily because of the hydrologic conditions created by the laterally confining, impermeable, floodplain-marsh clay deposits.

ENGINEERING ANALYSIS OF THE LANDSLIDES

Very little information concerning engineering properties of the earth materials had been obtained in the study area prior to this analysis. The SDHPT drilled and partially logged two exploration holes near the overpass and three holes along the Interstate 45 centerline. Samples were taken and tested triaxially for subgrade classification, but no shear-strength tests were performed on undisturbed material from the slopes. Atterberg limits were determined for materials obtained from auger borings along the east frontage road.

The problem is that most of the previous engineering work is generally qualitative or not applicable to the material properties of the slopes. Furthermore, information which was readily available during the site exploration was neglected. Therefore, the engineering section of this study is concerned with identifying the earth materials according to the Unified Soil Classification System and with quantitatively measuring the shear-strength parameters contained in the Mohr-Coulomb effective stress failure theory, $\tau = C' + (\sigma_n - u) \tan \phi'$ (variables explained in App. 1).

Engineering Tests

The soils in the slopes consist of overconsolidated clay, silt, and sand. Atterberg limit determinations on samples obtained from coring show that most of the material in the slopes can be classified as CL (low-plasticity clay), CH (high-plasticity clay), or ML soil (low-plasticity silt). Grain-size analyses of the sand layers show that these can be classified as SP soils (poorly graded sand). To a geologist the sand is well sorted.

During construction of the underdrains, in situ vane shear tests were performed on the slide plane material and on the light gray clay. Vane shear tests are comparable to the unconsolidated-undrained direct shear test, and data obtained by such field testing can be used in a short-term stability analysis. A vane shear strength of 190 psf (0.92 kg/cm²) was measured for the slide plane material, and a strength of 650 psf (3.17 kg/cm²) was measured for the light gray clay.

Six undisturbed samples were prepared for direct shear testing (Lambe, 1951), three from the slide plane material and three from the light gray clay. The purpose of this testing was to determine the peak and the residual shear strengths of the materials.

Samples were first consolidated and then sheared under drained conditions. Figures 9 and 10 show the shear stress versus shear box displacement curves for each sample tested. Peak strengths occur at the peak of each curve after relatively little displacement; with continued movement, the shear strengths are reduced significantly to the residual shear-strength values.

Three undisturbed samples of the slide plane material were tested triaxially using the consolidated-undrained method with pore-pressure measurements. Figure 11 shows the Mohr-Coulomb failure envelope and Mohr's circles of the specimens at failure using total and effective principal stresses. The plot gives a ϕ' value of 16° and C' value of 187 psf (0.91 kg/cm²).

Discussion of Test Results

Atterberg limits indicate the general material properties of the soils, and therefore can be useful in a reconnaissance investigation. The lower few feet of drill hole CE-2 and the upper part of CE-3 penetrate the stratigraphic horizon which contains the slide plane (Fig. 12). In terms of soil classification this stratigraphic horizon is high-plasticity clay (CH). Therefore, one should expect this material to exhibit relatively low shear-strength properties, it should be relatively impermeable, and it should have the capability of volume change under

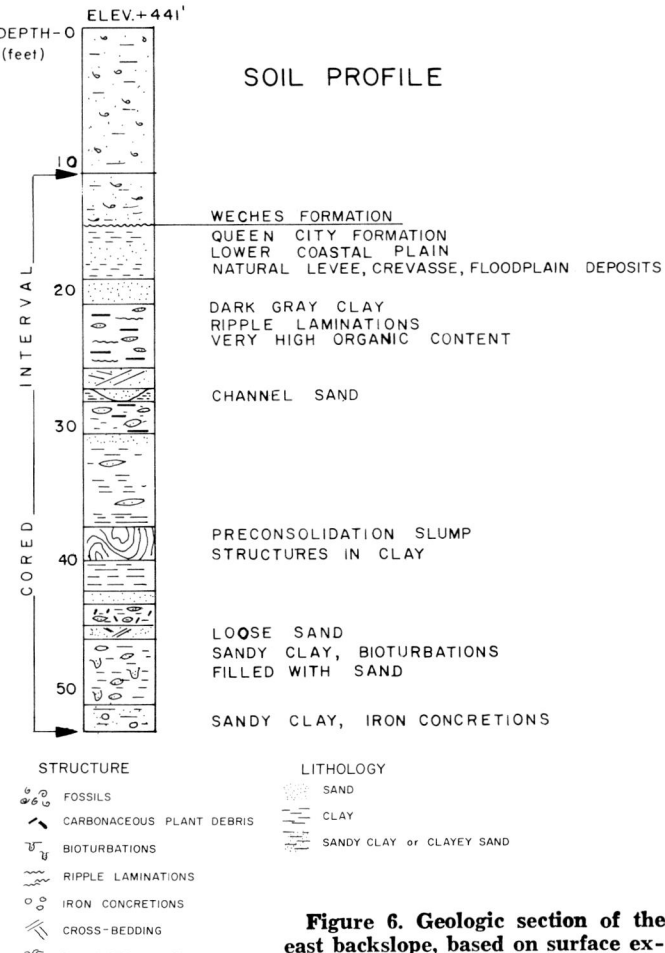

Figure 6. Geologic section of the east backslope, based on surface exposures, trenches, and core holes CE-1, CE-2, and CE-3.

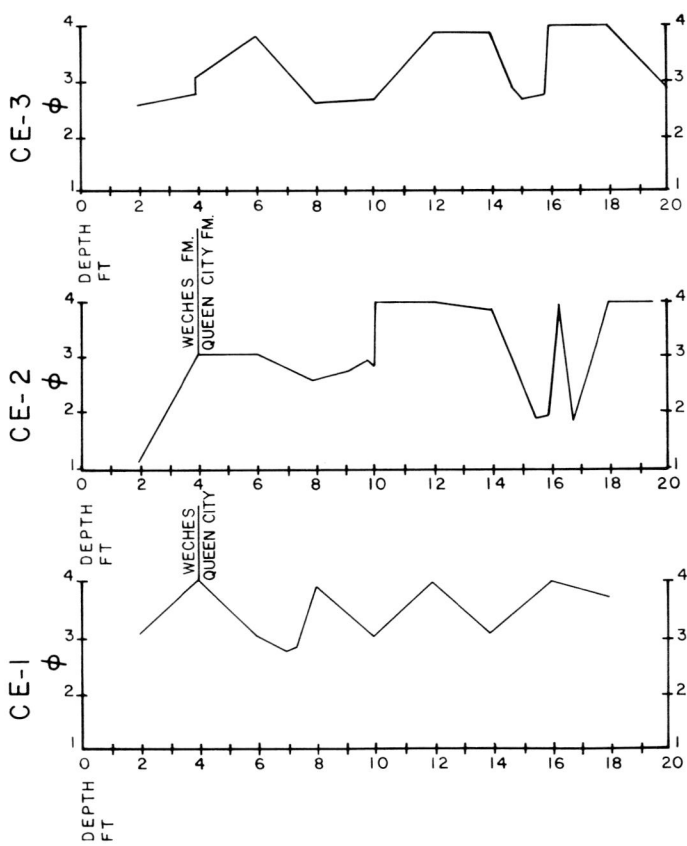

Figure 7. Grain size (ϕ) variation with depth in core holes CE-1, CE-2, and CE-3. Note the Weches–Queen City contact in CE-1 and CE-2.

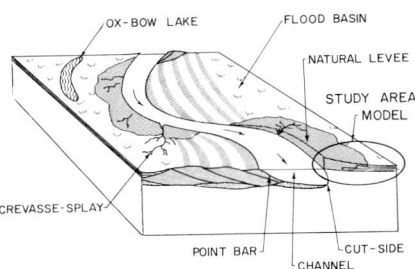

Figure 8. Depositional model of a meandering stream showing the landslide model to be within the natural levee, crevasse-splay, and flood-basin deposits (after Reineck and Singh, 1973).

the right environmental conditions. Water flowing through the sand units can be absorbed by the enclosing clay to reduce its strength further. Fluctuations in the water table may also create high pore-water pressure within the sand lenses. Artesian conditions and perched water tables are known to exist in both backslopes, as noted in the drill-hole logs.

When dealing with a stability analysis in overconsolidated clay, the investigator is faced with deciding which type of shear-strength test to use for design purposes. After a landslide has occurred, the material involved in the slide no longer responds according to its peak strength values but to some lower value called the residual strength. Residual strengths can be characterized by the equation (Skempton, 1964)

$$\tau_R = C'_R + (\sigma_n - u) \tan \phi'_R.$$

Skempton's test results show that C'_R (the effective residual cohesion) is usually very small or zero, and ϕ'_R (the effective residual angle of internal friction) decreases when going from peak to residual strengths. Therefore, residual strength is considerably lower than the peak shear strength of overconsolidated clay, and residual strengths must be used when analyzing a surface that has already failed.

Figure 13 shows the relationship between normal stress (σ_n) and shear stress (τ) determined from direct shear testing of the gray clay. Note that almost all of the cohesion is lost, and the angle of internal friction is reduced by about 5.5° when the gray clay reaches residual strength. The decrease in cohesion and in the angle of internal friction is probably caused by two factors: (1) Increase in water content (due to the expansion of the voids during shear) decreases cohesion. (2) Reorientation of clay particles parallel to the direction of shear decreases the angle of internal friction. Slickensides were observed on the shear surfaces of the specimens, which indicates a reorientation of the clay particles.

In going from peak to residual shear strength in the slide plane material (Fig. 14), the cohesion also goes to nearly zero, but the angle of internal friction does not change. These changes probably occur because the clay particles of the slide plane material were already oriented parallel to the direction of movement of the shear box, and the loss in strength then is due only to a reduction in cohesion.

Field observations revealed that the slide plane occurred within the gray clay. This relationship is also indicated by the equal residual shear strengths of the gray clay and the slide plane material. (Figs. 13 and 14).

The effective stress parameter ϕ' from the triaxial results is 2° less than ϕ_R determined from the direct shear tests, and C' from triaxial testing is greater than C_R of the direct shear tests. Bishop and Bjerrum (1960) found that in heavily overconsolidated clay the consolidated-drained test usually gave high ϕ values, possibly because of the work done by the increase in volume during shear in the drained test and the smaller strain at failure. It is odd, however, that

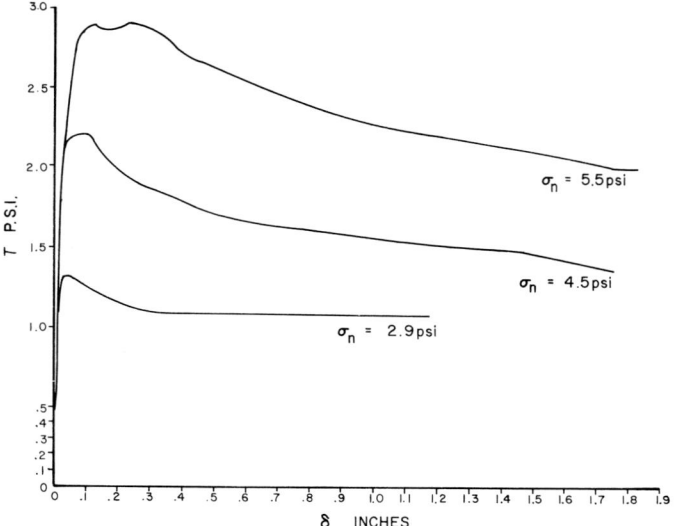

Figure 9. Shear stress (τ) versus shear box displacement (δ) plots of the slide plane material using the consolidated-drained method in a direct shear box.

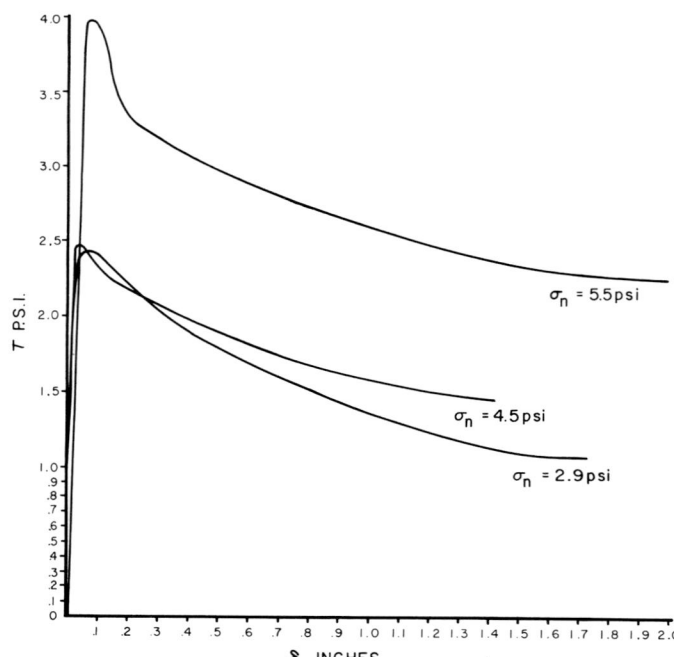

Figure 10. Shear stress (τ) versus shear box displacement (δ) plots of the gray clay using the consolidated-drained method in a direct shear box.

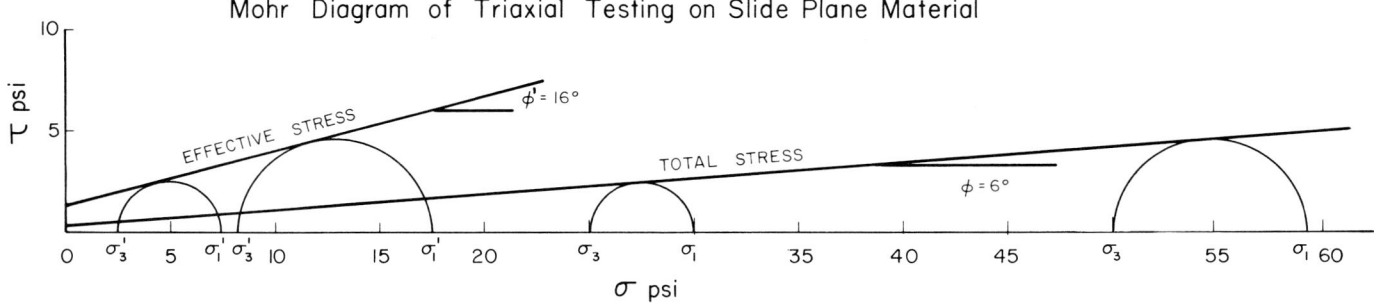

Figure 11. Consolidated-undrained tests on the slide plane material showing total and effective stress envelopes.

ϕ_R, determined on a specimen which failed along a plane parallel to the direction of applied shear stress (direct shear test), is greater than ϕ', determined on a sample which failed along a plane oriented at an angle of $45° - \phi'/2$ to the deviator stress (triaxial test). It seems that the opposite would be true for these two loading geometries.

The final step in understanding the stability problem is the actual slope-stability analysis using the data determined from the laboratory and field testing. As discussed in this section, different types of shear-strength tests give varying results for the shear-strength parameters of a material. Therefore, it is necessary to perform stability analyses using the data of the in situ shear tests, the direct shear tests, and the triaxial tests to determine the mode of slope failure because the shear-strength parameters from these three methods give three different safety factors.

Slope-Stability Analysis

The Bishop (1954) method of slices was applied to the failure surface shown in Figure 15 which represents the simplified section through the failed part of the east back-slope near station 631. The results of the analysis using shear-strength data from the various tests on the gray clay are given in Table 1. Table 2 shows the same results from tests performed on undisturbed samples of the slide plane material.

On the basis of data from Table 1, it is evident that the

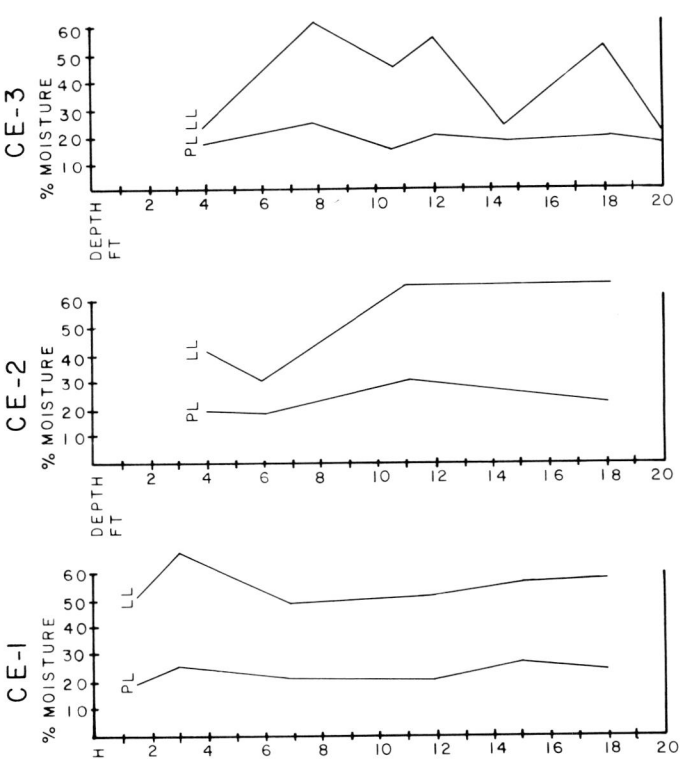

Figure 12. Variation of plastic and liquid limits (PL and LL) with depth in core holes CE-1, CE-2, and CE-3.

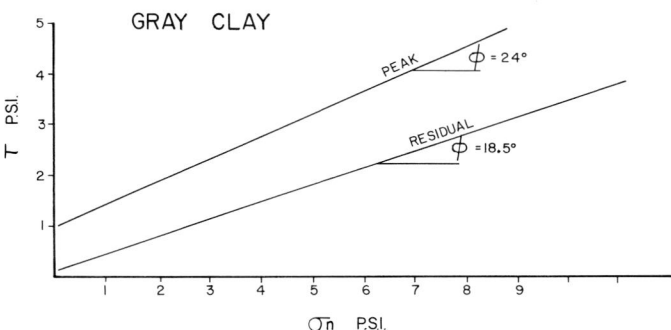

Figure 13. Normal stress versus shear stress plot of the gray clay showing peak and residual strength conditions.

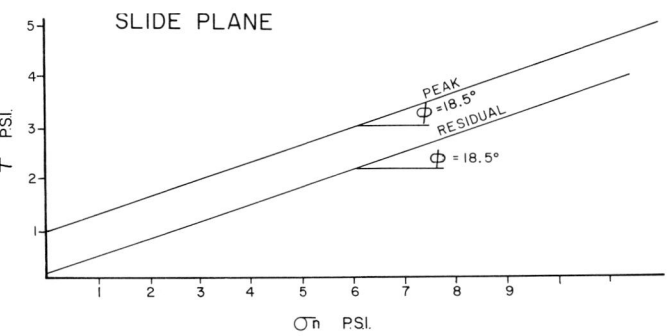

Figure 14. Normal stress versus shear stress plot of the slide plane material showing peak and residual strength conditions.

initial failures occurred either because the pore-water pressure in the slopes was very high or because the material was near its residual strength at the time of failure. If it is assumed that the initial failure occurred when the overconsolidated clay was near its peak strength, high pore-water pressure ($r_u = 0.58$) was necessary to cause failure. This indicates that the initial failure occurred under the undrained (or possibly partially drained) conditions that resulted from excavation and not under conditions of long-term, steady-state seepage. The presence of springs and ponded water before construction, the observation of "weeping" sand in both slopes after construction, and the water levels in drill holes support the pore-water pressure explanation. Thus, the west backslope probably failed as a result of (1) a decrease in the shear strength due to increased pore-water pressure and (2) a decrease in physical support caused by excavation that undercut the slope. Because the east slope failure occurred two years later, it is likely to have resulted from increased pore-water pressure and a concomitant softening of the clay.

On the basis of the r_u values in Table 2, one can see that only small increases in the pore-water pressure ($r_u = 0.1$ to 0.25) in the failed slope are sufficient to reactivate sliding. Continued movement of the slide masses in both slopes has been in response to periods of heavy rainfall which has increased the pore-water pressure in a material at or near its residual shear strength.

ENGINEERING GEOLOGY OF THE CENTERVILLE LANDSLIDES

The absolute solution to the problem of a potential landslide is total avoidance, but avoidance requires recognition of the potential problem. Even so, avoidance may not be the most economic solution. If this is the case, the stability

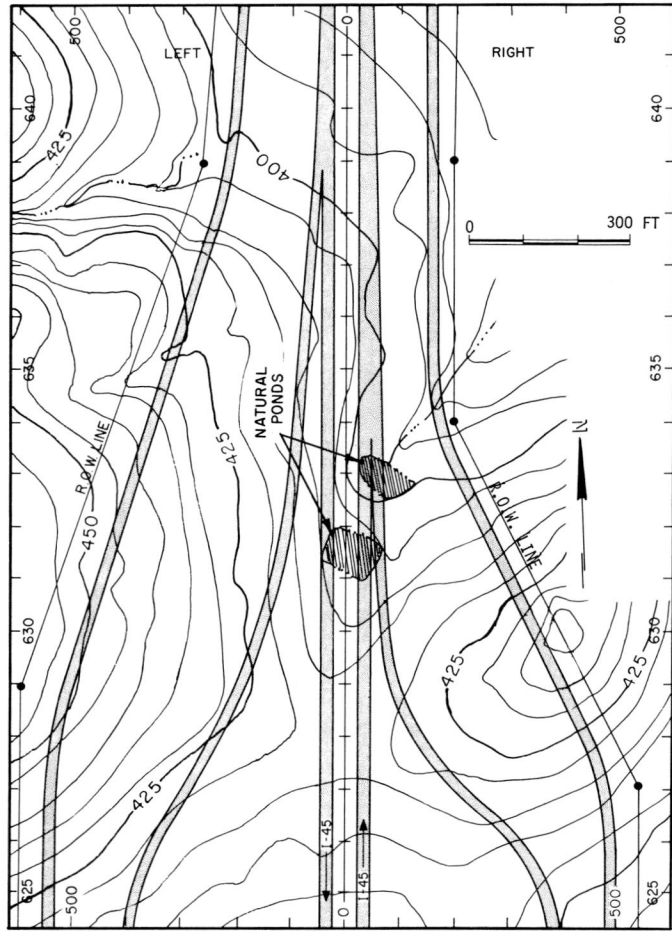

Figure 16. Preconstruction topographic map; note the two ponds and the drainage pattern. (Shaded areas, proposed highways and frontage roads; R.O.W. line, right of way.)

problem should be dealt with during the design phase, but once again, any solution calls for recognition of potentially unstable areas during preconstruction investigations.

In the case of the Centerville landslides, the stability problem was not recognized until after initial slope failure occurred in the west backslope. The slopes had been cut according to the design manual specifications (1:4 or 25%, the desirable maximum backslope), which take into consideration backslope maintenance, automobile safety (in the case of a car's running off the road), and economics along with slope stability. In most instances throughout the state of Texas, the design manual slopes are stable; therefore, it is usually economical to design slopes without performing stability analyses. However, preconstruction topography (Fig. 16) and aerial photographs (Fig. 17) of the study area show ponded water near a ground-water seep in the proposed east backslope. This one bit of information should have led to a thorough examination of the geologic and engineering properties of the materials in the proposed highway cuts.

The stability problem was created during the construction of the backslopes by oversteepening of the slopes combined

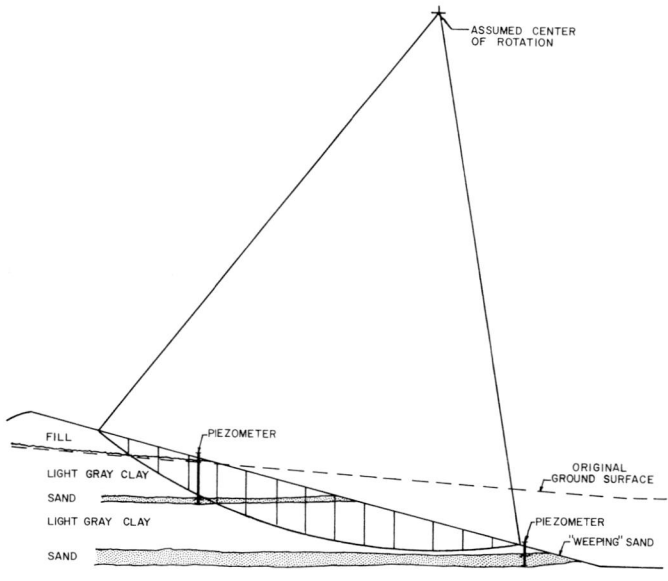

Figure 15. Cross section of the east backslope at station 631 showing the approximate failure surface and slices used in the stability analysis.

Figure 17. Preconstruction aerial photograph of the study area in 1960. The scale is the same as that of the preconstruction topographic map (Fig. 16). The proposed centerline of I-45 is shown (photo by U.S. Department of Agriculture, Soil Conservation Service).

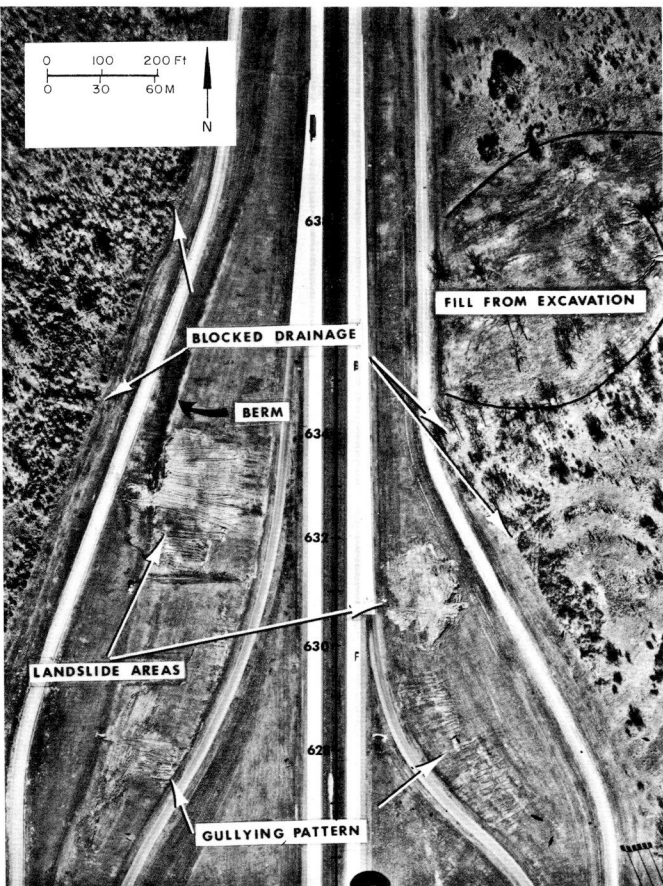

Figure 18. Aerial photograph of the study area in 1970. The landslide scars are from the 1968 failures. Note the berm and the blocked drainage paths. This photo is at the same scale as Figure 16 and Figure 17 (photo by U.S. Department of Agriculture, Soil Conservation Service).

with high pore-water pressure. Furthermore, field investigations of the area revealed areas of blocked drainage and ponded water (Fig. 18). Surface water which should have been directed away from the cut slopes was blocked by highway construction and drained into the ground; this situation contributed to subsequent failures.

Post-facto investigations included in this study show that an understanding of the local geology provides a sufficient basis on which to predict potentially unstable slope conditions. The Queen City Formation is a fluvial deposit in the study area. Data from core holes, auger holes, and trenches indicated that the stratigraphy consists of natural levee clay and crevasse-splay sand layers which grade southward into floodplain-marsh organic clay deposits. The downdip impermeable floodplain deposits help to create perched water tables within the natural levee and crevasse-splay deposits, especially during times of high rainfall. For this reason, the landslides are located opposite each other in the north end of the cut. Thus, the delineation of such stratigraphic relationships is a valuable tool for locating potential problem areas.

Atterberg limit determinations can be very useful for reconnaissance work and would have helped to pinpoint the potential problem once the warning signals on the map and aerial photo had been noticed. In the study area the Atterberg limits indicate the low shear strengths of the materials of the backslopes. According to the Unified Soils Classification, most of the soils in the area can be classified as high-plasticity clay and silt and poorly graded sand (well-sorted sand).

The results of the Bishop (1954) method of stability analysis indicate that the cut slopes would have been stable were it not for the hydrologic conditions created by the stratigraphic relationships. The low-strength, natural levee clay has sufficient strength to stand on a 1:4 slope (25%) with zero pore-water pressure, but water levels within the auger holes are indicative of perched water tables and artesian conditions. Now that failure has already occurred, only a slight increase in pore-water pressure (from $r_u = 0$ to $r_u = 0.1$) leads to unstable conditions.

In the west backslope the strata dip at a low angle toward the cut, a situation which is favorable to a buildup of pore-water pressure within the slope. Thus, failure occurred in the west backslope during construction because of the partial

TABLE 1. SAFETY FACTOR AS A FUNCTION OF THE r_u VALUE AND THE TYPE OF TEST PERFORMED ON THE LIGHT GRAY CLAY

Type of test	C (psf)	C (kg/cm³)	Φ (°)	r_u	F
Direct shear, residual	0	0	18.5	0	1.37
Direct shear, residual	0	0	18.5	0.25	1.0
Direct shear, peak	144	0.70	24.0	0	2.37
Direct shear, peak	144	0.70	24.0	0.58	1.0
Torsional vane shear	650	3.17	0	..	2.15

Note: Variables defined in Appendix 1.

TABLE 2. SAFETY FACTOR AS A FUNCTION OF THE r_u VALUE AND THE TYPE OF TEST PERFORMED ON THE SLIDE PLANE MATERIAL

Type of test	C (psf)	C (kg/cm²)	Φ (°)	r_u	F
Direct shear, residual	0	0	18.5	0	1.37
Direct shear, residual	0	0	18.5	0.25	1.0
Direct shear, peak	144	0.70	18.5	0	1.87
Direct shear, peak	144	0.70	18.5	0.45	1.0
Triaxial, consolidated, undrained	187	0.91	16.0	0	1.16
Triaxial, consolidated, undrained	187	0.91	16.0	0.1	1.02
Torsional vane shear	190	0.92	0	..	0.77

Note: Variables defined in Appendix 1.

removal of the physical support and the increased pore-water pressure. In the east backslope, the strata dip obliquely away from the cut; therefore, pore-water pressure is not likely to be as high as in the west slope. Failure of the east backslope occurred two years later owing to excessive pore-water pressure, caused by heavy rains, and concomitant progressive weakening of the clay.

CONCLUSIONS AND RECOMMENDATIONS

The following conclusions are considered justifiable on the basis of field observations and the analysis of data gathered during the course of this study:

1. The east and west backslopes are composed of the upper Queen City Formation overlain by the Weches Formation as mapped by H. B. Stenzel in 1938.

2. The Queen City Formation in the study area is composed of interbedded clay and sand deposited as natural levee, crevasse-splay, and floodplain-marsh deposits.

3. The occurrence of the landslides is due to the hydrologic conditions created within the natural levee clay by the laterally confining, impermeable floodplain clay.

4. In the west backslope, failure occurred during construction under undrained conditions upon excavation of the cut.

5. Failure of the east backslope occurred owing to excessive pore-water pressure and progressive weakening of the clay.

6. Continued slope failure has occurred because residual shear strengths are lower than initial ones and increases in pore-water pressure due to high rainfall.

The following recommendations are considered justifiable on the basis of this study:

1. Any department of highways and public transportation should require the presence of an engineering geologist during all coring or augering operations to supervise and log all drill holes. In the Centerville case, much information was lost during preconstruction operations because of incomplete logging.

2. A detailed stability analysis should be performed on suspicious-looking areas prior to construction to prevent economic losses due to landsliding. After all, prevention is the best solution. It is evident that in this case, the significance of the rock and soil types and their orientations, the natural drainage, and the natural springs and ponds was not realized before highway construction.

3. Maintenance vehicles on the backslopes contribute to the slope erosion and create ruts which collect water. Therefore, the slopes should be planted with phreatophytes such as cottonwood trees and low shrubs or vines rather than with grass. This will act to minimize maintenance traffic on the slopes, it will help to keep the slopes dry, and the deep roots may increase the stability of the slopes.

ACKNOWLEDGMENTS

This paper is based on the Master of Science thesis by James H. Clary. We wish to thank Robert E. Long and the Texas State Department of Highways and Public Transportation for the drill rig, crews, and equipment; Bob Bigham and Soil Mechanics, Inc., for the use of sampling equipment; Robert R. Berg for his review of the stratigraphic interpretation; and Wayne A. Dunlap for his review of the stability analyses. We also thank Paul Hilpman and Donald R. Coates for their review comments of this manuscript.

APPENDIX 1. LIST OF SYMBOLS

C = cohesion, general case (psf).
C_R = residual cohesion (psf).
C' = effective cohesion (psf).
C'_R = effective residual cohesion (psf).
F = safety factor, ratio of resisting forces to driving forces.
δ = displacement of direct shear box (in.).
σ_n = normal stress (psf).
τ = shear stress (psf).
τ_R = residual shear stress (psf).
ϕ = angle of internal friction, general case (°).
ϕ' = effective angle of internal friction (°).
ϕ_R = residual angle of internal friction (°).
ϕ'_R = effective residual angle of internal friction (°).
u = pore-water pressure (psf).
r_u = pore-water pressure parameter = $u/\gamma h$.
γ = unit weight of the soil material (lb/ft³).
h = height of soil column (ft).

REFERENCES CITED

Bishop, A. W., 1954, The use of the slip circle in the stability analysis of slopes: Geotechnique, v. 5, p. 7–17.

Bishop, A. W., and Bjerrum, L., 1960, The relevance of the triaxial test to the solution of stability problems: Oslo, Norwegian Geotech. Inst., no. 34, 56 p.

Lambe, T. W., ed., 1951, Soil testing for engineers: New York, John Wiley & Sons, 165 p.

National Weather Service, 1960–1976, Climatological data for Texas: U.S. Dept. Commerce, Natl. Oceanic and Atmos. Adm.

Plummer, F. B., 1933, Cenozoic systems in Texas: Geology of Texas: Univ. Texas at Austin, Bur. Econ. Geology, no. 3232, v. 1, p. 517–818.

Reineck, H. E., and Singh, I. B., 1973, Depositional sedimentary environments with reference to terrigenous clastics: New York, Springer-Verlag, 437 p.

Skempton, A. W., 1964, Long-term stability of clay slopes: Geotechnique, v. 14, p. 81–102.

Stenzel, H. B., 1938, The geology of Leon County, Texas: Univ. Texas at Austin, Bur. Econ. Geology, no. 3818, 294 p.

Wendlandt, E. A., and Knebel, G. M., 1929, Lower Claiborne of east Texas with special reference to Mt. Sylvan dome and salt movements: Am. Assoc. Petroleum Geologists, v. 13, p. 1347–1375.

MANUSCRIPT RECEIVED BY THE SOCIETY SEPTEMBER 7, 1976
MANUSCRIPT ACCEPTED SEPTEMBER 17, 1976

Printed in U.S.A.

17
Three major California freeway landslide areas

F. BEACH LEIGHTON
Leighton & Associates, 17975 Sky Park Circle, Irvine, California 92714

ABSTRACT

Three geologic case histories of freeway landslides in California offer significant guideposts in the sound economic planning of freeways. The three case histories are (1) San Diego Freeway, Bel Air, Los Angeles (Interstate 405; cut-slope and ancestral block glides); (2) Pomona Freeway (State Highway 60; cut slope block glides in areas previously unslid); and (3) Golden State Freeway, West Fork, Liebre Gulch (Interstate 5; fill slope in area of previous sliding).

The San Diego Freeway along Sepulveda Canyon in the Bel Air area of Los Angeles has been beset with landslide problems in the Santa Monica Formation (Triassic-Jurassic) since its construction in 1961–62. Because this area is predominantly slate, proper positioning of this stretch of the freeway could have eliminated much remedial grading and redesign.

A 1.6-km stretch of Pomona Freeway west of its junction with State Highway 71 has been subject to landsliding of the Puente Formation (upper Miocene) since construction. Freeway cuts had to be flattened appreciably and new rights-of-way established.

The Liebre Gulch landslide along Interstate 5 occurred when road fill was placed at the head of an undetected ancestral landslide.

These three case histories not only underline that the public right-of-way for freeways has not been given adequate geologic attention, but that it represents one of the most neglected opportunities in the application of geology to highway selection, design, construction, and maintenance.

INTRODUCTION

Geologic factors have substantially affected the construction and maintenance of freeways in California. Freeway problems of a geologic nature are principally matters of stability, drainage, and difficulty of excavation. Although the preventive landslide measures are generally known, detailed geology has commonly awaited the grading stage when exposures are more abundant and complete. The emphasis in this paper is on the need to assess to a greater extent the geologic stability of highway location and design in advance of grading operations.

The case histories and types of landslides dealt with are categorized as follows: cut-slope glide (previously unslid area) — Pomona Freeway, Pomona, and San Diego Freeway, Bel Air; Cut-slope block glide (area of old landslide) — San Diego Freeway, Bel Air; and fill slope (area of old landslide) — West Fork, Liebre Gulch, Los Angeles County.

In each landslide case cited, investigations were undertaken following landsliding to determine the origin of the landslide and events leading to the movement. Other landslides in the same areas testified to similar conditions of landsliding and the lack of predesign investigations to take these costly landslides into consideration.

SAN DIEGO FREEWAY ALONG SEPULVEDA CANYON IN THE BEL AIR AREA, LOS ANGELES

The San Diego Freeway, along Sepulveda Canyon between Mulholland Drive and Sunset Boulevard, traverses an extensively deformed section of the Santa Monica Mountains (Fig. 1). Mass movements in freeway cuts began as soon as the freeway was excavated before 1963.

The Santa Monica Mountains in this area represent a broad anticline that has been severely faulted and profusely intruded by irregularly shaped thin and elongated masses of diorite to quartz diorite (Hoots, 1931; Metropolitan Water District of Southern California, 1966–1967). Inasmuch as the gentle plunge of the Santa Monica anticline is westerly, the oldest bedrock unit of the mountains, the Santa Monica Formation, has been extensively exposed by relative uplift and erosion in the vicinity of Sepulveda Canyon. Here in the core of the anticline, slopes along the freeway expose this formation — a thick section of clay- and silt-rich sedimentary

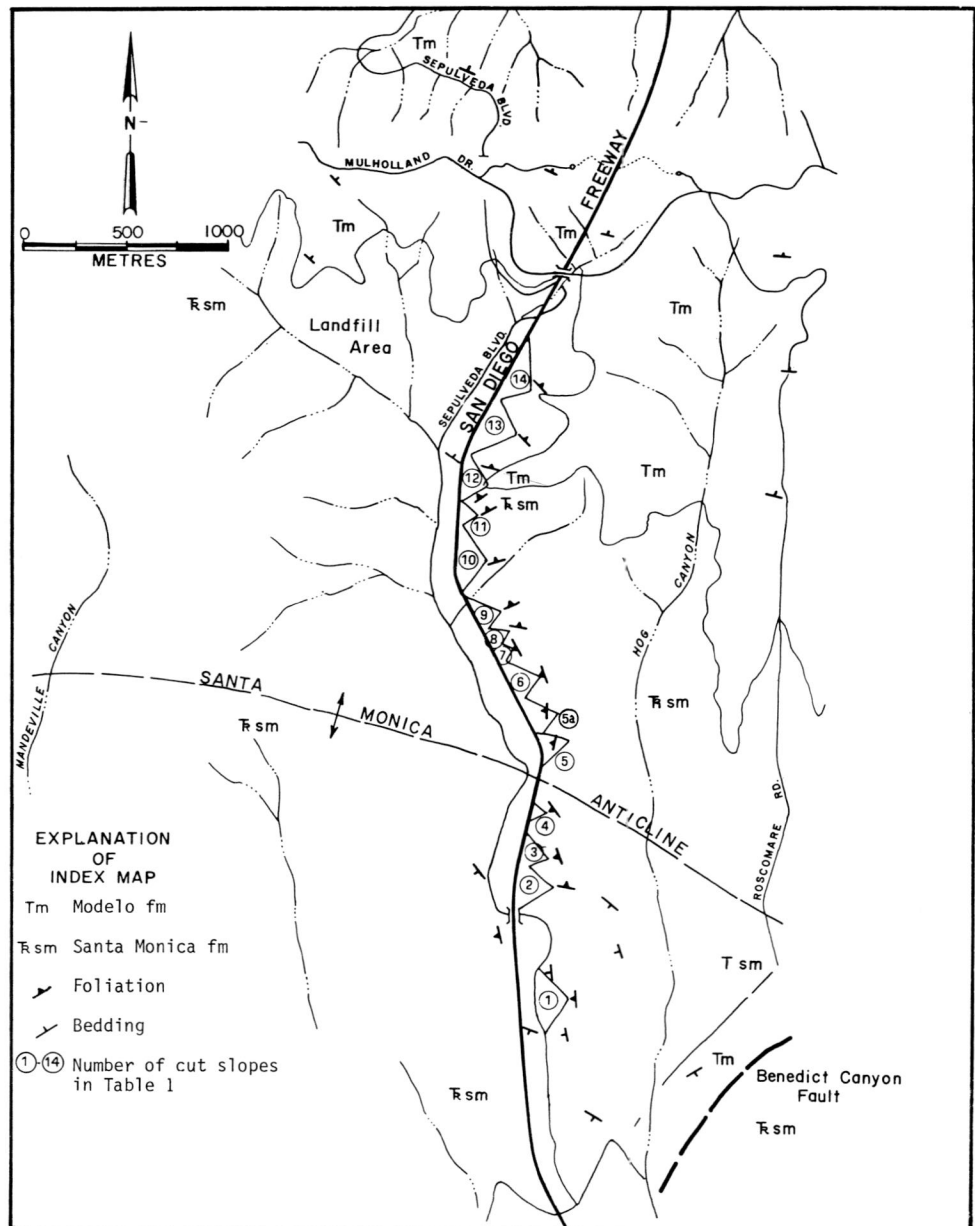

Figure 1. Index map of the San Diego Freeway along Sepulveda Canyon in the Bel Air area of Los Angeles. The 15 numbered cut slopes (including 5a) were mapped at 1:240 to assess slope stability for the Right-of-Way Department, California State Division of Highways.

rocks that has been mildly metamorphosed to chiefly slate, metagraywackes, and metasandstones. All the slides considered in this paper were developed in the Santa Monica Formation, although some have developed both to the south and north in the overlying Modelo Formation.

The chief structural features that control both previous sliding and potential sliding are foliation, fracture cleavage, joints, faults, shear zones, and folds. All of these structures were mapped on a scale of 1:240 in the 15 cuts shown in Figure 1.

Those structures most important in determining the geologic stability of sizeable cut slopes in this area are foliation and fracture cleavage. When wet, persistent foliation and fracture-cleavage surfaces with adverse orientations become slippery and slopes fail. Areas of unstable bedrock along the freeway are those in which the planar structures are at a flatter dip angle than the cut angle and those in which a large amount of overburden or active or old landslide debris are present. It was possible to distinguish three types of slope failure: (1) natural slope failures (chiefly block glides), many of which are ancient; (2) artificially induced slope failures; and (3) potential slope failures.

Table 1 differentiates those cuts in which old slope failures, cut-slope failures, and potential slope failures were mapped and indicates the angle of excavation of each cut. Figure 2, a geologic map of cut slope 1 that was flattened from 1:1 to 1:1.5 (vertical:horizontal), illustrates one of the cut slopes that evidenced all three types of slope failure. Geologic data were plotted on an essentially vertical plane intersecting the toe of each cut; thus benches and terraces are represented by lines, not bands. On the basis of geologic conditions observed, a recommendation was made that building setbacks upslope be controlled by projecting a 1:3 imaginary plane (approximately 18° from the horizontal) from the toe of existing unstable cuts.

A marked contrast exists in the stability of the east and west slopes of Sepulveda Canyon. Active and old landslides are concentrated on the east sides of the canyon where unfavorably oriented structural features are prominent. Had the freeway been placed more toward the west side, slopes on the east side could have been reduced in height and

TABLE 1. MAJOR CUT SLOPES ALONG PART OF SAN DIEGO FREEWAY, SEPULVEDA CANYON

Cut No.	Angle of excavation (vertical: horizontal)	Old slope failures	Cut slope failures	Potential slope failures
1	1:1.25–1:1.5	x	x	x
2	1:1.5			
3	1:1			x
4	1:1			
5	1:1.5	x	x	x
5a	1:1.5			x
6	1:1.5		x	x
7	1:1		x	x
8	1:1			
9	1:1.5			x
10	1:1.5			
11	1:1			
12	1:1.5			x
13	1:1.5			x
14	1:1.5			

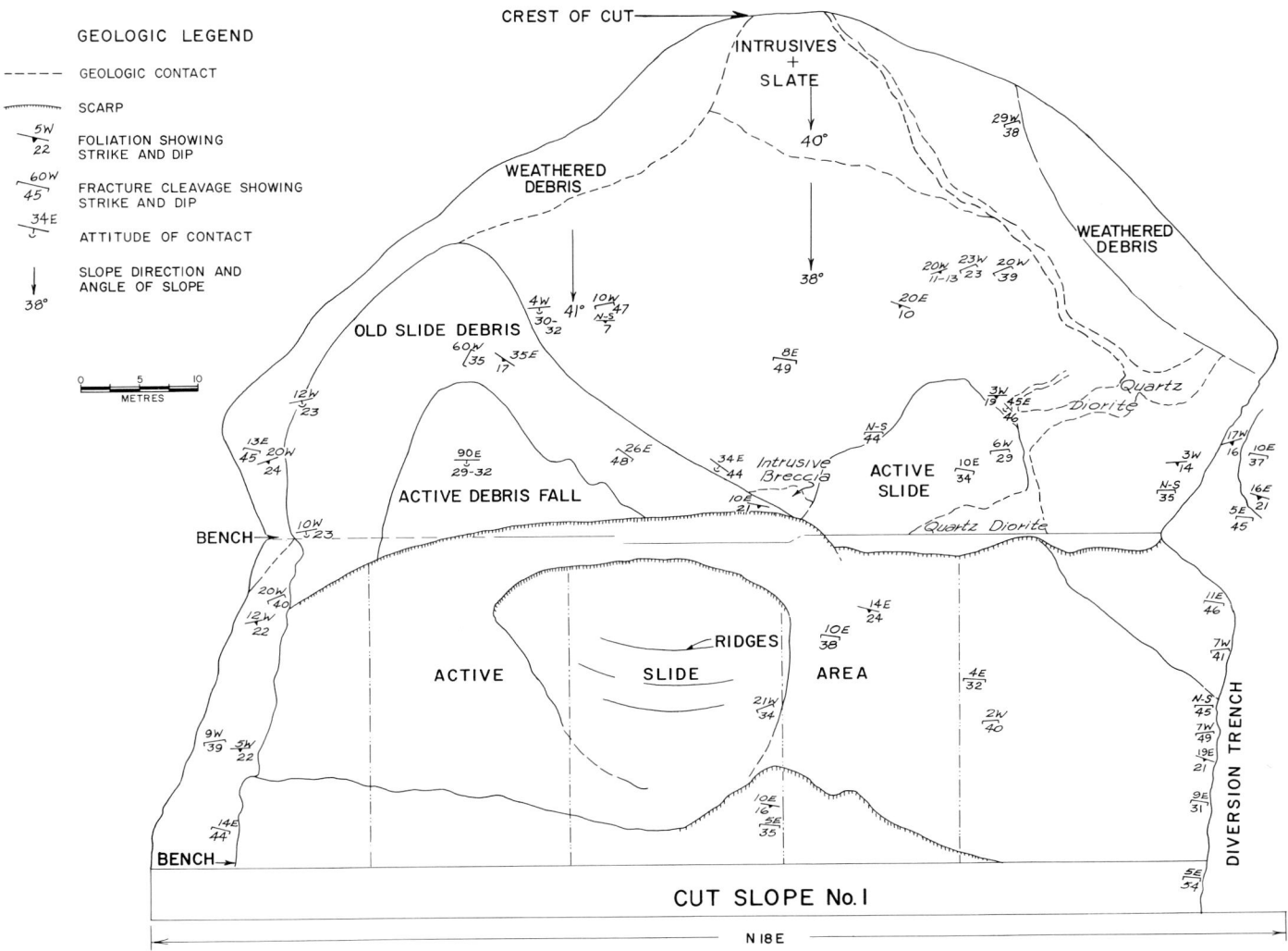

Figure 2. Geologic map of cut slope 1 (Fig. 1, index map). The 14 other cut slopes were mapped.

steepened in slope angle, saving much time, remedial grading, and redesign.

POMONA FREEWAY LANDSLIDES, POMONA AREA, COUNTY OF LOS ANGELES

The history of landsliding along the Pomona Freeway extends back to its creation in 1966-67 when a major landslide, occurring first in 1967 and again in 1969, blocked the freeway for hours and pushed a car broadside through the chain-link center divider. Fortunately no one was injured, but major traffic snarls ensued. This case history is already documented (Leighton, 1969).

Fifteen landslides and numerous popouts occurred between 1970 and 1971 along another east-west stretch of the Pomona Freeway located within Phillips Ranch near Pomona, California. The nine cut slopes along this part of the freeway extend about 1.6 km westward of the junction with State Highway 71, as shown in Figure 3. As soon as the cut slopes were excavated as designed at 1:1.5 and 1:2, six failed (cut slopes 3, 5, 6, 7, 8, and 9) because of unsupported bedding surfaces in a folded and broken siltstone-shale member of the Puente Formation. These slides ranged from popouts less than 3 m in length to bedrock slides over 60 m in length and 4.5 to 6 m in maximum thickness. Figure 4 is a cross section through cut slope 7. Most slides and potential failures were on the north-facing side of the freeway because of the dip-slope conditions on that side; all were the result of freeway excavation in bedrock with no evidence of previous sliding.

Original state right-of-way lines differed appreciably from those shown to be geologically necessary. For remedial purposes, the California State Division of Highways had recommended that the slopes be trimmed to 1:2 on most cut slopes, but studies for the developer show that the redesign, in most cases, should range between 1:3, 1:4, and 1:5, as shown in Figure 5. Geologic mapping of the cut slopes indicated that safe future foundation setbacks or trimlines would extend as much as 75 m beyond the designed freeway right-of-way. Cut slope 7 at 1:2 had to be redesigned and regraded at 1:4 to remove the slide debris.

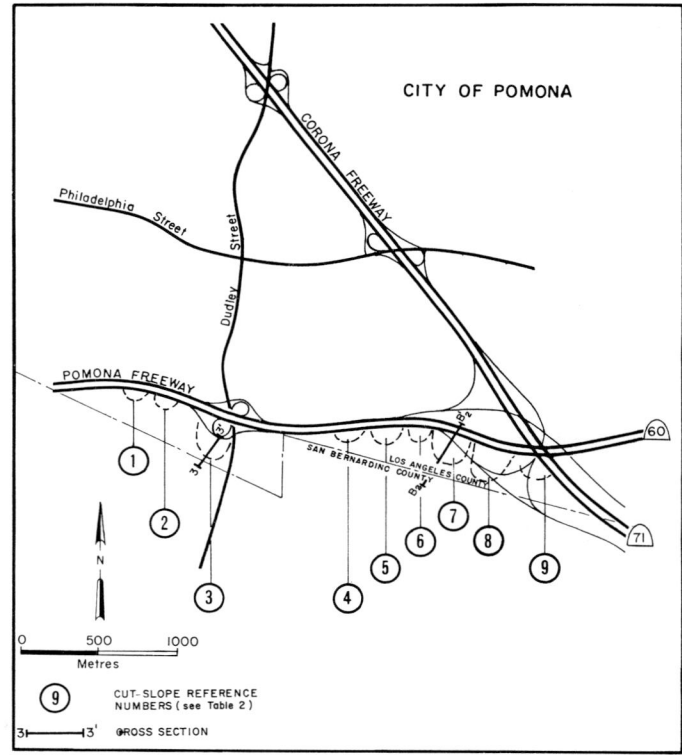

Figure 3. Index map of nine north-facing cut slopes along the Pomona Freeway subject to landsliding and potential failure.

Detailed geologic evaluation of the principal seven major-problem cut slopes is shown in Table 2.

Landslides along the freeway had been anticipated from recent geologic mapping, including subsurface exploration, on both sides of the freeway. This information was available to the California State Division of Highways, but the freeway cuts had been designed on the basis of a much earlier geologic

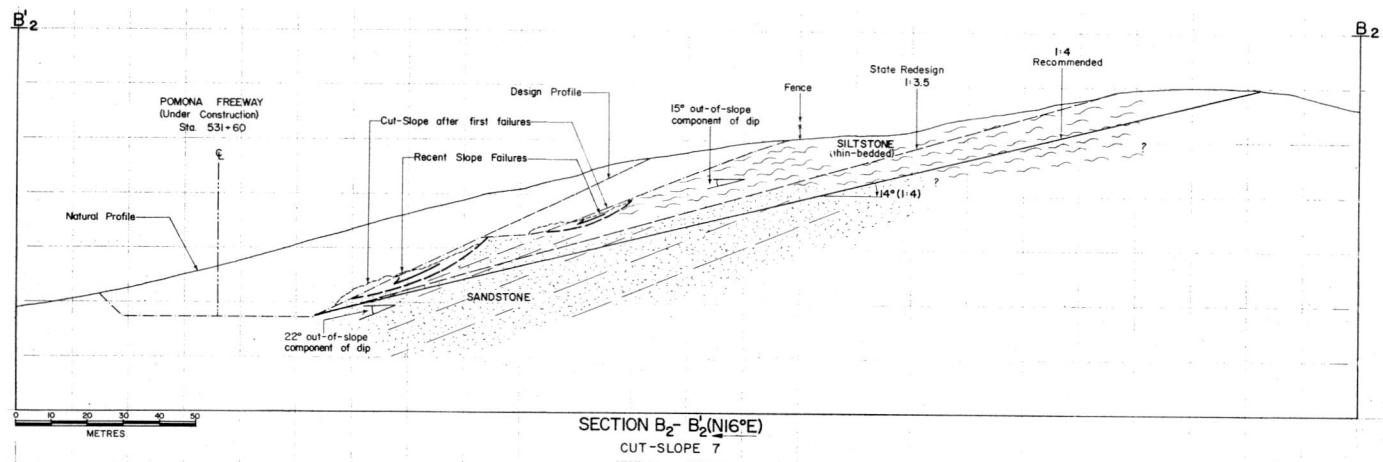

Figure 4. Section through cut slope 7 of Pomona Freeway cut showing its relationship to freeway construction and geologic structure (Fig. 3).

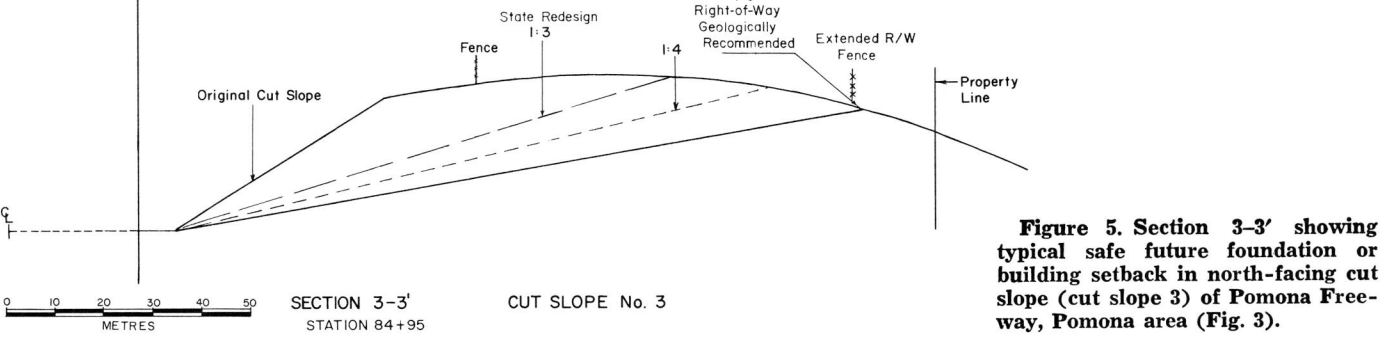

Figure 5. Section 3–3' showing typical safe future foundation or building setback in north-facing cut slope (cut slope 3) of Pomona Freeway, Pomona area (Fig. 3).

examination conducted by the division and there was no review of the recent geologic mapping. Had the hazardous cut slopes been buttressed at a slope angle of 1:1.5 or 1:2, rather than flattened to a safe angle (1:3–1:5), much valuable property could have been preserved.

During the period of debate regarding the extent of slope flattening necessary, one landslide increased in size nearly threefold, half blocking a freeway ramp. In order to remove the slide debris, the slope was trimmed to 1:4, underlining the need for a 1:5 foundation setback. Opening of the freeway was delayed three months.

WEST FORK, LIEBRE GULCH LANDSLIDE ON INTERSTATE ROUTE 5 (GOLDEN STATE FREEWAY)

A major landslide occurred on April 12, 1967, on the east side of the West Fork, Liebre Gulch (Fig. 6). This landslide involved failure of a fill embankment on Interstate 5 and broke pipelines in older landslide deposits. Figure 7 shows geologic relationships and Table 3 presents the chief parameters of the slide.

Grading of this section of the freeway, which was still

TABLE 2. GEOLOGIC PARAMETERS OF MAJOR–PROBLEM CUT SLOPES

Pomona Freeway, Pomona area (stations 457+00 to 545+00)				
Location by station no.	Designed cut-slope angle	Approximate height of cut (in metres) and direction of cut faces	Geologic conditions	Geologic recommendations
457–462	1:1.5	26± north	Out-of-slope 6°–20° in siltstone	Will require, in west half, 1:3 cut or stability equivalent; 1:5, or stability equivalent in east half of cut; if unbuttressed, safe residential setback will be approximately 75 m beyond present chain-link fence in 1:5 portion, 20–25 m in 1:3 part
469–478	1:1.5	30± northeast	Anti-dip slope at northwest end of cut. Adverse, 5°–13°, center and southeast end of cut. Massive sandstone	Will require 1:3 to 1:4 cut or stability equivalent; if unbuttressed, safe residential setback will be 30–68 m beyond link fence
Off-ramp A 483–486	1:1.5	26± north	Station 485–486+ is anti-dip slope. Most of cut (483–485) has 2°–12° out-of-slope with some sharp folds with out-of-slope plunge	Will require 1:5 cut or stability equivalent; if unbuttressed, will require residential setback of 15–68 m beyond present link fence
503–508.5	1:1.5	29± northwest	Out-of-slope 6°–17°. Folded, faulted sandstone and siltstone	Will require 1:3 to 1:5 cut or stability equivalent; if unbuttressed, safe residential setback will be approximately 38–68 m beyond link fence
510–513	1:1.5	18± northwest	Out-of-slope 6°±. Faulted sandstone and siltstone	Will require 1:3 to 1:5 cut or stability equivalent; if unbuttressed, safe residential setback will be 8–30 m from fence
526.5–535	1:1.5	44± north-northeast	Out-of-slope 18°–28° in siltstone	Will require 1:3 cut or stability equivalent; if unbuttressed, safe residential setback will be 68 m beyond fence
535–545	1:2	17± north-northeast	Dips range from in-to-slope to 32° out-of-slope. Folded and broken siltstone with out-of-slope dips, 10°–30°	Will require 1:3 to 1:5 cut or stability equivalent; if unbuttressed, safe residential setback will be approximately 10–38 m beyond present top of slope

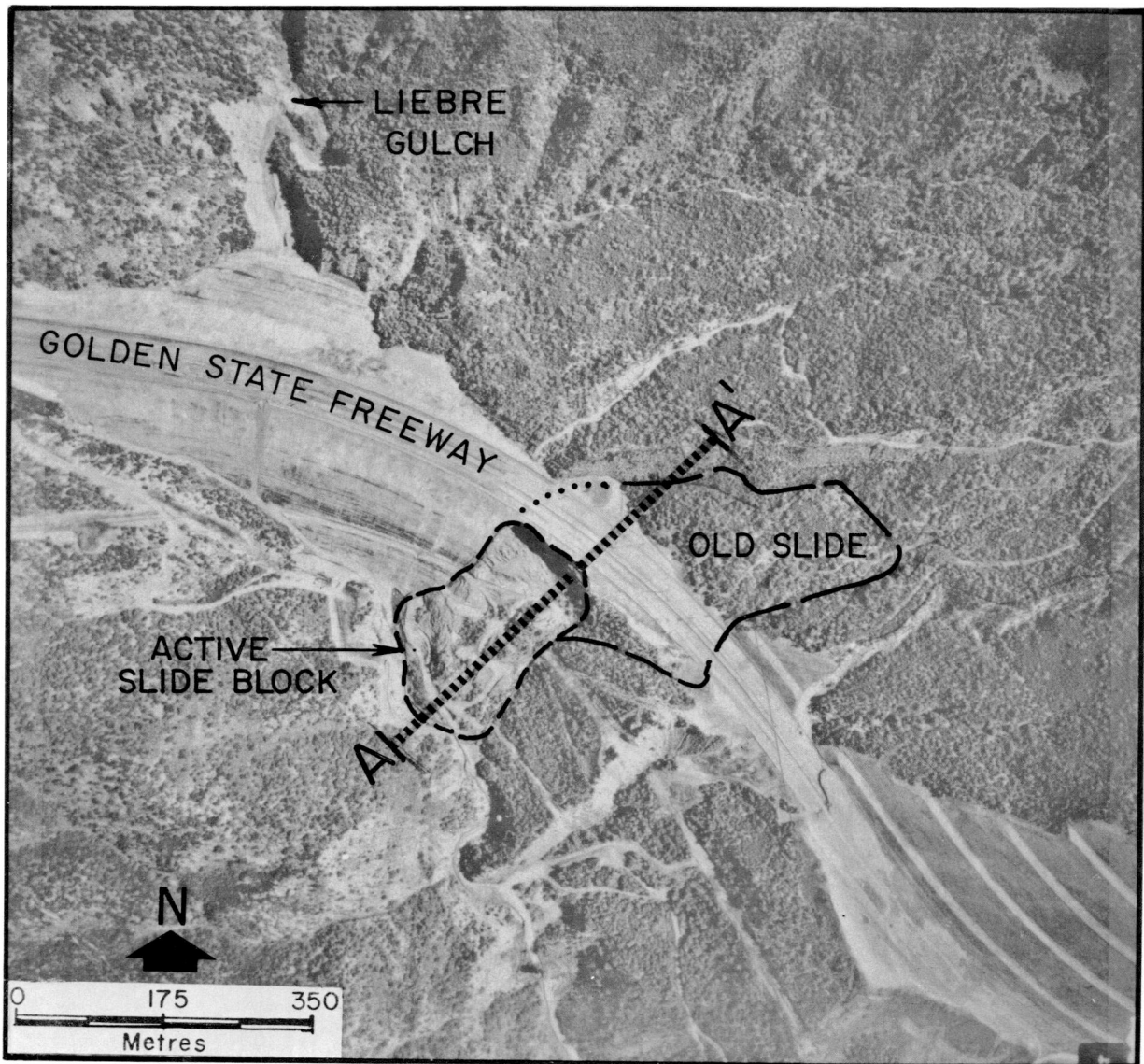

Figure 6. Vertical aerial photograph of Liebre Gulch slide, Golden State Freeway, which occurred April 12, 1967.

unopened at the time of sliding, was completed in March 1967. The only other man-made development in the slide area involved installation of three oil pipelines. These lines were buried about one metre deep except at drainage crossings where they were exposed. All lines were broken and bent by the slide, releasing some oil that appeared at one stage to endanger a nearby lake and reservoir.

The factors controlling the slide were as follows:

1. Previous Slide History. Boring logs, geologic mapping, and examination of preslide (1967) aerial photographs incicated the presence of deposits from one or more older landslides beneath and adjacent to the slide. This old slide material slakes rapidly in water, is extremely fragmented, and tends to behave as a mush of wet concrete when moving on the nearly planar bedrock surfaces below.

2. Type of Bedrock. Highly fractured and well-bedded siltstone and shale.

3. Attitude of the Bedding. Dip averages 20° to 30° in the same direction as the slope of the land (essentially a dip slope).

4. Steepness of the Slope. Ground slope was at the same angle as the dip of the beds; undercutting by running water in West Fork Liebre Gulch produced oversteepening at the base of the potential slide area and probably contributed to reactivation of the prehistoric slide.

5. Fault-Zone Alignment. A fault zone, trending about 65°

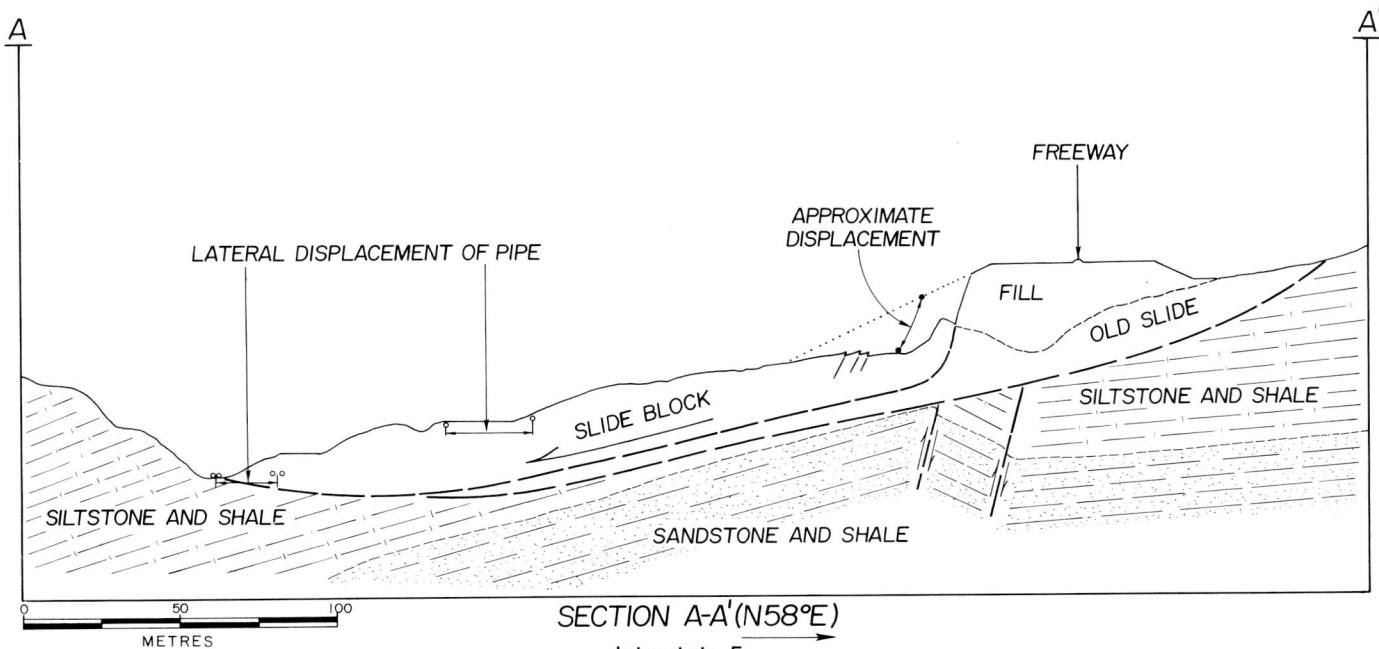

Figure 7. Section A–A' of Liebre Gulch slide, Golden State Freeway (Fig. 6).

to the strike of the regional bedding, dips about 80° in the same direction as the bedding and lies directly beneath the head slide scarp. Bearing, position, and inclination of the westernmost fault plane were identical to those of the slide scarp.

6. Addition of Water. The rains of early April 1967 served as a triggering device for the slide. Precipitation measured at station 409 on West Fork Liebre Gulch for that period was April 1, 2.16 cm; April 4, 3.05 cm; April 7, 0.76 cm; and April 12, 1.78 cm.

7. Fill Surcharge at Head of Slide. The placement of fill at the head of the potential slide block acted as a driving force as well as helping to define the position of the head slide scarp. The freeway fill north of the former slide probably acted as a bulwark, directing the slide to the southwest rather than to the west and northwest in the direction of bedding dip and that of older slides.

CONCLUSIONS

Three phases of preconstruction geological studies that lie at the heart of terrain investigations for freeways (and other important transportation arteries) and that should be incorporated into route selection, highway design, construction estimates, and contracts are:

1. Regional reconnaissance, which would accomplish the rating of multiple alignments geologically, on the basis of photogeology and review of geologic maps of area.

2. Detailed mapping and subsurface exploration, which would allow for design modifications, identification of local problems, and recommendations for modifications in road design and rights-of-way.

3. Review of construction plans and subsequent design revisions, which would describe and compute quantities of unsuitable earth materials for removal, blasting yardage, and on-site construction materials; geological design of remedial measures would be presented.

The routing and design of all three slide areas could have benefited from more intensive application of geology in each

TABLE 3. CHARACTERISTICS OF WEST FORK, LIEBRE CANYON LANDSLIDE

Time of slide	April 12, 1967
Classification	Bedrock slump (involving reactivation of portion of old slide, and failure of fill at head of slide)
Volume	±612,000 m³
Maximum width	165 m
Maximum length	240 m
Difference in elevation between crown and toe	63 m
Direction of movement	S56°W (70° from dip direction)
Maximum vertical displacement (at the head)	24 m
Maximum horizontal displacement	19.5–27 m (as determined from position of displaced pipelines)
Average vertical depth to basal slide surface	18 m
Pre-slide slope angles	15°–18°
Average inclination of basal slide surface	14°
Amount of rotation near head of slide mass	20° (as measured in vertical plane)

of the three phases. Major design modifications during grading could have been avoided, the width of the eventual freeway rights-of-way minimized, and the costs of construction and maintenance reduced.

ACKNOWLEDGMENTS

I gratefully acknowledge the aid of Lawrence Cann and David Adams on the Pomona and San Diego Freeway slides, of Walter Reiss on the Liebre Gulch slide, and of Lawrence Cann, Donald Coates, Donald Nichols, and Richard Lung for review of the text.

REFERENCES CITED

Hoots, H. W., 1931, Geology of the eastern part of the Santa Monica Mountains, Los Angeles County, California: U.S. Geol. Survey Prof. Paper 165-C, 134 p.

Leighton, F. B., 1969, Landslides, *in* Olson, R. A., and Wallace, M. M., eds., Geologic hazards and public problems: Washington, D.C., U.S. Office Emergency Preparedness, p. 97–132.

Metropolitan Water District of Southern California, 1966–1967, Geologic map and section, Sepulveda Tunnel area (L-1082), prepared under the direction of R. J. Proctor: scale 1:12,000.

MANUSCRIPT RECEIVED BY THE SOCIETY SEPTEMBER 7, 1976
MANUSCRIPT ACCEPTED SEPTEMBER 17, 1976

PART 5
Environmental Planning

18

Slope-stability studies in the San Francisco Bay region, California

TOR H. NILSEN
EARL E. BRABB
U.S. Geological Survey, 345 Middlefield Road, Menlo Park, California 94025

ABSTRACT

An extensive program of slope-stability studies has been concluded in the San Francisco Bay region in California. Work to date has resulted in the publication of estimates of landslide damage; an estimated-landslide-abundance map of the region; new slope maps prepared by photo-mechanical processes; photointerpretive maps of landslide, colluvial, and other surficial deposits; and maps of relative slope stability. These studies indicate that landsliding is a major slope-erosion process in the region, that the damage resulting from landsliding is very great, and that additional development in the upland parts of the region should not be undertaken without careful evaluation of slope stability.

INTRODUCTION

The U.S. Geological Survey has concluded a 5-yr investigation of the slope stability characteristics of the San Francisco Bay region. This work was part of a larger, more broadly based study done in cooperation with the U.S. Department of Housing and Urban Development. San Mateo County also provided some additional funding. The larger study, entitled "The San Francisco Bay Region Environment and Resources Planning Study," was initiated in the spring of 1970. Elements of the project include new topographic mapping; geologic, geophysical, and seismologic investigations; hydrologic studies; and regional planning studies. The overall scope and organization of the larger study has been outlined in the program design (U.S. Geological Survey and U.S. Department of Housing and Urban Development, 1971). The purpose of the present paper is to outline the types of slope-stability studies that have been undertaken in this project and to list the publications now available.

The San Francisco Bay region includes the nine counties that border San Francisco Bay, a total land area of about 18,000 km² (7,000 mi²), with a population exceeding 5 million (Fig. 1). It lies primarily within the Coast Ranges but includes part of the western Sacramento Valley. The great variety of climatic conditions, vegetation, topographic situations, and geologic conditions makes the area attractive for future growth. Population has been confined primarily to the flatlands surrounding San Francisco Bay and its adjacent waterways, the city of San Francisco, and some of the larger inland valleys. At present, however, development is spreading rapidly into adjacent hillside areas, where landsliding is becoming an increasing problem.

The geology of the region is very complex. Many different types of rocks (Schlocker, 1968, 1971) and numerous active faults (Brown, 1970; Brown and Lee, 1971) are present, and the structural and tectonic history has been complicated. The following characteristics of the San Francisco Bay region contribute to the widespread landsliding: (1) steep, irregular slopes; (2) abundant and seasonally intense rainfall; (3) extensive human activity, including logging and the grading and cutting of slopes; (4) an abundance of rock types that are very susceptible to sliding, including extensively crushed and fractured Franciscan mélange complexes and poorly consolidated upper Tertiary to Holocene sediments; (5) thick unconsolidated colluvial deposits and thick weathered zones on steep slopes; (6) abundant expansible clay soils; and (7) frequent high-level seismic activity.

Previous geologic studies have focused primarily on mapping bedrock units rather than on slope stability. Private engineering geology consultants have examined the slope-stability characteristics of many small parcels of land in detail, but little of this work has been published, and few regional studies have been undertaken. Some of the earlier studies that have been concerned with landslide features and slope-stability characteristics in the region, and which have provided much necessary data and insight for the current studies, include those by Kachadoorian (1956), Schlocker and others (1958), Bonilla (1960a, 1960b, 1971), Radbruch (1957, 1969), Radbruch and Weiler (1963), Kojan (1968), Harding (1969), Clague (1969), Pampeyan (1970), Rogers

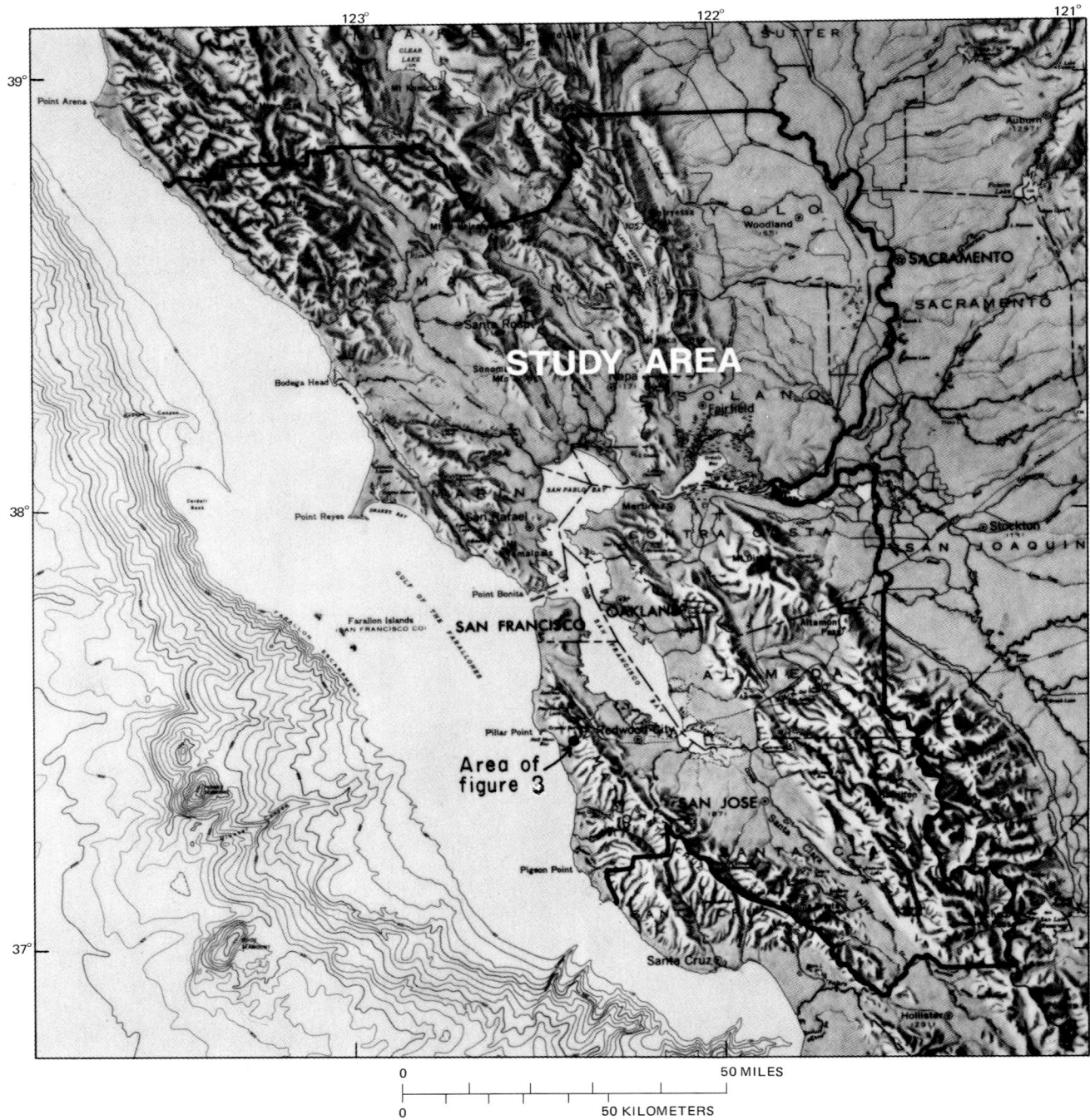

Figure 1. Index map of the San Francisco Bay region, showing area of Figure 2. Depth contour interval, 100 fm.

(1971), and Waltz (1971). Within most of the region, however, information regarding the distribution of landslide deposits was very meager at the time that this slope-stability study began. New work has been directed toward the preparation of slope-stability maps at a regional scale (1:125,000) for the entire bay region. To attain this goal, the acquisition of much new data and much new mapping have been necessary, including the mapping of landslide deposits, bedrock geology, active faults, slopes, recent landslides, other surficial deposits, and engineering properties of bedrock and soil units.

The maps and reports from the study have been published in a three-part series: (1) "Basic Data Contributions," which are based on the initial data-gathering phase of the study; (2) "Technical Reports," which are derived from the basic data; and (3) "Interpretive Reports," which are nontechnical and will be addressed primarily to governmental policy makers and the non-technically oriented private sector. To date, 71 Basic Data Contributions, 6 Technical Reports, and

7 Interpretative Reports have been released. Those that relate to slope-stability studies are noted separately in the list of references at the end of this chapter. All of these publications and a current list of publications from the study can be obtained from the Public Inquiries Office, U.S. Geological Survey, 555 Battery Street, San Francisco, California 94111.

ESTIMATED COSTS OF LANDSLIDE DAMAGE

Before the study began, it was known that the yearly cost of landslides in the bay region was great, but no reliable figures were available. As part of the overall study of slope stability, we have attempted to estimate in several ways the costs to the community. This type of information is useful not only in pointing out the necessity of slope-stability studies prior to construction but also to inform various local governmental agencies of the costs involved in developments that did not consider geologic factors.

Costs related to landsliding are generally very difficult to ascertain, primarily because few public agencies keep accurate records about landslide damage. Data on the costs of damage to structures, re-evaluation of land or housing for tax purposes, and expenditures for road, sidewalk, sewer, and railroad repairs are available from some public agencies, but much damage goes unreported, and many other costs are impossible to determine. Indeterminable costs include those for litigation, salaries of firemen and policemen, detours, and a host of other problems, many of them intangible.

Taylor and Brabb (1972) compiled data on landslide costs from governmental and private agencies in the San Francisco Bay region for the rainy season of 1968–1969, dividing the costs into public, private, and miscellaneous categories and enumerating costs to each county separately. They found that public and private costs totalled more than $25 million, and this figure is probably minimal.

A similar report for the same area during the 1972–1973 rainy season was prepared by Taylor and others (1975). The total economic loss for the second test period was about $10 million.

Nilsen and Brabb (1972) investigated a group of landslides in the northeastern part of the city of San Jose that had damaged public and private facilities during a 10-yr period. Their report shows the locations of landslide deposits and damaged buildings and streets and also summarizes some of the costs of the landslides to the city. During a 5-yr period, San Jose spent more than $750,000 to study one of the landslides, to attempt to control the movement, and to maintain roads across it.

ESTIMATES OF LANDSLIDE ABUNDANCE

Radbruch and Wentworth (1971) made preliminary estimates of the relative abundance of landslides in the bay region by comparing types of earth materials in the region, amount of rainfall, and degree of slope. They divided the region into six categories, ranging from least abundant to most abundant landslides. Their map is an approximation, because when it was made, little information about the distribution of landslides was available. Nevertheless, it provides a reasonably good overview of the distribution of landslides in the region.

Taylor and Brabb (1972) and Taylor and others (1975) included in their reports a map (scale 1:500,000) showing all landslides that they investigated for the 1968–1969 and 1972–1973 rainy seasons. The map gives an impression of the distribution of structurally damaging landslides in the region and of how many new ones might be expected from development of some hilly regions where Radbruch and Wentworth's (1971) map shows landslides to be abundant.

INVENTORY OF LANDSLIDE DEPOSITS

Photointerpretive maps at (scale 1:62,500) showing the distribution of landslide deposits have been prepared for nearly the entire bay region (Fig. 2). The techniques enable recognition of a large number of landslide deposits in most areas and have shown that landsliding is one of the major erosional processes in the bay region. The techniques depend upon recognition of scarps, anomalous bulges and lumps, hummocky topography, ridgetop trenches and fissures, terraced slopes, abrupt slope changes, altered stream courses, discontinuous drainage patterns, closed depressions, springs, and anomalous color, texture, shade, vegetation, and bedrock patterns. The landslide inventory maps were prepared by Brabb and Pampeyan (1972b), Nilsen (1971, 1972a, 1972b, 1972c, 1972d, 1973a, 1973b), Sims and Nilsen (1972), Sims and Frizzell (1976), Frizzell (1974), Frizzell and others (1974), Wentworth and Frizzell (1975), Wright and Reid (1975), Dwyer and others (1976), and Cooper Clark and Associates (1975).

Although the type of movement, date of most recent activity, and nature of the landslide materials are not indicated, the maps are nevertheless useful for land-use planning. Most landslide activity consists of renewed movement of older landslide deposits. The maps can therefore be used as a general guide to problem areas, and, because they provide a regional picture of the past history of landslide activity, they are also useful for site investigations by consulting engineering geologists. Construction activities can alter the marginally stable character of older landslide deposits and induce renewed movement. The inventory of landslide deposits provided by these maps has been used in conjunction with other data to prepare derivative maps showing relative slope stability.

ISOPLETH MAP

The landslide inventory maps for the southern San Francisco Bay region have been generalized and quantified on an isopleth map by Wright and Nilsen (1974). Isopleth maps show the distribution of landslide deposits by contours (isopleths) drawn through points representing equal percentages of landslide deposits within a unit area. The contour format is easy to combine with other quantified map data

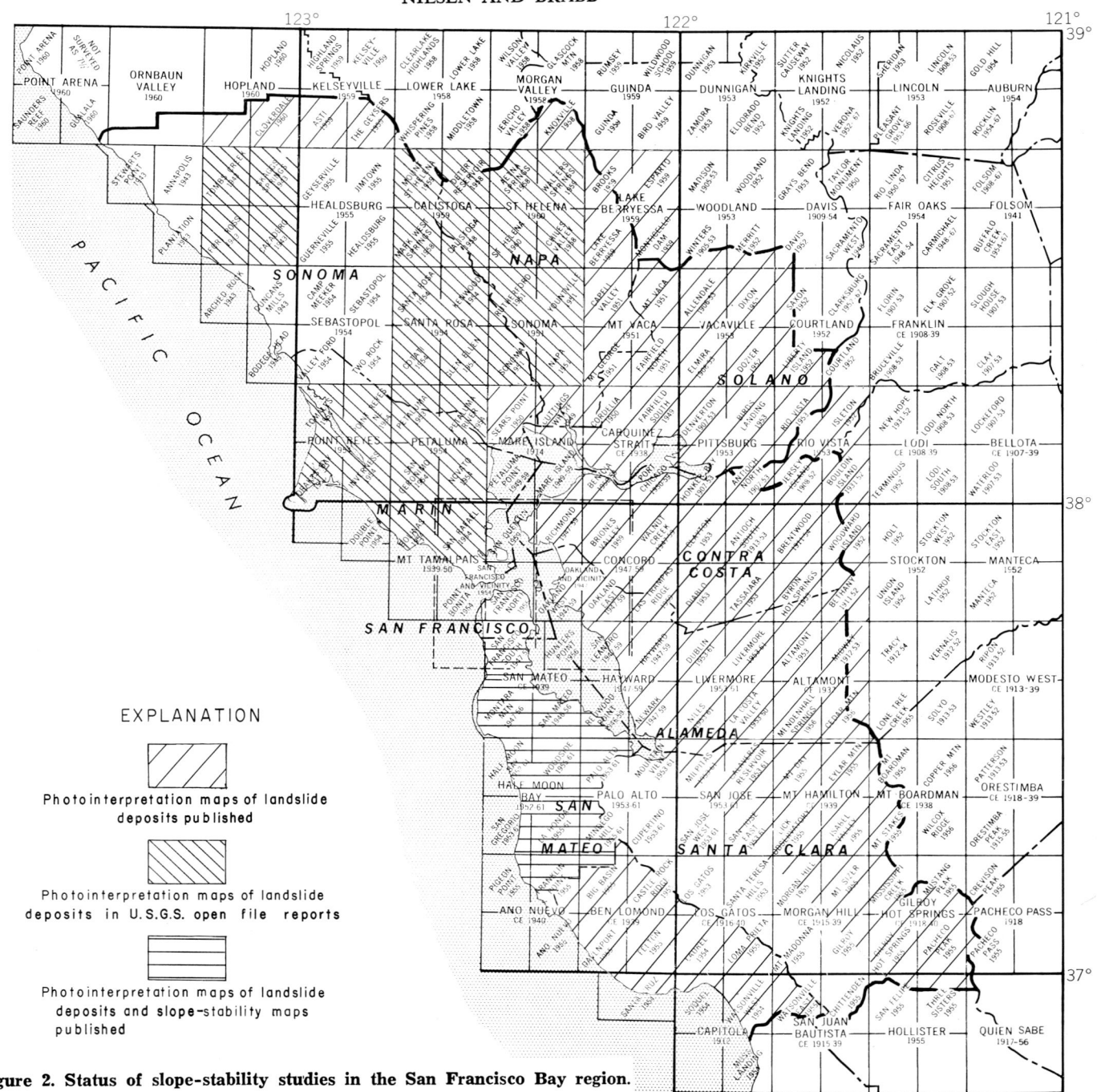

Figure 2. Status of slope-stability studies in the San Francisco Bay region.

for the preparation of derivative maps, such as those suitable for regional planning. A description of the method is provided by Wright and others (1974).

SLOPE MAPS

Slope maps of selected areas within the San Francisco Bay region have been prepared by a new, experimental technique utilizing contour negatives of the U.S. Geological Survey's 1:24,000 topographic map series. The contour lines are diffused photomechanically in measured increments to produce selected slope intervals (in percent): 0 to 5, 5 to 15, 15 to 30, 30 to 50, 50 to 70, and greater than 70. These six intervals are currently being used for many other studies and were selected after discussions with project geologists and city, county, and regional planners and engineers. Complete coverage of the nine San Francisco Bay counties with slope maps at scales of 1:62,500 and 1:125,000 has been completed.

Color transparencies as well as black-and-white prints have been combined several different ways to determine the best way to show slope intervals. For San Mateo County, a color transparency at a scale of 1:62,500 was superimposed on geologic and landslide maps to make a landslide-susceptibility analysis (see discussions under Slope Stability). Inexpensive diazo prints were also prepared from positives of the color-separation negatives and were used to provide slope information for a coastline study by regional planners. Black-and-white composites as well as transparencies with dark shades of color have so far proved unsatisfactory for slope-stability studies, because the contour base or other information needed for the study, such as the identification of the rock unit, cannot easily be seen.

Minor errors are introduced by the photomechanical process, especially along narrow ridge crests and valley floors, where slope categories tend to be too high. The errors can be taken into account when the contour base is examined in conjunction with the slope data.

SLOPE STABILITY

Geologic maps emphasizing engineering properties of rock units, including slope stability and shear strength, have been published for only a few parts of the bay region. Maps showing only slope stability, however, are rare. A. M. Johnson and S. D. Ellen (unpub. data) and A. M. Johnson and A. U. Lobo-Guerrero (unpub. data) made slope-stability maps of small areas in the town of Portola Valley. These maps were the first in the San Francisco Bay region to depict slope stability in a cartographic format that could be readily understood by urban planners.

A report by Blanc and Cleveland (1968) established a methodology for preparing slope-stability maps in a small area of southern California. Brabb and others (1972) prepared a landslide-susceptibility map of San Mateo County based in part on the methodology of Blanc and Cleveland (1968) and in part on data provided by Bonilla (1960a, 1960b). A geologic map, landslide map, and slope map (Fig. 3) were used to prepare the landslide-susceptibility map. All three maps are at a scale of 1:62,500. Part of the geologic map, compiled by Brabb and Pampeyan (1972a), is shown in Figure 3A, The landslide map (Fig. 3B) was prepared by Brabb and Pampeyan (1972b), largely by photointerpretation. The slope map (Fig. 3C) was prepared experimentally by the U.S. Geological Survey specifically for the slope-stability studies. The susceptibility of slope-material units was averaged over the region. Their map differs from others in that the landslide failure record for each rock unit in six different slope categories was systematically measured and used in the analysis. Slope and strength of the rock unit seem to be the principal factors controlling slope stability in San Mateo County — other factors, such as structural control, soil type, rainfall, climate, and vegetation, appear to be averaged through time and space so that their relative influence, at least on a regional basis, is minimal.

The landslide-susceptibility map (Fig. 3D) was prepared by the following method: the percentage of area within 35 geologic map units in San Mateo County that are covered by landslide deposits was estimated by use of a grid overlay with a grid size of 0.01 mi^2 at the map scale. The geologic map units were then arbitrarily grouped in six classes, from class I with 0% to 1% area covered by landslide deposits to class VI with 54% to 70%. The class numbers express, therefore, the relative susceptibility of the geologic map units to landsliding, from I (very low) to VI (very high). The landslide deposits themselves are shown as a separate class, L. Each geologic map unit was then further evaluated to determine which slope categories were critical for the formation of landslides. If few or no landslide deposits formed on low slopes, the class number for the geologic map unit was reduced. Thus, a geologic map unit might be assigned the classification III in comparison with the landslide susceptibility of other geologic map units; in low-slope areas where few failures occurred, a classification of II or I would be assigned.

One advantage of this method is that it provides the regional planner with a way of evaluating the area between landslide deposits on the basis of the average failure record of the geologic map unit throughout a county. The method also compensates for stable formations on steep slopes and unstable formations on low solpes. The principal disadvantages are that (1) the landslide susceptibility of flat areas adjacent to unstable slopes is underestimated and (2) the map does not express in absolute terms the likelihood that any given slope will fail in any given period of time. Field studies by engineering geologists are, of course, required for evaluating any particular site.

EARTHQUAKES

The general problems of landslides forming during earthquakes and the effects and history of landsliding in the San Francisco Bay region were reviewed by Nilsen and Brabb (1975). Earthflows and other types of landslides that formed during the 1906 (magnitude 8.3) and 1957 (magnitude 5.3) earthquakes resulted (in 1906) in several fatalities and major damage to man-made structures.

Ground failures associated with liquefaction are closely related to landsliding. The liquefaction process has been discussed by Youd (1973), and a map of liquefaction potential for the southern San Francisco Bay region has been prepared by Youd and others (1975).

The predicted geologic effects, including landsliding and liquefaction, for a magnitude 6.5 earthquake along the San Andreas fault have been discussed by Borcherdt and others (1975). The predicted effects are shown in schematic form along a profile extending from the mountainous area near the San Andreas fault to the margins of San Francisco Bay.

CORRELATIONS

The correlation between recent landslides and slope, rainfall, and ancient landslide deposits has been described for

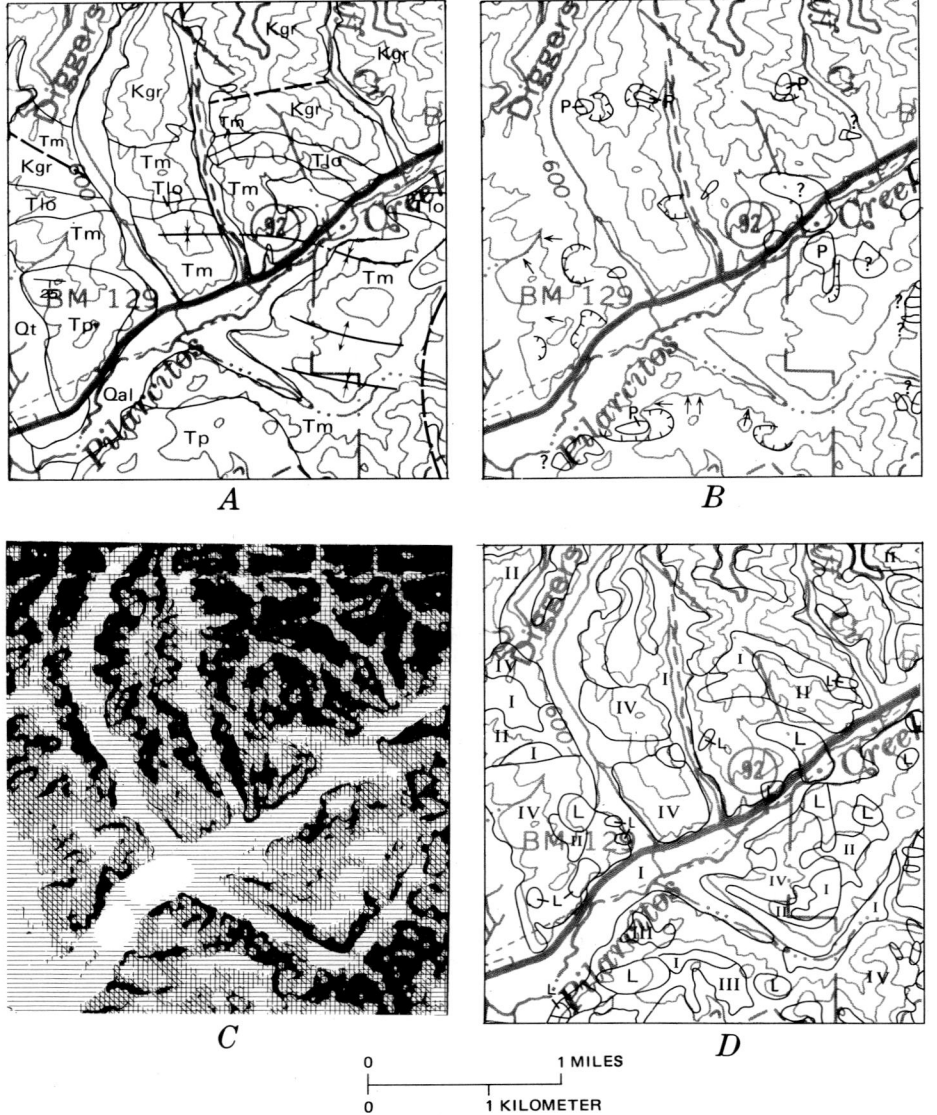

Figure 3. Materials used for a landslide-susceptibility map of San Mateo County, California. The area shown is near the town of Half Moon Bay, about 25 km south of San Francisco (see Fig. 1).

A. Geologic map. Includes granitic rocks of Cretaceous age (Kgr), shale and sandstone of Miocene and Pliocene age (Tp, Tm, Tlo), terrace deposits (Ot), and alluvium (Oal). Heavy dashed lines are faults; thinner lines with arrows are fold axes.

B. Landslide map. Small arrows show landslide deposits 15 to 150 m in size, and the other lines show landslides larger than 150 m. Hachured lines show landslide scarps. P, probable landslide deposit; ?, questionable landslide deposit.

C. Slope map. The darker the tone, the greater the slope.

D. Landslide-susceptibility map. The higher the roman numeral, the greater the susceptibility to landsliding. Landslide deposits (L) are shown as a separate category (the highest).

Contra Costa County by Nilsen and Turner (1975), for Alameda County by Nilsen and others (1976), and for the entire San Francisco Bay region by Nilsen and others (1977). In Contra Costa County, most of the landslides that caused damage to manmade structures from 1950 to 1971 occurred on pre-existing landslide deposits. Most of the landslides occurred during or immediately after storm periods in which more than 7 in. (18 cm) of rain fell, particularly if the ground was already wet from previous storms. Amounts of rain required to generate abundant landslides are smaller in the spring than in the fall. The pattern of rainfall is more important than the total amount — long periods of relatively continuous rainfall produce more landsliding than short discrete storms separated by dry periods.

In Alameda County, 85% of the landslides occurred on slopes greater than 15%. More than $5 million worth of damage was caused by landsliding in 1968–1969; this cost averages out for that year to about $400 per developed acre of land, or $100 per dwelling unit, on slopes greater than 15%.

In the entire San Francisco Bay region, 335 landslides were recorded for the 1968–1969 rainy season, and 411 for the 1972–1973 rainy season. Of these recorded landslides, 55% of those in 1968–1969 and 69% of those in 1972–1973 took place on or near underlying, pre-existing landslide deposits (Nilsen and others, 1977). With regard to slope, 74% of the 1968–1969 landslides and 80% of the 1972–1973 landslides took place on slopes steeper than 15% (8.5°). In addition, more than 60% of the landslides in both rainy seasons took place in either overlying soils or within certain geologic formations that have been previously mapped as being prone to slope failures. Large numbers of landslides were observed to have occurred during storm periods yielding more than 6 to 8 in. of rain (15 to 20 cm) in areas where 10 to 15 in. of rain (25 to 38 cm) had previously accumulated during the same rainy season.

USE BY CITIES AND COUNTIES

Studies by Kockelman (1975, 1976) indicate that landslide and other reports prepared as part of the San Francisco Bay Region Environment and Resources Planning Study have been used by all counties and by 90% of the cities, primarily for geologic hazard studies; for seismic safety, public safety, conservation, and open-space elements of general plans; for general reference; and for the preparation and review of environmental impact reports and statements. For example, the landslide inventory (Brabb and Pampeyan, 1972b) and the landslide-susceptibility map (Brabb and others, 1972) for San Mateo County have been used to determine where and when geologic analyses are needed; to help formulate policies and programs to reduce landslide risks; and to integrate landslide hazard data into the decision-making process. Perhaps the most innovative use of these maps was in the preparation of a San Mateo County ordinance relating landslide susceptibility to the density of dwelling units — only one dwelling unit per 40 acres is allowed in areas highly susceptible to landsliding, and even that one dwelling unit cannot be constructed without site studies by engineering geologists.

CONCLUSIONS

Slope-stability studies in the San Francisco Bay region are probably unique in scope and amount of area covered — landslide and slope maps have been prepared for almost the entire 18,000-km^2 (7,000-mi^2) area. New methods for preparing landslide-susceptibility maps and liquefaction maps have been developed. Economic analyses provide, for the first time, some objective estimates of the enormous cost of landslide damage in a large region. A prediction of landslide damage during an earthquake has been attempted. A correlation between rainfall and the formation of landslides has been established, providing the opportunity for landslide forecasts similar to tornado warnings. Other correlations show that most landslide damage occurred on pre-existing landslide deposits where the slope is greater than 15%. Some of the uses of the slope-stability maps by cities and counties are unprecedented.

Much more could be done. A second generation of landslide maps showing the kind of movement involved and an estimate of the age of the landslide deposit would be helpful in preparing more accurate slope-stability maps. More accurate maps of surficial deposits are needed, especially in wooded areas where small landslides cannot be easily detected on aerial photographs. More landslides should be dated and an attempt made to correlate climatic changes, seismic events, and landslide formation. These and other studies would provide greater knowledge of slope processes and the potential for greatly reducing the risk of building in landslide terrain.

ACKNOWLEDGMENTS

We are grateful to R. F. Yerkes and D. R. Coates for reviewing the manuscript.

REFERENCES CITED

Reports that have been released as "Basic Data Contributions" of the San Francisco Bay Region Environment and Resources Planning Study are noted in brackets after the formal reference citation "BDC," followed by the number of the publication.

Blanc, R. P., and Cleveland, G. B., 1968, Natural slope stability as related to geology, San Clemente area, Orange and San Diego Counties, California: California Div. Mines and Geology Spec. Rept. 98, 19 p.

Bonilla, M. G., 1960a, Landslides in the San Francisco South quadrangle, California: U.S. Geol. Survey Open-File Rept., 44 p.

———1960b, A sample of California Coast Range landslides: U.S. Geol. Survey Prof. Paper 400-B, art. 66, p. B149.

———1971, Preliminary geologic map of the San Francisco South and part of the Hunters Point quadrangles: U.S. Geol. Survey Misc. Field Studies Map MF-311, scale 1:24,000 [BDC 29].

Borcherdt, R. D., Brabb, E. E., Joyner, W. B., Helley, E. J., Lajoie, K. R., Page, R. A., Wesson, R. L., and Youd, T. L., 1975, Studies for seismic zonation of the San Francisco Bay region—Predicted geologic effects of a postulated earthquake: U.S. Geol. Survey Prof. Paper 941A, p. A88–A95.

Brabb, E. E., and Pampeyan, E. H., compilers, 1972a, Preliminary geologic map of San Mateo County, California: U.S. Geol. Survey Misc. Field Studies Map MF-328, scale 1:62,500 [BDC 41].

———1972b, Preliminary map of landslides in San Mateo County, California: U.S. Geol. Survey Misc. Field Studies Map MF-344, scale 1:62,500 [BDC 42].

Brabb, E. E., Pampeyan, E. H., and Bonilla, M. G., 1972, Landslide susceptibility in San Mateo County, California: U.S. Geol. Survey Misc. Field Studies Map MF-360, scale 1:62,500 [BDC 43].

Brown, R. D., Jr., 1970, Faults that are historically active or that show evidence of geologically young surface displacements, San Francisco Bay region; a progress report, October 1970: U.S. Geol. Survey Misc. Field Studies Map MF-331, scale 1:250,000 [BDC 7].

Brown, R. D., Jr., and Lee, W.H.K., 1971, Active faults and preliminary earthquake epicenters (1969–1970) in the southern part of the San Francisco Bay region: U.S. Geol. Survey Misc. Field Studies Map MF-307, scale 1:250,000 [BDC 30].

Clague, J. J., 1969, Landslides of southern Point Reyes National Seashore: California Div. Mines and Geology Mineral Inf. Service, v. 22, no. 7, p. 107–110, 116–118.

Cooper Clark and Associates, 1975, Preliminary map of landslide deposits in Santa Cruz County, California, in Seismic safety element: Santa Cruz, Calif., Santa Cruz County Planning Dept.

Dwyer, M. J., Noguchi, N., and O'Rourke, J., 1976, Reconnaissance photointerpretation map of landslides in 24 selected 7.5 minute quadrangles in Lake, Napa, Solano and Sonoma Counties, California: U.S. Geol. Survey Open-File Map 76-74, scale 1:24,000.

Frizzell, V. A., Jr., 1974, Reconnaissance photointerpretation map of landslides in parts of the Hopland, Kelseyville, and Lower Lake 15-minute quadrangles, Sonoma County, California: U.S. Geol. Survey Misc. Field Studies Map MF-594, scale 1:62,500 [BDC 66].

Frizzell, V. A., Jr., Sims, J. D., Nilsen, T. H., and Bartow, J. A., 1974, Preliminary photointerpretation map of landslide and other surficial deposits of the Mare Island and Carquinez Strait 15-minute quadrangles, Contra Costa, Marin, Napa, Solano and Sonoma Counties, California: U.S. Geol. Survey

Misc. Field Studies Map MF-595, scale 1:62,500 [BDC 67].

Harding, R. C., 1969, Landslides—A continuing problem for Bay area development, *in* Danehy, E. A., ed., Urban environmental geology in the San Francisco Bay region: Assoc. Eng. Geologists Spec. Pub., San Francisco sec., p. 65–74.

Kachadoorian, R., 1956, Engineering geology of the Warford Mesa Subdivision, Orinda, California: U.S. Geol. Survey Open-File Rept. 13 p.

Kockelman, W. J., 1975, Use of USGS earth science products by city planning agencies in the San Francisco Bay region, California: U.S. Geol. Survey Open-File Rept. 75-276, 110 p.

——1976, Selected applications of SFBRS earth science products to county planning in the San Francisco Bay region, California: U.S. Geol. Survey Open-File Rept., 131 p.

Kojan, E., 1968, Mechanics and rates of natural soil creep: Eng. Geologists and Soil Engineers Symp., 5th, Pocatello, Idaho, 1967, Proc., p. 233–253.

Nilsen, T. H., 1971, Preliminary photointerpretation map of landslide and other surficial deposits of the Mount Diablo area, Contra Costa and Alameda Counties, California: U.S. Geol. Survey Misc. Field Studies Map MF-310, scale 1:62,500 [BDC 31].

——1972a, Preliminary photointerpretation map of landslide and other surficial deposits of parts of the Altamont and Carbona 15-minute quadrangles, Alameda County, California: U.S. Geol. Survey Misc. Field Studies Map MF-321, scale 1:62,500 [BDC 34].

——1972b, Preliminary photointerpretation map of landslide and other surficial deposits of the Byron area, Contra Costa and Alameda Counties, California: U.S. Geol. Survey Misc. Field Studies Map MF-338, scale 1:62,500 [BDC 38].

——1972c, Preliminary photointerpretation map of landslide and other surficial deposits of the Mount Hamilton quadrangle and parts of the Mount Boardman and San Jose quadrangles, Alameda and Santa Clara Counties, California: U.S. Geol. Survey Misc. Field Studies Map MF-339, scale 1:62,500 [BDC 40].

——1972d, Preliminary photointerpretation map of landslide and other surficial deposits of parts of the Los Gatos, Morgan Hill, Gilroy Hot Springs, Pacheco Pass, Quien Sabe, and Hollister 15-minute quadrangles, Santa Clara County, California: U.S. Geol. Survey Misc. Field Studies Map MF-416, scale 1:62,500 [BDC 46].

——1973a, Preliminary photointerpretation map of landslide and other surficial deposits of the Concord 15-minute quadrangle and the Oakland West, Richmond and part of the San Quentin 7½-minute quadrangles, Contra Costa and Alameda Counties, California: U.S. Geol. Survey Misc. Field Studies Map MF-493, scale 1:62,500 [BDC 57].

——1973b, Preliminary photointerpretation map of landslide and other surficial deposits of the Livermore and part of the Hayward 15-minute quadrangles, Alameda and Contra Costa Counties, California: U.S. Geol. Survey Misc. Field Studies Map MF-519, scale 1:62,500 [BDC 59].

Nilsen, T. H., and Brabb, E. E., 1972, Preliminary photointerpretation and damage maps of landslide and other surficial deposits in northeastern San Jose, California: U.S. Geol. Survey Misc. Field Studies Map MF-361, scales 1:12,000 and 1:24,000 [BDC 45].

——1975, Studies for seismic zonation of the San Francisco Bay region—Landslides: U.S. Geol. Survey Prof. Paper 941A, p. A75–A87.

Nilsen, T. H., and Turner, B. L., 1975, Influence of rainfall and ancient landslide deposits on recent landslides (1950–71) in urban areas of Contra Costa County, California: U.S. Geol. Survey Bull. 1388, 18 p.

Nilson, T. H., Taylor, F. A., and Brabb, E. E., 1976, Recent landslides in Alameda County, California (1940–71); an estimate of economic losses and correlations with slope, rainfall and ancient landslide deposits: U.S. Geol. Survey Bull. 1398, 21 p.

Nilsen, T. H., Taylor, F. A., and Dean, R. M., 1976, Natural conditions that control landsliding in the San Francisco Bay region—An analysis based on data from 1968–69 and 1972–73 rainy seasons: U.S. Geol. Survey Bull. 1424, 35 p.

Pampeyan, E. H., 1970, Geologic map of the Palo Alto 7½ minute quadrangle, San Mateo and Santa Clara Counties, California: U.S. Geol. Survey Open-File Map, scale 1:24,000 [BDC 2].

Radbruch, D. H., 1957, Areal and engineering geology of the Oakland West quadrangle, California: U.S. Geol. Survey Misc. Geol. Inv. Map I-239, scale 1:24,000.

——1969, Areal and engineering geology of the Oakland East quadrangle, California: U.S. Geol. Survey Geol. Quad. Map GQ-769, scale 1:24,000.

Radbruch, D. H., and Weiler, L. M., 1963, Preliminary report on landslides in a part of the Orinda Formation, Contra Costa County, California: U.S. Geol. Survey Open-File Rept., 35 p.

Radbruch, D. H., and Wentworth, C. M., 1971, Estimated relative abundance of landslides in the San Francisco Bay region, California: U.S. Geol. Survey Open-File Map, scale 1:500,000 [BDC 11].

Rogers, T. H., 1971, Environmental geologic analysis of the Santa Cruz Mountain study area, Santa Clara County, California: Sacramento, California Div. Mines and Geology, 64 p.

Schlocker, Julius, 1968, The geology of the San Francisco Bay area and its significance in land use planning: Berkeley, Calif., Assoc. Bay Area Govts. Suppl. Rept. IS-3, 47 p.

——1971, Generalized geologic map of the San Francisco Bay region, California: U.S. Geol. Survey Open-File Map, scale 1:500,000 [BDC 8].

Schlocker, Julius, Bonilla, M. G., and Radbruch, D. H., 1958, Geology of the San Francisco North quadrangle, California: U.S. Geol. Survey Misc. Geol. Inv. Map I-272, scale 1:24,000.

Sims, J. D., and Frizzell, V. A., Jr., 1976, Preliminary photointerpretation of landslide and other surficial deposits of the Mount Vaca, Vacaville and parts of the Courtland, Davis, Lake Berryessa and Woodland 15-minute quadrangles, Napa and Solano Counties, California: U.S. Geol. Survey Misc. Field Studies Map MF-719, scale 1:62,500.

Sims, J. D., and Nilsen, T. H., 1972, Preliminary photointerpretation map of landslide and other surficial deposits of parts of the Pittsburg and Rio Vista 15-minute quadrangles, Contra Costa and Solano Counties, California: U.S. Geol. Survey Misc. Field Studies Map MF-322, scale 1:62,500 [BDC 35].

Taylor, F. A., and Brabb, E. E., 1972, Map showing distribution and cost by counties of structurally damaging landslides in the San Francisco Bay region, California, winter of 1968–69: U.S. Geol. Survey Misc. Field Studies Map MF-327, scale 1:1,000,000 [BDC 37].

Taylor, F. A., Nilsen, T. H., and Dean, R. M., 1975, Distribution and costs of landslides that have damaged manmade structures during the rainy season of 1972–1973 in the San Francisco Bay region, California: U.S. Geol. Survey Misc. Field Studies Map MF-679, scale 1:1,000,000.

U.S. Geological Survey and U.S. Department of Housing and Urban Development, 1971, Program design 1971, San Francisco Bay region environment and resources planning study, PB2-06826: Springfield, Va., U.S. Dept. Commerce Natl. Tech. Inf. Service, 123 p.

Waltz, J. P., 1971, An analysis of selected landslides in Alameda and Contra Costa Counties, California: Assoc. Eng. Geologists Bull., v. 8, no. 2, p. 153–163.

Wentworth, C. M., and Frizzell, V. A., 1975, Reconnaissance landslide map of parts of Marin and Sonoma Counties, California: U.S. Geol. Survey Open-File Map 75-281, scale 1:24,000.

Wright, R. H., and Nilsen, T. H., 1974, Isopleth map of landslide deposits, southern San Francisco Bay region, California: U.S. Geol. Survey Misc. Field Studies Map MF-550, scale 1:125,000 [BDC 63].

Wright, R. H. and Reid, G. O., 1975, Photointerpretive map of landslides and surficial deposits of northernmost Napa County, California: U.S. Geol. Survey Misc. Field Studies Map MF-677, scale 1:24,000.

Wright, R. H., Campbell, R. H., and Nilsen, T. H., 1974, Preparation and use of isopleth maps of landslide deposits: Geology, v. 2, p. 483–485.

Youd, T. L., 1973, Liquefaction, flow, and associated ground failure: U.S. Geol. Survey Circ. 688, 12 p.

Youd, T. L., Nichols, D. R., Helley, E. J., and Lajoie, K. R., 1975, Studies for seismic zonation of the San Francisco Bay region—Liquefaction potential: U.S. Geol. Survey Prof. Paper 941A, p. A68–A74.

MANUSCRIPT RECEIVED BY THE SOCIETY SEPTEMBER 7, 1976
MANUSCRIPT ACCEPTED SEPTEMBER 17, 1976

Printed in U.S.A.

19
Landslides in West Virginia

PETER LESSING
ROBERT B. ERWIN
West Virginia Geological Survey, P.O. Box 879, Morgantown, West Virginia 26505

ABSTRACT

The West Virginia Geological Survey, under contract from the Appalachian Regional Commission, is mapping landslides and slide-prone areas around urban localities in the State. Recent and old landslides and rockfalls are initially identified on low-altitude panchromatic and, where available, black and white infrared photography and checked by extensive field investigations. They are mapped on 1:24,000-scale topographic maps, and slide-prone areas are delineated based on slide occurrence, slope, and certain rock and soil types. Results to date indicate an approximate density of five slides/mi^2 (2.6/km^2). These slides correlate well with the occurrence of red shale bedrock, slopes of 15% to 45%, concave topographic slopes, and certain actions of man. Particular emphasis is placed on U.S. Public Law 92-234 and on our present inability to predict exactly when, where, or how severe a future landslide could be or to what degree "wet flow" will be involved.

A final report, including landslide mapping for all 28 7.5-minute quadrangles, will discuss other aspects of this study including statutory responsibilities and sociological and psychological problems of forced relocation. The report and maps will provide decision makers with necessary information for land-use planning policies, construction and development projects, and legal and legislative actions.

INTRODUCTION

This paper is a progress report based on a detailed landslide mapping program being performed by the West Virginia Geological Survey under an Appalachian Regional Commission contract. Areas under investigation include seven urban centers (Morgantown, Fairmont, Clarksburg, Charleston, Huntington, Parkersburg, and Wheeling) and a part of the Ohio River valley (Fig. 1). These urban areas cover 23 7.5-minute quadrangles and parts of 15 others along the Ohio River, or approximately 1,300 mi^2 (3,367 km^2). All of the

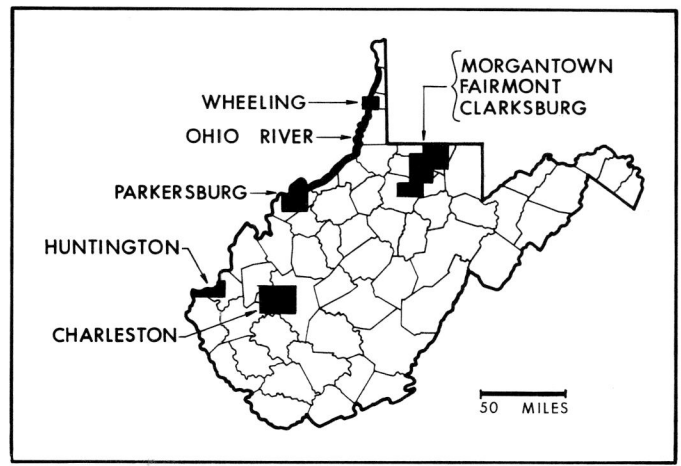

Figure 1. Index map of West Virginia landslide study.

quadrangles, at a scale of 1:24,000, will be included with the final report (Lessing and others, 1976).

We have reviewed extensive literature dealing with landslide descriptions (Ladd, 1935; Sharpe, 1938) and many excellent reports covering correction (A. W. Martin Associates, Inc., 1975; Eckel, 1958; Leach, 1968; Long and Stinnett, 1969), methodology (Eckel, 1958; National Academy of Sciences, 1974), mechanics (Terzaghi, 1950; Zaruba and Mencl, 1969), and regional mapping in the Appalachian Plateau (Briggs and others, 1975). Only rarely are regional maps produced with a report expressly designed for the nongeologist even though the nongeologist constantly makes decisions concerning land use, grading permits, housing developments, subdivision regulations, landslide insurance, and legal interpretations.

Consequently, our concern for the nongeologist will be evident in this paper, and in the final report we will also present advice to homeowners, buyers, and builders; outcomes of court cases; correction methods; governmental responsi-

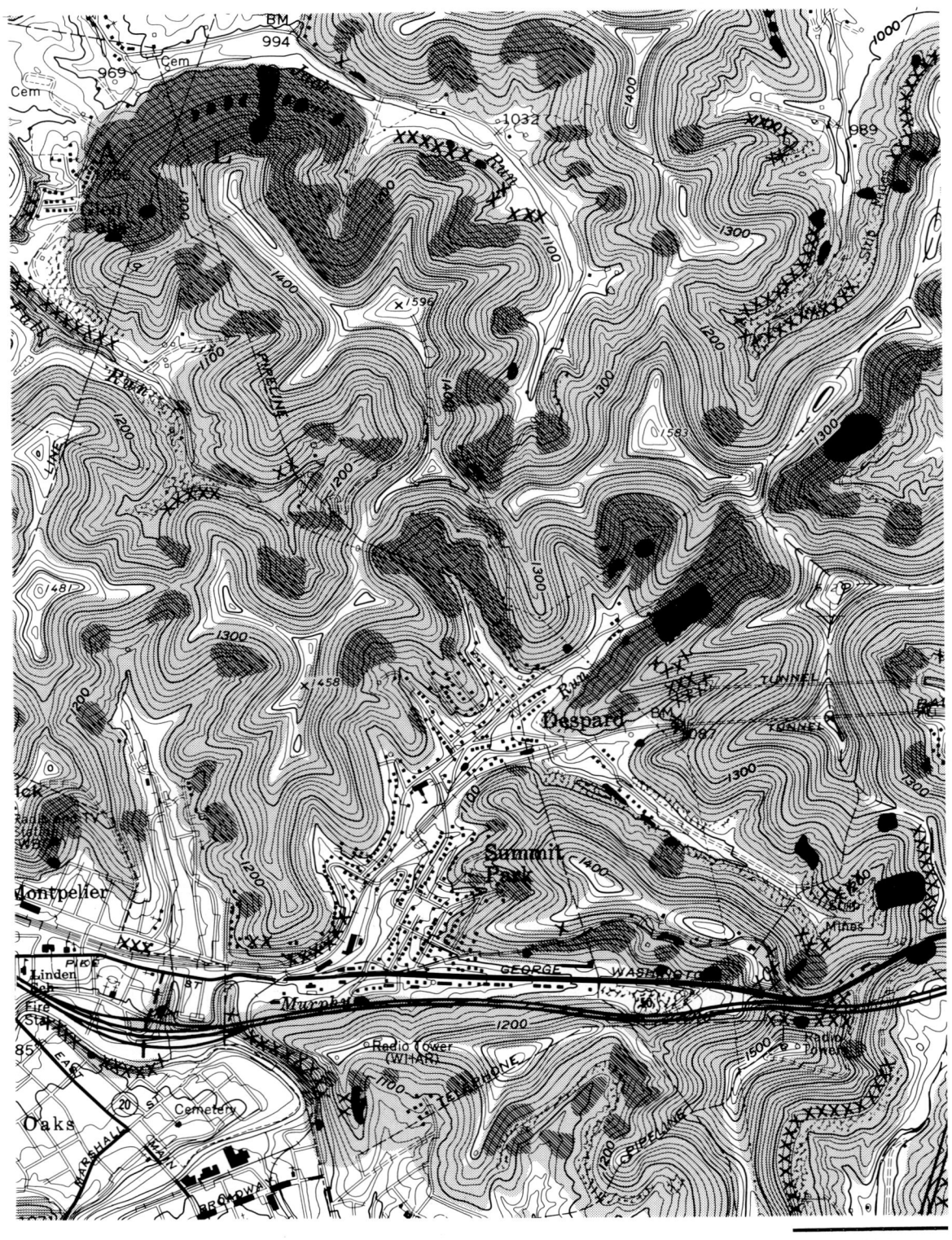

Figure 2. Landslide map of part of the Clarksburg 7.5-minute quadrangle showing old landslides (pattern), recent landslides (black), rockfalls (XXX pattern), and slide-prone areas (gray). Scale is 1:24,000. Large landslide in upper left is shown as stereo pair on Figure 7.

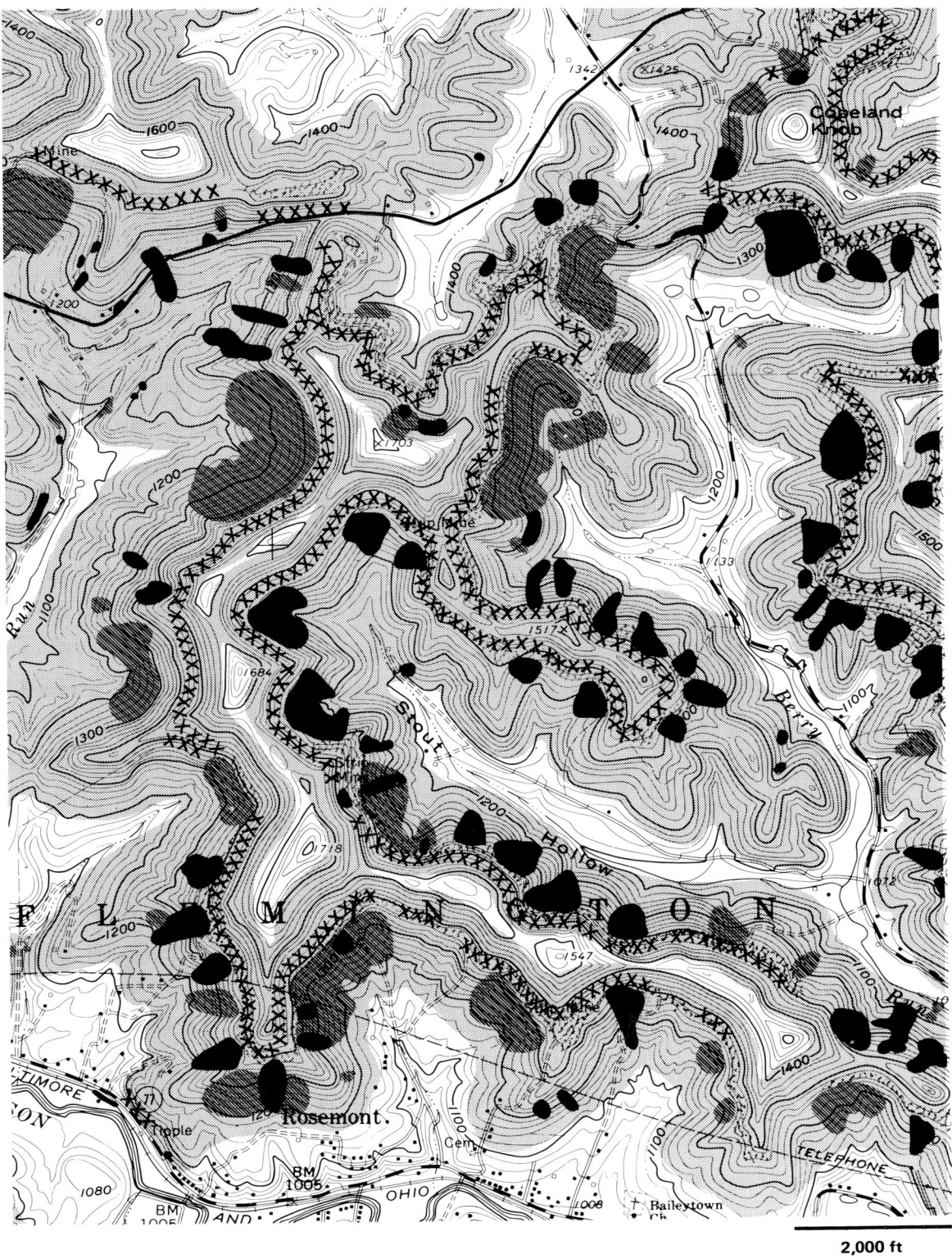

Figure 3. Landslide map of part of the Rosemont 7.5-minute quadrangle showing old landslides (pattern), recent landslides (black), rockfalls (XXX pattern), and slide-prone areas (gray). Scale is 1:24,000. Most of the recent slides are associated with mine-spoil banks from surface mining.

Figure 4. Recent slump and earthflow on relatively gentle slope, Fairmont West 7.5-minute quadrangle. (Photo by Kulander.)

Figure 5. Bedrock and soil that have fallen on secondary road, Morgantown North 7.5-minute quadrangle. (Photo by Art Jordan.)

bilities; and personal, psychological, and sociological effects of landslides.

We have not attempted to develop a new landslide classification scheme, nor have we designated specific types of landslides on our maps (Figs. 2 and 3). However, we have found that the majority involve only soil with rotational slumps at the head and earthflows at the toe (Fig. 4). Individual examples of debris slides, rotational slumps, debris avalanches, earthflows, and mudflows can also occur. Bedrock is only involved in a minor number of landslides (Fig. 5) and, of course, in all rockfalls (Fig. 6).

AERIAL PHOTOGRAPH INTERPRETATIONS AND FIELD INVESTIGATIONS

The landslide maps shown on Figures 2 and 3 are produced by combining aerial-photo interpretation and extensive field studies. An individual 7.5-minute quadrangle involves approximately three to five days of aerial-photo work and two to four weeks of field investigation. Interpretations from aerial photos precede and greatly augment field work, and photos are also checked periodically during and after field investigations.

The two stereo pairs shown in Figures 7 and 8 illustrate rather obvious landslides. Most slides, however, are much more subtle and require detailed aerial-photo interpretation and extensive field checks in order to map them with any degree of confidence. We have found that "leaves-off" (late fall to early spring) photography is far superior to "leaves-on" (late spring to early fall) no matter what the vintage of photography. Normally, low-altitude (1:20,000 to 1:31,000) black and white photos are used because of their availability from the U.S. Geological Survey and Agricultural Stabilization and Conservation Service. Certain areas of West Virginia are also covered by low-altitude black and white infrared photography which augments the standard photography. High-altitude black and white and color infrared photography hold considerable promise, but lack the detail obtainable from low-altitude stereo pairs.

Detailed field investigations are mandatory for locating recent landslides (Fig. 9), recording small slips (Fig. 10), and confirming or denying aerial-photo interpretations. In forest-covered areas, where aerial-photo interpretations are more questionable, field examinations have revealed many landslides.

USE AND LIMITS OF MAPS

The landslide maps that will accompany the final report of our study must be considered the major contribution of this project (Figs. 2 and 3). We have purposely mapped the urban areas primarily because landslides in these areas will affect the most people and will cause high financial losses.

The maps are all at a scale of 1:24,000 (1 in. = 2,000 ft) and consequently lack the detail that may be required by a user at a specific site. However, on the bottom of each map we have stressed that the information is intended as a general

Figure 6. Rockfall (above arrow) caused by undercutting of incompetent shale, Fairmont East 7.5-minute quadrangle. (Photo by Lessing.)

Figure 7. Stereo pair of aerial photographs showing recent and older landslides, Clarksburg 7.5-minute quadrangle. Scale is 1:20,000. (Photos by Agricultural Stabilization and Conservation Service, Oct. 22, 1967.)

guide and should not be used as a substitute for detailed geological engineering and on-site investigations. In other words, these maps have been designed for planning purposes as general indices of risk and represent a pictorial view of what has happened and what can be expected. Also, we have simplified a rather complex geologic situation and presented it in a manner and style which nongeologists can comprehend. For these reasons, there are no actual borders or black lines delineating landslide areas; thus, the gradational nature of the landslide categories is emphasized.

The delineation of "slide-prone areas" and "relatively stable ground" will have major short- and long-term significance, making it advantageous to explain in more detail what these two terms mean, how they were designated, and the limitations imposed. A standard and very common procedure for delineating "slide-prone areas" is based on the frequency and distribution of prior landslides, assuming that the historical record of landslides in an area is indicative of slide potential in the future.

Thus, "slide-prone areas" on the accompanying maps were based in large part on the occurrence of recent and old landslides. No attempt was made to include rockfalls within "slide-prone areas," except those directly associated with landslides. In addition, the slope of the ground surface, the type(s) and structure of rock underlying the region, and the type(s) of soil on the slope contributed to our final interpretation. Finally, the expertise and judgment of the geologists are also reflected in the designation "slide-prone areas." The lack of these same situations (landslide occurrences, steep slopes, unstable bedrock and soil types) automatically delineates "relatively stable ground."

We cannot at present estimate the "probability" of a landslide occurrence based on statistical data. We are also not in a position to indicate the "safety factor" for any given slope, which is based on geotechnical laboratory tests and mathematical computations. Such statistical, mathematical, and geotechnical studies can only be undertaken for designs of specific facilities or limited geographic areas because of the time and financial resources required. However, the systematic compilation of these special geotechnical tests and studies will be beneficial in determining the actuarial rate zones (M, N, O, or P) to be used on flood insurance rate maps in accordance with the 1973 Flood Disaster Protection Act (U.S. Public Law 92-234, section 1914.2). These special

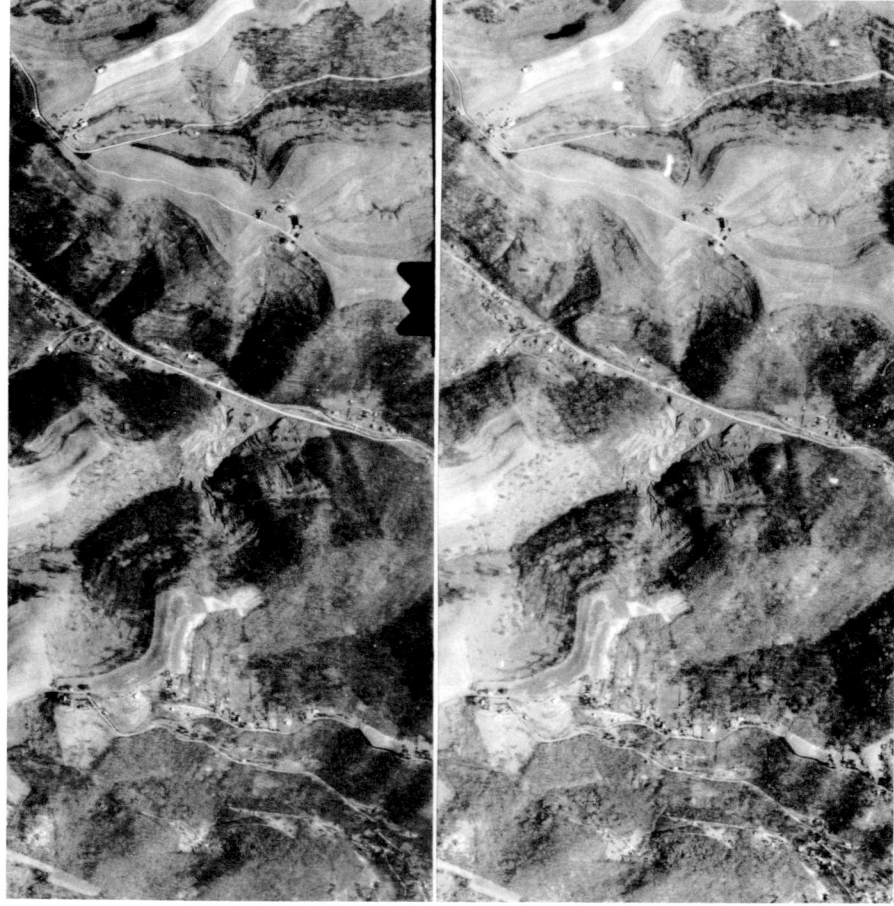

Figure 8. Stereo pair of aerial photographs showing recent landslide associated with surface mining, Wheeling 7.5-minute quadrangle. Scale is 1:24,000. (Photos by U.S. Geological Survey, Nov. 13, 1966.)

geotechnical studies are necessary to delineate the precise boundaries of areas subject to landslides; however, because such data are generally lacking for West Virginia, the boundaries of the "slide-prone areas" on the accompanying maps are only approximate and represent our very best estimate of the areas that will be subject to movement in the future.

This same reasoning applies to areas that are designated "relatively stable ground." It is entirely possible that natural or man-induced landslides could occur in these regions, although we judge these areas to have a low potential for such events.

LANDSLIDE CAUSES AND CORRELATIONS

Landslides constitute a continuous series of events that rarely can be attributed to one definite cause. Normally, only the final result is seen: the failure of a slope. The causes that finally bring the material down begin with the depositional environment of the rock many millions of years ago. Subsequent structural adjustments, erosion, weathering, groundwater flow, and the continuous pull of gravity all contribute to the unstable situation just prior to movement. The final event, *but not the sole cause,* may be triggered by such factors as a heavy rainfall, poor excavation techniques, or an earthquake. Several authors have written extensively on the causes of landslides; the interested reader is especially referred to Terzaghi (1950), Eckel (1958), and Zaruba and Mencl (1969).

The more common natural situations that may foster rockfalls and landslides in West Virginia and the Appalachian Plateau are incompetent rock (Fig. 9), primarily the red shales of the Conemaugh and Monongahela Groups (Pennsylvanian) and Dunkard Group (Pennsylvanian-Permian); considerable areas of moderate to steep slope, mainly within the range of 15% to 45% (Fig. 11); abundant rainfall or occasional heavy rains that saturate the soil; soil with a high clay content (for example, Upshur and Vandalia of U.S. Department of Agriculture Soils Series) and colluvial soil near the base of many slopes (Fig. 9); concave topographic slopes that concentrate surface runoff and ground water; and structural situations where rocks dip in the same direction as the slope or where bedrock fractures strike approximately parallel to slope surfaces.

Man-induced causes of rockfalls and landslides include excavation on slopes or on old landslides, particularly exca-

Figure 9. Recent landslide adjacent to excavation, Clarksburg 7.5-minute quadrangle. (Photo by Lessing.)

Figure 11. Major recent landslide in area of older landslides, New Martinsville 7.5-minute quadrangle. (Photo by Wilson.)

Figure 10. Recent slump and earthflow, Morgantown North 7.5-minute quadrangle. (Photo by Lessing.)

Figure 12. Damaged garage foundation caused by landslide that was activated by excavation and a leaky sewer pipe, Morgantown North 7.5-minute quadrangle. (Photo by Lessing.)

vation at the toe which removes support (Figs. 9 and 12); creation of road cuts that expose red shale to weathering and promote undercutting of more competent rock that will create rockfalls (Fig. 6); surface mining whereby spoil banks composed of poorly consolidated, incompetent material place excess weight on the crown, oversteepen slopes, and inhibit drainage (Figs. 3 and 8); faulty drainage caused by poor septic tank operation, leaky water and sewer pipes, paving, downspouts, and undrained shallow springs, and building construction on poorly designed fill (Fig. 13).

In the previous paragraphs, natural and man-induced causes have been separated only for clarity. Many of the landslides we have investigated show much evidence of natural processes being activated, accelerated, or aggravated by man. Also, many of the causes discussed separately are commonly found to simultaneously contribute their "fair share" to a given landslide. For example, the slide pictured in Figure 9 resulted from a combination of incompetent red shale, a relatively steep slope, and excavation of the toe. Figure 12 shows the result of a leaky sewer pipe, excavation, a relatively steep slope, and incompetent soil.

In order to characterize unstable slope conditions, a simple random sample consisting of 100 landslides was selected from 2,115 landslides mapped in the Morgantown-Clarksburg area of northern West Virginia (Fig. 1). Eight variables were determined from the maps for each landslide in the sample: age, size, morphology, topographic form, slope orientation, proximity to other slides, reactivated sliding in pre-existing slides, and association with mine-spoil banks. The scale of measurement for each variable was nominal, with the exception of size which was measured at an ordinal scale. Frequency distributions for the eight variables measured for the 100 landslides are shown in Figure 14. Although care must be taken in drawing firm conclusions from these statistics, some interesting tentative inferences can be made. The inferences concerning the variables are estimates of the percentage of landslides in northern West Virginia possessing the given characteristics with a bound placed on the error of estimation.

Figure 13. Part of parking lot that is slipping in spite of piling, Clarksburg 7.5-minute quadrangle. (Photo by Lessing.)

For the sake of brevity, the formulas and calculations are not included; the reader is referred to Mendenhall and others (1971).

Landslide age is categorized as either recent or old with recent slides being characterized by fresh cracks and other obvious evidence of recent movement not observed on older slides. Older slides constitute approximately 85% of the slides mapped with a bound on the error of estimation of ±7%.

The size of the landslide in the sample is characterized as small (approximately 1 acre or less; 1 acre = 2.47 ha [hectare]), medium (1 to 6 acres), large (6 to 23 acres), very large (23 to 92 acres), or giant (greater than 92 acres). Medium-size slides constitute approximately 57% of those mapped with a bound on the error of estimation of ±10%.

The morphology of the landslides sampled is characterized as equidimensional, elongate perpendicular to the slope, elongate parallel to the slope, crescent-shaped, or some complex combination. Approximately 52% of the slides mapped are equidimensional with a bound on the error of estimation of ±10%.

The topographic form of the slip slope for each landslide is characterized as planar, concave, convex, or a combination. Approximately 69% of the landslides mapped occur on concave slopes with a bound on the error of estimation of ±9%.

The orientation of the slip slope for each landslide in the simple random sample is characterized as one of eight compass directions. Visual inspection of the frequency distribution of slope orientation reveals no striking preferred orientation. The most frequently occurring orientation is northeast (approximately 18% of the landslides) with a bound on the error of estimation of ±8%.

The landslides sampled are categorized as either associated or not associated with mine-spoil banks. Association merely indicates the occurrence of a landslide either in or directly above (or below) a mine-spoil bank. Association does not necessarily imply a causal relationship. Approximately 8% of the slides mapped are associated with mine-spoil banks with a bound on the error of estimation of ±5%.

Each landslide in the sample is classed as either proximate to other slides (within 1,000 ft or 305 m) or distant (greater than 1,000 ft). Approximately 86% of the landslides mapped occur within close proximity to other slides with a bound on the error or estimation of ±7%.

The landslides in the simple random sample are classed as reactivated if recent sliding has taken place in an older, pre-existing slide. Approximately 10% of the landslides involve reactivation of old slides with a bound on the error of estimation of ±6%.

Based on our simple random sample of 100 of the 2,115 landslides mapped on seven 7.5-minute quadrangles covering 385 mi² (997 km²) in northern West Virginia, it appears that the most common characteristics of landslides in the Morgantown-Clarksburg area are that they are older, equidimensional slides, involving one to six acres of ground, and they occur on concave slopes in close proximity to other landslides. The statistics available at present as well as their analyses are far from complete. Work is currently in progress (Woodfork and Lessing, 1976) on multivariate statistical analyses of more complete data. It appears that a more precise and reliable model for unstable slope conditions can be developed that will provide a meaningful appraisal of unstable slope conditions and risk factors.

FEDERAL FOOD INSURANCE

U.S. Public Law 92-234 is entitled the "Flood Disaster Protection Act of 1973" and is basically an amendment to U.S. Public Law 90-448, the "Housing and Urban Development Act of 1968." Most people, if they are not familiar with the official name, at least are aware of part of the content — flood insurance subsidized by the Federal Government. What people are not usually aware of is that the law also provides insurance coverage for "mudslides."

Section 1901.1 provides some definitions:

"Mudslide" means a general and temporary movement down a slope of a mass of rock or soil, artificial fill, or a combination of these materials, caused or precipitated by the accumulation of water on or under the ground.

"Mudslide area" or "mudslide-prone area" means an area characterized by unstable slopes and land surfaces, whose history, geology, soil and bedrock structure, and climate indicate a potential for mudslides.

"Mudslide area having special mudslide hazards" means a mudslide area with a high potential for mudslides.

"Mudslide area management" means the operation of an overall program of corrective and preventive measures for reducing mudslide damage, including but not limited to emergency preparedness plans, mudslide control works, and land use and control measures.

"Land use and control measures" means zoning ordinances, subdivision regulations, building codes, health regulations, and other applications and extensions of the normal police power, to provide standards and effective enforcement provisions for the prudent use and occupancy of flood-prone and mudslide areas.

The point in parading this legislative jargon before you is to show that the definition, although labeled "mudslide," in

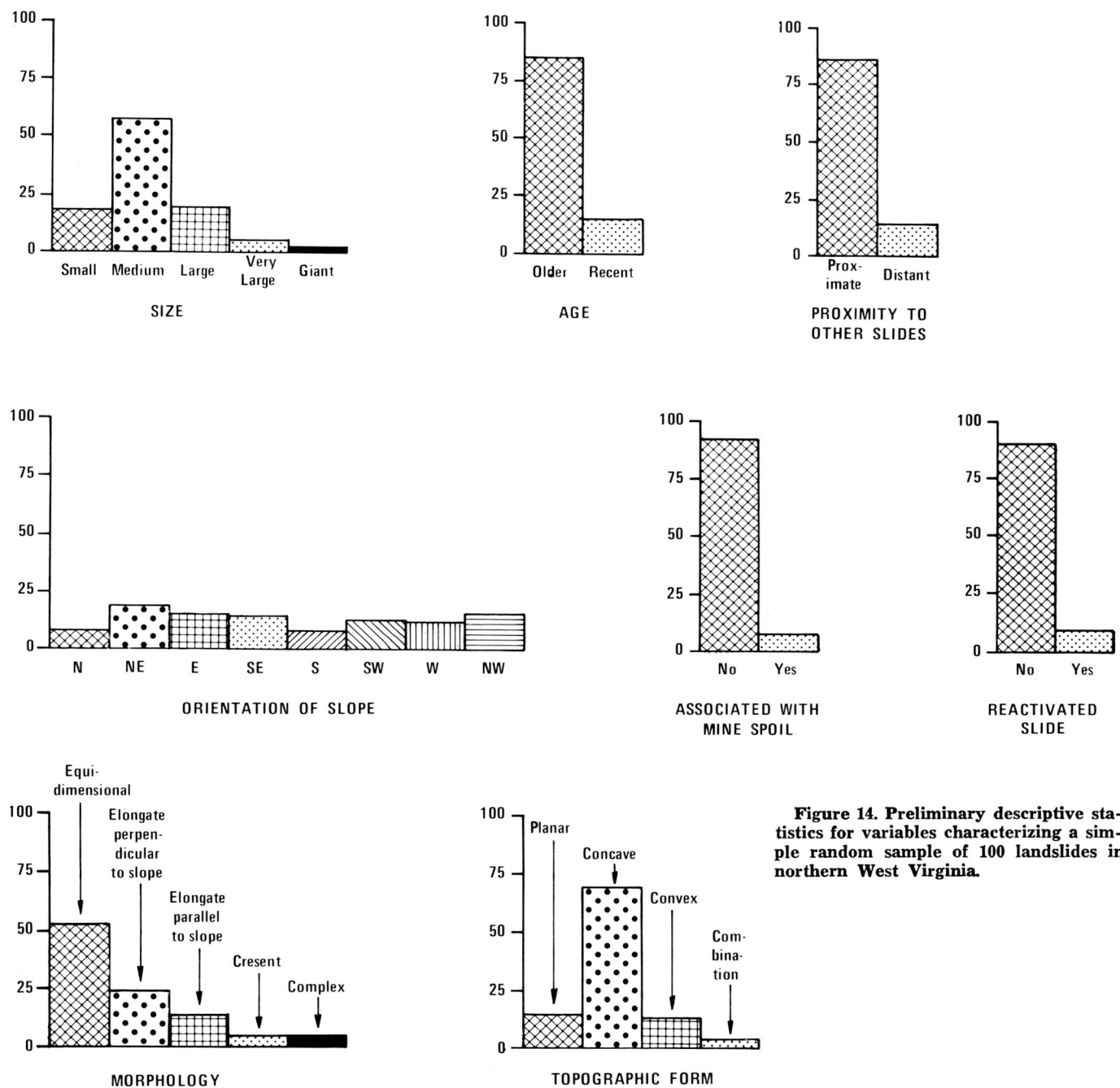

Figure 14. Preliminary descriptive statistics for variables characterizing a simple random sample of 100 landslides in northern West Virginia.

fact defines landslide. In our experiences, as well as others' (Eckel, 1958; Terzaghi, 1950; National Academy of Science, 1974), the vast majority of landslides are activated or aggravated by water. The National Academy of Sciences adds (1974, p. 14),

Because water is the proximate cause of most landslides and significant quantities of it are involved in almost all landslides, use of the term *mudslide,* or any similar term, to distinguish among landslides on the basis of cause and amount of wet flow does not provide a useful basis for determining, before sliding occurs, that a particular landslide will be the type intended to be covered by the National Flood Insurance Program. Thus, for purposes of the federal insurance program, the most practical approach to delineating areas within communities that are susceptible to *mudslides* requires delineation of areas that are susceptible to *landslides* without regard to the degree of wet flow likely to be involved.

The problem of insurance coverage has been compounded in West Virginia by the action taken by HUD to make Stonewood and McMechen eligible for compensation under the 1973 law. The precedent set in these two cases would suggest that landslides are indeed covered. Although there was a volume of mud involved in both locations, most of each area was affected by a landslide.

The National Academy of Science (1974, p. 3) has expressed the potential problem extremely well:

Because water is the proximate cause of most landslides and because significant quantities of water are involved in almost all landslides, it is difficult to see how a refinement such as is intended by use of the term *mudslide* can be used in a practical way in forecasting which areas are prone to sliding. Therefore, while recognizing that this issue is somewhat outside the scope of its assigned task, the Panel does believe it should call attention to its judgment that use of the term *mudslide* or any similar term to restrict insurance coverage to a certain type of landslide involving wet flow probably will result in extensive and expensive litigations in which policyholders will contend—and most often establish—that wet flow was involved in some way in the particular landslide at issue.

Further complications arise from the designated actuarial rate zones for a community, which are as follows: M = area of special mudslide hazards; N = area of moderate mudslide hazards; O = area of minimal mudslide hazards; and P = area of undetermined, but possible mudslide hazards. Not only does this type of designation require prediction, but it also requires predicting the *severity* of a future mudslide hazard. At the present time we can only indicate "slide-prone areas"; we can not predict when, where, how severe, or what the proportion of wet flow will be.

SUMMARY

While realizing that the geologic data gathered during this study are important, we have stressed the equally important implications to the nongeologist. Our contention is that the geologic information concerning landslides (and all geologic hazards) must be presented in an understandable format so that the nongeologist can use it effectively to make decisions. He must be able to comprehend the maps and reports, so that the information on them can be applied to land-use planning policies, construction and development, and legal and legislative actions. Only in this way can we realistically expect to decrease the losses caused by landslides.

ACKNOWLEDGMENTS

We thank Byron Kulander, Bruce Wilson, Stuart Dean, and Stanley Woodring, without whom this project could not have been accomplished. Financial assistance from the Appalachian Regional Commission is acknowledged. Cartography and photography by Daniel Barker and Raymond Strawser are greatly appreciated. Larry Woodfork, Byron Kulander, and Robert Behling critically reviewed the manuscript and offered helpful suggestions. Larry Woodfork kindly contributed preliminary statistical data.

REFERENCES CITED

A. W. Martin Associates, Inc., 1975, Review of construction methods for use in landslide-prone areas: Dept. Planning and Development, Allegheny Co., Pa., 57 p.

Briggs, R. P., Pomeroy, J. S., and Davies, W. E., 1975, Landsliding in Allegheny County, Pennsylvania: U.S. Geol. Survey Circ. 728, 18 p.

Eckel, E. B., 1958, ed., Landslides and engineering practices: Highway Research Board, Spec. Rept. 29, 232 p.

Ladd, G. E., 1935, Landslides, subsidence, and rock-falls: Am. Railway Eng. Bull., v. 37, no. 377, 72 p.

Leach, R. C., 1968, The problem and correction of landslides in West Virginia, *in* Erwin, R. B., ed., Proc. 19th annual highway geology symposium: West Virginia Geol. Survey Circ. 10, p. 11–33.

Lessing, P., Kulander, B. R., Wilson, B. D., Dean, S. L., and Woodring, S. M., 1976, West Virginia landslides and slide-prone areas: West Virginia Geol. Survey Environ. Geol. Bull. 15.

Long, D. C., and Stinnett, B. C., 1969, Landslide recognition and control on West Virginia highways, *in* Wang, J.W.H., and Fisher, S. P., eds., Proceedings — A symposium on landslides: Athens, Ohio Univ., p. 98–127.

Mendenhall, W., Ott, L., and Schaeffer, R. L., 1971, Elementary survey sampling: Belmont, Calif., Duxbury Press, 247 p.

National Academy of Sciences, 1974, Methodology for delineating mudslide hazard areas: Washington, D.C., 107 p.

Sharpe, C.F.S., 1938, Landslides and related phenomena: New York, Cooper Square Pub., Inc., 137 p. (1968 reprint).

Terzaghi, K., 1950, Mechanism of landslides, *in* Paige, S., chairman, Application of geology to engineering practice [Berkey volume]: Boulder, Colo., Geol. Soc. America, p. 83–123.

Woodfork, L. D., and Lessing, P., 1976, Regional appraisal of unstable slope conditions in the Appalachian plateau province of West Virginia: Assoc. Eng. Geologists, Program with Abs., 19th ann. mtg., Philadelphia, Pa., p. 36.

Zaruba, Q., and Mencl, V., 1969, Landslides and their control: New York, American Elsevier Pub. Co., 214 p.

MANUSCRIPT RECEIVED BY THE SOCIETY SEPTEMBER 7, 1976
MANUSCRIPT ACCEPTED SEPTEMBER 17, 1976

20
Landslides at Sardis in western Turkey

GERALD W. OLSON
Department of Agronomy (Soil Survey), Cornell University, Ithaca, New York 14853

ABSTRACT

Geomorphological and archaeological observations were related during pedologic field studies at Sardis to elucidate the role of landslides in the past history of several millenniums at the ruined city. Examinations of deep archaeologic trenches and 107 soil samples and descriptions from 55 sites indicated that the landslide materials could be readily identified by their mixed nature, in contrast to stratified unmixed conglomerate and alluvial deposits and unstratified eolian mantles. Photographs illustrate the geomorphic features that correlate landslides with ruins. Physical and chemical analyses indicated that high silt content (to 57%), high saturation percentage (to 84%), and numerous mica flakes contribute to liquefaction of the soil mass when it becomes wet. Abnormal amounts of phosphorus in subsoils provided a chemical marker of ancient occupation areas disrupted by landslides. Cations and pH give indications of weathering and seepage water movement in landforms, and organic matter content is an index of erosion and soil management affecting landslides. Features like soil structure and slope enable us to make predictions about future slides. A few months before the field study began in 1970, an earthquake with associated landslides near Sardis killed 1,100 people and damaged or destroyed 15,000 buildings. Conservation of monuments and landforms at Sardis would help to exploit the historical and tourist potential of the place and to preserve evidence of many of the geological and archaeological correlations still buried beneath the landslide deposits, for discovery by future scientists.

INTRODUCTION

Sardis is a ruined city about 75 km east of Izmir in western Turkey, probably first settled in the second or third millennium B.C. Landslides and earthquakes (along with droughts, floods, fires, famines, periods of prosperity, invasions, and sieges) have played a prominent role in the history of the city and in the recurrent destruction and preservation of artifacts and archaeological monuments. Sardis was the capital of the ancient kingdom of Lydia, the western terminus of the Persian royal road described by Herodotus, a center for administration under the Roman Empire, and the metropolis of the province of Lydia in later Byzantine times. The place is unique for its geologic setting, giving it a strategic military location, a critical position on a main highway between the Anatolian plateau and the Aegean coast, and ready access to the wide fertile plain of the Gediz River valley.

Sardis was referred to in the *Iliad* (as Hyde), mentioned by the Greek poet Alcman about 650 B.C., and listed as one of the Seven Churches of Asia in the Book of Revelation in the Bible. It was captured by the Cimmerians in the seventh century B.C., by the Persians in the sixth century B.C., by the Ionian Greeks in the fifth century B.C., and by Antiochus the Great in the third century B.C. The ancient wealth of Sardis is still reflected in our familiarity with the phrase "rich as Croesus," in reference to the Lydian king who reigned there from 561 to 547 B.C. Alexander the Great admired the fortifications on the Sardis acropolis in 334 B.C., and Sardis was the scene of the final conference between the Roman generals Brutus and Cassius before they marched north to their deaths at Philippi in 42 B.C. The city was largely destroyed by landslides and earthquakes in A.D. 17 but was rebuilt under Tiberius. It was again mostly destroyed by the Sassanian Persians in A.D. 616, captured by the Arabs in A.D. 716, and taken by the Turks and retaken by the Byzantines in the period A.D. 1090–1098. The fort on the acropolis was handed over to the Turks about 1315, and Sardis was incorporated into the Ottoman Empire in 1425. The latest conflicts at Sardis were battles of the Greco-Turkish war in the first quarter of the twentieth century.

FIELD STUDY

The field study was conducted during the summer of 1970, as part of the field work to characterize the soils in the physical environment at Sardis (Hanfmann, 1961; Olson and Hanfmann, 1971). This report, however, concentrates on the observations and data relevant to geological and engineering aspects of landslides and does not include most of the pedological information. Because the study was designed to look at pedological and archaeological soils materials in great detail, it is different from most other geological studies of landslides. We collected 107 soil samples, along with soil-profile descriptions, from 55 sites to characterize the soils and their archaeological and historical significance. Table 1 summarizes the locations and topographic and geologic characteristics of each site. Generally a pit was dug at each site (Fig. 1) in a characteristic geomorphic landform unit (Fig. 2), and the soils and geologic materials were described and sampled at each site. Samples were not collected at sites 6, 24, 26, and 38 because of the large gravel content, and numerous other archaeological excavations (some to depths of 10 m or more) were examined for their geological and archaeological correlations. Most of the sites and samples were in landslide materials, but a few eolian, alluvial, and other samples were included for comparative purposes.

Figure 3 is a topographic map of Sardis, with a superimposed grid plan for the location of the sites listed in Table 1 in relation to the excavated ruins. Apparently, manmade structures at one time or another have covered nearly all of the area shown on this map. They extended southward up the narrow valley of the Pactolus River and spread out into the large Gediz River valley to the north. Although probably only a very small proportion of the area of the ancient city has been excavated (Hanfmann and Waldbaum, 1975), the floor plans of some of the large structures show clearly in Figure 3. A Roman gymnasium complex at location 1 (Fig. 3) has a floor plan about 175 m by 125 m; a synagogue (location 2) has a floor plan about 100 m long. A row of 30 shops at location 3 constituted an ancient shopping center. The Greek temple of Artemis at location 17 had 64 marble columns almost 20 m high and more than 2 m in diameter. The Roman bath at location 28 is of similar large proportions. The Greek theater at location 26 had seating capacity for 20,000 persons. The Byzantine city wall (remnants at locations numbered 9) has been traced for 4,100 m and was apparently about 3 m thick and 10 m high. A row of imposing mounds, more than 1 km long, stretches from location 1 eastward, marking the remains of sizeable buildings yet unexcavated. The population of Sardis was considerable at several periods in the city's history, probably numbering as high as 100,000 persons in Roman times. Boundaries of the more rural areas of influence of Sardis extended to the present ruins of other ancient cities located tens of kilometres away in the large valley to the east, north, and west of the site; smaller associated settlements extended southward up the narrow Pactolus Valley and into the mountains. Numerous burial mounds at Bin Tepe about 10 km north of Sardis include the one attributed to the Lydian king Gyges and built about 650 B.C.; the mound is about 50 m high and 230 m in diameter. More than 1,100 Lydian chamber tombs were excavated in the necropolis across the Pactolus River west from Sardis. Piers supported a Roman aqueduct apparently as high as 30 m from the ground in places (Butler, 1922), bringing water from the mountains northward into Sardis.

Figure 2. Pit location in geomorphic landform unit at Sardis. Peak of acropolis and stratification of undisturbed sediments are visible in upper middle part of photograph. Landforms across wadi (dry drainageway) have been much affected by numerous landslides, and soil materials in foreground have also been moved by gravity. View is to southeast from the northwest slopes of acropolis.

Figure 1. Description and sampling in pit dug in landslide materials at Sardis. Notice platy structure inclined downslope. Near Pyramid Tomb on northwest slopes of acropolis.

RECORDS OF LANDSLIDES

Sardis is unique for the wealth of its historical and archaeological records, many of which can be related indirectly to landslides and earthquakes. A recent bibliography of Sardis lists almost 100 pages of reference citations. Numerous ancient authors wrote about Sardis; excavations sponsored by Princeton University from 1910 to 1914 and in 1922 produced several volumes; and since 1958 Harvard and Cornell Universities have been involved in excavation and restoration at Sardis. The recent excavations have confirmed many of the earlier records; books and publications in prepa-

TABLE 1. DESCRIPTION OF COLLECTED SOIL SAMPLES AND STUDY SITES AT SARDIS IN WESTERN TURKEY

Site no.	Location (m from grid base)		Elevation (m)	Slope (%)	Sample no.	Depths (cm)	Description of sampled and observed soil materials
1	S1300	W100	135	8	1, 2, 3, 4	10, 35, 75, 100	Eolian (1), alluvial (2), and landslide materials (3, 4) in soil profile with artifacts
2	S1300	W300	120	1	5	5	Sandbar of recent deposition in Pactolus Riverbed
3	S1100	W1100	375	3 to 10, convex	6, 7	5, 15	Surface soil (6) and subsoil (7) on top of Byzantine barracks
4	S1000	E800	400	160, convex ridge	8, 9, 10, 11	10, 20, 40, 60	Eolian materials blown upslope from acropolis conglomerate face
5	S1200	E0	150	12	12, 13, 14, 15	10, 30, 60, 100	Soil horizons in landslide materials with artifacts
6	S1200	E500	215	15	..	..	Gravelly water-worked torrent deposits
7	S1000	E900	340	170	16	5	Stratum in conglomerate at base of acropolis cliff
8	S1100	E500	240	20	17, 18, 19	10, 25, 100	Surface soil (17), subsoil (18), and substratum materials (19) from landslide of about A.D. 668 to 867.
9	S1300	E500	200	60	20, 21, 22, 23	3, 10, 25, 50	Soil profile under pines in landslide materials representing landform stabilized by vegetation
10	S800	E0	185	9	24, 25, 26	5, 30, 75	Eolian surface soil (24), distorted horizon (25), and landslide materials (26) with platy structure inclined downslope
11	S1300	E700	235	10	27	10	Alluvium in recent narrow terrace in ravine bottom
12	S900	E400	235	30	28, 29, 30	10, 15, 100	Eolian surface (28), subsoil in landslide materials (29), and hardpan (30) in conglomerate with mottles indicating seepage
13	S900	E600	310	3 to 5, convex	31	10	Shallow soil formed on top of ruined wall of "flying tower"
14	S1100	E300	195	45	32, 33, 34	10, 30, 130	Soil horizons in landslide materials in toeslope of steep colluvial fan
15	S1100	E200	180	15	35	75	Alluvial substratum from island in dry wadi
16	S900	W100	140	15	36, 37, 38, 39	5, 25, 75, 130	Soil profile with artifacts in colluvial bowl facing necropolis
17	S700	E100	150	20	40, 41, 42	5, 25, 65	Soil profile with platy structure inclined downslope in colluvial bowl-fan across ravine from Pyramid Tomb
18	S900	E200	175	15	43	5	Recent alluvium in narrow terrace in dry wadi
19	S700	E200	165	30	44, 45	30, 130	Landslide materials with artifacts beside a Roman wall
20	S700	E0	170	45, sharp convex	46	80	Hill crest of landslide materials with strong coarse plates inclined downslope
21	S1100	E1000	410	160, ridge crest	47, 48, 49, 50	5, 20, 45, 75	High point of acropolis crest with updraft eolian deposits over gravelly conglomerate
22	S900	E600	320	30	51, 52, 53, 54	5, 25, 60, 90	Soil profile with artifacts on tilted mesa below "flying towers"
23	S900	E700	275	Overhang	55	5	Stratum at base of cliff below "flying towers"

(Continued on next page.)

TABLE 1. (*Continued*)

Site no.	Location (m from grid base)		Elevation (m)	Slope (%)	Sample no.	Depths (cm)	Description of sampled and observed soil materials
24	S600	E200	195	50, ridge crest	..	..	Gravelly remnant ridge of acropolis conglomerate with sloping strata
25	S1200	W200	130	Level foundation	56	5	Soil under foundation of Lydian altar buried by landslides
26	S700	E400	180	15	..	..	Torrent deposits in colluvial concave valley bowl
27	S800	E500	255	85, ridge crest	57, 58, 59	5, 20, 90	Remnant ridge crest of stratified acropolis conglomerate
28	S600	E500	190	20	60, 61, 62	5, 50, 110	Soil profile about 1,500 yr old in colluvial materials with artifacts over a Roman wall
29	S700	E500	200	50	63	105	Colluvium in steep terrace face
30	S900	E500	250	70	64	5	Surface soil (64) under pines in vegetation-stabilized profile with Roman bricks
31	S900	E700	260	15	65	125	Torrent deposit in head of valley bottom; sample taken from 2 mesh/cm screenings of gravelly materials
32	S800	E700	275	80, sharp crest	66, 67	5, 80	Profile with eolian deposition (66), erosion remnant (stone line), with salt encrustations and platy structure (67) inclined downslope
33	Bin Tepe (about 10 km north of Sardis)		100	1	68	5	Surface soil formed from limestone at base of Gyges mound
34	Bin Tepe		140	15	69	5	Surface soil at top of artificial Gyges mound built about 650 B.C.
35	Bin Tepe		100	Inside mound	70	Inside mound	Sample from tunnel at ground level inside Gyges mound
36	Lake shore (about 15 km north of Sardis)		70	20	71	5	Clay weathered from limestone at shore of lake (Marmaragolu)
37	S800	E800	290	30	72	115	Gravelly torrent deposits with artifacts in head of valley bottom above ruins of Roman stadium
38	S700	E1000	270	30	..	..	Gravelly torrent deposits with Roman bricks in valley bowl below tunnels in acropolis
39	S900	E1100	365	7	73, 74, 75	5, 30, 80	Soil profile in eolian deposits (73, 74) and acropolis conglomerate (75) on mesa promontory
40	S400	E1500	165	35	76	105	Gravelly landslide materials in peninsula ridge ending abruptly in braided gravelly wadi floor with fallen cut stone
41	S1200	E1000	325	30	77, 78	5, 25	Colluvium on slope below end of acropolis wall
42	S900	E800	350	35	79	10	Crest of end of ridge spur of colluvium with many Roman bricks in soil profile
43	N200	E1500	105	2	80, 81, 82	15, 30, 75	Aluvium about 6 m above foundation base of Roman "CG" complex
44	N200	E1400	105	2	83, 84, 85	10, 30, 75	Soil in alluvial plain raised about 6 m in 1,500 yr
45	S100	W100	110	7	86, 87, 88	5, 30, 95	Soil profile in landslide materials originally covering House of Bronzes excavation area

(*Continued on next page.*)

TABLE 1. (*Continued*)

Site no.	Location (m from grid base)	Elevation (m)	Slope (%)	Sample no.	Depths (cm)	Description of sampled and observed soil materials
46	Deep strata near S100 W100	100	7	89, 90, 91	200, 400, 600	Landslide materials (89) with Roman pipes over alluvial flood deposits (90) over Lydian occupation layer (91)
47	N200 W100	110	2	92, 93, 94	10, 25, 70	Alluvium covering Roman pipe of 18-cm diam at 110-cm depth
48–49	S1200 E200	195	55	95, 96, 106, 107	5, 5, 50, 125	Surface topsoil (95) and soil profile in sideslope (96, 106, 107) of anthropically shaped mound
50	S300 W300	115	6	97, 98, 99, 100	10, 25, 115, 250	Landslide colluvium (97, 98, 99) with Roman pipe of 18-cm diam installed at 70 cm overlying stratified alluvial deposits (100) dated about 600 B.C.
51	N300 E700	100	1, concave	101	5	Wet seep spot in alluvial plain near Roman complex "C"
52	S100 E200	115	5	102	300	Landslide materials in mudbrick quarry
53	N300 E500	115	3 to 7, convex	103	10	Topsoil on mound formed from buildings destroyed by earthquakes
54	N500 E800	105	1, concave	104	5	Bare salty spot in cotton field
55	S900 E100	165	70 before landslide, 45 after landslide	105	5	Recent unvegetated landslide; this soil has excellent platy structure inclined downslope; soil plates persisted from larger plates into tiny plates even when material was passed through screen of 2 mesh/cm

ration and in press will continue to make available many records of events at Sardis which can be further correlated in the future with the observations and data on landslides presented in this paper. Recent publications that summarize the history and archaeology of the city include books by Foss (1976), Hanfmann (1972), Pedley (1968), and Hanfmann and Waldbaum (1975).

Many of the archaeological and historical records from Sardis have direct references to earthquakes and landslides. Campbell (1971) included lists of some of the important earthquakes in Turkey. Ancient literary sources (Pedley, 1972, p. 10, 17) contain numerous references to "the kind of changes which overtook this land frequently" and "great storm and earthquake" occurrences. Pedley (1968, p. 15) gave a quotation recorded by the Roman historian Tacitus about the earthquake of A.D. 17:

That year twelve famous cities of Asia fell by earthquake in the night, so that the destruction was all the more unforeseen and fearful. Nor was there the means of escape usual in such disaster, by rushing out into the open country, for there people were swallowed up by the yawning earth. Vast mountains, it is said, collapsed, what had been level ground seemed to be raised aloft, and fires blazed amid the ruin. The calamity fell most fatally on the inhabitants of Sardis, and attracted them the largest share of sympathy. The emperor promised ten million sesterces, and remitted for five years all they paid to the exchequer or the emperor's purse.

Even the archaeological monuments themselves bear witness to the earthquakes and landslides at Sardis. A system of wooden tie beams was incorporated into the top of the Church E foundation walls (location 11 in Fig. 3), apparently as a precaution against earthquake damage. Mitten (1966, p. 61–62) has summarized the effects of the earthquake of A.D. 17 upon the thinking of the later engineers who attempted to design reconstructions in the vicinity around location 1 in Figure 3 to withstand the geologic instabilities:

The fear of a second devastating earthquake that dominated the thinking of the architects and engineers who rebuilt Sardis is shown by the elaborate precautions taken to strengthen the foundations of the new public buildings which were erected during the first and second centuries A.D. In the gymnasium-synagogue complex, foundation walls over a meter thick were sunk to a depth of a least five meters, with intervening spaces laced by a network of buried crosswalls. The floor of the south hall of the gymnasium was laid upon two meters of rubble in concrete. In some places, large open courts were built upon an underlying system of parallel barrel vaults, which may have served as drains or cisterns as well as supports; such vaults run east-west under the entrance court to the gymnasium, and north-south under the

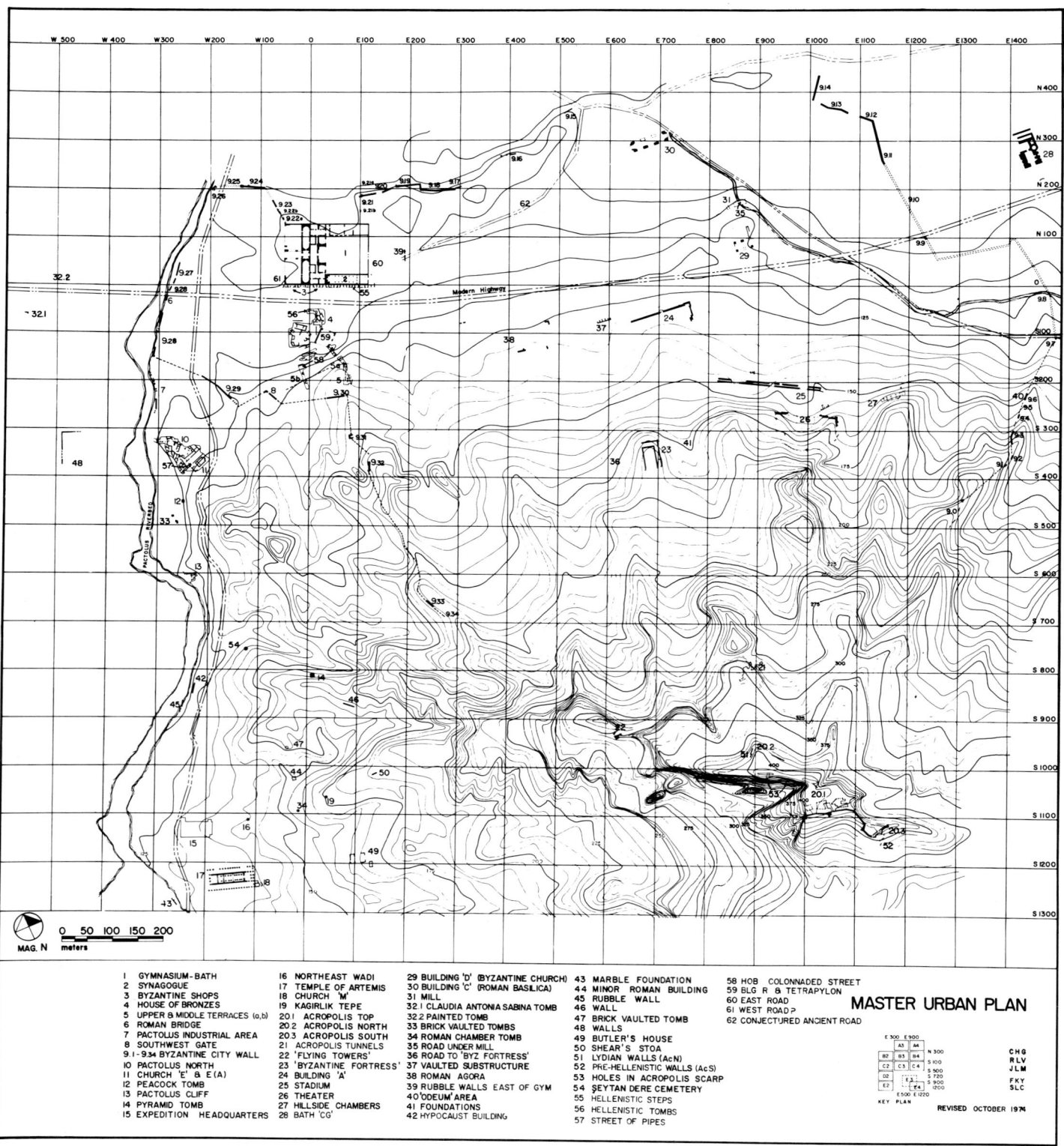

Figure 3. Topographic map of landforms and ancient ruins at Sardis, with superimposed location grid. Numbers indicate locations, and legend gives names of archaeological excavations and monuments. Contour interval is 5 m; elevations range from about 100 m to more than 400 m. Large valley of Gediz River lies to north of this map area; Tmolus mountain range is to south. Across Pactolus River to west is necropolis, which contained more than 1,100 Lydian chamber tombs. Map made by Sardis Expedition Staff of Harvard and Cornell Universities.

palaestra and the main hall of the synagogue. The whole area north of the modern highway, which perhaps originally sloped steeply north into the plain, appears to have been raised several meters to form a huge platform or terrace, upon which the gymnasium complex was built.

GEOLOGY

The Princeton University excavation team at Sardis included a geologist as a consultant, and many geologic interpretations are given in the massive Volume I and other volumes reporting on the 1910–1914 excavations (Butler, 1922). A recent volume (Campbell, 1971) summarizes the geology of Turkey; several of the chapters in the book deal with landslides and earthquakes related to the region in which Sardis is located. A report on a survey of Sardis (Hanfmann and Waldbaum, 1975) gives excellent information about the history, topography, and ecology of the city as well as the archaeological details. A soil survey (Topraksu, 1971) is available for part of the irrigation project area in the Gediz Valley below Sardis; it has information on the geology, climate, topography, and land use and includes maps at a scale of 1:50,000. All of these reports and many other publications provide a considerable amount of information about the geologic and physical environment at Sardis and show how the natural environment has interacted with and influenced human activities through the millenniums. Unfortunately, no modern geological or geomorphological studies of landslides by geologists have been undertaken at Sardis within the area shown in Figure 3; some electrical resistivity measurements and investigations of the great marble quarries near Sardis have been done recently (Hanfmann and Waldbaum, 1975), but these studies are not directly relatable to the landslides discussed in this paper.

The region around Sardis is an intensely deformed and faulted geologic area; it is part of the 40% of Turkey subject to destructive earthquakes. Since 1938 more than 50,000 people have been killed and several hundred thousand buildings demolished or severely damaged by earthquakes in Turkey; in the Sardis region, 350 earthquakes have been reported since the eleventh century A.D. (Campbell, 1971). On March 28, 1970, an earthquake of magnitude 6.9 to 7.8 on the Richter scale was centered at Gediz near Sardis; in this quake 1,100 lives were lost, more than 1,000 people were injured, and more than 15,000 buildings (mostly dwellings) were badly damaged or collapsed completely. Numerous landslides accompanied the Gediz earthquake, particularly in some of the unconsolidated sediments that were saturated with water and had the ability to deform by liquefaction processes.

The Gediz River valley north of Sardis is a graben that includes infillings of colluvial and alluvial materials. The faults at the boundaries of the graben are currently active. Deposits of Quaternary, Tertiary, and Cretaceous age crop out near Sardis and include a wide range of lithologies such as volcanic rocks, phyllite, schist, gneiss, marble, quartzite, limestone, sandstone, and conglomerate. Thus, large quantities of rock for building materials are close to Sardis. The area shown on the map in Figure 3, however, does not have any outcrops of consolidated bedrock. The acropolis at Sardis is composed entirely of crumbling Tertiary conglomerate. William Warfield, consulting geologist for Butler (1922, p. 176–177) noted that the acropolis is composed of fine grains of kaolin, muscovite, biotite, and quartz, with quantities of pebbles; flat pebbles of schist usually about 10 cm in diameter but frequently much larger or smaller; irregular pebbles of gneiss, often slightly angular, but always water-worn; and quartz and pebbles of all sizes up to 20 cm (but usually not more than 10 cm) in diameter. He said that the beds are mostly uniform and massive, but about every 3 or 4 m a 30-cm-thick layer of well-sorted fine material occurs. The bottom beds contain less coarse material and are better sorted, averaging 50 cm in thickness. The beds are cemented with lime. The X-ray patterns of clay minerals in the acropolis conglomerate include peaks indicating quartz, plagioclase, albite, epidote, muscovite, and kaolinite (D. Kamilli, 1976, personal commun.). Numerous mica flakes reduce the shearing resistance of the soil mass owing to their low coefficient of friction in landslides, especially when the material is saturated with water and in a state conducive to movement as a flowing liquid. Figure 4 is a view of the bedding of the conglomerate of the acropolis. The view is from a point about at S1300 E900 (Fig. 3), looking northward toward the steepest slopes extending from S1000 E700 to E1000 on the topographic map. The view in Figure 4 shows a cap of gravel-free sand at the top of the acropolis ridge; the sand apparently was eroded from the steep slope faces by strong winds and carried upward to be deposited in the sparse vegetation growing on the ridge. Landslide deposits and alluvial deposits occupy the lower slope (Fig. 3). Typically, the deepest footslope deposits are stratified sands and gravels with inclusions of Lydian artifacts; landslide materials cover the flood deposits in many places around the base of the acropolis (Fig. 5) and are associated with the later Roman and Byzantine structures.

The geomorphology and topography at Sardis are very complex because of numerous landslides that have created a motley array of terrain features. Figures 6 through 9 show general views of some Sardis landforms whose topographic

Figure 4. Conglomerate bedding of acropolis. View is from south. Holes visible in conglomerate below crests are associated with rooms constructed as part of habitations or defensive networks. These holes are at locations numbered 53 in Figure 3.

Figure 5. Stratified alluvial sand and gravel covered by landslide materials in ravine on northwest slopes below acropolis. Shovel is about 1 m long. Notice cut marble blocks at stream bottom; these fell from ancient buildings and were carried downslope by slides and torrents.

Figure 6. Acropolis at Sardis, viewed from north. Piers in foreground are remains of Roman building on alluvial sediments. Wall in background marks location of Roman civic center on landslide materials. Acropolis, first fortified in eighth century B.C., had in Lydian and Persian times a triple system of defenses admired by Alexander the Great. Lands around Sardis are now used mainly for grazing of goats and sheep and for crops including cotton, eggplant, figs, grapes, melons, olives, peppers, sesame, tomatoes, and wheat.

relations are illustrated in Figure 3. Figure 6 is a view from the Gediz Valley, with the ruins at location 30 in Figure 3 in the foreground. The highest point in Figure 6 is the acropolis peak, shown also in a view from the opposite side in Figure 4. The geologic advantage of the acropolis for defense is well illustrated; the footslopes of landslide deposits are also easy to see in Figure 6. Figure 7 is the view in the opposite direction, showing the landslide spurs radiating outward from the acropolis, with their associated wadis or small valleys. Figures 8 and 9 are views of the acropolis west face, including the landslide covering the temple of Artemis (Fig. 8) and the landslide, erosion, and deposition surfaces of the steep slopes (Fig. 9). The great and small human events at Sardis through the millenniums have been considerably influenced by these slopes and their geologic instabilities; much information about the history of the place is still buried under the landslides and flood deposits.

ENVIRONMENT

The environment at Sardis is also conducive to landslides, particularly where human activities have had a degrading influence. Sardis has a Mediterranean-type climate, with a mean annual precipitation of about 50 cm and a monthly average temperature of 16.7°C (Hanfmann and Waldbaum, 1975). Summers are dry; it rarely rains in July and August. Precipitation is concentrated in violent downpours from late

Figure 7. Gediz valley and landslide slopes at Sardis, viewed toward north from acropolis. Soil materials in foreground have been moved by gravity from upper slopes. Radiating landslide-spur remnant ridges are composed of some original stratified undisplaced conglomerate and some landslide deposits. Narrow valley bottoms between radiating ridges have gravelly water-worked torrent deposits. Footslopes are mostly landslide materials. Valley in background has alluvium. Some large Roman ruins can be seen all along modern road running east-west at base of acropolis. Gymnasium-bath complex, under reconstruction in this photograph, is visible in upper left-hand corner.

October into March, when earthquakes are also likely to occur; the saturated soils in a critical unstable state flow readily into landslides during the wet winter periods. Both increased saturations and the seepage of ground waters contribute to instability of slopes. The original vegetation was forest, but the trees have been cut and the steep slopes overgrazed for millenniums. Good evidence for the man-induced environmental abuse is found in the historical and archaeological records. The Lydians, Greeks, and Romans all had goats and other animals prone to overgraze when populations exceeded the carrying capacity of the land; dumps with animal bones serve as evidence of the high animal numbers supporting the human populations. Much timber was also cut for the massive urban constructions. The Greek

Figure 8. Acropolis at Sardis, viewed from west. Slopes in foreground of photograph were used as Lydian burial ground; about 1,100 chamber tombs were excavated here. Ruins at lower middle right are Greek temple of Artemis, probably built mostly before third century B.C. and overwhelmed by later landslides. By ninth century A.D., temple was covered by landslides to depth shown by dark bands on columns; it was excavated in 1910–1914 by Princeton University expedition. Walled compound with trees is Cornell-Harvard expedition headquarters.

Figure 9. Pactolus Valley, with landslides and eroded hillslopes, at Sardis, viewed toward west from acropolis. This view shows slide that covered temple of Artemis, gravelly torrent deposits in wadis, and erosion of stratified remnant acropolis conglomerate. Narrow valley of Pactolus River has small floodplain of gold-bearing alluvium; this was source of wealth of Lydian king Croesus. Modern Turkish village of Sart Mustafa is in upper right corner.

temple of Artemis at location 17 in Figure 3 had wooden beams in the roof, some of which were of large dimensions to cover spans of 8.5 m, with outside coverings of marble tiles (Butler, 1925). The huge Roman baths (locations 1 and 28 in Fig. 3) also required large amounts of wood for heating the water, so the original forests were severely depleted as demands for wood increased. Overgrazing and overcutting continues at Sardis to the present day and is threatening the remaining archaeological monuments with increasing hazards of erosion and landslides.

GEOLOGY AND ARCHAEOLOGY CORRELATIONS

Some of the potentials for correlation of geological and archaeological information at Sardis can be illustrated by details from some of the excavated locations. Debate about some of the finer points of dating by various archaeologists continues, but many of the major areas of knowledge about Sardis have been verified by past and recent discoveries. The temple of Artemis, for example, was buried by about 10 m of landslide materials from the acropolis. The geomorphic facts of the event are well established. Volume I on the 1910–1914 excavations at Sardis (Butler, 1922) contains many photographs of the location before excavations started and records also of the geologic observations as the unearthing continued. Modern views of the acropolis (Fig. 10) and the temple (Fig. 11) confirm the earlier accounts. Volume II on the Princeton excavations (Butler, 1925) is devoted to the temple of Artemis, and documents the dates of the construction and destruction of the temple. Apparently (Butler, 1925), the location was hallowed ground for a long time, with old foundations of a temple or some other structure

Figure 10. Highest point on acropolis at Sardis, viewed from east. Much of stratified conglomerate has crumbled from these slopes and moved in landslides to wadis and valleys. Temple of Artemis is in valley in middle left. Part of ancient wall still clings to acropolis, at right.

Figure 11. Landslide above temple of Artemis, viewed from east. This landslide covered temple to depth of about 10 m. Most of temple not buried was broken up and used for construction materials in other buildings. Pits in necropolis across Pactolus River are excavations from uncoverings of Lydian chamber tombs. Geologic materials of necropolis are similar to those of acropolis at Sardis.

that was destroyed. Another temple was erected and probably mostly completed before the year 400 B.C., because "between this date and 350 B.C. a number of copper coins were dropped, accidentally or intentionally, into cracks between the base of the cult-statue and the surrounding pavement" (Butler, 1925, p. 140); these coins were dated by archaeologists with considerable precision. Butler (1925, p. 142) stated that "this temple was never completed in all its finer details, though it was roofed over with marble tiles at an early day, and gifts were continually made to Artemis." Then (p. 142), "it appears that the work was nearing completion at the beginning of the first century after Christ, there being nothing left to do but to finish off the mouldings and carve away the lifting-bosses, when the great earthquake of 17 A.D. occurred." In this catastrophe (p. 142) "the columns of the front row seem to have been overthrown" and they were later replaced by new shafts, which (p. 143) "were given a slight increase of diameter, so that the old capitals do not fit them as well as they had fitted the shafts of the other columns of the peristyle." The colossal head of the Empress Faustina discovered at the location indicated (p. 143) "clear evidence that the restoration of the temple was complete, or nearly so, at the time of the empress' death in 141 A.D."

Apparently, the temple of Artemis was not deeply buried in the A.D. 17 earthquake or else was excavated at an early date, because Butler (1925, p. 11) stated that "it is evident that the temple was not deeply buried in the fourth, fifth, and sixth centuries, when the little church just outside the southernmost columns of the east front was built and in use." The ground level at that time is verified by the church foundation resting on soil at the southeast angle of the temple, which by then had risen only 1.4 m (p. 11). At one place in the excavations (p. 12) "was a sack of 216 coins dating between the years 569 and 615 A.D." apparently deposited there while parts of the temple were being broken up to provide building materials for other constructions. The dating of the final most massive landslide is given in this description by Butler (1925, p. 13):

Figure 12. Soil profile in landslide materials at site 1 (Table 1) near temple of Artemis (location 17 in Fig. 3) at Sardis. Most of lower part of profile consists of mixed landslide materials that covered much of temple by ninth century A.D. Top of shovel marks later water-deposited (stratified) layer of loam. Thin layer of gravelly loamy sand was deposited on top of loam by more recent rapidly flowing water. The 15-cm layer at top of exposure is probably mostly wind deposited. Shovel is about 1 m long.

Soon after the middle of the seventh century A.D. a great change came upon the temple. There must have been another great landslip from the Acropolis; for the levels rise suddenly to within a meter or more of the depth of earth in which the temple was found in 1910 at the east end, diminishing toward the river as before. In all this mass of debris no antiquities were found, and not a coin dating between the years 668 and 867 — a period of two hundred years — while coins of the succeeding centuries, from 867 to 1400 A.D., were all found on the higher levels. The little church at the southeast angle of the temple was overwhelmed and almost completely buried, and a few fallen details of the temple which would soon have been broken up were saved by being deeply hidden.

The landslide above the temple of Artemis was described and sampled at several locations in this study, including sites 1, 5, and 8 listed in Table 1 at locations S1300 W100, S1200 E0, and S1100 E500 (Fig. 3). Figure 12 shows the nature

Figure 13. Pyramid Tomb (upper left) in landslide materials at Sardis. Tomb was apparently built in 547 B.C. or shortly thereafter, nearly buried by landslides, excavated by Butler about 1914, nearly buried again by landslides, excavated in 1961, again nearly buried by landslides, and again excavated in 1969. This photograph, taken in 1970, illustrates how colluviation will likely bury structure once again. Trenches at bottom of slope are deep archaeological excavations.

Figure 14. Pyramid Tomb, showing how tomb, excavated in 1969, was again being buried in 1970. Much of soil covering it has moved from upper slopes in one year. If tomb is to be kept exposed as an archaeological monument, protective structures must be built in unstable soils above to prevent or reduce slippage.

of the landslide materials in the lower part of the exposed pit. The landslide materials above the temple of Artemis have been mixed considerably during the slides, but some remnants of sandy material with less gravel can be observed in "pockets" distorted from the original stratification (see Fig. 4). Geologic evidence in these archaeological excavations can be used to easily distinguish the landslide materials (which are mixed) from the alluvial deposits (stratified) and the eolian (windblown) nonstratified materials. Any acropolis conglomerate unmoved by landslides can also be readily identified by the distinctive nature of the original strata characteristics.

The Pyramid Tomb[1] (location 14 in Fig. 3) is another example of an archaeological excavation where landslides can be related to the ruins. Figure 13 illustrates the general landslide nature of the tomb location, Figure 14 gives a closer view of the recent soil movements around the tomb, and Figure 15 shows the distinctive nature of the deep landslide disturbances of the original acropolis strata to their present position below the tomb. The archaeological monuments serve well as "markers" to show the ancient and modern landslide movements. Slope stabilization and erosion control will be necessary in the future if the ruins are to be protected from landslide and erosion damage and observed by the public.

DISCUSSION OF DATA

Tables 2 and 3 list the data from some of the analyses of the samples collected at the sites at Sardis listed in Table 1. The sampling sites were located mostly in the higher elevations of landslide materials in the lower (southern) parts of the Sardis area shown in Figure 3, but they included some sites in the lower Pactolus Valley and the Gediz Valley and the footslopes. For comparative purposes, samples were included from Bin Tepe (an area of Lydian burial mounds about 10 km north of Sardis) in a loess and limestone area and the clayey lakeshore of Marmaragolu (lake), where raw materials might have been obtained for some of the pottery manufacturing at Sardis. Each site examination included detailed soil-profile descriptions (Olson, 1971, 1976), and the laboratory analyses included standard physical and chemical determinations by both the Ankara and Cornell soil laboratories (Greweling and Peech, 1965; Soil Survey Staff, 1967). Many sites have variable materials, including acropolis conglomerate strata, landslide materials, alluvium, and loess; even the alluvial deposits in the large Gediz Valley have probably been mainly deposited from water-transported landslide materials. The objective of the study was to characterize the soil materials affecting the history of the place, so sites

---

[1]The Pyramid Tomb is a distinctive structure at Sardis attributed to the Persian occupation. In 547 B.C. a Persian nobleman named Abradatus was killed in the battle when the Persian king Cyrus defeated the Lydian king Croesus. Xenophon reports that the nobleman's wife Pantheia killed herself over his body. Cyrus built for them a tomb "high above the Pactolus" (Hanfmann, 1972, p. 92). The tomb may have served as a model for Persian architecture such as the tomb of king Cyrus at Pasargadae in Persia, where Lydian captives from Sardis were taken after the fall of Croesus.

Figure 15. Deformed strata in deep trench below Pyramid Tomb at Sardis. These sandy strata with and without gravel are distinctly different from undisturbed strata of acropolis conglomerate or recent alluvial deposits. These are characteristic landslide materials at Sardis, but apparently they have not moved very far or else mixing would have been more complete, as in Figure 12. Shovel is about 1 m long.

and samplings were selected in landforms associated with the ruins. Although artifacts were not deliberately sought out in these diggings, they were often discovered in the pits more or less accidentally. Sites 46 and 47 listed in Table 1, for example, were initially assumed to have landslide materials and alluvium relatively undisturbed by man, but tile pipes were found to cross the pit areas as the digging progressed. These geologic and pedologic examinations helped to indicate that the ancient ruins are more extensive than could be determined from visible features and that much of the extent of Sardis lies buried beneath soil surfaces that appear to be undisturbed. Some of the pits (for example, at site 22 listed in Table 1) had artifacts, indicating that these sites might be fruitful locations for future archaeological excavations.

Table 2 gives some of the physical analyses of the soil samples. The saturation percentage, field capacity, and wilting percentage are particularly relevant to the water content and strength state of the materials for movement as a liquid in landslides. The saturation percentage is determined using the weight of water (based on dry weight of soil) at which all of the continuous void spaces are filled. Field capacity is measured by placing soil samples under 0.33-b vacuum tension and is an approximation of the amount of water a soil will hold when it has drained for several days after a heavy rain. The wilting percentage is measured at 15-b tension and is an estimate of the water in a soil when plants wilt from drought. The moisture percentages, of course, are related to the amount of sand, silt, and clay in a soil sample and to the texture class. The particle size analyses in Table 2 have been put into the texture classes used by the U.S. Department of Agriculture (Olson, 1976). Nearly all of the samples from the area shown in Figure 3 are high in sand and low in clay and in the loamy sand–sandy loam–loam texture classes. Gravel, of course, was excluded from the laboratory analyses but was described in the notes on the soil profiles (Olson, 1971). Only samples 101 and 105 (Table 2) have clay loam texture. Sample 101 was collected from a wet seep spot in the Gediz Valley, and sample 105 was from a recent landslide (Table 1); sample 105, however, has a relatively high content of silt that has a high percentage of tiny mica flakes. Sample 105 from the recent landslide has one of the highest saturation percentages (77%) of all of the samples. Sample 67 (Table 2), from a pit with strong platy structure inclined downslope (Table 1), had silty clay loam texture, a saturation percentage of 69, and 57% silt content (highest of all the samples). Only sample 71 from the lakeshore about 15 km north of Sardis has a clay texture; this material was one of the likely clay sources for the pottery makers at Sardis.

The chemical data in Table 3, of course, are closely related and supplementary to the physical data in Table 2 and the site characteristics listed in Table 1. The organic-matter content is a good indicator of the vegetative cover protecting the soil from erosion and landslides. The pH is influenced by the weathering and leaching of the soil and by the effects of vegetation. Phosphorus levels indicate past and present human activities; high levels occur where human or animal wastes were deposited. Phosphorus is fixed to soil particles so that it is not translocated very far in the undisturbed soil over time. Thus, where landslide materials have high phosphorus levels deep in the soils, that phosphorus was probably carried down from upslope in a slide and buried. Feces and urine of grazing animals would currently increase phosphorus levels in surface soils, and fertilizer applications would also raise the phosphorus contents in the surface horizons (but not in the subsoils). The exchangeable cation status indicates a great deal about the nature of the soil material, its weathering status, relative rates of movement in landscapes, clay mineralogy, and so on. One would expect mobile cations like Na^+ and K^+ to move farther and faster in geomorphic landscapes than Ca^{++} and Mg^{++}, which are relatively less soluble in leaching waters.

Organic matter (Table 3) is generally at a low level at Sardis in subsoils and substrata. The percentage 0.2 in samples 16 and 55 is typical of the acropolis conglomerate beds in the cliffs. The 10.8% organic matter in sample 20 under pines is the extreme high of the range; this amount represents the soil condition under natural vegetation, where the landforms are relatively stabilized by vegetation — as the Sardis area was before people began cutting the trees and allowing animals to overgraze the land. The 5.1% organic matter of sample 64 represents a recovery stage of the soil in an area where the land surface had been disturbed in

TABLE 2. PHYSICAL ANALYSES OF SOIL SAMPLES FROM SARDIS IN WESTERN TURKEY

Sample no.	Saturation*	Field capacity†	Wilting§	Sand (%)	Silt (%)	Clay (%)	Texture#	Sample no.	Saturation*	Field capacity†	Wilting§	Sand (%)	Silt (%)	Clay (%)	Texture#
1	37	12	4	70	22	8	sl	55	48	13	5	57	34	9	sl
2	58	20	6	45	44	11	l	56	33	12	6	70	15	15	sl
3	33	14	4	70	20	10	sl	57	44	16	6	54	31	15	sl
4	30	9	2	79	16	5	ls	58	42	14	5	56	30	14	sl
5	39	12	3	76	20	4	sl	59	41	14	4	66	23	11	sl
6	50	21	7	53	34	13	sl	60	41	18	7	62	24	14	sl
7	40	20	6	62	27	11	sl	61	39	18	7	58	24	18	sl
8	46	10	6	80	14	6	ls	62	38	18	4	59	28	13	sl
9	39	10	4	82	13	5	ls	63	35	11	4	64	24	12	sl
10	38	7	4	81	13	6	ls	64	59	19	8	65	27	8	sl
11	29	8	3	78	15	7	ls	65	25	7	2	80	15	5	ls
12	42	20	5	56	32	12	sl	66	50	18	5	47	44	9	l
13	35	18	12	59	26	15	sl	67	69	34	12	14	57	29	sicl
14	33	20	16	59	26	15	sl	68	70	42	19	33	41	26	l
15	38	18	13	56	32	12	sl	69	50	28	13	51	28	21	l
16	34	9	4	73	21	6	sl	70	44	25	9	50	29	21	l
17	35	14	6	73	19	8	sl	71	58	34	17	25	28	47	c
18	27	11	6	75	15	10	sl	72	24	7	3	86	11	3	ls
19	25	12	6	74	17	9	sl	73	49	24	6	64	26	10	sl
20	84	33	14	48	40	12	l	74	38	19	7	64	25	11	sl
21	52	22	8	43	39	18	l	75	46	25	5	51	39	10	l
22	50	21	8	39	41	20	l	76	34	17	4	62	30	8	sl
23	40	25	15	44	40	16	l	77	33	12	3	72	21	7	sl
24	44	24	6	46	36	18	l	78	29	12	4	68	25	7	sl
25	28	8	1	81	16	3	ls	79	39	19	7	65	24	11	sl
26	49	27	9	80	16	4	ls	80	50	21	4	56	35	9	sl
27	30	11	2	32	44	24	l	81	42	17	4	60	34	6	sl
28	45	18	5	73	23	4	sl	82	38	26	6	45	38	17	l
29	41	16	5	56	31	13	sl	83	49	28	6	47	41	12	l
30	49	15	5	54	35	11	sl	84	38	23	8	54	35	11	sl
31	46	15	6	64	29	7	sl	85	46	28	9	44	44	12	l
32	41	12	4	66	26	8	sl	86	46	27	7	54	34	12	sl
33	34	12	4	64	26	10	sl	87	44	25	10	58	27	15	sl
34	33	10	4	69	22	9	sl	88	46	26	10	52	30	18	sl
35	26	5	2	81	14	5	ls	89	42	22	8	58	27	15	sl
36	40	14	5	64	27	9	sl	90	30	5	4	89	8	3	s
37	30	13	4	63	25	12	sl	91	38	20	7	65	27	8	sl
38	33	13	5	62	26	12	sl	92	35	21	6	54	32	14	sl
39	37	16	7	57	33	10	sl	93	41	23	8	47	37	16	l
40	31	11	4	70	24	6	sl	94	44	28	9	52	31	17	l
41	48	21	5	46	41	13	l	95	35	13	4	63	29	8	sl
42	47	22	6	42	41	17	l	96	36	9	5	76	18	6	sl
43	38	7	3	73	22	5	sl	97	50	24	11	55	33	12	sl
44	33	14	5	63	24	13	sl	98	58	30	17	50	33	17	l
45	33	16	5	64	24	12	sl	99	30	12	4	72	22	6	sl
46	63	27	9	29	46	25	l	100	44	7	2	76	20	4	ls
47	46	10	5	74	18	8	sl	101	68	33	12	30	39	31	cl
48	38	10	5	80	14	6	ls	102	41	23	5	49	39	12	l
49	36	8	5	81	13	6	ls	103	34	21	6	62	27	11	sl
50	35	7	4	78	14	8	ls	104	49	25	6	51	34	15	l
51	40	18	7	66	22	12	sl	105	77	29	10	22	49	29	cl
52	38	14	6	66	20	14	sl	106	36	10	4	69	22	9	sl
53	36	14	4	64	22	14	sl	107	31	12	4	70	24	6	sl
54	33	14	6	61	23	16	sl								

Note: Analyses by Ankara laboratory.
*Percent H_2O at saturation.
†Percent H_2O at 0.33-b tension.
§Percent H_2O at 15-b tension.

#Designated according to U.S. Department of Agriculture classes: ls = loamy sand, sl = sandy loam, l = loam, cl = clay loam, sicl = silty clay loam, c = clay.

TABLE 3. CHEMICAL ANALYSES OF SOIL SAMPLES FROM SARDIS IN WESTERN TURKEY

Sample no.	Organic matter (%)	pH	P (ppm)	Exchangeable cation status (meq/100 g)					Sample no.	Organic matter (%)	pH	P (ppm)	Exchangeable cation status (meq/100 g)				
				CEC*	Ca^{++}	Mg^{++}	K^+	Na^+					CEC*	Ca^{++}	Mg^{++}	K^+	Na^+
1	1.1	7.3	90	7.31	6.09	0.97	0.29	0.24	55	0.2	7.2	4	16.41	14.85	1.02	0.15	2.43
2	1.0	7.6	9	11.90	9.98	1.47	0.40	0.24	56	0.5	8.6	118	8.95	5.90	2.36	0.39	0.20
3	0.6	7.8	45	6.81	5.15	0.98	0.31	0.24	57	1.0	8.6	6	10.42	9.58	0.52	0.31	0.14
4	0.4	7.8	65	4.93	4.42	0.16	0.12	0.25	58	0.6	8.7	4	10.94	6.38	4.21	0.21	0.14
5	0.4	8.0	3	5.15	4.63	0.14	0.15	0.32	59	0.3	8.6	6	7.16	5.90	0.98	0.17	0.21
6	3.5	6.9	300	15.37	13.73	1.05	0.24	0.19	60	2.5	6.7	52	11.77	7.28	2.40	0.39	0.16
7	1.7	7.4	355	14.75	13.52	0.93	0.16	0.19	61	1.7	6.8	105	12.05	8.90	2.52	0.36	0.17
8	2.3	7.2	55	8.84	8.00	0.20	0.39	0.19	62	0.8	7.9	75	8.41	6.91	1.22	0.29	0.19
9	1.4	7.3	74	8.49	7.44	0.43	0.29	0.19	63	0.4	7.2	2	8.18	5.14	2.52	0.22	0.17
10	0.9	7.4	78	7.72	6.49	0.69	0.36	0.14	64	5.1	7.5	64	16.79	15.04	1.04	0.71	0.10
11	0.5	7.4	72	6.83	5.58	0.65	0.31	0.17	65	0.4	7.8	58	5.29	4.50	0.57	0.16	0.12
12	2.1	6.5	48	8.57	7.19	0.63	0.45	0.19	66	0.9	7.8	5	10.28	8.21	1.63	0.21	0.16
13	0.9	6.8	110	9.21	7.80	0.85	0.41	0.19	67	0.4	8.1	2	17.87	12.56	5.70	0.37	0.35
14	1.1	6.9	280	11.35	9.64	0.88	0.59	0.17	68	7.8	7.2	218	30.56	29.28	1.12	0.07	0.18
15	2.0	7.4	375	13.71	12.38	0.89	0.34	0.19	69	3.9	7.6	12	22.15	20.33	1.27	0.16	0.19
16	0.2	8.4	4	6.06	4.57	0.72	0.12	0.47	70	0.2	7.4	12	16.44	13.95	2.11	0.22	0.32
17	1.9	5.4	5	8.58	5.05	0.48	0.37	0.17	71	0.5	7.9	30	23.35	19.56	3.77	0.10	0.20
18	0.6	5.8	4	8.25	5.39	0.76	0.39	0.18	72	0.3	8.1	90	5.54	4.84	0.54	0.15	0.14
19	0.4	6.0	18	8.28	4.95	0.98	0.27	0.22	73	3.8	5.9	6	11.74	6.86	0.58	0.47	0.17
20	10.8	6.4	17	27.24	23.79	2.50	0.70	0.20	74	0.8	6.1	10	9.87	6.56	0.81	0.29	0.14
21	3.2	6.0	8	17.81	12.21	0.74	0.50	0.19	75	0.6	7.4	98	10.25	8.73	1.01	0.24	0.17
22	1.2	4.6	1	15.76	8.17	0.38	0.32	0.27	76	0.6	7.9	3	9.82	8.78	0.86	0.14	0.18
23	0.6	6.5	1	14.14	11.09	2.51	0.23	0.19	77	1.5	6.5	2	7.51	4.98	0.49	0.19	0.16
24	1.1	7.5	2	10.68	9.09	1.08	0.42	0.19	78	0.9	7.0	2	7.32	5.88	1.25	0.21	0.14
25	0.1	7.9	2	2.78	1.81	0.69	0.11	0.19	79	3.6	6.3	32	13.80	9.57	0.85	0.42	0.13
26	0.6	7.9	1	14.19	12.16	1.52	0.32	0.31	80	1.2	7.8	48	8.38	7.40	0.58	0.35	0.15
27	0.3	7.9	4	3.81	3.10	0.52	0.11	0.19	81	0.7	7.6	10	7.07	6.10	0.83	0.16	0.15
28	2.5	5.4	1	10.35	6.95	0.52	0.38	0.20	82	1.6	7.9	19	11.47	9.76	1.36	0.24	0.14
29	0.9	5.8	1	11.41	8.28	0.95	0.29	0.18	83	1.6	7.6	42	8.96	7.30	0.86	0.64	0.15
30	0.3	6.7	3	9.53	8.27	0.73	0.18	0.19	84	1.2	8.0	20	8.08	7.64	0.47	0.34	0.15
31	4.1	6.9	72	15.65	14.47	0.53	0.20	0.24	85	0.8	8.1	8	11.60	10.86	0.29	0.24	0.13
32	1.5	7.0	6	9.05	7.88	0.59	0.23	0.19	86	2.9	7.5	390	11.97	10.67	1.50	0.71	0.15
33	0.9	7.1	2	8.31	7.27	0.77	0.23	0.17	87	1.6	8.0	104	12.96	12.12	0.28	0.62	0.18
34	0.2	7.7	5	9.57	9.39	0.02	0.15	0.17	88	1.2	8.1	112	13.70	13.10	0.48	0.31	0.21
35	0.2	7.4	28	4.53	2.81	1.42	0.25	0.12	89	1.0	8.2	91	11.24	10.34	0.55	0.24	0.18
36	2.5	5.9	5	8.71	5.80	0.46	0.50	0.11	90	0.2	8.6	105	6.85	4.05	2.32	0.26	0.14
37	0.4	6.7	6	7.05	5.09	1.46	0.60	0.06	91	0.5	8.5	255	8.33	7.75	0.67	0.08	0.23
38	0.4	6.7	6	7.57	6.56	0.70	0.26	0.10	92	1.6	8.0	285	10.66	9.78	0.33	0.48	0.16
39	0.5	7.6	14	8.69	8.11	0.21	0.23	0.18	93	1.6	8.0	262	12.68	11.64	0.50	0.42	0.17
40	0.7	7.7	50	5.28	3.98	0.84	0.32	0.17	94	1.1	9.1	105	11.74	10.13	0.62	0.40	0.25
41	0.8	7.9	55	9.35	8.59	0.39	0.30	0.17	95	0.3	8.2	7	10.47	9.56	0.43	0.24	0.14
42	0.4	8.2	6	7.78	4.33	2.45	0.78	0.28	96	1.3	6.3	3	9.10	5.05	0.76	0.24	0.14
43	0.4	7.9	4	6.26	3.24	2.76	0.14	0.14	97	3.1	7.6	600	12.89	10.86	0.74	1.22	0.16
44	0.8	7.6	18	9.13	7.23	1.46	0.30	0.13	98	2.5	7.7	675	15.88	13.88	0.43	1.43	0.17
45	0.4	7.8	11	7.56	5.40	1.85	0.22	0.17	99	0.7	8.3	135	5.09	3.39	1.08	0.39	0.19
46	0.4	8.1	2	15.21	11.41	2.46	0.21	1.41	100	0.2	9.4	32	7.21	2.54	1.06	3.42	0.32
47	1.8	7.4	56	9.98	7.37	2.46	0.24	0.09	101	2.6	7.9	70	17.94	15.07	2.05	0.59	0.34
48	1.4	7.5	40	9.74	7.43	1.96	0.24	0.09	102	0.2	9.5	65	10.15	4.60	0.55	2.81	2.12
49	0.9	7.5	42	8.83	8.45	0.14	0.17	0.16	103	2.0	7.8	98	11.17	10.39	0.37	0.93	0.19
50	0.7	7.5	38	8.85	6.19	2.51	0.12	0.14	104	1.2	8.0	31	13.52	10.94	0.55	0.44	1.67
51	2.3	5.3	7	11.89	8.05	0.21	0.32	0.17	105	0.5	8.0	1	17.93	14.68	1.96	0.58	0.85
52	1.4	5.7	31	10.01	7.21	0.41	0.22	0.11	106	0.5	6.9	1	9.87	6.50	0.34	0.20	0.16
53	0.9	6.4	240	9.37	8.80	0.22	0.24	0.11	107	0.2	8.2	5	8.69	7.68	0.69	0.19	0.17
54	0.7	6.7	292	10.68	8.87	1.45	0.21	0.12									

Note: Analyses by Ankara and Cornell laboratories.
*Cation exchange capacity.

antiquity. Organic matter does not generally persist in subsoils when buried in the Sardis climate regime. Thus, organic matter in most of the pits decreases abruptly with depth. Exceptions are observed in samples 80 to 82 and 92 to 94, where subsoil alluvial deposits were derived from topsoils high in organic matter eroded from upslope — and that higher level of organic matter persisted to the present day. Sample 68, from Bin Tepe north of Sardis, was formed from limestone under a grass climax vegetation; these kinds of soils are typically higher in organic matter than most of the eroded soils at Sardis.

The pH of soils at Sardis (Table 3) is typically lowest in surface horizons and increases with depth in the profile, due to leaching. The acropolis conglomerate materials, exemplified by samples 16 and 55, have pH ranges around 8.4 and 7.2. Some of the samples with the higher pH values from the lower elevations (sample 100, 9.4; sample 102, 9.5; sample 104, 8.0) are probably enriched by increasing salinity and alkalinity from soluble salts (Na^+ and K^+) and other materials moving downslope in seepage waters. The high salt content is detrimental to crop growth in some spots in the Gediz Valley.

Phosphorus levels in soils of Sardis are very meaningful for identification of landslides associated with ancient occupation areas. Subsoil samples with high phosphorus are indicative of landslides that buried the early soil surfaces where accumulations of wastes caused phosphorus enrichments. Thus, the landslide subsoils in Table 3 samples 3, 4, 14, 15, 53, 54, 61, 62, 87, and 88, with high phosphorus contents, correlate very well with the observations identifying the landslide deposits at sites 1, 5, 22, 28, and 45 in Table 1. The original undisturbed acropolis conglomerate strata have very low levels of phosphorus (4 ppm in samples 16 and 55).

The exchangeable cations (Table 3) correlate well with the nutrients in the soils and the weathering and leaching of the soils. The cations are related to the organic matter, pH, and phosphorus and to the erosion, landslides, and ancient and recent land use practices. Cations in the acropolis strata are variable (compare samples 16 and 55), so that all of the past interactions affecting cations cannot be precisely known or interpreted. Cation exchange capacity, Ca^{++}, and Mg^{++} increase with depth in some soil profiles and decrease with depth in others. The mobile cation sodium (Na^+) is generally lowest on the highest most leached site on the landscape (site 21, Table 1; samples 47 through 50, Table 3) and highest in the lower parts where seepage waters break out and evaporate (site 52, sample 102; site 54, sample 104). Complete and perfect interpretations of the chemical analyses cannot be made because a complete and perfect record of all the landslides, erosions, and land use shifts is not available. The chemical analyses, however, support what is known about the physical properties (Table 2) and the landscape characteristics (Table 1) and will be of value in future correlations with further archaeological and geological investigations at Sardis.

PREDICTIVE IMPLICATIONS

Observations and data on past landslides at Sardis help in interpretation of the history and archaeology of the place but are most valuable in helping to predict future causes and effects of landslides. Figure 16 is a view of a recent landslide at Sardis — one of many. Strong platy structure (Fig. 17) is characteristic of the soils in these locations and is probably a relic flow impression in the soil material. This structure is apparently formed in the soils as they approach the liquid state when the soil strength is reduced because of seepage

Figure 16. Recent landslide scar at Sardis. Many of these slips are being initiated at present time by overgrazing of vegetation and by timber cutting. Exclusion of animals and reforestation would help greatly in preservation of ruins at Sardis. Site 55, where sample 105 was collected, is within slipped materials at middle of this landslide.

Figure 17. Left (western) scarp of recent landslide scar at Sardis shown in Figure 16. Notice typical platy structure inclined downslope characteristic of landslide-prone soils (at left near shovel). Soil structure to right of shovel has been severely disrupted from its bedding planes by recent slide. Shovel is about 1 m long.

forces or liquefaction. The structure is amazingly persistent (Fig. 18); tiny plates persist even when the large fragments are dried, smashed, and screened. I had the opportunity to observe soil failure into the liquid state of soil flowing one day at Sardis during a rainstorm. I sought shelter from the storm within one of the many Lydian chamber tombs, and within minutes after the rain started, small liquid flows of conglomerate were moving into the mouth of the tomb. The rainstorm did not last very long, but the event demonstrated clearly that the soils readily move under the stress of flowing ground and surface waters.

In Figure 3, nearly all of the steepest slopes in the higher elevations with contour lines closest together have had landslides fall from them. The midslopes in the ravines and wadis in Figure 3 have several metres of landslide deposits overlying flood deposits (Fig. 5); the footslopes in the lower elevations have as much as 10 m of landslide deposits over flood deposits. In the area shown in Figure 3, probably at least 1,000,000 m^2 have landslide deposits several metres thick; probably at least 500,000 m^2 have deposits as much as 10 m thick. In a conservative estimate, based on pedological field work over the entire area (Fig. 3), probably at least 10,000,000 m^3 of materials have moved and been redeposited in identifiable landslides. In the future, the general vicinity of sites 4, 7, 21, 27, 32, and 55 on steep slopes (Table 1) stretching in a band from location S900 E100 through S800 E500, S800 E700, S1000 E800, and S1000 E900 to S1100 E1000 (Fig. 3) is almost certain to produce major slides; the areas with platy soil structure inclined downslope (Table 1) in a broad band including sites 10, 17, 20, 32, and 55, stretching from S800 E0 through S700 E0, S700 E100, and S800 E700 to S900 E1000, are also likely to slide, even where the slopes are not so steep (Table 1). Although no single factor can be considered absolutely diagnostic for landslide prediction, the data in Tables 1, 2, and 3 are strong indicators

Figure 18. Fragment from recent landslide at Sardis shown in Figure 16. Soil has strong platy structure, which persists even when soil fragments are broken up. Scale intervals are 1 cm; when material was broken up and passed through a screen, plates persisted to thicknesses of 1 to 3 mm.

Figure 19. Roman bath "CG" at Sardis (location 28 in Fig. 3). This building was almost covered with more than 6 m of sediments deposited above its original base. Alluviation is closely related to destruction of forest, overgrazing, and increased numbers of landslides and floodings. Many of the alluvial deposits are probably derived from landslide materials that have fallen from acropolis.

Figure 20. Blocks fallen from ancient buildings into ravine on north slopes of acropolis at Sardis. Peak of acropolis is visible at top of photograph, and trail of cut stones, bricks, and other artifacts litter bottom of dry wash.

Figure 21. One of "flying towers" at Sardis (location 22 in Fig. 3). Extensive engineering conservation measures would be needed to save this monument from next major earthquake.

Figure 22. Acropolis fortifications at Sardis (location 20.3 in Fig. 3). Conglomerate foundations of this structure are crumbling and eroding, but walls can still be saved with prompt execution of engineering designs for conservation and protection. Structure was probably built in Byzantine time, and conglomerate has eroded back from foundation base within last few centuries.

of future movements at each site (Table 1) and in each layer (Tables 2, 3). The platy soil horizon from which sample 46 was collected (site 20 in Table 1), for example, appears to be a likely candidate for sliding, with 63% saturation, 46% silt (Table 2), only 0.4% organic matter (Table 3), 15.21 meg/100 g cation exchange capacity, and 1.41 meg/100 g sodium. On the other hand, sites 9 and 30 (Table 1), at locations S1300 E500 and S900 E500 (Fig. 3), have steep slopes (60% and 70%) but are relatively stabilized under pines, with a high organic-matter content (10.8% and 5.1%, Table 3) that somewhat counteracts a high saturation percentage (84% in sample 20, Table 2); this stabilized situation would change rapidly if the pines were cut and the areas overgrazed. All the data in Tables 1, 2, and 3 are included for all the sites and samples to assist future workers at Sardis in correlating possible additional studies of past and predicted slides.

Many of the landslides at Sardis have effects far beyond the site itself, in providing debris and alluvium to deposits far downstream (Fig. 19). Floodings and alluvial deposits are even observed in the modern cities in the area (J. A. Scott, 1976, personal commun.). Many of the Sardis monuments have fallen in earthquakes and landslides (Fig. 20), and others will continue to fall (Fig. 21). Many others, however, can still be saved (Fig. 22), with good conservation and reconstruction techniques. Most important of all, the events recorded in the soils at Sardis should help geologists, engineers, architects, pedologists, and many others to predict and plan better for the landslides and earthquakes in the future.

MANAGEMENT RECOMMENDATIONS

No study or work at Sardis has been done to attempt to significantly reduce the hazards of soil erosion and landslides, and even the national reforestation and soil conservation programs in Turkey have been relatively small scale compared with those in some other countries. The hazards of neglect and environmental abuse in mountain environments, however, are being increasingly documented in many places (Eckholm, 1975). At Sardis, in fact, land abuse and overgrazing are becoming more intense as the human population pressures increase and as farmers are forced farther up the slopes onto poorer and poorer lands. Precious trees are currently being cut from steep slopes for firewood and other uses. Sheep and goats and other animals have overgrazed the acropolis at Sardis for millenniums, but only recently have many of the steep slopes been planted to grapes, small grains, and other erosive crops. The movements of farmers upslope are apparently a result of the rising value and scarcity of land and increasing demands for farm produce for export and domestic use. Comparison of modern photographs with the photographs taken by Butler in 1910–1914 show how the grass was then lush in places (Butler, 1922, p. 1, 17) where it is now overgrazed (Fig. 11), and how steep slopes then only grazed (p. 15, 29 in Butler) are now both intensively farmed and overgrazed (Figs. 6, 8). The numerous recent landslides observed during the field study are apparently a reflection of the land abuses and are caused at least partly by the overgrazing and the forest destruction.

The ruined city of Sardis deserves preservation and protection because of its historic importance as verified by past records, and it also deserves protection because of the archaeological discoveries still to be made in excavations there. The area should be made into a national park and protected from grazing and farming by exclusion of those activities. If the land were allowed to revegetate, the forest would be quickly re-established; the trees within the protected compound shown in Figure 8 have grown to that height in about 10 years. Vegetation is probably the most feasible protection to prevent and reduce the landslips. Some critical ruins, like those shown in Figures 14, 21, and 22, need special slope and embankment stabilization with expensive engineering grading and structures to prevent slippage; joint archaeologic, engineering, and geologic studies would be necessary to formulate landform designs to save these monuments. Some landscape design has been done by architects and archaeologists at Sardis in restoration of a few of the monuments, and some stream channelization has helped to divert torrents where monuments were seasonally threatened with streambank erosion; but more commonly, tombs, walls, and other artifacts have been deliberately reburied by archaeologists after excavation to prevent vandalism. Money expended for development and preservation at Sardis could be quickly regained by promotion and encouragement of the tourist potential of the place — many foreign and national visitors already are attracted to Sardis for its historic and esthetic qualities. If extensive soil-conservation measures are not taken, the monuments will continue to fall in earthquakes and landslides accelerated by the environmental abuse. This soil-geology-archaeology study, however, along with others that might be conducted later, could assist appreciably in improving Sardis for its role in education for the future as well as in understanding the past.

CONCLUSIONS

This paper illustrates the importance of interdisciplinary work in studying a problem. Geologists do not normally examine 55 soil pits in a landslide area, but this approach in this study gave a broad areal aspect to the research and allowed examination of factors such as soil structure and soil phosphorus which contribute to knowledge about landslides. Pedologists usually do not have deep trenches with datable artifacts available for examination, but archaeological diggings at Sardis made these available for geologic and pedologic correlations with archaeology. And, of course, archaeologists normally do not have the services of geologists and pedologists and other specialists to the extent that might be ideal, but the Sardis research is unique in the diversity of the excavators and their findings. One hopes that in the future many more interdisciplinary studies of this type can be initiated.

ACKNOWLEDGMENTS

I thank G.M.A. Hanfmann, M. Ozuygur, K. W. Flach, G. Holmgren, R. E. Bentley, M. G. Cline, T. Greweling, S. W. Jacobs, D. C. Marshall, and L. L. Stewart for their assistance. Travel was supported by Humanities Grant No. 111-70-3966 for Archaeological Exploration of Sardis through Harvard University, and other contributions to the expenses of the project were borne by the Department of Agronomy of Cornell University and the Sardis Expedition Fund of Harvard University. This presentation is the result of the suggestions and encouragements of D. R. Coates, who also reviewed the manuscript. The work was also assisted and reviewed by J. A. Scott and D. A. Sangrey.

REFERENCES CITED

Butler, H. C., 1922, Sardis: Vol. I — The excavations 1910–1914: Leyden, Holland, E. J. Brill Ltd., 213 p.
——1925, Sardis: Vol. II — Architecture — The Temple of Artemis: Leyden, Holland, E. J. Brill Ltd., 146 p.
Campbell, A. S., 1971, Geology and history of Turkey: Petroleum Explor. Soc. Libya, Ann. Field Conf. No. 13, 511 p.
Eckholm, E. P., 1975, The deterioration of mountain environments: Science, v. 189; p. 764–770.
Foss, C., 1976, Byzantine and Turkish Sardis: Cambridge, Mass., Harvard Univ. Press, 216 p.
Greweling, T., and Peech, M., 1965, Chemical soil tests (rev. ed.): Cornell Univ. Agricultural Experiment Station, New York State College of Agriculture, Bull. 960, 59 p.
Hanfmann, G.M.A., 1961, Excavations at Sardis: Sci. American, v. 204, p. 124–135.
——1972, Letters from Sardis: Cambridge, Mass., Harvard Univ. Press, 366 p.
Hanfmann, G.M.A., and Waldbaum, J. C., 1975, A survey of Sardis and the major monuments outside the city walls: Cambridge, Mass., Harvard Univ. Press, 368 p.
Mitten, D. G., 1966, A new look at ancient Sardis: Biblical Archaeologist, v. 29, p. 38–68.
Olson, G. W., 1971, Descriptions, notes, maps, and data on soils of Sardis, Turkey — A collection of materials for relating the soils environment to the Lydian, Greek, Roman, and Byzantine ruins and to the ancient civilizations at Sardis: Cornell Univ. Agronomy Mimeo. 71-1, 150 p.
——1976, Criteria for making and interpreting a soil profile description — A compilation of the official United States Department of Agriculture procedure and nomenclature for describing soils: Kansas Geol. Survey, Bull. 212, 47 p.
Olson, G. W., and Hanfmann, G.M.A., 1971, Some implications of soils for civilizations: New York's Food and Life Sciences Quart., v. 4, p. 11–14.
Pedley, J. G., 1968, Sardis in the age of Croesus: Norman, Univ. Oklahoma Press, 146 p.
——1972, Ancient literary sources on Sardis: Cambridge, Mass., Harvard Univ. Press, 96 p.
Soil Survey Staff, 1967, Soil survey laboratory methods and procedures for collecting soil samples (Soil Survey Inv. Rept. 1, Soil Conservation Service): Washington, D.C., U.S. Govt. Printing Office, 50 p.
Topraksu, 1971, Gediz ovasi topraklari [Soil survey of Gediz River valley area]: Ankara, Turkey, Soil Survey Agency Rept. 8, 109 p. and 21 soil map p.

MANUSCRIPT RECEIVED BY THE SOCIETY SEPTEMBER 7, 1976
MANUSCRIPT ACCEPTED SEPTEMBER 20, 1976

Printed in U.S.A.

Index

Aggradation, 89, 92
Air cushion theory, 16
Alaska, 10
    Anchorage, 10
    Copper River delta, 137
    Gulf of Alaska, 19, 137–148
    Kayak Trough, 137, 139–140, 143
    Kenai Lake, 10
    Ketchikan, 23
    Seward, 10
    Sherman landslide, 10, 15
    Whittier, 10
Alluvial fans, 85, 88–89, 93, 96, 98, 110–111, 160, 162
    age, 110
    relation to landslides, 99, 104–107
    stability influences, 109–110
Appalachian Regional Commission, 245
Arizona
    Pima Mine, 12
Asia, 7
Atterberg limits, 207–208, 221

Batholiths, 127
Bedload, 119
Bedrock, 11, 20, 86, 115, 127, 248
    susceptibility to failure, 127–128
Bentonite, 14, 46
Bimodal flows, 53
Block glides, 5, 172, 226
Block stabilization, 193
Bonneville Power Administration, 77
Bootlegger Cove Clay, 10, 23
Brazil, 7
    Floresta Creek Valley, 7
    Ipe, 26
    Laranjeiras district, 7
    legislation of hillslopes, 23–24
    Rio de Janeiro, 7, 129–130
Bread crust surface, 134
British Columbia, 32, 36–38, 85–111
    climate, 99, 102–103, 107–109
    drainages, 93–94
    Fraser Canyon, 85–111
    Fraser River fault zone, 92–93, 95
    glaciation, 88–89, 93
    Hells Gate Bluffs area, 100–102
    landslide locations, 98–99
    physiographic and geologic setting, 86, 88–89, 92–94, 110
    regional slope stability factors, 95, 97–98
    structures, 95, 97
Buttresses, 25, 182–183

Calcareous deposits, 164
California, 23, 225–232, 235–241
    Alameda County, 23, 240
    Blackhawk landslide, 10, 15, 155
    Coachella Valley, 155, 159, 163
    Contra Costa County, 23, 240
    Division of Highways, 228
    Emerald Bay landslide, 128
    geology, 225–226, 235
    Golden State Freeway, 229–231
    Imperial Valley, 155
    inventory of deposits, 237
    isopleth maps, 237–238
    Lake Tahoe landslide, 18, 128
    landslide abundance, 237
    landslide causes, 8
    Los Angeles, 23–24
        grading ordinances, 3, 23
    Palo Verde Hills, 12
    Portuguese Bend landslide, 12, 24
    San Andreas fault, 155, 158, 239
    San Diego Freeway, 225–228
    San Francisco Bay region, 3–4, 6, 13, 23, 235–241
    San Jacinto fault, 158
    San Mateo County, 13, 23
    San Pablo, 9
    Santa Clara County, 23
    Sears Point, 11
    Sepulveda Canyon, 225–227
    slope failure, 226–227
    slope maps, 238–239
    Toro Canyon fault, 159
    Van Norman Reservoir, 17
    Ventura, 17, 24
Canada, 25, 29–53, 59–67, 85–111
    Alberta, 35, 43
    Bearpaw Formation, 39–43, 45
    Cordilleran region, 29–38
    Downie Creek slide, 36–37
    Frank landslide, 12, 30–36
    Grierson Hill landslide, 44
    Hope slide, 30–36
    Hudson Bay, 46
    Interior Plains, 29, 38–39, 44

Labrador, 46, 59–67
landslide costs, 22
landslide types, 29
Lower Mackenzie Valley, 29, 52
Maligne Lake landslide, 31, 33–35
Morgan Creek block slide, 44–46
Northwest Territories, 52
Nicolet slide, 51–52
Ontario, 47–48, 51
Peace River, 40–41
Quebec, 51
Saskatchewan, 38–42, 44
South Nation River earthflow, 17, 47–49, 51
St. Jean Vianney liquefaction flow, 17, 49
St. Lawrence Lowland, 29, 46, 49
St. Thuribe liquefaction flow, 17
Stalk Lake landslide, 31, 33–35
Toulnustouc landslide, 49
Turtle Mountain, 35–36
Canadian National Railway, 85
Canadian Pacific Railway, 44, 46
Catastrophic geology, 3
Champlain Sea, 48–49
China, 10
Clays, 133–136
    expandable, 115
    sensitive, 46–47, 50
Colombia, 130
Colorado, 13
    Loveland Basin landslide, 25
    San Juan landslides, 14
    Silver Mountain landslide, 14
Columbia River Gorge, 69–83
    Bonneville landslide area, 74–77
    geologic setting, 69, 72–73
    landslide types, 12
    Oregon shore landslides, 77, 79
    Skamania landslide area, 74
    Washougal landslide area, 73
    Wind Mountain landslide area, 75, 77–78, 80–82
Consolidated undrained method, 217
Consolidation
    curves, 43
    tests, 198
Corestones, 128
Creep, 36, 113–114, 117, 119
    definition, 113
    gravity, 44
    movement rate, 114, 119
    occurrence, 114
    relation to earthflow, 117
Czechoslovakia, 6, 13

Dams
    Bonneville, 69, 74
    Gardiner, 39, 43–44
    Grand Coulee, 12, 23, 26
    Lookout Point, 11
    Vaiont, 12
Debris, 165
    currents, 123
    dams, 123, 133
    flow, 6
    forms, 30
Debris avalanches, 6, 10, 123, 128–130
    damages, 7–8
    occurrence, 14, 107
Desert varnish, 164, 167
Dewatering, 181–182
Disasters, 3
Double jeopardy, 3, 21
Drainage anomalies, 149–150
Dry flow, 6

Earthflow, 6, 10, 32, 46–52, 113, 117, 130, 158, 205, 248
    moisture influence, 121
    movement rate, 114, 120–122
    occurrence, 114
    seasonal deformation, 121
Earthquakes, 8–11, 17, 32, 77, 128, 130, 166, 239, 259, 261–262
    Alaska, 10, 23, 147
    Borrego Mountain, 158
    Himalayan, 11
    Indian, 9
    Kwanto, 19
    Lisbon, 8
    New Madrid, 9
    Peruvian, 128
    San Fernando Valley, 10, 17
    San Francisco, 9
    Sardis, 255, 264
    St. Lawrence Valley, 50
    Yakutat, 147
Elastic rebound, 42–43, 45
Electric strain gauges, 191
Electro-osmosis, 25
Embankment, 190
Engineering and environmental implications, 35, 44–45, 50–51
Engineering geology, 79, 82, 97, 130, 135, 155, 172–195, 205–222, 225–232
    tests, 217–220
England, 8
    Axmouth Landslip, 8
Erosion, 88–89, 95, 111, 114, 119, 123, 129, 152, 163, 176
    fluvial, 41
    lateral, 97–98, 110
    toe, 38, 40–41, 48, 93
Exploratory drilling, 191
Extensometers, 191

Falls, 5
Faults, 9, 88–89, 103, 150, 158
Fiji, 19
Fill surcharge, 231
Floodplain, 123
Floods, 123, 168
Flows, 6
Flowslide, 6, 46–52
    bottleneck form, 47
    case histories, 51–52
    characteristics, 50
    failure mechanisms, 50
    geological conditions, 49–50
    implications, 50–51
    morphologic features, 47–48
    occurrence, 47
    rate of movement, 50
    stratigraphy, 48–49
Fluidization, 15–16
Forests, 117

clear cutting, 119
deforestation, 11, 117, 263
Fractures, 38, 209–211
France, 13

Geomorphology, 31, 33, 50, 63, 99, 113, 122, 206, 263
Glacial loading, 40, 202
Glaciation processes, 20, 35, 42, 48–49, 53, 59–61
Glaciolacustrine sediments, 15, 20–21, 198, 201
Grading ordinances, 3–4
Granite, 127–130
Granitoid, 127–130
Gravity creep, 44
Great Northern Railway, 23
Ground water, 43–45, 128–130, 200, 203, 250
artesian, 43–45, 177–179
Grouting, 25, 184
Grus, 127, 129
Gulf of Alaska, 137–148
geologic setting, 137, 139
glaciation, 139
submarine slide, 139

Hazards, 19–20, 51, 127, 130, 136, 148–149, 153, 168, 241, 254, 271
definition, 3
flooding, 21
rating, 51
reduction, 136
Hawaiian Islands, 4, 19
Herodotus, 255
Highways, 213, 225–232
cause of landslides, 185
defoliation, 203
effects on stability, 107, 113, 213–222
impacts, 11
Hillside benching, 25
Hillslope development, 23
Hong Kong, 8, 130
Horizontal drains, 24, 79, 178
Hurricane Camille, 8

Incompetent rock, 250
Indonesia, 4
Interceptor trench, 24
Iran
Saidmarreh landslide, 15
Italy, 11, 13, 23
Vaiont, 12, 17

Japan, 7, 129
Honshu, 129
Kobe, 8
Kure, 8
legislation, 24
Mt. Fuji, 7
Ohsawakuzure slide, 7
Sagami Wan, 19
Tokyo, 8
Joints, 43, 159, 166, 187, 189, 191
roughness, 66
tear, 175
tension, 63, 179
water pressure, 65

Karst, 149–150

Labrador-Ungava, 59–67
deglaciation, 60
geology, 60
glaciation, 62
joints, 60–61, 64, 66
Labrador trough, 60
Schefferville area, 59–60, 66
slope failure mechanisms, 60–62
Lake Albany Clays, 133–136
cause of slides, 135–136
instability, 134
length of slide travel, 134
problems, 133–136
varved character, 133
Land abuse, 271
Landslide-prone terrain, 20, 39, 133–134
Landslides
avoidance methods, 24
categories, 9
causes, 7–14, 32, 51–52, 77, 98–99, 129, 150, 172–173, 175, 187, 189, 220–222, 230–231, 235, 240, 250–252
chemical treatment, 184
classification, 6–7, 9
costs, 21–23, 237, 240
criteria for, 5
damages, 4, 7, 8–12, 18, 51–52, 69, 77, 107, 129, 134, 171, 185, 214, 237, 240, 255
definition, 3–5
deposits, 237, 239, 261
engineering remedies, 24
excavation methods, 24
external factors, 7
features, 20
hydraulic ram mechanism, 82
internal influences, 15
inventory system, 21
legal affairs, 23–24
length of travel, 166
management, 271–272
mechanisms, 201–203
movement rate, 5
occurrence, 99, 240
planning, 19–26
postglacial, 99, 102, 110
prediction, 269–271
prevention, 24–26, 222
recognition, 19–21, 237
reduction, 135–136
remedial methods, 189–195, 214
stabilization 6, 29, 181–184
susceptibility map, 239
terminology, 4–6
types, 5–6
vegetation treatment, 184
water control methods, 24
Landslip, 4
Lateral spreading landslides, 5
Laurentide Ice Sheet, 64
Law of equifinality, 20–21
Leda clay, 17
Lewis and Clark expedition, 69
Lime, 26, 184
Liquefaction, 239, 270

flow, 6, 8, 17–18, 46–52
Loess, 133
    flow, 10
Lunar avalanches, 16
Lydia, 255

Man, 11–12, 24, 85, 107, 109, 111, 130, 250–251, 262
Martinez Mountain rock avalanche, 155–168
    age, 167–168
    archaeology, 158
    characteristics, 163–164
    failure, 164–166
    faulting, 160–161
    geologic setting, 155–158
    location, 161
    mechanisms of movement, 166–167
    morphometry, 161
    movement, 161
    slide fault, 160
    structural relationships, 158–161
    Torro Canyon fault, 160
Mass movement (wasting), 43, 113–123
    debris, 109
    factors involved, 115, 117
    processes, 115, 117
    rates, 119
Mazama ash, 110, 114
Mining operations, 12
Mohr-Coulomb theory, 216
Montana
    Hebgen Lake, 10
    Madison rock avalanche, 10, 16
Moon, 16
Mountains
    Blue Ridge, 128
    Caribou, 39
    Cascade, 86, 92, 97, 103, 113–123
    Chugach, 139
    Coast, 86, 92, 97, 103, 235
    French Alps, 30
    Klamath, 129–130
    New Guinea, 130
    Pamir, 15
    Peruvian Andes, 10
    Rocky, 43
    San Bernardino, 15, 155
    San Jacinto, 155
    San Juan, 13–14
    Santa Monica, 225
    Santa Rosa, 155
    Sierra Nevada, 128
    Skeena, 35, 37
    Teton, 14
Mudflows, 5, 7
Mudslides, 252–254
Mylonite, 186–187, 194

Netherlands, 17
Nevada, 9
    Las Vegas Valley, 128
    Slide Mountain rockfall avalanche, 128
New Guinea, 13
New Hampshire, 185–195
    geology, 186–187
    Highway Department, 185, 187
    Interstate 93, 185
    landslide cause, 187, 189
    Woodstock landslide, 185–195
New York, 21, 133–136
    Albany, 134, 136
    Haverstraw, 12
    Hudson River valley, 136
    Lake Albany sediments, 133–134
    Tioughnioga valley, 14
    Troy, 134, 136
New Zealand, 8, 171–184
    Forest Service, 172
    geology, 172
    Lands and Survey Department, 176
    State Highway 1, 171
    Utiku landslide, 171–184
Norway, 18, 25, 65–66

Ohio, 197–203
    Cuyahoga River valley, 197
    geology, 198
    hydrology, 199–200
    morphology, 200–201
    slope stability problems, 197
    Stumpy Basin slide, 197
    vegetation, 200–201
Oregon, 9, 113–123
    climate, 115
    Coyote Creek site, 117–119
    dimensions of movements, 137, 148
    forest cover, 115
    geology and soils, 115
    Lookout Creek earthflow, 119–120
    South Umpqua Experimental Forest, 117
Overburden pressure, 202
Overconsolidated materials, 38–39, 45, 205–213
Overgrazing, 263, 271

Pacific Ocean, 18, 127
Panama Canal, 79
Pennsylvania, 20
    Belle Vernon, 25
    Cowanesque Valley, 3
Permafrost
    slope failures, 52–53
    terrain, 32
Peru, 10–11, 13
    Cerro Condor-Seneca rockslide avalanche, 128
    Chimbote, 11
    Mount Huascaran, 10, 128
    Ranrahirca, 10
    Yungay, 10
Petroleum development, 52, 137, 147–148
Philippines, 149–153
    Bagacay landslide, 150–151
    faulting, 151–152
    geology, 149
    geomorphology, 150–152
    Maqueda Bay, 149–150, 152
    Samar Island, 149–153
    strip plain, 150–152
Piezometer, 198, 202
Piles, 25
Pipelines, 137, 147, 230
Piping, 15, 21, 130

Poisson ratio, 64
Popouts, 228
Pore water pressure, 7, 25, 42, 46, 50, 128, 186, 202–203, 213, 218, 220–222
Portugal, 8
Precipitation, 77
    excessive, 7–8, 32, 129–130, 231, 240
Presplitting, 190–192

Quick clay, 5, 46

Railroad cuts, 109
Rebound, 32, 36, 38, 43, 65, 88
Regolith, 3, 8, 12, 20, 127–130
Regrading, 24
Residual strength, 218
Restraining structures, 25
Retaining walls, 25, 186, 190, 215
Retrogressive slope failure, 30, 38–46, 48, 52
    character, 44
    engineering implications, 44
    geologic conditions, 41–44
    morphologic factors, 38
    rate of movement, 44
    stratigraphy, 39
Richter scale, 10
Riprap, 109
Rivers
    Chenango, 13–14
    Columbia, 12, 23, 37–38, 75
    Cowanesque, 20
    Cuyahoga, 198
    Fraser, 85–111
    Hautapu, 172–173, 175–176, 180, 183
    Hudson, 12, 133, 136
    Llanganuco, 11
    Lookout Creek, 120–122
    Magdelena, 18
    Mantara, 13
    Mississippi, 9
    Morgan Creek, 45
    Murderers Creek, 136
    Nicolet, 51
    Ohio, 9
    Pactolus, 264
    South Saskatechwan, 38–39, 41, 43
    Tioughnioga, 13–14
    Thunder Creek, 53
    Trowbridge Creek, 21
    Tubig, 149–150, 152
    Ulot, 149–150, 152
    Washougal, 73
Rock avalanche, 6, 15–17, 30–37, 155–168
    characteristics, 32
    definition, 30
    morphologic features, 15, 30–31
    rate of movement, 32
    theories of motion, 15–17, 166–167
Rock bolts, 25, 183, 191–194
Rockfalls, 10, 75, 98–99, 107, 109, 191, 194, 205, 211, 245, 248
Rock flour, 133
Rock mass geometry, 62
Rockslides, 5, 7, 13, 128, 185–195
Rooting strength, 117
Rotational slides, 39, 50, 130, 134, 180

Rumania, 26
Safety factor, 180–181, 193–194, 222, 249
Sand boils, 179
Sand dikes, 75
Sand dunes, 133
Saprolite, 114, 127, 130
Sediment yield, 114, 119, 123
Seismic profiles, 139, 142–145
Seismic triggering, 155
Sensitive sediments, 3, 6, 38–47, 50, *see also* Flowslides
Septic tanks, 12, 176
Shearing resistance, 43, 63, 65
Shearing stress, 202
Shear keys, 25, 183
Shear strength, 45, 65–66, 130, 165, 175, 179–181, 220
Sheeting, 65
Sheet slide, 5
Sink holes, 150
Skin flows, 52
Slab slide, 5, 10
Slide-prone areas, 249–250
Slides, 5
Slopes, 109
    failure mechanisms, 61, 95
    regrading, 24
    stability, 110–111, 219–221, 239
Slumping, 5, 114, 117, 119, 130, 205, 209, 211
Soils, 115, 130, 217, 248, 256–266, 269
    overconsolidated, 199
    stabilization, 11
    strength, 116, 202, 269
Solifluction lobes, 21
Springs, 117, 178, 251
St. Lawrence Region, 17
Storm waves, 147
Stress
    history, 207
    horizontal, 64–65, 202
    lateral, 202
    residual, 64–65
    vertical, 64
Sturzstrom, 16
Subaqueous slide, 6, 18–19
Submarine slide, 137–148
    character, 137
    morphology, 139
    sediments, 139
    triggering mechanisms, 146
Subsidence cracks, 172
Subsurface water, 24
Surface mining, 251
Sweden, 64, 67
Switzerland, 18
    Elm landslide, 12, 16
    Lake Zurich, 18

Taiwan, 8
Talus, 60, 64, 77, 79, 89
Tanzania, 8
Terraces, 48, 93, 123, 175
Teton National Forest, 14
Texas, 205–222
    Austin Chalk, 210–211
    Balcones fault, 206

Bosque Escarpment, 207, 211
 Centerville, 213
 Centerville landslide, 220–222
 Del Rio Clay, 205–210
 Department of Highways and Public Transportation, 214
 geology, 206–207, 215–216
 Interstate 45, 213–214
 Queen City Formation, 215
 rock properties, 207–208
 slope failures, 209–211
 South Bosque shale, 205–210
 types of slope failure, 205
 Waco, 205, 211
 Weches Formation, 215
Thermokarst, 53
Timber, 113
Toppling, 63
Tree splitting, 122
Triaxial tests, 194, 216–219
Trimlines, 228
Turkey, 255–272
 acropolis, 261, 263
 archaeology, 259, 263–265
 Artemis temple, 264
 environment, 262–263, 265
 geology, 261–263
 history, 255–256
 landslide record, 256, 259
 Pactolus Valley, 263
 Pyramid Tomb, 265–266
 Sardis, 255–272
Typhoons, 8, 149
 Tanogama, 8

Underconsolidated clay slide, 6
Underdrains, 214, 217
United States, 22
 Corps of Engineers, 3, 20, 69, 74–75, 77
 Department of Housing and Urban Development, 235
 Geological Survey, 137, 235
 Federal Housing Administration, 185, 187
 Flood Disaster Protection Act of 1973, 23, 245, 249, 252–253
 Housing and Urban Development Act of 1968, 252
 National Environmental Policy Act of 1969, 3–4
 National Flood Insurance Program, 253
Unloading, 24, 38, 64, 109
Urban development, 11, 155, 241, 245–254, 262

Vane shear tests, 217
Vegetation, 24, 110, 133, 184, 197, 199, 272
Vertical drains, 24
Virginia, 8

Washington, 69–83
 Bonneville landslide, 69
 Emmons Glacier, 15
 Franklin D. Roosevelt Lake, 12
 Little Tahoma Peak landslide, 15
 Marcus, 23
 Mt. Ranier, 23
Water table, 199, 201, 218
 control methods, 24
 level, 12
 perched, 221
Water yield studies, 117
Weathering, 41, 89, 115, 127–130, 255
 chemical, 65–66
Wedge separation, 63–64
West Virginia, 245–254
 Geological Survey, 245
 landslide types, 248, 252
 mapping, 248–249
 Morgantown-Clarksburg area, 251–252
Wyoming
 Teton National Forest, 14

TA
705
R4
v.3
1977

JAN 19 1978